FISH DISEASES

FISH DISEASES

Editor
Italo Naggi

scitus
academics

Fish Diseases

Edited by **Italo Naggi**

This Book is Printed on Acid-Free Paper. ♾

ISBN: 9781681179490

Library of Congress Control Number: 2017949093

SCITUS Academics LLC,
616, Corporate Way, Suite 2, 4766,
Valley Cottage, NY 10989

www.scitusacademics.com

Notice

Printed in United States of America.

Preface

It is generally accepted that all outbreaks of a communicable disease are the result of interaction between the host, the pathogen, and the environment. This is particularly true in regard to fishes, which are coldblooded and utilize oxygen dissolved in water. The environment of the open sea is very stable, but in the inland waters and in fish farms extreme water temperatures, low dissolved oxygen contents, presence of fish catabolic products, and general pollution often produce stresses which contribute to outbreaks of infectious diseases. Fish suffer ill health from a variety of reasons. The best way to prevent the disease is to maintain a healthy environment in the aquarium for all times to avoid the stress which is one of the most important factor for many fish diseases. Stress may be associated with the temperature variation, deterioration of water quality, stocking density, insufficient or excess supply of food and poor environmental conditions. In some chronic diseases, such as corynebacterial kidney disease, timely use of chemoprophylaxis with sulfonamides may prevent, or reduce, losses very significantly. Fungal infections are among the most common diseases seen in temperate fish. Because fungal spores are found in all fish ponds and create problems in stressed fish. Poor water quality can also lead to an increase in fungal infections in an otherwise healthy fish population. Most fungal infections only attack the external tissues and only few fungal infections that will infect the internal organs of fish. The treatment of aquatic fish diseases has assumed major importance in this last decade especially in view of the boom in aquatic fish farming.

This book focuses on fish diseases and disorders, which is an outcome of a tremendous volume of research conducted on parasites, especially those that cause morbidity and mortality in fish. It covers theoretical and practical information to identify, treat and cure the disease. The aim of this book is to help students, researchers, and aquarists to recognize signs of illness so they can identify and resolve problems.

Table of Contents

CHAPTER 1

Fish Food and Fish Diseases

Bassleer G

Pathologist and CEO for Bassleer Biofish, Herselt, Belgium

ABSTRACT

Without too technical information we will explain what we experienced during our 40+ years of feeding ornamental fish at wholesale level. In the early 90s, during our daily practice, we had 1 goal: using a complete food for ornamental fish that would provide high quality fish within 3-7 days, depending on the origin and quality of the fish. Next to the basic ingredients, we have learned in the past 25 years that many other additional feed additives can play an important role to control the health of our fish: to help the immune (defense) system of the fish, to assist in faster repair after damage, to prevent or control infections, etc. In general, with the recent studies and experience (especially in aquaculture) we learned to understand the health benefits of a functional fish food= a food that gives more benefits than just 'feeding fish', a food that can be seen as a "nutricament"= nutritional medication with products from nature, avoiding (preventing) the use of medications.

Keywords: Ornamental fish; Omega-3 DHA fatty acids; Nutricaments; Functional food

WHAT CAN BE DIFFERENT AS FISH FOOD?

This is interesting content that our competitors would like to read although we practice common sense after many years of trial and error while working with millions of ornamental fish **(Figure 1)**.

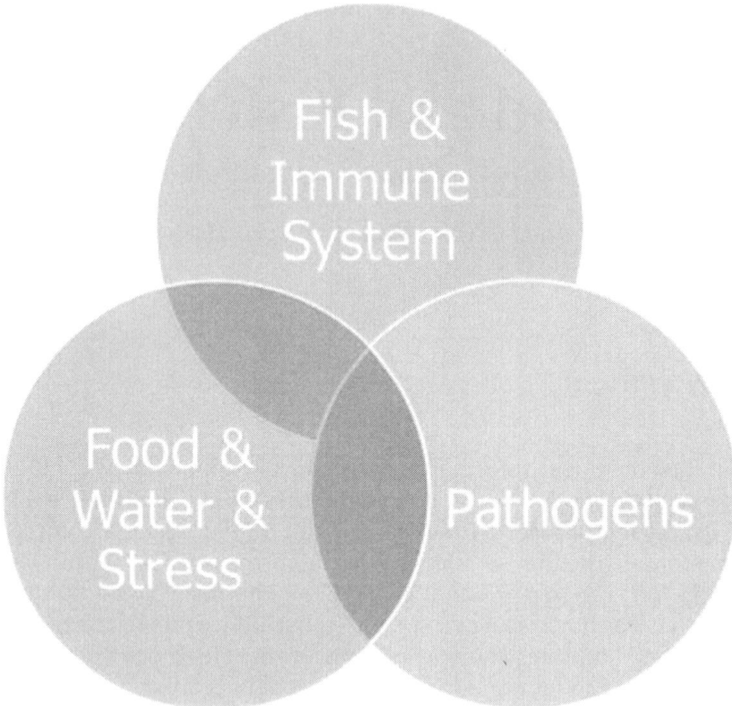

Figure 1: Hippocrates already said 2000 years ago: 'Let the food be your medicine'.

The main raw ingredient we recommend in its production is fresh wild Scandinavian fish (herring, mackerel, anchovy) which is very rich in proteins and essential Omega-3 DHA fatty acids; besides, this food also contains fresh mollusks and crustaceans such as squids, octopus, shrimps, prawns and krill, as well as fish oils, cereals, yeasts and algae.

In general, the composition of a food for most of our aquarium fish (freshwater and marine) is ideal when high % of digestible proteins, many essential amino acids and fatty acids, vitamins A+B+C+D, trace elements and minerals. In this article, we will not go into detail about the function of each essential food molecule. More important is to share my experience as fish health specialist. During my many years of work in 4 contents of the world, as fish pathobiologist, we did not only look at just feeding the fish with good ingredients but we added also extra health beneficial natural products to make it a functional food (nutricaments) [1] **(Figure 2)**.

Figure 2: Scandinavian fishes.

We also give a large preference to granulate (pelleted) food, (see further for more reasons why this is better for your fish).

When we produce granulate food the majority of ingredients are mixed and then cold-pressed; then it is cooked for a very short time at only 70°C to eliminate potential virus and bacteria. Once food is cold but still a little humid, the vitamins and nutricaments (such as algae, immune stimulants, acai, herbs, aloe vera, garlic, etc.) are added so that none of their exceptional nutritional values are lost.

On the other hand it is outmost important that a food has enhanced palatability=many fish come from specific breeding facilities or from the wild and need to learn to eat a 'new' kind of food. We have put a lot of attention (especially by using wild fish) that the fish 'smell/taste' the granulates very quickly.

Through our 40 years' experience between millions of fish we learned to understand the problems that the large variety of our ornamental fish encounter/face in our industry: handling, packing, shipping, acclimating, netting, mixing of species, competition and many other stress factors.

Next to the selection of good quality fish and shrimp (mostly Scandinavian) we added extra natural products (to produce functional feed or nutricaments) that help in the prevention of bacterial, viral, fungal and parasitic infections or faster repair/recovery after disease.

At the same time we try to optimize digestive and metabolic efficiency: less waste (Important=less pollution of the aquarium water) and better growth and coloration of the fish.

The investment in good food contributes to less problems with fish, less diseases, less use of medications (especially antibiotics), less work, less costs, etc., and higher quality of fish and higher survival rate. In the fish business this means=more profit! From my personal experience I have seen companies going broke (out of business) because they never wanted to invest in food and they did spend lots of money in medications, water changes, excessive working hours...combined with high losses, complaints (DOAs) from customers of course this results in no profit.

What about functional fish food for the health of fish (nutricaments)?

WIKIPEDIA explains="*Functional food is a food given an additional function (often one related to health-promotion or disease prevention) by adding new ingredients or more of existing ingredients*".

As explained above, we are specialized in using nutricaments for the wellbeing of our fish! This has become an important issue in 21st Century because the use of antibiotics and other medications will eventually become very much restricted.

We make our function fish food by coating natural products around each granulate, after production of the granulate, so the quality of beneficial ingredients are guaranteed!

These natural products have a functional purpose with health benefits for the fish: we use different categories: probiotics, prebiotics and phytotherapeutics.

Probiotics

"Probiotics are microorganisms are believed to provide health benefits when consumed" We recommend the use of Pedioccocus acidilactici since it has been proven to have great success on health & growth of fish! (Reference Nobel prize laureate Élie Metchnikoff & Beneficial effect of yoghurt! In recent years also 'Yakult' for humans.)

Since we mix the bacteria (as coating) around each granulate we can do this perfectly during our production process.

Those beneficial bacteria, introduced by eating the food, change the content of the intestinal flora in a very positive way. The intestinal flora has a great impact on the resistance against infections: a weak gut flora creates a weak defense against infections (the same counts for all animals and humans). Since 2014 many positive scientific reports have been published **(Figure 3)**.

Figure 3: Lab culture probiotics.

The probiotics are incorporated in our fish food can be fed permanently.

Prebiotics

"Prebiotics are substances that induce the growth or activity of microorganisms (e.g., bacteria and fungi) that contribute to the wellbeing of their host. The most common example is in the gastrointestinal tract, where prebiotics can alter the composition of organisms in the gut microbiome. However, in principle it is a more general term that can refer to other areas of the body as well. As a functional food component, prebiotics, like probiotics, are conceptually intermediate between foods and drugs". Very common used in aquaculture & our industry are Yeast Extracts (from *Saccharomyces cerevisae*) with a high content of Betaglucan as it is effective in all mammals as it is in fish and birds with its immune-enhancing properties established in a number of fish species. It has applications for prophylactic and therapeutic use. Beta-glucan may be used as an alternative to antibiotics and vaccines for protecting farmed fish against microorganisms, or microparasitic disease. It can also be used together with vaccines to improve the effectiveness and it is non-toxic to fish. Beta-glucan products have been in practical use for many years in animal feeds all over the world to improve health of aquaculture species, pets and farm animals. Pretreatment definitely increases survival rates with enhanced phagocytes function and the inhibition of cellular injury. An overall enhancement of immune response can be achieved by the use of beta glucan. This may in turn allow fish the opportunity to effectively combat disease **(Figure 4)**.

Using beta glucan as a dietary supplement carries the potential to significantly impact the quality of health as well as longevity. At the same time administration in the food of bioflavonoids and vitamin C (from Citrus fruit) with anti-oxidant and anti-inflammatory activity, especially in the intestine (gut)! This works synergetic with the glucans. We recommend this kind of immune-stimulant food during acclimation, before shipping, after unpacking, after stress situations and during disease/treatment (also during treatment fish should be fed!) [1].

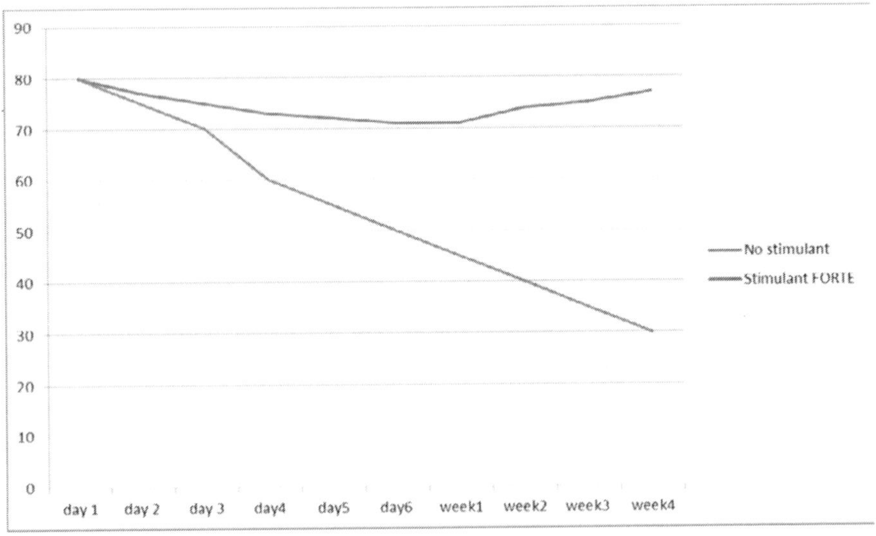

Figure 4: Growth or activity of microorganisms with and without stimulate forte.

Phytotherapeutics or phytotherapy

"Phytotherapy is the use of plants and plant extracts for attempted therapeutic purposes". In our experience, with trial and error during my 40 years of working between millions of fish and with exchange of data with experts in the aquaculture, we have used different kind of FRESH plant (phyto) materials. We cannot regard them as medicines but more as 'aid' during treatment or for faster repair/recovery after disease.

Hereby we give you a short overview of some we have used in our work:

+ Chlorella: from the freshwater algae *Chlorella pyrenoidosa*

+ Garlic: from *Allium sativum*

+ Moringa oliveira

+ Grapefruit seed extract

+ Aloe vera

+ Thymus (thyme)

+ Mentha (peppermint)

+ Artemisia (mugwort)

+ Stellaris (chickweed)

+ Açai: fruit berries from Euterpe oleracea (palmtree)

+ Sophora flavescens (ingredient matrine)

We prefer granulate (pelleted) fish food: but why granulated fish food?

We use granulate food because of different reasons, that makes it different than frozen, fresh or flake food

1. Granulated food can be processed in such way that it guarantees no loss of quality of the ingredients (Processing of flake food at 140°C destroys or changes many interesting ingredients)
2. The ingredients (fish, shrimp, vegetables, etc.)are selected and prepared without a risk of introducing (transmitting) diseases (a big risk to introduce unnecessary potential harmful bacteria with you feed frozen or live food);
3. The process of granulated food allows us to coat (each ranulate) with natural products so their ingredients, that have a health benefit, remain fully active;
4. Granulated food can be produced in different sizes, well adapted to the different sizes of the mouths of our aquarium fish;
5. The granulates remain for many hours stable in the water: not falling apart and still ready to eat for the 'slow' eaters;
6. We prefer, for most fish, a sinking granulate food: slowest sinking smaller granulates and faster sinking larger granulates... adapted for the different fish ages, sizes and behavior;
7. Due to high nutritional content, the quantity of food to be used is much less in comparison with non-granulated fish food;
8. The cost of feeding granulate food is lower than most ordinary flake, frozen or live feed **(Figure 5)**.

Figure 5: Granulate (pelleted) fish food.

What problems of fish health can be prevented with food?

Overfeeding: It is problem nr.1 for many hobbyists

- Most people give too much of cheap, inadequate food. Then they have to give large quantities of poor quality food to make fish somewhat happy(?).
- Also some people think that fish are always hungry because they are 'begging' for food. No, this is conditioning behavior of the fish like Pavlov's –dog.
- Please just feed 1× or maximal 2×/day with a high quality food (do not save on expenditure of fish food since it is a very small cost...for fish with a rather small mouth; since most fish hobbyists have also a cat or dog they can compare it with the much more expensive bags of cat or dog food they have buy! See further down) [2].

Lack of food: Lack of good feeding practice because of competition between the species or mix with other fish (e.g., Clown loach mixed with other fast-eating fish like Rainbow fish: see photo) the fish is not able to reach (or eat) the food. (e.g., Pleco used as 'cleaner fish' not getting any food: see photo). This can be prevented, as explained above, by using good granulated food **(Figure 6)**.

Poor food quality: By poor ingredients, lack of high nutritional value, is a common problem for wild fish that lack good food sources (many are too skinny) and also a problem in farming fish when using very cheap, low grade food (some use chicken food!). These kinds of fish are very weak, no cure is possible and wholesaler or shops have daily or weekly losses. See 2 examples in these photographs of Pearl gourami and Swordtail with pale, fatty organs (especially the liver). All these "bad feeding" lead eventually to bad organs and a failing immune system: the fish becomes weak and gets infections, diseases which are difficult to cure. Some blame it on a bad medication, but it all goes back to a failing nutritional approach and because the fish is too weak to coop with the infection **(Figure 7)**.

Figure 6: Fishes with lack of good feeding practice.

Figure 7: Fishes Poor food quality.

Is feeding high quality food expensive?

Our goal: learn to understand that your investment in fish food is a very small cost in comparison with the benefits of more healthy fish AND less medications, less diseases, less losses, less spoiled water, etc.

The price of food mainly depends on its ingredients: it is obviously cheaper to produce food with fishmeal, a high content of cereals and a low content of fish oils than producing high quality food; This latter one is produced using fresh wild fish as main ingredient, is very rich in healthy, good digestible fish oils with high content of Omega-3 and low content of cereals, just the required quantity to have food consistent enough. But the price of a food being higher does not necessarily mean that feeding fish with it is more expensive, since the amount of food to be used depends on the energy-giving value and digestibility of the food itself. Thus digestibility of carbohydrates is only 40%, whilst that of proteins and Omega-3 fatty acids from fresh fish is over 90% (digestibility is the amount of food that is assimilated by the organism and is not eliminated as waste through excrements). On the other hand, fish oils' energy value is more than twice of carbohydrates'. All this means that pellets with' high energy value and digestibility, the user saves around 50-60% of the amount he would need to use if feeding with other conventional foods. Therefore, the cost of feeding the fishes with a high quality food is considerably lower than using other food with a cheaper price. Besides, higher digestibility and usefulness of this food makes the pollution in aquaria produced by excrements is notably reduced and above all, fishes will be healthier and stronger, and less prone to catch diseases [3].

What about Food Safety? Animal welfare? Environmental and Ecological care?

The supply and production of the food sources and supplements of fish feed can be controlled, standardized and certified with the assistance of several organizations which are working close with the fish food industry: e.g., GlobalGAP, IFFO, ISO and FAVV [4].

a) Standards of Bio safety and sustainable and animal welfare=GLOBAL G.A.P. The GLOBALG.A.P. Aquaculture Standard sets strict criteria for: Legal compliance, Food safety, Workers' occupational health, safety and welfare, Animal welfare, Environmental and ecological care.

b) Sustainable marine products supply=IFFO is the international 'not for profit' organization that represents and promotes the fishmeal, fish oil and wider marine ingredients industry worldwide. We are globally respected and regularly represent the industry at international forums, as well as holding observer status at the UN Food and Agriculture Organization (FAO) and the EU Commission and Parliament.

c) Basic ingredients can have ISO-Certified production: ISO22000

d) Food safety Controlled by local, national Food Safety Authority such as FAVV in Belgium

The FAVV is an executive body and has jurisdiction over the entire territory of Belgium. In accordance with its legal obligations the FASFC is responsible for laying down, implementing and enforcing measures.

REFERENCES

1. Lim C, Webster C (2001) Nutrition and Fish Health
2. Stoskopf(1993)Fish Medicine.
3. Bassleer G (2006)The New illustrated guide to fish diseases.
4. Bassleer G (2011) The practical guide to fish diseases: 104.

CHAPTER 2

Columnaris Disease in Fish: A Review with Emphasis on Bacterium-Host Interactions

Annelies Maria Declercq[1*], Freddy Haesebrouck[2], Wim Van den Broeck[1], Peter Bossier[3] and Annemie Decostere[1]

[1]*Department of Morphology, Faculty of Veterinary Medicine, Ghent University, Merelbeke, Belgium*

ABSTRACT

Flavobacterium columnare (*F. columnare*) is the causative agent of columnaris disease. This bacterium affects both cultured and wild freshwater fish including many susceptible commercially important fish species. *F. columnare* infections may result in skin lesions, fin erosion and gill necrosis, with a high degree of mortality, leading to severe economic losses. Especially in the last decade, various research groups have performed studies aimed at elucidating the pathogenesis of columnaris disease, leading to significant progress in defining the complex interactions between the organism and its host. Despite these efforts, the pathogenesis of columnaris disease hitherto largely remains unclear, compromising the further development of efficient curative and preventive measures to combat this disease. Besides elaborating on the agent and the disease it causes, this review aims to summarize these pathogenesis data emphasizing the areas meriting further investigation.

1.1 History and taxonomy

Flavobacterium columnare (*F. columnare*), the causative agent of columnaris disease, belongs to the family *Flavobacteriaceae*[1, 2, 3]. Columnaris disease was first described by Davis among warm water fishes from the Mississippi River [4]. Although unsuccessful in cultivating the etiological agent, Davis described the disease and reported large numbers of slender, motile bacteria present in the lesions [4]. Upon examining a wet mount preparation of these lesions, column-like structures formed by these bacteria were evident. The organism was hence named *Bacillus columnaris* and the disease elicited columnaris disease. Over the decades the taxonomic status of the pathogen has changed several times since the pioneering work of Davis [4]. In 1944, Ordal and Rucker were the first to isolate the bacterium from a natural outbreak of columnaris disease among sockeye salmon (*Onchorhynchus nerka*) [5]. A diluted culture medium was used to grow the bacterium. Based on cellular morphology, they identified the bacterium as a myxobacterium. Organisms classified in the order *Myxobacteria* are long, thin Gram-negative rods that are motile on agar media by a creeping or flexing motion. They have a life cycle composed of vegetative cells, microcysts (resting cells), and fruiting bodies, or only vegetative cells and microcysts [6]. Ordal and Rucker reported that the myxobacterium from columnaris disease produced both fruiting bodies and microcysts and named the organism *Chondrococcus columnaris*[5]. Garnjobst assigned the bacterium to the family *Cytophagaceae* as *Cytophaga columnaris* after isolating a pathogenic bacterium resembling *Chondrococcus columnaris* morphologically, but not producing microcysts [7]. Bernardet and Grimont reclassified the organism and placed it in the family *Cytophagaceae* and the genus *Flexibacter*, assigning it as *Flexibacter columnaris*[8]. Finally, in 1996, the bacterium received its current name, *Flavobacterium columnare*, based on DNA-rRNA hybridization data and protein and fatty acid profiles [3]. In 1999, the *F. columnare* cluster was subdivided in three genomovars based on differences in 16S rRNA sequences, restriction fragment length polymorphism (RFLP) and DNA-DNA-hybridization [9].

1.2 Morphology and biochemical characteristics

The morphological and biochemical characteristics of *F. columnare* are summarized in Table 1. For a full biochemical profile of *F. columnare*, the reader is referred to Bernardet and Bowman [1].

Table 1 Morphological and biochemical characteristics of *F. columnare* (adapted from[1, 3, 10, 11, 12])

Characteristic	**Description**
Growth condition	Strictly aerobic
Gram-stain	Gram-negative
Morphology	Long, slender gliding rods of 4 to 10 µm and 0.3 to 0.5 µm wide. In aging cultures spheroplasts may occur
Capsule	Described to be absent [10] or present [11] depending on the adopted strain
Congo red absorption	Present due to an extracellular galactosamine glycan in the mucus and the production of flexirubin-type pigments
H_2S-production	Present
Degradation of crystalline cellulose	Absent
Degradation of complex acidic polysaccharides of connective tissue	Present

1.3 Epidemiology

F. columnare is distributed worldwide in fresh water sources and may infect many different wild and cultured freshwater fish species, such as (but not limited to) carp, channel catfish, goldfish, eel, perch, salmonids and tilapia [1, 12, 13, 14, 15, 16, 17]. This disease also assails many tropical freshwater aquarium fish [12, 14]. In the channel catfish (*Ictalurus punctatus*) industry in the United States, *F. columnare* is the second most prevalent bacterium, after *Edwardsiella ictaluri*, to cause disease and mortality [18, 19], with yearly losses estimated at 30 million dollars [20]. This organism can also be part of the bacterial microbiota of freshwater fish, eggs and the rearing waters the fish live in [21]. Fish may reside in a clinically healthy carrier status harbouring an isolate remaining from a previous outbreak of columnaris disease and in this way act as an infection source for other fish [22, 23]. Fujihara and Nakatani reported that rainbow trout surviving a *F. columnare* infection can release up to 5×10^3 colony forming units/mL/h of viable bacteria into tankwater [22]. The gills were shown to be the major release site of this pathogen. Dead fish would be able to spread the disease at a higher transmission rate compared to living fish [24].

Several studies have indicated the potential for *F. columnare* to survive for extended periods in water. Survival was demonstrated to be influenced by physical and chemical characteristics of the surrounding water. Fijan indicated that *F. columnare* can survive up to 16 days at 25°C in hard, alkaline water with a high organic load [25]. Soft water with 10 mg/L $CaCO_3$, especially when acid or with a low organic content, does not provide a favorable environment for the organism [25]. Chowdhury and Wakabayashi determined that calcium, magnesium, potassium and sodium ions all are important for long-term survival of *F. columnare* in water [26]. Ross and Smith found that survival of *F. columnare* in static, sterile river water was directly related to temperature, with a higher survival percentage at lower temperatures [27]. The bacterium can keep its infectivity in lake water in laboratory conditions for at least five months [28]. *F. columnare* is also capable of surviving in sterile river mud [6]. Apparently, mud slurry often contains sufficient nutrients to maintain viability of *F. columnare* longer than sterile river water. In this case, however, the percentage survival of *F. columnare*seeded into mud seems to be higher at 25°C than at 5°C. Temperatures below 5°C are even detrimental to *F. columnare* cells in mud. *F. columnare* also grows well on particulate fish feed [29]. When surviving outside the host, *F. columnare* can change from a virulent to a less virulent form with an altered colony morphology, probably to save energy [28]. It has been suggested that *F. columnare* strains at fish farms originate from environmental waters and that the farm environment and practices may select for virulent strains that cause outbreaks in the farm [28, 30].

2. THE DISEASE

2.1 Clinical signs, histopathology, ultrastructural features and haematology

F. columnare causes acute to chronic infections and typically affects the gills, the skin and fins. The clinical manifestation of columnaris disease amongst others is dependent on the virulence of the eliciting strain. In a study of Rucker et al., the strains of low virulence induced slow progressive infections at water temperatures above 21°C and caused massive tissue damage before death occurred [31]. Strains of high virulence caused fulminating infections and killed young salmon (*Salmo salar*) in 12 to 24 h at 20°C. Ordinarily, these fish did not show gross tissue damage at the time of death [31]. The same was found in a study of Pacha and Ordal [32] and Foscarini [33]. The gross pathology observed in the fish experimentally infected with strains of *F. columnare* of high virulence was usually very limited. Apparently, death occurred before gross external manifestations of the disease appeared. However, some of the last fish to die did show macroscopically visible signs [34]. Besides the virulence of the strain being a determinant factor, in coldwater and temperate fish, age also seems to have an important impact on the severity of the clinical signs. In young fish, the disease develops acutely and mostly damages the gills (Figure 1). In adults, the disease may adopt an acute, subacute or chronic course. When the disease course is acute or subacute in adult fish, yellowish areas of necrotic tissue can appear in the gills ultimately resulting in complete gill destruction (Figure 2) [1, 32, 35].

In chronic cases, it takes longer before gill damage appears and skin lesions may develop as well [1, 32, 35]. On the body, small lesions start as areas of pale discolorations of the skin, which usually are surrounded by a zone with a distinct reddish tinge. This mostly begins at the base of the dorsal fin. Fin deterioration then occurs, starting from the lesion at the base of the fin and progressing to the outer edge, the opposite to normal finrot. The lesions then begin extending laterally from their common location at the base of the dorsal fin to encircle the fish resembling a "saddle-back" (Figure 3). The disease therefore is referred to as "saddle-back disease" [1, 32, 36]. Finrot is also often present [1, 37].

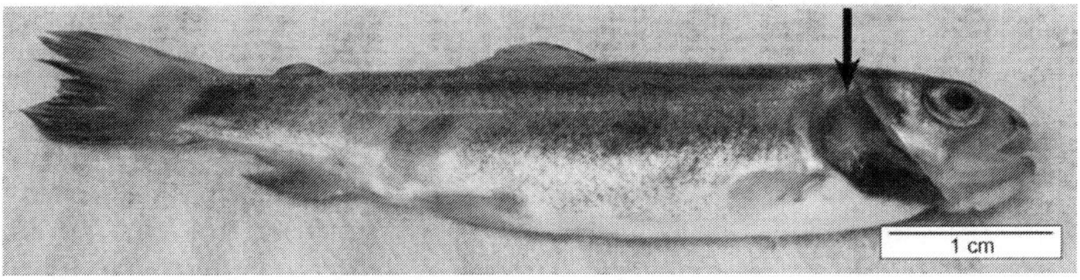

Figure 1 ***F. columnare*** **elicited gill lesions in rainbow trout (*Oncorhynchus mykiss*) fry (operculum removed).** In young fish, the disease is mostly acute and the gill is the major site of damage. The lesions are exhibited by pale necrotic areas (arrow, bar = 1 cm).

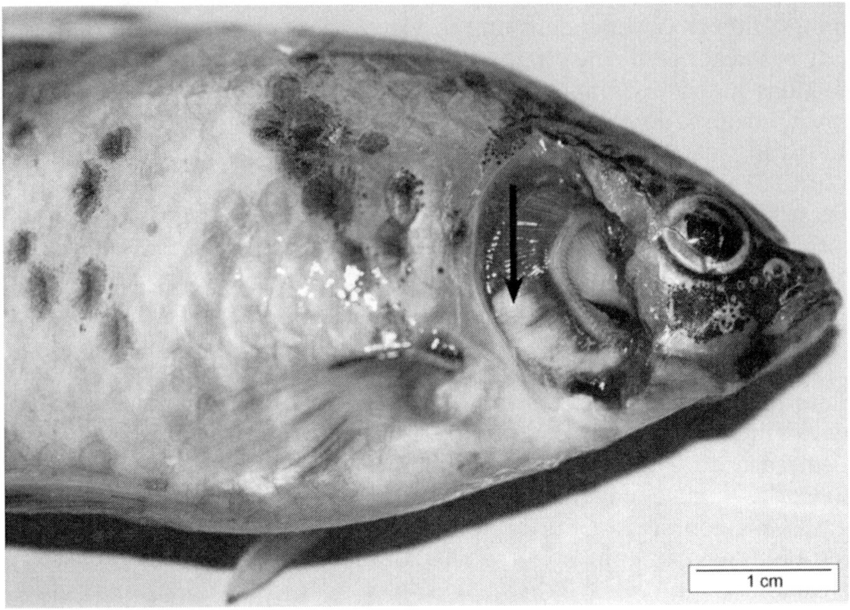

Figure 2 Gill lesions in a shubunkin (*Carassius auratus*) (operculum removed) caused by *F. columnare*. Yellowish-white areas of degeneration are visible in the ventral part of the first gill arch (arrow). When the *F. columnare* infection spreads rapidly throughout the gill lamellae, the fish may die in a short period of time without any other apparent lesions. Respiratory distress, caused by damage to the gills, appears to be the cause of death (bar = 1 cm).

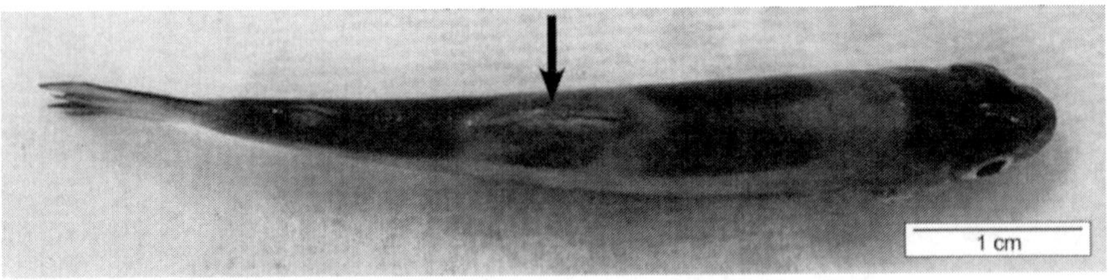

Figure 3 *F. columnare* induced a saddleback lesion (arrow) in rainbow trout fry. The lesion is visible as a discoloration starting around its common location at the base of the dorsal fin and extending laterally to encircle the fish resembling a saddle. Hence, the descriptive term "saddle-back" is often used and the disease is denoted as "saddle-back disease" (bar = 1 cm).

In rainbow trout, the area around the adipose fin may become dark and show erosions. These lesions expand to the peduncle, hence the name "peduncle disease" [1]. The lesions may progress cranially and caudally and even into the deeper skin layers, exposing the musculature leading to deep ulcers (Figures 4 and 5) [1, 35]. The lesions typically are covered with yellowish-white mucus [1].

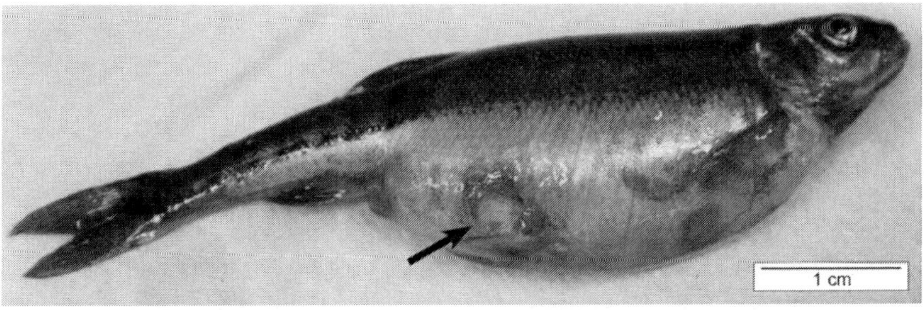

Figure 4 Skin ulceration (arrow) in a minnow (*Phoxinus phoxinus*) caused by *F. columnare.* The lesion has progressed into deeper skin layers, exposing the musculature. The edge of the ulceration displays a distinct reddish tinge and its centre is covered with yellowish-white mucus (bar = 1 cm).

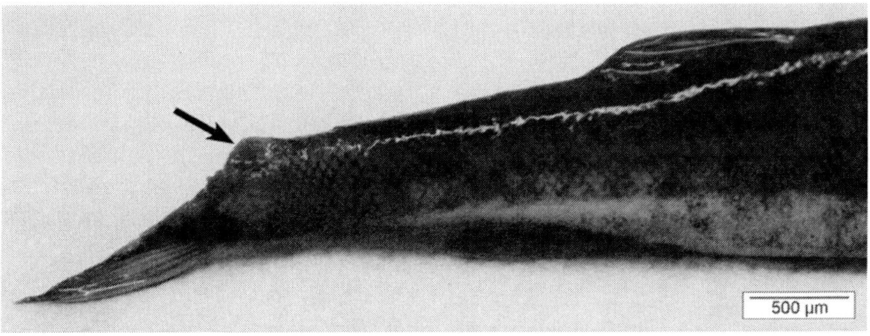

Figure 5 Tail fin erosion in a minnow caused by *F. columnare.* Columnaris disease led to complete disappearance of the upper half of the fin and exposing the underlying musculature (arrow, bar = 500 µm).

Ulceration of the oral mucosa also occurs, resulting in mouthrot. These mouth lesions are more lethal than are the skin lesions, since the painful oral lesions render the fish anorectic and lead to death due to starvation.

Moreover, the disease spreads easily to the mandible and the maxilla. Secondary infections with fungi or other bacteria may deteriorate the situation and can be seen together with the filamentous bacteria [38]. In tropical fish, this clinical sign led to the disease being termed "cotton wool disease" or "mouth fungus" [1]. Finrot can also be present in tropical fish [37]. Lesions can be restricted to local skin discoloration, with or without ulceration, and degeneration of underlying muscle fibers [1, 12, 39].

The skin or gills need to be abraded, for bacteria to enter the bloodstream and cause systemic infections [40]. However, Hawke and Thune have isolated *F. columnare* from internal organs without any external lesions appearing [18]. Foscarini described that the pathological changes to the gill structure caused by columnaris disease go hand in hand with cardiac alterations [33]. The first day after infection, bradycardia was noticed with the formation of hyperplastic gill lesions. Degenerative processes of the lamellae in the following days resulted in a compensatory tachycardia. This research suggests that the interaction between the impaired gill vascular blood circulation and the cardiac changes could result in the death of the fish [33].

Light microscopic examination of the affected gill tissue reveals the disappearance of the normal structure of primary and secondary filaments (Figure 6) [32, 33, 38, 41]. In the initial phase, proliferation of epithelial cells of the gill filaments can be accompanied by an increase of mucous cells [33]. The proliferating tissue can occlude the space between adjacent gill lamellae. In more advanced stages, the occlusion can be total causing the gill lamellae to be completely surrounded by the propagating tissue. Congestion of gill lamellae occurs due to accumulation of blood masses and inflammatory cell infiltration can be noticed. Edema causes lifting of the surface epithelium of gill lamellae from the underlying capillary bed. In more advanced stages of the disease, fusion of gill lamellae and/or gill filaments appears [32, 33, 38, 41]. Complete clubbing of gill filaments can finally result in circulatory failure and extensive internal hemorrhage [33]. Moreover, huge clusters of *F. columnare* can be found on the cell surface and/or in between necrotic sites (Figure 6, detail).

Columnaris disease may cause acute ulcerative dermatitis extending into the hypodermis and the muscle. Waterlogging can be present. The latter appears when the osmotic barrier is broken and thus forces water into the tissues, leading to severe dermal edema. Rupture of pigment cells with the loss of melanocytes can also be seen. Columns or hay-stack-like aggregates of bacteria can be gathered between dermal collagen fibers. The typical long and slender bacterial cells can easily be noted upon inspecting hematoxylin and eosin (H&E) or Giemsa stained sections from affected tissue where they appear bluish-purple and blue, respectively [38]. When all these changes occur rapidly, they may proceed to severe necrosis and sloughing of the epidermis [32, 38].

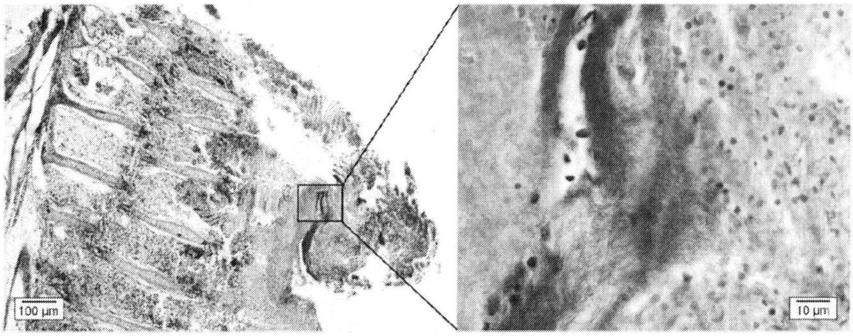

Figure 6 Gill section of a koi carp infested with *F. columnare*. In the left figure, extensive loss of branchial structures is visible. This is an advanced stage of the disease in which the filaments and the lamellae have fused and the gill epithelium is destroyed (H&E, bar = 100 µm). Complete clubbing of gill filaments may finally result in circulatory failure and extensive internal hemorrhage. A detail of this is depicted in the right figure, where *F. columnare*bacteria are visible as long, slender, purplish structures in between remnants of the gill tissue (H&E, bar = 10 µm).

Scanning electron microscopic (SEM) pictures of affected gill arches reveal the presence of rod-shaped bacterial cells, approximately 0.3-0.5 µm wide and 3-10 µm long (Figure 7). These long, thin bacteria adhere on the surface of the gills and appear to be aggregated rather than evenly distributed across the gill epithelium [42, 43]. Transmission electron microscopic (TEM) examination of gill tissue shows numerous long, slender bacteria in close contact with the gill tissue (Figure 8).

Bullard and McElwain were the first to describe the ultrastructural features of saddleback lesions associated with experimental infections of *F. columnare* in channel catfish and zebrafish (*Danio rerio*) using SEM [44]. Channel catfish skin lesion samples had margins typified by shed epidermal cells and lesion centers that exhibited a multitude of rod-shaped bacterial cells, approximately 3-10 µm long and 0.3-0.5 µm wide, intermixed with cellular debris. Zebrafish skin lesion samples displayed a multitude of rod-shaped bacterial cells and exhibited similar ultrastructural changes. Scales were missing or, when present, denuded of epidermis.

Figure 7 Gill of a rainbow trout infested with *F. columnare.* This scanning electron microscopic (SEM) picture of an affected gill arch reveals the presence of long, thin, rod-shaped bacterial cells, approximately 0.3-0.5 µm wide and 3-10 µm long (SEM, bar = 5 µm).

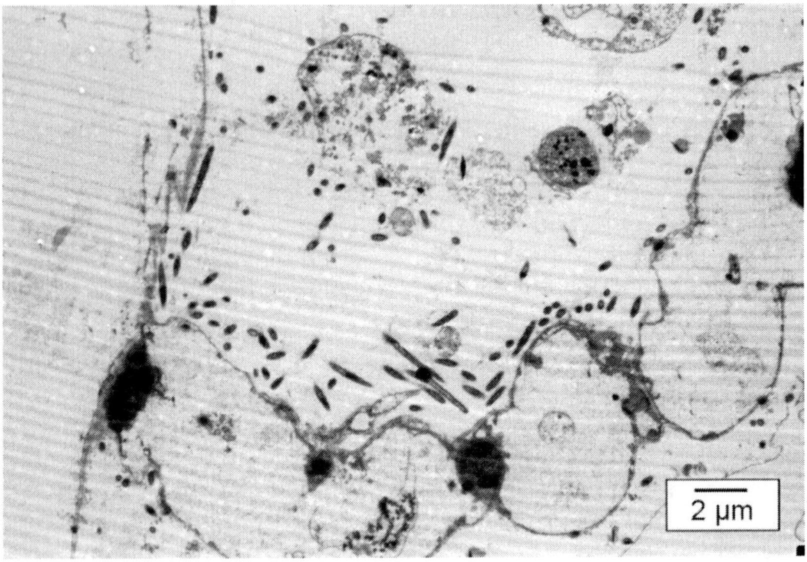

Figure 8 Gill lamellae of a koi carp following infection by *F. columnare.* This transmission electron microscopic picture clearly demonstrates the loss of structure of the gill lamellae. The *F. columnare* bacteria are lining up around the remnants of these lamellae, and are closely associated with the gill epithelium (TEM, bar = 2 µm).

In a cutaneous columnaris challenge model in koi carp, significant changes in blood parameters were observed in the infected fish [45]. For the hematologic parameters, a significant decrease was noted in Packed Cell Volume (PCV), haemoglobin concentration, red blood cell count, mean corpuscular volume and absolute lymphocyte counts. As for the biochemical parameters, marked hyponatremia, hypochloridemia and hyperglycemia were observed. Calcium and magnesium levels dropped only slightly and total serum protein and albumin-like protein concentrations decreased mildly. Alkaline phosphatase (ALP), aspartate aminotransferase (AST), lactate dehydrogenase (LDH) and creatine kinase (CK) secretions were significantly increased [45]. In a study by Řehulka and Minařík blood parameters from brook trout suffering from an acute, natural outbreak of *F. columnare* also revealed anemia, but levels of mean corpuscular volume and mean corpuscular haemoglobin were higher [15]. Total protein levels fell much below physiological parameters in infected fish. Calcium concentrations were reduced significantly. Blood Urea Nitrogen (BUN) measurements were much higher compared to normal levels. The catalytic activity of AST, alanine aminotransferases and LDH reached multiples of normal values. In contrast, ALP concentrations decreased. Hypoglycemia was noted. No data were shown on sodium or chlorine [15].

2.2 Diagnosis

Timely detection of this pathogenic agent is important to prevent its spreading and to reduce the economic loss to fish farmers.

The isolation of *F. columnare* is possible from external lesions, provided that the samples are taken from the edge of recent lesions [1]. *F. columnare* requires low nutrient media. No growth of *F. columnare* is reported on trypticase soy agar (TSA), nutrient agar or Marine 2216 agar [1]. In 1944, *F. columnare* was first cultured on Cytophaga agar, a nutrient poor medium [5]. Since then, several media, including Shieh [46] and TYES [47] have been developed in an attempt to improve growth of the bacterium. Growth does not occur in media that contain NaCl concentrations of over 0.5% or that have a pH lower than six [1]. Depending on the strain, the bacterium grows between 15 and 37°C [9, 12], with an optimal growth occurring between 25°C and 30°C. Colonies appear after 24 to 48 h of incubation [12]. Glucose does not improve growth. Considering that *F. columnare* is easily overgrown by contaminating bacteria, selective media have been developed based on inherent resistance of *F. columnare* to different antimicrobial agents, including polymyxin and neomycin [6, 8, 25]. Decostere et al. added one µg/mL tobramycin to modified Shieh medium, demonstrating this to be an effective selective supplement for isolating *F. columnare*[41].

F. columnare largely displays two colony types on solid media; smooth and rhizoid [48]. Kunttu et al. characterized a third, rough colony morphology variant [49]. The rhizoid colony variants were assigned virulent and moderately adherent, the non-rhizoid rough colony variants non-virulent and highly adherent, and the smooth colony variants non-virulent and poorly adherent [49]. Colonies of *F. columnare* are notorious for their sometimes strong adherence to the agar. This trait is also exhibited in broth cultures where yellow, filamentous clumps of bacterial cells can form a thick ring at the surface of a glass recipient [12, 50]. Adherence may be lost after several subcultures. Colonies can be recognized by their distinctive yellow pigmentation. The yellow color is due to the production of flexirubin pigments [1]. A method to identify the typical protein profile of *F. columnare* is whole-cell protein analysis [3]. Genomovar ascription has been performed using 16S-restriction fragment length polymorphism to divide *F. columnare* into three genomovars I-III [9]. Compared to RFLP, single strand conformation polymorphism (SSCP) would have an improved resolution power in the study of intraspecies diversity in *F. columnare*[51]. Polymerase chain reaction (PCR) has gained interest for definitive identification of *F. columnare* based on the selective amplification of the 16S ribosomal RNA gene using species-specific primers [52, 53]. This technique gives a conclusive identification of the organism within a few hours and eliminates the need for biochemical testing which is laborious and sometimes inconclusive [52].

Besides isolation, other methods may be used for detecting this pathogen. These include serological methods such as the enzyme-linked immunosorbent assay [54] and the fluorescent antibody test [55, 56]. Both have proven to be efficient and rapid for diagnosing columnaris disease. A loop-mediated isothermal amplification method (LAMP) for rapid detection of *Flavobacterium columnare* from infected fish organs (gills, skin and head kidneys) was established in channel catfish [57]. PCR may also be adopted, having the advantage of being able to detect very low levels of *F. columnare*[52, 53]. In addition, Panangala et al. have developed a TaqMan-based real-time PCR targeting a 113 bp nucleotide region of the chondroitin AC lyase gene of *F. columnare*[58]. This PCR is specific, sensitive and reproducible for the detection and quantitation of *F. columnare* in tissues (blood, gills and kidney) of infected fish. Moreover, real-time PCR-based methods are distinctively more advantageous than conventional PCR as they eliminate the need for detection of amplified products by gel electrophoresis, thus reducing costs, time and labour [58]. Length heterogeneity PCR (LH-PCR) is based on comparing naturally varying lengths of 16S rDNA PCR products between bacterial groups [59]. LH-PCR has also proven effective in detecting *F. columnare* in fish tissue [60].

3. PATHOGENESIS

Various reports have noted the difference in virulence among *F. columnare* strains [11, 12, 40, 42, 61, 62, 63, 64, 65]. In the last decade, various studies have attempted to elucidate the pathogenesis of columnaris disease. Although significant progress has been made, still many questions remain unanswered on how this pathogenic organism elicits disease, and information concerning the bacteriological events preceding disease and death is scarce. Li et al. related this lack of knowledge to the absence of an efficient molecular manipulation system for *F. columnare*, especially a plasmid-based inframe knockout system [66], despite two recent reports on the establishment of genetic manipulation system for the bacterium [67, 68]. Li et al. identified the type I restriction-modification system (R-M system) in *F. columnare* to improve electroporation efficiency and suggested that it would be of significant interest to examine the composition and diversity of R-M systems in strains of *F. columnare* in order to set up a suitable genetic manipulation system for the bacterium [66]. Only very recently, the full genome of *F. columnare* ATCC 49512 was sequenced [69].

3.1 Colonization

The colonization of the fish tissue is to be regarded as a complex multistep process which can be subdivided into the stages of attraction, adhesion and aggregation, requiring a step-by-step analysis. The exact factors mediating colonization have however not yet received the full attention they merit and to date remain largely unidentified.

The research group of Klesius et al. demonstrated by means of the traditional capillary tube method that the mucus from the skin and gills of catfish promotes chemotaxis of *F. columnare*[63]. The gliding motility of this bacterium is well known [1, 12, 41]. Indeed, the observation of a drop of bacteria grown in broth under a phase contrast microscope shows the slowly forward and backward gliding of *F. columnare*. Pate and Ordal described that the location of fibrillar structures spanning the gap between the outer membrane and the mucopeptide layer might play a role in the gliding motility of the *F. columnare* bacterial cells [70]. Although Klesius et al. acknowledged that they were not able to fully define the role that chemotaxis plays in the virulence of *F. columnare*, the chemotactic response of the more virulent genomovar ll isolates suggested that chemotaxis could be associated with virulence [63]. LaFrentz and Klesius developed a culture independent method to quantify the chemotactic response of *F. columnare* to skin mucus using blind-well chemotaxis chambers [71] generating similar results as stated above [63]. At least three carbohydrate-binding receptors (D-mannose, D-glucose and *N*-acetyl-D-glucosamine) associated with the capsule of *F. columnare* might be involved in chemotactic responses [62]. The gliding motility gene *gld* H was found to be significantly ($P < 0.001$) upregulated in *F. columnare* as soon as five minutes post-exposure to catfish mucus. When pretreated with D-mannose, there was no upregulation of gliding motility genes [62].

The ability to adhere is a prerequisite for the successful colonization of the host tissue. Decostere et al. performed a series of both in vivo and in vitro experiments and found that a highly virulent strain adhered more readily to the gill tissue of black mollies (*Poecilia reticulata*) than did the low virulence strain [11, 12, 42]. This research group underscored that the adhesion of *F. columnare* to the gill tissue constitutes an important step in the pathogenesis of columnaris disease. Bader et al. adopted an adhesion defective mutant of *F. columnare* in immersion challenge trials and found that the mortality was reduced with 75% and occurred 24 h later compared to the strains that still possessed the adhesion capacities, confirming the findings of the former research group [48]. In channel catfish genomovar II is considered to be more virulent than genomovar I [51, 72]. Challenge of rainbow trout with genomovar I and II isolates of *F. columnare* demonstrated a difference in the cumulative percent mortalities (CPM), with the genomovar II isolates inducing significantly higher CPM [73]. Olivares-Fuster et al. compared adhesion of *F. columnare* genomovar I and II strains to the skin and gill of channel catfish and the gill of zebrafish (*Danio rerio*) [43]. At 0.5 h post-challenge, both strains adhered to the gill of channel catfish at comparable levels, but significant differences in adhesion were found in later datapoints over time. They concluded that particular strains of *F. columnare* exhibit different levels of adhesion to their fish hosts and that adhesion to fish tissues is not sufficient to cause columnaris disease. The same statement was previously made by Klesius et al. [63].

Kunttu et al. have shown that colony morphology affects the adhesion capacity of *F. columnare* to polystyrene and put forward the hypothesis that there is a link between virulence and rhizoid colony morphology [49, 74]. The formation of different colony morphologies could be caused by changes in the cell surface components of the bacteria. It was also found that *F. columnare* changes colony morphology during experimental long-term storage in lake water, indicating the importance of different colony morphologies to bacterial survival [64, 74, 75]. Arias et al. did note that long-term starvation appears to decrease cell fitness and resulted in loss of virulence [75]. However, whether adhesion impairment was the cause of the observed virulence loss was not investigated.

F. columnare produces two types of mucus. The first one is an acidic polysaccharide and is made visible by ruthenium red staining. Another type of mucus is a basic, partially acetylated polygalactosamine, which cannot be stained with ruthenium red. Pate and Ordal described a capsular material which coated the surface of the bacterial cell that could be stained with ruthenium red [70]. They stated that the ruthenium red-positive material was probably an acid mucopolysaccharide that might be involved in the adhesive properties of the cells. The exact role of the mucopolysaccharides was however not further illuminated in this study. Decostere et al. demonstrated that the adherence capabilities of a highly virulent *F. columnare* strain to the gill were significantly reduced following treatment of the bacteria with sodium metaperiodate or incubating them with D-glucose, *N*-acetyl-D-glucosamine, D-galactose and D-sucrose [11]. Treatment with pronase or trypsin did not cause any significant inhibition of adhesion [11]. The same research group noted that the highly virulent strain had a thick capsule with a regular and dense appearance, whereas the capsule of the low virulent strain was much thinner. This made them to speculate that a lectin-like carbohydrate substance incorporated in the capsule might be partially responsible for the adhesion to the gill tissue.

Sun et al. conducted the first transcriptomic profiling of host responses to columnaris disease following an experimental immersion challenge by using illumina-based RNA-sequencing expression profiling [76]. A rhamnose-binding lectin (RBL) was detected as by far the most highly upregulated gene observed in their differentially expressed set, with expression increasing 105-fold by four hours following infection. This upregulation dramatically decreased at the later verified timepoints (24 h and 48 h), suggesting the importance of this gene during early infection events [76]. Immunohistochemical staining with antisera against an RBL in rainbow trout revealed the presence of these RBLs in mucous cells of the gill and in various cells related to innate immunity [77]. This expression pattern is consistent with the finding of *F. columnare* bacterial cell aggregates around goblet cells in common carp [11] and catfish [43] following experimental challenge with a highly virulent strain [76]. Beck et al. identified two distinct catfish families with differential susceptibilities to columnaris disease, with the one family completely resistant and the other susceptible [78]. In the susceptible family, an acute and robust upregulation in catfish RBL was observed following challenge that persisted for at least 24 h. After exposure of the catfish to different doses of the putative RBL ligands L-rhamnose and D-galactose, these sugars were found to protect the fish against columnaris disease, likely through competition with *F. columnare* binding of host RBL. Moreover, RBL expression was upregulated in fish fasted for 7 d (as compared to fish fed to satiation daily), yet expression levels returned to those of satiated fish within 4 h after re-feeding. These findings highlight putative roles for RBL in the context of columnaris disease and reveal new aspects linking RBL regulation to feed availability [79]. Further studies are needed to further pinpoint the (various) factor(s) responsible for mediating adhesion to the fish tissue.

The adhesion to the fish tissue was shown to be impacted by various environmental parameters. Using a gill perfusion model [61, 80], Decostere et al. noted that the adhesion of a highly virulent strain to the gill tissue was enhanced by a number of factors, including immersion of the gill in divalent ion water, the presence of nitrite or organic matter, and high temperatures [61]. The positive effect of high temperatures on the adhesion has also been demonstrated by Kunttu et al. [49]. Adhesion seems to decrease in vitro as salinity goes up [79]. Expression of adhesins by bacteria is regulated on two different levels. One is directed by environmental sensing and transcription of specific regulatory elements, and the other is a random switching on and off of adhesion genes by submission of the bacterial population to unpredictable environmental conditions [81]. If and how both phenomena are established in *F. columnare* remains to be elucidated.

An aggregative adhesion pattern of a highly virulent *F. columnare* strain onto gill tissue is a distinct feature in both in vivo and organ culture experiments [35, 42, 53, 61]. This results in an irregular gill surface covered by a thick mat consisting of numerous clumps of *F. columnare* bacterial cells, most likely impeding oxygen uptake and causing death of the fish. These microcolonies are not observed when a low virulence strain is used. Biofilm formation capacity was demonstrated in vitro for one *F. columnare* strain when exposed to mucus [67]. These features point towards biofilm formation potentially being an important stage in the pathogenicity of *F. columnare*, warranting its further investigation.

The developmental switch to the biofilm state is commonly regulated by quorum sensing. Quorum sensing (QS) is a mechanism of gene regulation in which bacteria coordinate the expression of certain genes in response to the presence or absence of small signal molecules. In many Gram-negative bacteria, the signal molecule is an *N*-acylhomoserine lactone (AHL) [82]. Additionally, a signaling molecule known as autoinducer-2 (AI-2) may also be employed [83]. Wagner-Döbler et al. found short-chain AHL-type activity in *Flavobacterium*sp., but no AHL-presence could be confirmed using gas chromatography-mass spectrometry (GS-MS) [84]. Using light chromatography-mass spectrometry (LC-MS), Romero et al. described the presence of short-type AHL activity in the culture media of nine *Tenacibaculum maritimum* strains, biofilm-forming members of the phylum *Bacteroidetes*, formerly referred to as "*Cytophaga-Flavobacterium-Bacteroides*" group [82]. To our knowledge, so far, QS by AHL or AI-2 has not yet been demonstrated in *F. columnare*.

3.2 Exotoxins, bacteriocins and endotoxins

It is known that polysaccharide degradation in combination with the secretion of various extracellular enzymes participate in the destruction of skin, muscle and gill tissue [1], enhancing pathogenicity. In culture, *F. columnare* produces an enzyme that degrades chondroitin sulfates A and C and hyaluronic acid, the complex polysaccharides of connective tissue. This so-called chondroitin AC lyase acts specifically on a group of acidic mucopolysaccharides found primarily in animal connective tissue [40]. No correlations were found between host origin, geographic distribution, and amount of enzyme produced by different isolates [85]. AC lyase is alleged to play a role in the virulence of *F. columnare*[49, 65]. Though high AC lyase activity solely would not be enough to induce virulence in *F. columnare* strains, both high AC lyase activity and gliding motility of the bacteria would be needed for *F. columnare* to be virulent [49]. Proteases also contribute to damaging the tissue or enhancing invasive processes [40]. Newton et al. isolated and partially characterized proteases of 23 isolates of *Flavobacterium columnare* derived primarily from channel catfish raised in the southeastern United States [86]. The bacterial isolates were divided into two groups according to the apparent molecular masses of proteases after zymographic resolution by non-reducing, non-denaturing sodium dodecyl sulfate–polyacrylamide gel electrophoresis (SDS–PAGE) with gelatin as the protease substrate. Isolates of group one, produced two proteases with apparent molecular masses of 53.5 and 58 kilodaltons. The isolates of group two, revealed three proteases with apparent molecular masses of 44.5, 48 and 59.5 kilodaltons. All isolates degraded gelatin and casein. Seven out of 23 isolates degraded elastin. More protease was produced in a medium with low nutrients and salts than in media with higher concentrations of nutrients. Moreover, a sharp increase was seen in protease production during the first 24 h of incubation and these levels dropped only slightly in the remaining days of the experiment [86].

Bacteria not only need to enter the host tissue, they also need to eliminate competitive bacteria. Different strains of *F. columnare* release specific, non-transmissible, bactericidal substances equivalent to colicins of *Escherichia coli* into the environment to reduce competition from other bacterial strains [87]. It was postulated that cells of *F. columnare* also possess multiple specific receptors for the bacteriocins, and, consequently, the cells are sensitive only to those bacteriocins for which the cell possesses receptors [87]. Antagonism of *Pseudomonas sp.* MT5 against *F. columnare* bacteria was found to be very strong in agar assays [88]. However, antagonistic baths of the *Pseudomonas* bacterial strain could not yet prevent nor treat a *F. columnare* infection following experimental challenge [23].

Differences in LPS composition between highly and low virulent strains of *F. columnare* retrieved from channel catfish have been detected [89]. Analysis of LPS by immunoblotting revealed that an avirulent mutant of a *F. columnare* isolate lacked the high molecular weight components of LPS present in virulent isolates. Based on these differences of LPS and total protein profiles, the research group was able to discriminate the attenuated mutant from other *F. columnare* strains [89]. Kunttu et al. determined LPS-profiles of different colony morphology variants of *F. columnare* in rainbow trout [49]. Colony morphology variants of the same strain produced a similar single LPS band. However, there were size differences between different strains. Both research groups used different LPS extraction and detection methods, rendering comparison of the obtained results impossible [49].

3.3 Interaction with the fish immune system

When encountering an infection, the first system activated is the innate immune system. It is assumed that the surface mucus layer, as a first physical-immunological barrier, plays an important adhesive role and that it is part of the innate host resistance of fish to disease [43]. Antibacterial characteristics of the fish mucus against *F. columnare* have been demonstrated. In an experimental challenge trial of cutaneous columnaris disease in koi carp [45], lesions were noted on locations where mucus had been removed. After incubating *F. columnare*inoculated agar plates, a lower number of colonies was counted on plates to which mucus was added. Fluorescent microscopy of a stain-based bacterial viability assay also revealed a higher number of dead bacteria in *F. columnare* cultured with mucus [45]. Staroscik and Nelson compared growth, biofilm formation, extracellular protease production and changes in protein expression of a highly virulent *F. columnare* strain cultured in media supplemented with juvenile Atlantic salmon (*Salmo salar* L.) skin mucus with the same media without mucus [67]. Interestingly, and in contrast to the reasoning resulting from the above mentioned studies, salmon surface mucus promoted the growth of *F. columnare*, induced the bacteria to grow as a biofilm and increased extracellular protease activity [67]. This might indicate that skin mucus of different fish species responds differently to the bacteria, or that it is the *F. columnare* strain which is critical in determining the antibacterial capacities of the mucus.

The fish skin in itself also forms a barrier against pathogens. Aranishi et al. immersed Japanese eel (*Anguilla japonica*) in a *F. columnare* suspension to monitor dermal nonspecific stress responses [90]. They found that cathepsins B and L activities in the infected fishes increased more than 1.5-fold over their initial values over a 48 h period, along with a 4.5-fold increase in bacteriolytic activity. These cathepsins likely participate in bacteriolysis associated with Japanese eel skin and their activities may represent an important nonspecific response of eels [90].

Some studies reveal that *F. columnare* might be able to avoid parts of the immune system. In a study by Ourth and Bachinski, the catfish alternative complement pathway (ACP) was inhibited by large amounts of sialic acid contained by Gram-negative bacterial pathogens, including *F. columnare*[10]. Sialic acid seemed to be the determining factor for the pathogenicity of *F. columnare*, as very little or no bactericidal activity was produced against this bacterium by the catfish ACP. Furthermore, the latter greatly increased after removal of sialic acid with neuraminidase [10]. Another recurrent feature in *F. columnare* infections, is the lack of an inflammatory response as observed upon inspecting affected tissues microscopically. This resulted in the hypothesis that *F. columnare* triggers the endogenous programmed cell death machinery of immune cells to evade the immune system. Do Vale et al. have already proven that apoptosis can be a very powerful pathogenic strategy by inducing lysis of phagocytic cells [91]. Sun et al. speculated that negative regulation of one of the central innate immune signaling pathways NF-κB, may be the result of immune evasion or manipulation by *F. columnare* via secreted toxins [76]. Furthermore, high levels of inducible nitric oxide synthases (iNOS), apoptotic-promoting interferon (IFN) and other members of oxidative stress responses and apoptotic pathways, such as caspase 8 and G3BP1 (Rasputin), were observed following a *F. columnare* infection. Rasputin plays an inhibitory role in negative regulation of apoptosis and was highly downregulated at all examined time-points [76]. This could explain why hardly any inflammatory cells appear in early infections with *F. columnare* as was described by Morrison et al. [36].

Ourth and Wilson demonstrated that *F. columnare* is resistant to the bactericidal action of serum of non-immunized catfish via the ACP [92]. Ourth and Bachinski confirmed this finding [93], but also demonstrated that the classical, antibody-mediated complement pathway is highly effective in killing *F. columnare*[10]. The virulence of *F. columnare* can also be influenced by transferrin [94]. The survival time of fish experimentally challenged with *F. columnare* by intraperitoneal injection with iron-free human transferrin (Sigma) was reduced when iron was injected prior to exposure. The effect of iron was only evident in one of the two strains examined when the challenge was delivered via immersion. These results indicate that iron depletion may limit the virulence of *F. columnare* more in systemic infections than in external infections. This hypothesis is supported by data indicating that administration of transferrin prior to challenge increased survival after challenge by injection but had little or no effect on bath-challenged fish [94]. However, the results found were not consistent for all strains tested and no statistical data were presented.

Several immunization experiments adopting different administration routes, have proven that fish can be protected from subsequent *F. columnare* infections by activating the adaptive immune system [20, 22, 95, 96, 97, 98]. High agglutinin titers and good protection were obtained in trout following subcutaneous or intraperitoneal injection with heat-killed *F. columnare* cells [22]. Schachte and Mora obtained a high agglutinin titer in channel catfish by intramuscular injection of heat-inactivated cells of the pathogen, but the actual level of protection was not examined [99]. Becker and Fujihara reported that rainbow trout injected with heat-killed cells produced an agglutinating titer of 1:5120, and that 60-70% of the trout later survived an injection of 10^6live *F. columnare* cells [96]. Tilapia (*Oreochromis niloticus* (L.)) could mount a significant humoral response in plasma and cutaneous mucus to *F. columnare* after intraperitoneal immunization with formalin-killed sonicated cells in Freund's complete adjuvant [97]. Protection levels were not investigated in the latter study. Protection was obtained after oral immunization with heat-killed or formalin-killed cells of *F. columnare* in the fish feed of three-month-old coho salmon (*Oncorhynchus kisutch*) [22, 100]. However, the protection as observed in these studies did not go hand-in-hand with high agglutinin titers. Bath-immunization with a bacterin was shown to protect carp against experimental challenge but antibodies against *F. columnare* were not detected in sera from immunized fish [101]. Song's work with bacterins demonstrated unequivocally that immersion vaccination could result in high levels of protection, but field test results were inconsistent [102]. Song was able to demonstrate that there was cross-protection between isolates and that there may be a common protective antigen among the strains tested [102]. Polyvalent vaccines have also been tested. Using intraperitoneal injections of a combination of formalin-killed *F. columnare, Aeromonas salmonicida* and *Yersinia ruckeri* antigens, interference from *A. salmonicida* antigen was shown to suppress responses to two other antigens [100]. Commercially available oral and bathing vaccines have been successfully tested in largemouth bass (*Micropterus salmoides*) fry and salmon, respectively [95, 103]. Vaccination trials are further elaborated on below.

4. IMPORTANCE OF ENVIRONMENTAL FACTORS

Karvonen et al. described the effect of global warming on the prevalence of different fish parasites and bacteria [104]. *F. columnare* could be one of the many taking advantage of this phenomenon. Indeed, transmission of columnaris disease is more efficient in higher temperatures [30, 105]. Holt et al. found that when steelhead trout (*Salmo gairdneri*) or coho salmon (*O. kisutch*) experimentally infected with *F. columnare*were held in water at 12 to 20°C, mortality increased with temperature [106]. As stated above, adhesion to gill tissue of a highly virulent *F. columnare* strain is enhanced at increased temperature [61] and chondroitin AC lyase activity of this pathogen increases along with the temperature [107]. The influence of rearing density and water temperature in rainbow trout was studied by Suomalaien et al. [105]. Normal rearing densities with high temperatures (23°C) proved to increase both transmission rate of columnaris disease and mortality in the fish. Normal densities at low temperatures (18°C) did not affect mortality, but increased the transmission rate of columnaris disease [105].

Columnaris disease is furthermore influenced by water quality. Decostere et al. observed significantly higher bacterial titers on the gills when organic matter or nitrite were added to an organ bath when performing ex vivo trials with *F. columnare*[61]. They discussed that organic matter could concentrate nutrients to feed the bacteria and that degrading enzymes could be kept in close contact with the host tissue. Bacterial titers were furthermore markedly lower in gills placed in an organ bath with distilled water with or without 0.03% NaCl compared to the titers of gills suspended in Ringer solution or in formulated water containing divalent ions (magnesium and calcium) [61]. Morris et al. noted that the survival of the fish exposed to *F. columnare*significantly increased as unionized ammonia concentrations increased [108]. These results suggest that complex interactions can complicate prediction of the responses of fish to concurrent chemical stressors and bacterial pathogens [108]. Bandilla et al. described that co-infections of ectoparasites with *F. columnare*increased the susceptibility of rainbow trout to the bacterial pathogen [109]. Compared with single infections, the mortality was significantly higher and the onset of disease condition occurred earlier in fish which were concomitantly infected by the parasite *Argulus coregoni* and *F. columnare*[109].

5. EXPERIMENTAL TRIALS/PATHOGENICITY TESTS

A reliable and reproducible experimental infection model is crucial for studying bacterium-host interactions and evaluating the efficacy of both curative and preventive measures. Most researchers recognize, in experimental challenges with *F. columnare,* the fine and delicate balance and complex interplay between the bacterial cells, fish and environment in the successful reproduction of columnaris disease.

Hitherto, columnaris disease has been reproduced experimentally in black mollies (*Poecilia sphenops*) [12], channel catfish [43, 44, 110, 111], eel [94], golden shiner (*Notemigonus crysoleucas*) [112], koi carp [45], rainbow trout [49, 65, 105, 113], tilapia [94] and zebrafish [43, 44].

In the various experimental infection trials, several ways to both cultivate and harvest the bacterial cells for inoculation were adopted. Different media were used to cultivate the bacteria, including Hsu-Shotts or modified Hsu-Shotts medium [79, 114], Ordal's medium [115], Anacker and Ordal's medium also referred to as "Cytophaga medium" [94], *F. columnare* growth medium broth [115] and Shieh or modified Shieh medium [12, 23, 43, 44, 45, 105]. The bacteria were mostly grown on a shaker [43, 44, 45, 75, 109, 110, 111]. Temperatures and incubation times of the bacteria varied from 21°C to up to 30°C and from 24 to 48 h, respectively.

To reproduce columnaris disease experimentally, largely two inoculation routes were adopted, viz. bath (immersion) and injection. With regard to immersion challenges, mostly full-grown broth cultures were added to the water with water temperatures of the immersion water varying from 18°C to 30°C [12, 23, 42, 43, 44, 111]. Only in a minority of cases were the bacterial cells first harvested through centrifugation and consequently resuspended before being added to the inoculation bath [33, 45, 114]. Times during which the fish were bath exposed varied from 15 min [13, 72, 105] over 30 min [12, 23, 43, 44, 45, 109] to 45 min [42], one hour [79, 94] and longer [110, 111]. When the bacterial cells were administered through injection, then they were retrieved from centrifuged broth cultures following discard of the supernatant [12] or immediately taken from a full-grown broth culture without primary centrifugation [13, 58, 94]. A third inoculation route consisted of adding dead fish that were first injected intraperitoneally with a virulent *F. columnare* strain, to the aquarium water in which fish were immersed [20]. In some researches, the successful reproduction of columnaris disease depended on or the severity of the elicited disease increased by abrading the skin or gills of the fish [45, 94, 105] whereas in another study, clipping the fin had no effect on success after immersion challenge with *F. columnare*[12].

The infection route, which determines the way the bacterial cells are to be grown/collected, also defines the disease producing capacity of the adopted strains and the disease picture they elicit. Indeed, especially for low virulent strains, higher morbidity and mortality rates were noted after injection of fish as compared to inoculation through immersion [12]. Pacha and Ordal reported a similar finding [34]. They stated that contact with highly virulent strains induced infection and disease contraction more than intramuscular injection, whereas injection of low virulent bacteria more readily induced infection and disease contraction than did contact [34].

Environmental conditions can also highly influence morbidity and mortality rates during a challenge trial. To illustrate the dramatic effects of water temperature on the level of mortalities, the investigation of Holt et al. is particularly relevant [106]. This team challenged steelhead trout (*Oncorhynchus mykiss*), Chinook salmon (*O. tshawytscha*) and coho salmon with *F. columnare*, via the water-borne route. At a water temperature of 9.4°C, there were no mortalities attributable to *F. columnare*. By increasing the temperature to 12.2°C, mortality became as high as 4-20%; at 20.5°C, all the steelhead trout and coho salmon and 70% of the Chinook salmon died [106]. High (23°C) rearing temperatures also increased mortality significantly in rainbow trout compared to lower (18°C) temperatures [105]. Fish density at fish farms is also a key player in influencing mortality in an outbreak of columnaris disease as mortality rates started earlier and remained higher when fish were stocked at high densities [105]. Water flow is another important factor since it acts as a determining factor with regard to the contact time between the possibly present bacteria and the host tissue. High mortality rates were observed in elvers kept in standing water, while in aquaria with running water, mortality in elvers was reduced by half [94].

6. CONTROL

6.1 Preventive measures

Management plays a key role in the prevention of the disease. Cunningham et al. showed that some commonly recorded production variables (feed consumption, pond depth, ammonia levels and stocking events) were associated with columnaris disease outbreaks and, if monitored, could help identify "at risk" ponds prior to disease outbreaks [116]. Suomalainen et al. pointed out that reduction of fish density could be used in the prevention of columnaris disease especially if water temperature is high [105]. As lower rearing density can also decrease the transmission of ectoparasites and penetrating endoparasites, it could be an efficient tool in ecological disease management as a whole [105]. High nitrite levels and organic load can stimulate the adherence capacity of *F. columnare*[61], and therefore it is important to control these parameters as well. Furthermore, water treatment could aid in averting a bacterial outbreak. Conrad et al. reported that ozone treatment of water significantly reduced the numbers of added *F. columnare*, which could be a practical method of prevention [117]. Salt and acidic bath treatments could be used to disinfect water contaminated by *F. columnare*[23]. An in vivo immersion challenge of *F. columnare* in channel catfish and goldfish (*Carassius auratus* L.) revealed decreasing mortality as salinity goes up, with significantly lower and no mortalities when salinity reaches values of 1.0‰ and between 3 and 9‰, respectively [79]. If the fish can be adapted to salt levels of at least 1.0‰, this method could be used as a possible preventive measure in columnaris disease. Shoemaker et al. suggested that in the absence of natural food, juvenile channel catfish should be fed at least once every other day to apparent satiation in order to maintain normal physiological function and improve resistance to *F. columnare,* since deprivation reduced innate resistance of catfish to columnaris disease [118].

Besides optimizing and adjusting management practices, chemical agents can also be adopted as a preventive approach. Davis concluded that the development or intensification of columnaris disease could be prevented by treating the fish for 20 min in a copper sulfate ($CuSO_4$) bath at 37 mg/L (1:30 000) or by adding copper sulfate to pond water at 0.5 mg/L [4]. Dipping the fish one at a time in a 1:2000 copper-sulfate for one to two minutes was also proven to be effective in the prevention of the disease. Rogers suggested the addition of potassium permanganate ($KMnO_4$) to the water at 2 mg/L [119]. Darwish et al. also confirmed the prophylactic value of $KMnO_4$ at doses around 2 mg/L [120]. Prophylactic treatment of channel catfish with 15 mg/L chloramine-T reduced fish mortality from a *F. columnare* infection from 84–100% to 6–14% [121]. Thomas-Jinu and Goodwin demonstrated the efficacy of prophylactically given oxytetracycline against mortality in channel catfish and also reported zero mortality for the combination of sulphadimethoxine and ormetoprim in feed prior to bacterial challenge with four highly virulent strains of *F. columnare*[111].

Another method to prevent columnaris disease is through vaccination. Although vaccination trials have not always been successful, success rates have increased as knowledge on fish immunity and its role in the defense against bacterial diseases continues to expand. Bath immunization with a bacterin was shown to protect carp against experimental challenge, but no agglutinin could be found in sera from immunized fish [101]. Immersion of channel catfish in a bacterin, when performed each year, induced a significant decrease in mortality compared to unvaccinated fish [98]. Fujihara and Nakatani obtained protection against columnaris disease in 3-month-old coho salmon by oral immunization with heat-killed cells of *F. columnare* incorporated into fish feed [22]. Ransom proved that prolonged feeding (over three months) of formalin-killed bacteria provided high levels of protection [100]. Ourth and Bachinski proposed that strains containing sialic acid could serve as potential vaccine strains for columnaris disease [10]. As stated above, immunization with formalin-killed sonicated cells in Freund's complete adjuvant injected intraperitoneally in tilapia resulted in a significant systemic humoral response within two weeks and antibody levels almost tripled following secondary immunization [97]. At 10 weeks postimmunization, the mean antibody titer remained significantly elevated. Antibodies were also observed in cutaneous mucus of these fish at six and eight weeks postimmunization [97]. An attenuated immersion vaccine currently is registered for the use in channel catfish in the USA [20]. Fry between 10 to 48 days post hatch that were vaccinated through immersion achieved a relative percent survival (RPS) between 57 and 94% following *F.*

columnare challenge. This vaccine was also proven to be efficient in largemouth bass fry resulting in RPS values between 74 and 94%, depending on the vaccine dose [20]. Bebak et al. tested a commercial oral vaccine in largemouth bass fry [95]. Vaccinated fish had a 43% lower risk of death by *F. columnare* during the field trial [95]. An immersion vaccine consisting of a bacterin of *F. columnare*was also brought to the market in the USA as an aid in the prevention of columnaris disease in healthy salmonids of over three grams [103].

Probiotics appear to be a promising way in the prevention of different bacterial diseases in aquaculture [122]. Boutin et al. isolated different strains of commensal bacteria from the skin mucus of unstressed brook charr (*Salvelinus fontinalis*) which in vitro revealed antagonistic effects against *F. columnare*[123]. The strains were mixed and used to treat columnaris disease in an in vivo experiment. This resulted in a significant decrease of mortality indicating the potential use of these probiotic candidates in the efficient and durable management of columnaris disease. The immunostimulants β-glucan and β-hydroxy-β-methylbutyrate raised the levels of immune function parameters, but did not improve survival in challenge trials with *F. columnare* at any concentration of the stimulants used [113]. The research group of Sink et al. demonstrated that mortality rates in golden shiners fed high-fat diets with a dairy-yeast prebiotic were significantly lower after a challenge trial with *F. columnare*[112].

Recent studies have demonstrated genetic variation in resistance towards *F. columnare*[73, 78, 124]. Arias et al. presented experimental data on the susceptibility to columnaris disease of hybrid catfish (female channel catfish × male blue catfish (*I. furcatus*)) (C×B) [124]. Under experimental conditions, C×B hybrids were significantly more resistant to columnaris disease caused by a highly virulent strain of *F. columnare* belonging to genomovar II compared to channel catfish and blue catfish. Beck et al. also found one of the two investigated catfish families to be completely resistant towards *F. columnare* resulting in no mortality after inoculation with the bacteria [78]. These *F. columnare* resistant families could be of great financial importance in the catfish industry. Interestingly, LaFrentz et al. found that the two families that exhibited the highest CPM after *F. columnare* challenges, had the lowest CPM following *E. ictaluri* challenge, the latter being the most prevalent bacterium to cause disease and mortality in the catfish industry [73]. Further research on larger numbers of families is needed to determine whether there is any genetic correlation between resistance to *E. ictaluri* and resistance to *F. columnare*, the two leading bacterial diseases in the catfish industry [125].

6.2 Curative approach

Treatment of columnaris disease using antimicrobial agents has known different success rates. External treatments are possible only in early stages of the disease, when the infection is still superficial [6]. Drugs which have been used effectively in bath therapies are chloramphenicol [126], nifurpirinol [127, 128], nifurprazine [129, 130] and oxolinic acid [131, 132]. If the disease is in an advanced stage and/or signs of septicaemia are observed, it is necessary to administer antimicrobials in the feed. Oxytetracycline given orally for up to 10 days proved effective in early as well as advanced outbreaks of columnaris disease in Pacific salmon (*Salmo salar*) [6, 133]. Lack of success of orally administered oxytetracycline has also been reported [134]. Sulfonamides, such as sulfamerazine and sulfamethazine, can be used orally but would be less effective than other drugs [6, 135]. Nitrofuran can also be administered orally for 3 to 5 days [6, 96, 130]. Gaunt et al. demonstrated the efficacy of florfenicol in the feed against columnaris disease in channel catfish [136]. Darwish et al. also illustrated the clear benefit of florfenicol against a mixed infection of *A. hydrophila* and *F. columnare* in Sunshine bass (hybrid striped bass, *Morone chrysops* female × *Morone saxatilis* male) [115]. The excessive use of antimicrobial agents to withstand *F. columnare* has its negative attributes though [20]. These include possible allergic reactions elicited in the user after food contact [137]. Potential impacts on human health resulting from the emergence of drug-resistant bacteria and the associated risk of transfer of these resistant traits to the environment and human-associated bacteria are also a major concern [137]. Declercq et al. demonstrated the in vitro multiple resistance of *F. columnare* strains originating from ornamental fish toward several clinically important antibiotics, such as quinolones and tetracyclines [138]. The results obtained in this study appeal for less prudent use of antimicrobials especially in the ornamental fish industry and therefore urges to limit their use and to focus on the development of alternative curative and preventive measures against columnaris disease.

Besides resorting to antimicrobial agents, chemicals have also been adopted in the curative treatment of columnaris disease. In a study by Thomas-Jinu and Goodwin, the herbicide Diquat® (Zeneca Agricultural Products, Wilmington, DE, USA) was shown to significantly reduce channel catfish mortalities to zero percent after challenge with *F. columnare*[111]. The herbicide has also proven to be effective in the treatment of columnaris disease in salmonids [6, 139]. Copper sulfate [4] and potassium permanganate [119] are among the older chemicals used for treatment and prevention of columnaris disease in pond fishes. The organic load in water affects the efficacy of potassium permanganate, but methods are available to estimate that organic load and compensate by adjusting the level of the chemical [6]. Darwish et al. suggested that copper sulfate has clear therapeutic value against *F. columnare* infections in channel catfish when treated in an ultralow flow-through system during 4 h [110]. Thomas-Jinu and Goodwin on the other hand proved the inefficacy of this same chemical against columnaris disease, which might be due to the advanced stage of the experimental infection at the time of treatment [111]. An in vitro assay of commercial products containing peracetic acid was proven to be effective against *F. columnare* infection [140]. Laanto et al. reported some *Flavobacterium*sp. phage lysates to inhibit growth or lyse the bacterial cultures [141]. The authors recommended that the causative agent of this strong inhibition or lysis should be studied further for the possibility of developing antimicrobial agents. Prasad et al. also described the successful use of *F. columnare* phage FCP1 to combat columnaris disease in walking catfish (*Claries batrachus*) [142]. Phage treatment led to disappearance of gross symptoms, resulted in a negative bacteriological test, a detectable phage, and 100% survival in experimentally infected *C. batrachus*. The result of this study opens new perspectives for the treatment of columnaris disease elicited by antimicrobial resistant *F. columnare* strains [142].

7. CONCLUSION

Despite the worldwide importance of columnaris disease and the multitude of studies focusing on its causative agent, only snippets of the bacterium-host interactions have hitherto been exposed. This results in major knowledge gaps still prevailing on amongst others how the pathogen is able to establish and maintain a grip on the skin and gill tissue and elicit disease and mortality. Methods for rapid identification of the bacterium in the environment, host tissue or in culture have indeed evolved rapidly and some techniques permit to trace even a few bacterial cells. However, the rationale for the difference in virulence among *F. columnare* strains remains to be unraveled. Full genome sequencing of the reference strain has been an important step towards gaining insight into the pathogenesis of columnaris disease. Efficient molecular manipulation systems could be of great importance to further explore this domain. Genetic knockouts could in effect illuminate virulence factors involved in the pathogenesis of this disease. Although various models to experimentally induce columnaris disease are available, there is a lack of a standardized experimental model hampering the comparison of retrieved data in between research groups and the drawing of consistent conclusions. Consistency in the cultivation protocol of the bacteria in terms of incubation temperature, infection dose, exposure time and infection route is desirable. Extrapolation of a standardized model from one to another fish species is not possible rendering it necessary to set-up an experimental model for each fish sensitive species. Besides further elucidating the pathogenesis of columnaris disease, the development of efficient curative and effective preventive measures also needs to continue. With the upcoming antimicrobial resistance, focus should be laid on the search for preventive measures. Management plans trying to optimize or adjust fish densities and control water quality parameters are a first critical step in controlling columnaris disease. Vaccines, chemotherapeutics and probiotics could possibly have a promising future in the prevention and control of the disease as some studies have already demonstrated. Another interesting envisaged method to mitigate columnaris disease is the creation of a fish breeding program introducing *F. columnare* resistant fish. Data generated through pathogenesis studies will allow for the further development and optimization of cost-effective and environmentally friendly prevention methods of columnaris disease in the various susceptible fish species.

10. ACKNOWLEDGEMENTS

The authors thank Prof. Paul Simoens of the Department of Morphology of the Faculty of Veterinary Medicine at Ghent University for critically reviewing this manuscript. Prof. Koen Chiers of the Department Bacteriology, Pathology and Avian Diseases of the Faculty of Veterinary Medicine at Ghent University is gratefully acknowledged for his expertise and help with the interpretation of the fish pathology. Prof. Maria Cornelissen and Leen Pieters of the Department of Anatomy, Embryology, Histology and Medical Physics of the Faculty of Medicine and Health Sciences at Ghent University are accredited for their cooperation. Furthermore, we also give thanks to Lobke De Bels, Jurgen De Craene, Bart De Pauw and Patrick Vervaet for their technical assistance. The financial support by the Special Research Grant (Bijzonder Onderzoeksfonds, BOF, grant number 01Z06210) of Ghent University, Belgium is gratefully acknowledged.

REFERENCES

1. Bernardet JF, Bowman JP: The genus *Flavobacterium*. The Prokaryotes: A Handbook on the Biology of Bacteria: Volume 7: Proteobacteria: Delta and Epsilon Subclasses. Deeply Rooting Bacteria. Edited by: Dworkin M, Falkow S. 2006, New York: Springer Science+Business Media, LLC, 481-531.
2. Bernardet JF, Nakagawa Y, Holmes B: Proposed minimal standards for describing new taxa of the family *Flavobacteriaceae* and emended description of the family. Int J Syst Evol Microbiol. 2002, 52: 1049-1070. 10.1099/ijs.0.02136-0.
3. Bernardet JF, Segers P, Vancanneyt M, Berthe M, Kersters K, Vandamme P: Cutting a Gordian knot: emended classification and description of the genus *Flavobacterium*, emended description of the family *Flavobacteriaceae*, and proposal

of *Flavobacterium hydatis* nom. nov. (basonym, *Cytophaga aquatilis* Strohl and Tait 1978). Int J Syst Evol Bacteriol. 1996, 46: 128-148. 10.1099/00207713-46-1-128.

4. Davis HS: A new bacterial disease in freshwater fishes. United States Bureau of Fisheries Bulletin. 1922, 38: 37-63.
5. Ordal EJ, Rucker RR: Pathogenic myxobacteria. PSEBM. 1944, 56: 15-18.
6. Bullock GL, Hsu TC, Shotts EB: Columnaris disease of fishes. USFWS Fish Disease Leaflet. 1986, 1986: 1-9.
7. Garnjobst L: *Cytophaga columnaris* (Davis) in pure culture: a myxobacterium pathogenic to fish. J Bacteriol. 1945, 49: 113-128.
8. Bernardet J-F, Grimont PAD: Deoxyribonucleic acid relatedness and phenotypic characterization of *Flexibacter columnaris* sp. nov., norn. rev., *Flexibacter psychrophilus* sp. nov., nom. rev., and *Flexibacter maritimus* Wakabayashi, Hikida, and Masumura 1986. IJSB. 1989, 39: 346-354.
9. Triyanto , Wakabayashi H: Genotypic diversity of strains of *Flavobacterium columnare* from diseased fishes. Fish Pathol. 1999, 34: 65-71. 10.3147/jsfp.34.65.
10. Ourth DD, Bachinski LM: Bacterial sialic acid modulates activation of the alternative complement pathway of channel catfish (*Ictalurus punctatus*). Dev Comp Immunol. 1987, 11: 551-564. 10.1016/0145-305X(87)90044-9.
11. Decostere A, Haesebrouck F, Van Driessche E, Charlier G, Ducatelle R: Characterization of the adhesion of *Flavobacterium columnare (Flexibacter columnaris)* to gill tissue. J Fish Dis. 1999, 22: 465-474. 10.1046/j.1365-2761.1999.00198.x.
12. Decostere A, Haesebrouck F, Devriese LA: Characterization of four *Flavobacterium columnare* (*Flexibacter columnaris*) strains isolated from tropical fish. Vet Microbiol. 1998, 62: 35-45. 10.1016/S0378-1135(98)00196-5.
13. Figueiredo HCP, Klesius PH, Arias CR, Evans J, Shoemaker CA, Pereira DJ, Peixoto MTD: Isolation and characterization of strains of *Flavobacterium columnare* from Brazil. J Fish Dis. 2005, 28: 199-204. 10.1111/j.1365-2761.2005.00616.x.
14. Morley NJ, Lewis JW: Consequences of an outbreak of columnaris disease (*Flavobacterium columnare*) to the helminth fauna of perch (*Perca fluviatilis*) in the Queen Mary reservoir, south-east England. J Helminthol. 2010, 84: 186-192. 10.1017/S0022149X09990459.
15. Řehulka J, Minařík B: Blood parameters in brook trout *Salvelinus fontinalis* (Mitchill, 1815), affected by columnaris disease. Aquac Res. 2007, 38: 1182-1197. 10.1111/j.1365-2109.2007.01786.x.
16. Soto E, Mauel MJ, Karsi A, Lawrence ML: Genetic and virulence characterization of *Flavobacterium columnare* from channel catfish (*Ictalurus punctatus*). J Appl Microbiol. 2008, 104: 1302-1310. 10.1111/j.1365-2672.2007.03632.x.
17. Suomalainen L-R, Bandilla M, Valtonen ET: Immunostimulants in prevention of columnaris disease of rainbow trout, *Oncorhynchus mykiss* (Walbaum). J Fish Dis. 2009, 32: 723-726. 10.1111/j.1365-2761.2009.01026.x.
18. Hawke JP, Thune RL: Systemic isolation and antimicrobial susceptibility of *Cytophaga columnaris* from commercially reared channel catfish. J Aquat Anim Health. 1992, 4: 109-113. 10.1577/1548-8667(1992)004<0109:SIAASO>2.3.CO;2.
19. Wagner BA, Wise DJ, Khoo LH, Terhune JS: The epidemiology of bacterial diseases in food-size channel catfish. J Aquat Anim Health. 2002, 14: 263-272. 10.1577/1548-8667(2002)014<0263:TEOBDI>2.0.CO;2.
20. Shoemaker CA, Klesius PH, Drennan JD, Evans J: Efficacy of a modified live *Flavobacterium columnare* vaccine in fish. Fish Shellfish Immun. 2011, 30: 304-308. 10.1016/j.fsi.2010.11.001.
21. Barker GA, Smith SN, Bromage NR: Effect of oxolinic acid on bacterial flora and hatching success rate of rainbow trout, *Oncorhynchus mykiss*, eggs. Aquaculture. 1990, 9: 205-222.
22. Fujihara MP, Nakatani RE: Antibody protection and immune responses of rainbow trout coho salmon to *Chondrococcus columnaris*. J Fish Res Bd Canada. 1971, 28: 1253-1258. 10.1139/f71-191.

23. Suomalainen L-R, Tiirola M, Valtonen ET: Treatment of columnaris disease of rainbow trout: low pH and salt as possible tools?. Dis Aquat Organ. 2005, 65: 115-120.
24. Kunttu HMT, Valtonen ET, Jokinen EI, Suomalainen L-R: Saprophytism of a fish pathogen as a transmission strategy. Epidemics. 2009, 1: 96-100. 10.1016/j.epidem.2009.04.003.
25. Fijan FJ: Antibiotic additives for isolation of *Chondrococcus columnaris* from fish. Appl Microbiol. 1968, 17: 333-334.
26. Chowdhury MBR, Wakabayashi H: Survival of four major bacterial fish pathogens in different types of experimental water. Bangladesh J Microbiol. 1990, 7: 47-54.
27. Ross AJ, Smith CA: Effect of temperature on survival of *Aeromonas liquefaciens*, *Aeromonas salmonicida*, *Chondrococcus columnaris*, and *Pseudomonas fluorescens*. Progr Fish Cult. 1974, 36: 51-52. 10.1577/1548-8659(1974)36[51:EOTOSO]2.0.CO;2.
28. Kunttu HMT, Sundberg L-R, Pulkkinen K, Valtonen ET: Environment may be the source of *Flavobacterium columnare* outbreaks at fish farms. Environ Microbiol Rep. 2012, 4: 398-402. 10.1111/j.1758-2229.2012.00342.x.
29. Wakabayashi H: Columnaris disease. Bacterial diseases of fish. Edited by: Inglis V, Roberts RJ, Bromage NR. 1993, Oxford, England: Blackwell Scientific Publications, 23-39.
30. Pulkkinen K, Suomalainen L-R, Read AF, Ebert D, Rintamäki P, Valtonen ET: Intensive fish farming and the evolution of pathogen virulence: the case of columnaris disease in Finland. Proc Biol Sci. 2010, 277: 593-600. 10.1098/rspb.2009.1659.
31. Rucker RR, Earp BJ, Ordal EJ: Infectious diseases of Pacific salmon. Trans Am Fish Soc. 1953, 83: 297-312.
32. Pacha RE, Ordal EJ: Histopathology of experimental columnaris disease in young salmon. J Comp Path. 1967, 77: 419-423. 10.1016/0021-9975(67)90027-8.
33. Foscarini R: Induction and development of bacterial gill disease in the eel (*Anguilla japonica*) experimentally infected with *Flexibacter columnaris:* pathological changes in the gill vascular structure and in cardiac performance. Aquaculture. 1989, 78: 1-20. 10.1016/0044-8486(89)90002-1.
34. Pacha RE, Ordal EJ: Myxobacterial diseases of salmonids. A symposium on diseases of fishes and shellfishes. Edited by: American Fisheries Society special publication, Snieszko SF. 1970, 243-257. 5
35. Decostere A: *Flavobacterium columnare* infections in fish: the agent and its adhesion to the gill tissue. Verhandelingen - KAGB. 2002, 64: 421-430.
36. Morrison C, Cornick J, Shum G, Zwicker B: Microbiology and histopathology of "saddle-back" disease of underyearling Atlantic salmon, *Salmo salar.* J Bacteriol. 1981, 78: 225-230.
37. Decostere A, Haesebrouck F: Outbreak of columnaris disease in tropical aquarium fish. Vet Rec. 1999, 144: 23-24. 10.1136/vr.144.1.23.
38. Ferguson HW: Systemic Pathology of Fish. 2006, London: Scottian Press, 2
39. Michel C, Messiaen S, Bernardet J-F: Muscle infections in imported neon tetra, *Paracheirodon innesi* Myers: limited occurrence of microsporidia and predominance of severe forms of columnaris disease caused by an Asian genomovar of *Flavobacterium columnare.* J Fish Dis. 2002, 25: 253-263. 10.1046/j.1365-2761.2002.00364.x.
40. Dalsgaard I: Virulence mechanisms in *Cytophaga psychrophila* and other *Cytophaga*-like bacteria pathogenic for fish. Ann Rev Fish Dis. 1993, 3: 127-144.
41. Decostere A, Haesebrouck F, Devriese L: Development of a medium for the selective isolation of *Flavobacterium columnare* from diseased fish. J Clin Microbiol. 1997, 35: 322-324.
42. Decostere A, Haesebrouck F, Charlier G, Ducatelle R: The association of *Flavobacterium columnare* strains of high and low virulence with gill tissue of

black mollies (*Poecilia sphenops*). Vet Microbiol. 1999, 67: 287-298. 10.1016/S0378-1135(99)00050-4.
43. Olivares-Fuster O, Bullard SA, McElwain A, Llosa MJ, Arias CA: Adhesion dynamics of *Flavobacterium columnare* to channel catfish *Ictalurus punctatus* and zebrafish *Danio rerio* after immersion challenge. Dis Aquat Organ. 2011, 96: 221-227. 10.3354/dao02371.
44. Bullard SA, McElwain A, Arias CR: Scanning electron microscopy of "saddleback" lesions associated with experimental infections of *Flavobacterium columnare* in channel catfish, *Ictalurus punctatus* (*Siluriformes: Ictaluridae*), and zebrafish, *Danio rerio* (*Cypriniformes: Cyprinidae*). J World Aquacult Soc. 2011, 42: 906-913. 10.1111/j.1749-7345.2011.00527.x.
45. Tripathi NK, Latimer KS, Gregory CR, Ritchie BW, Wooley RE, Walker RL: Development and evaluation of an experimental model of cutaneous columnaris disease in koi *Cyprinus carpio*. J Vet Diagn Invest. 2005, 17: 45-54. 10.1177/104063870501700109.
46. Shieh HS: Studies on the nutrition of a fish pathogen, *Flexibacter columnaris*. Microbios Letters. 1980, 13: 129-133.
47. Holt RA: Cytophaga psychrophila, the causative agent of bacterial cold-water disease in salmonid fish. 1988, Corvallis, OR: Oregon State University
48. Bader JA, Shoemaker CA, Klesius PH: Production, characterization and evaluation of virulence of an adhesion defective mutant of *Flavobacterium columnare* produced by β-lactam selection. Lett Appl Microbiol. 2005, 40: 123-127. 10.1111/j.1472-765X.2004.01641.x.
49. Kunttu HMT, Jokinen EI, Valtonen ET, Sundberg L-R: Virulent and nonvirulent *Flavobacterium columnare* colony morphologies: characterization of chondroitin AC lyase activity and adhesion to polystyrene. J Appl Microbiol. 2011, 111: 1319-1326. 10.1111/j.1365-2672.2011.05149.x.
50. Bernardet JF: "*Flexibacter columnaris*": first description in France and comparison with bacterial strains from other origins. Dis Aquat Organ. 1989, 6: 37-44.
51. Olivares-Fuster O, Baker JL, Terhune JS, Shoemaker CA, Klesius PH, Arias CR: Host-specific association between *Flavobacterium columnare* genomovars and fish species. Syst Appl Microbiol. 2007, 30: 624-633. 10.1016/j.syapm.2007.07.003.
52. Darwish AM, Ismaiel AA, Newton JC, Tang J: Identification of *Flavobacterium columnare* by a species-specific polymerase chain reaction and renaming of ATCC43622 strain to *Flavobacterium johnsoniae*. Mol Cell Probes. 2004, 18: 421-427. 10.1016/j.mcp.2004.07.002.
53. Welker TL, Shoemaker CA, Arias CR, Klesius PH: Transmission and detection of *Flavobacterium columnare* in channel catfish *Ictalurus punctatus*. Dis Aquat Organ. 2005, 63: 129-138.
54. Shoemaker CA, Shelby RA, Klesius PH: Development of an indirect ELISA to detect humoral response to *Flavobacterium columnare* infection of channel catfish, *Ictalurus punctatus*. J Appl Aquaculture. 2003, 14: 43-52. 10.1300/J028v14n01_03.
55. Panangala VS, Shelby RA, Shoemaker CA, Klesius PH, Mitra A, Morrison EE: Immunofluorescent test for simultaneous detection of *Edwardsiella ictaluri* and *Flavobacterium columnare*. Dis Aquat Organ. 2006, 68: 197-207.
56. Speare DJ, Markham RJ, Despres B, Whitman K, MacNair N: Examination of gill from salmonids with bacterial gill disease using monoclonal antibody probes for *Flavobacterium branchiophilum* and *Cytophaga columnaris*. J Vet Diagn Invest. 1995, 7: 500-505. 10.1177/104063879500700413.
57. Yeh H-Y, Shoemaker CA, Klesius PH: Sensitive and rapid detection of *Flavobacterium columnare* in channel catfish *Ictalurus punctatus* by a loop-mediated isothermal amplification method. J Appl Microbiol. 2006, 100: 919-925. 10.1111/j.1365-2672.2006.02853.x.

58. Panangala VS, Shoemaker CA, Klesius PH: TaqMan real-time polymerase chain reaction assay for rapid detection of *Flavobacterium columnare*. Aquac Res. 2007, 38: 508-517. 10.1111/j.1365-2109.2007.01695.x.
59. Suzuki M, Rappe MS, Giovanni SJ: Kinetic bias in estimates of coastal picoplankton community structure obtained by measurements of small-subunit rRNA gene PCR amplicon length heterogeneity. Appl Environ Microbiol. 1998, 64: 4522-4529.
60. Suomalainen L-R, Tiirola MA, Valtonen ET: Effect of *Pseudomonas sp.* MT5 baths on *Flavobacterium columnare* infection in rainbow trout and on microbial diversity on fish skin and gills. Dis Aquat Organ. 2005, 63: 61-68.
61. Decostere A, Haesebrouck F, Turnbull JF, Charlier G: Influence of water quality and temperature on adhesion of high and low virulence *Flavobacterium columnare* strains to isolated gill arches. J Fish Dis. 1999, 22: 1-11. 10.1046/j.1365-2761.1999.00132.x.
62. Klesius PH, Pridgeon JW, Aksoy M: Chemotactic factors of *Flavobacterium columnare* to skin mucus of healthy channel catfish (*Ictalurus punctatus*). FEMS Microbiol Lett. 2010, 310: 145-151. 10.1111/j.1574-6968.2010.02060.x.
63. Klesius PH, Shoemaker CA, Evans JJ: *Flavobacterium columnare* chemotaxis to channel catfish mucus. FEMS Microbiol Lett. 2008, 288: 216-220. 10.1111/j.1574-6968.2008.01348.x.
64. Kunttu H: Characterizing the bacterial fish pathogen *Flavobacterium columnare* and some factors affecting its pathogenicity. Jyväskyla Stud Biol Environ Sci. 2010, 206: 1-69.
65. Suomalainen L-R, Tiirola M, Valtonen ET: Chondroitin AC lyase activity is related to virulence of fish pathogenic *Flavobacterium columnare*. J Fish Dis. 2006, 29: 757-763. 10.1111/j.1365-2761.2006.00771.x.
66. Li N, Zhang LQ, Zhang J, Liu ZX, Huang B, Zhang SH, Nie P: Type I restriction-modification system and its resistance in electroporation efficiency in *Flavobacterium columnare*. Vet Microbiol. 2012, 160: 61-68. 10.1016/j.vetmic.2012.04.045.
67. Staroscik AM, Nelson DR: The influence of salmon surface mucus on the growth of *Flavobacterium columnare*. J Fish Dis. 2008, 31: 59-69.
68. Zhang J, Zou H, Wang LF, Li N, Wang GT, Huang B, Nie P: Construction of two selective markers for integrative/conjugative plasmids in *Flavobacterium columnare*. Chin J Oceanol Limnol. 2012, 30: 269-278. 10.1007/s00343-012-1077-z.
69. Tekedar HC, Karsi A, Gillaspy AF, Dyer DW, Benton NR, Zaitshik J, Vamenta S, Banes MM, Gülsoy N, Aboko-Cole M, Waldbieser GC, Lawrence ML: Genome sequence of the fish pathogen *Flavobacterium columnare* ATCC 49512. J Bacteriol. 2012, 194: 2763-2764. 10.1128/JB.00281-12.
70. Pate JL, Ordal EJ: The fine structure of *Chondrococcus columnaris*. III. The surface layers of *Chondrococcus columnaris*. J Cell Biol. 1967, 35: 37-51. 10.1083/jcb.35.1.37.
71. LaFrentz BR, Klesius PH: Development of a culture independent method to characterize the chemotactic response of *Flavobacterium columnare* to fish mucus. J Microbiol Meth. 2009, 77: 37-40. 10.1016/j.mimet.2008.12.011.
72. Shoemaker CA, Olivares-Fuster O, Arias CR, Klesius PH: *Flavobacterium columnare* genomovar influences mortality in channel catfish (*Ictalurus punctatus*). Vet Microbiol. 2008, 127: 353-359. 10.1016/j.vetmic.2007.09.003.
73. LaFrentz BR, LaPatra SE, Shoemaker CA, Klesius PH: Reproducible challenge model to investigate the virulence of *Flavobacterium columnare* genomovars in rainbow trout *Oncorhynchus mykiss*. Dis Aquat Organ. 2012, 101: 115-122. 10.3354/dao02522.
74. Kunttu HMT, Suomalainen L-R, Jokinen EI, Valtonen ET: *Flavobacterium columnare* colony types: connection to adhesion and virulence?. Microb Pathog. 2009, 46: 21-27. 10.1016/j.micpath.2008.10.001.
75. Arias CR, LaFrentz S, Cai W, Olivares-Fuster O: Adaptive response to starvation in the fish pathogen *Flavobacterium columnare:* cell viability and ultrastructural changes. BMC Microbiol. 2012, 12: 266-276. 10.1186/1471-2180-12-266.

76. Sun F, Peatman E, Li C, Liu S, Jiang Y, Zhou Z, Liu Z: Transcriptomic signatures of attachment, NF-κB suppression and IFN stimulation in the catfish gill following columnaris bacterial infection. Dev Comp Immunol. 2012, 38: 169-180. 10.1016/j.dci.2012.05.006.
77. Tateno H, Yamaguchi T, Ogawa T, Muramoto K, Watanabe T, Kamiya H, Saneyoshi M: Immunohistochemical localization of rhamnose-binding lectins in the steelhead trout (*Oncorhynchus mykiss*). Dev Comp Immunol. 2002, 26: 543-550. 10.1016/S0145-305X(02)00007-1.
78. Beck BH, Farmer BD, Straus DL, Li C, Peatman E: Putative roles for a rhamnose binding lectin in *Flavobacterium columnare* pathogenesis in channel catfish *Ictalurus punctatus*. Fish Shellfish Immun. 2012, 33: 1008-1015.
79. Altinok I, Grizzle JM: Effects of low salinities on *Flavobacterium columnare* infection of euryhaline and freshwater stenohaline fish. J Fish Dis. 2001, 24: 361-367. 10.1046/j.1365-2761.2001.00306.x.
80. Decostere A, Henckaerts K, Ducatelle R, Haesebrouck F: An alternative model to study the association of rainbow trout (*Oncorhynchus mykiss* L.) pathogens with the gill tissue. Lab Anim. 2002, 36: 396-402. 10.1258/002367702320389053.
81. Ofek I, Hasty DL, Doyle RJ: Adhesins as Bacterial Cell Surface Structures: General Concepts of Structure, Biogenesis, and Regulation. Bacterial adhesion to animal cells and tissues. Edited by: ASM Press. 2003, Washington D.C.: ASM Press, 77-81.
82. Romero M, Avendaño-Herrera R, Magariños B, Cámara M, Otero A: Acylhomoserine lactone production and degradation by the fish pathogen *Tenacibaculum maritinum*, a member of the *Cytophaga-Flavobacterium_Bacteroides* (CFB) group. FEMS Microbiol Lett. 2009, 304: 131-139.
83. Camilli A, Bassler BL: Bacterial small-molecule signaling pathways. Science. 2006, 311: 1113-1116. 10.1126/science.1121357.
84. Wagner-Döbler I, Thiel V, Eberl L, Allgaier M, Bodor A, Meyer S, Ebner S, Henning A, Pukall R, Schulz S: Discovery of complex mixtures of novel long-chain quorum sensing signals in free-living and host-associated marine *Alphaproteobacteria*. Chem Bio Chem. 2005, 6: 2195-2206. 10.1002/cbic.200500189.
85. Griffin BR: Characteristics of a chondroitin AC lyase produced by *Cytophaga columnaris*. T Am Fish Soc. 1991, 120: 391-395. 10.1577/1548-8659(1991)120<0391:COACAL>2.3.CO;2.
86. Newton JC, Wood TM, Hartley MM: Isolation and partial characterization of extracellular proteases produced by isolates of *Flavobacterium columnare* derived from channel catfish. J Aquat Anim Health. 1997, 9: 75-85. 10.1577/1548-8667(1997)009<0075:IAPCOE>2.3.CO;2.
87. Anacker RL, Ordal EJ: Studies on the myxobacterium *Chondrococcus columnaris*. II. Bacteriocins. J Bacteriol. 1959, 78: 33-40.
88. Tiirola M, Valtonen ET, Rintamaki-Kinnunen P, Kulomaa MS: Diagnosis of flavobacteriosis by direct amplification of rRNA genes. Dis Aquat Organ. 2002, 51: 93-100.
89. Zhang Y, Arias CR, Shoemaker CA, Klesius PH: Comparison of lipopolysaccharide and protein profiles between *Flavobacterium columnare* strains from different genomovars. J Fish Dis. 2006, 29: 657-663. 10.1111/j.1365-2761.2006.00760.x.
90. Aranishi F, Mano N, Nakane M, Hirose H: Epidermal response of the Japanese eel to environmental stress. Fish Physiol Biochem. 1998, 19: 197-203. 10.1023/A:1007746514851.
91. Do Vale A, Marques F, Silva MT: Apoptosis of sea bass (*Dicentrarchus labrax* L.) neutrophils and macrophages induced by experimental infection with *Photobacterium damselae* subsp. *piscicida*. Fish Shellfish Immun. 2003, 15: 129-144. 10.1016/S1050-4648(02)00144-4.
92. Ourth DD, Wilson EA: Bactericidal serum response of the channel catfish against Gram-negative bacteria. Dev Comp Immunol. 1982, 6: 579-583. 10.1016/S0145-305X(82)80044-X.

93. Ourth DD, Bachinski LM: Bactericidal response of channel catfish *Ictalurus punctatus* by the classical and alternative complement pathways against bacterial pathogens. J Appl Ichtyol. 1987, 3: 42-45. 10.1111/j.1439-0426.1987.tb00449.x.
94. Kuo S-C, Chung HY, Kou GH: Studies on artificial infection of the gliding bacteria in cultured fishes. Fish Pathol. 1981, 15: 309-314. 10.3147/jsfp.15.309.
95. Bebak J, Matthews M, Shoemaker C: Survival of vaccinated, feed-trained largemouth bass fry (*Micropterus salmoides floridanus*) during natural exposure to *Flavobacterium columnare*. Vaccine. 2009, 27: 4297-4301. 10.1016/j.vaccine.2009.05.026.
96. Becker CD, Fujihara MP: The bacterial pathogen, Flexibacter columnaris and its epizootiology among Columbia river fish. 1978, Washington, D.C.: American Fisheries Society
97. Grabowski LD, LaPatra SE, Cain KD: Systemic and mucosal antibody response in tilapia, *Oreochromis niloticus* (L.), following immunization with *Flavobacterium columnare*. J Fish Dis. 2004, 27: 573-581. 10.1111/j.1365-2761.2004.00576.x.
98. Moore AA, Eimers ME, Cardella MA: Attempts to control *Flexibacter columnaris* epizootics in pond-reared channel catfish by vaccination. J Aquat Anim Health. 1990, 2: 109-111. 10.1577/1548-8667(1990)002<0109:ATCCEI>2.3.CO;2.
99. Schachte JHJ, Mora EC: Production of agglutinating antibodies in the channel catfish (*Ictalurus punctatus*) against *Chondrococcus columnaris*. J Fish Res Board Canada. 1973, 30: 116-118. 10.1139/f73-017.
100. Ransom DP: Immune responses of salmonids: (a) oral immunization against Flexibacter columnaris. (b) effects of combining antigens in parenterally administered polyvalent vaccines. 1975, Corvallis, OR: MS Thesis. Oregon State University
101. Liewes EW, Van Dam RH: Procedures and application of the in vitro fish leukocyte stimulation assay. Dev Comp Immunol. 1982, 2: 223-232.
102. Song Y-L: Comparison of Flexibacter columnaris strains isolated from fish in North America and other areas of the Pacific Rim. PhD Thesis. 1986, Corvallis, OR: Oregon State University
103. AFS-FCS (2011):*Guide to using drugs, biologics, and other chemicals in aquaculture. American fisheries society fish culture section.*http://www .fws.gov/fisheries/aadap/AFS-FCS%20documents/GUIDE_FEB_2011.pdf,
104. Karvonen A, Rintamäki P, Jokela J, Valtonen ET: Increasing water temperature and disease risks in aquatic systems: climate change increases the risk of some, but not all, diseases. Int J Parasitol. 2010, 40: 1483-1488. 10.1016/j.ijpara.2010.04.015.
105. Suomalainen L-R, Tiirola M, Valtonen ET: The influence of rearing conditions on *Flavobacterium columnare* infection of rainbow trout. J Fish Dis. 2005, 28: 271-277. 10.1111/j.1365-2761.2005.00631.x.
106. Holt RA, Sanders JE, Zinn JL, Fryer JL, Pilcher KS: Relation of water temperature to *Flexibacter columnaris* infection in steelhead trout (*Salmo gairdneri*) coho (*Oncorhynchus kisutch*), and chinook (*O. tshawytscha*) salmon. J Fish Res Board Can. 1975, 32: 1553-1559. 10.1139/f75-182.
107. Suomalainen L-R, Reunanen H, Ijäs R, Valtonen ET, Tiirola M: Freezing induces biased results in the molecular detection of *Flavobacterium columnare*. Appl Environ Microbiol. 2006, 72: 1702-1704. 10.1128/AEM.72.2.1702-1704.2006.
108. Morris JM, Snyder-Conn E, Foott JS, Holt RA, Suedkamp MJ, Lease HM, Clearwater SJ, Meyer JS: Survival of lost river suckers (*Deltistes luxatus*) challenged with *Flavobacterium columnare* during exposure to sublethal ammonia concentrations at pH 9.5. Arch Environ Contam Toxicol. 2006, 50: 256-263. 10.1007/s00244-004-0194-x.
109. Bandilla M, Valtonen ET, Suomalainen L-R, Aphalo PJ, Hakalahti T: A link between ectoparasite infection and susceptibility to bacterial disease in rainbow trout. Int J Parasitol. 2006, 36: 987-991. 10.1016/j.ijpara.2006.05.001.

110. Darwish AM, Mitchell A, Straus DL: Evaluation of a 4-h static copper sulphate treatment against experimental infection of *Flavobacterium columnare* in channel catfish (*Ictalurus punctatus*). Aquaculture Res. 2011, 43: 688-695.

111. Thomas-Jinu S, Goodwin AE: Acute columnaris infection in channel catfish, *Ictalurus punctatus* (Rafinesque): efficacy of practical treatments for warmwater aquaculture ponds. J Fish Dis. 2004, 27: 23-28. 10.1046/j.1365-2761.2003.00504.x.

112. Sink TD, Lochmann RT, Goodwin AE, Mareceaux E: Mortality rates in golden shiners fed high-fat diets with or without a dairy-yeast prebiotic before challenge with *Flavobacterium columnare*. N Am J Aquaculture. 2007, 69: 305-308. 10.1577/A06-078.1.

113. Kunttu HMT, Valtonen ET, Suomalainen L-R, Vielma J, Jokinen IE: The efficacy of two immunostimulants against *Flavobacterium columnare* infection in juvenile rainbow trout (*Oncorhynchus mykiss*). Fish Shellfish Immun. 2009, 26: 850-857. 10.1016/j.fsi.2009.03.013.

114. Bader JA, Nusbaum KE, Shoemaker CA: Comparative challenge model of *Flavo bacterium columnare* using abraded and unabraded channel catfish, *Ictalurus punctatus* (Rafinesque). J Fish Dis. 2003, 26: 461-467. 10.1046/j.1365-2761.2003.00479.x.

115. Darwish AM, Bebak JA, Schrader KK: Assessment of Aquaflor®, copper sulphate and potassium permanganate for control of *Aeromonas hydrophila* and *Flavobacterium columnare* infection in sunshine bass, *Morone chrysops* female x *Morone saxatilis* male. J Fish Dis. 2012, 35: 637-647. 10.1111/j.1365-2761.2012.01393.x.

116. Cunningham FL, Jack SW, Hardin D, Wills RW: Pond-Level risk factors associated with columnaris disease on Mississippi commercial catfish farms. J Aquat Anim Health. 2012, 24: 178-184. 10.1080/08997659.2012.675932.

117. Conrad JF, Holt RA, Kreps TD: Ozone disinfection of flowing water. Prog Fish-Cult. 1975, 37: 134-136. 10.1577/1548-8659(1975)37[134:ODOFW]2.0.CO;2.

118. Shoemaker CA, Klesius PH, Lim C, Yildirim M: Feed deprivation of channel catfish, *Ictalurus punctatus* (Rafinesque), influences organosomatic indices, chemical composition and susceptibility to *Flavobacterium columnare*. J Fish Dis. 2003, 26: 553-561. 10.1046/j.1365-2761.2003.00489.x.

119. Rogers WA: Principal diseases of catfish: how to identify and fight them. Fish Farming Ind. 1971, 2: 20-26.

120. Darwish AM, Mitchell AJ, Straus DL: Evaluation of potassium permanganate against an experimental subacute infection of *Flavobacterium columnare* in channel catfish, *Ictalurus punctatus* (Rafinesque). J Fish Dis. 2009, 32: 193-199. 10.1111/j.1365-2761.2008.01015.x.

121. Riley TA: Treatment of Flavobacterium columnare and toxicity of chloramine-T in channel catfish. 2000, Ictalurus punctatus. MS thesis: Auburn University, Auburn, Alabama

122. Merrifield DL, Dimitroglou A, Foey A, Davies SJ, Baker RTM, Bøgwald J, Castex M, Ringø E: The current status and future focus of probiotic and prebiotic applications for salmonids. Aquaculture. 2010, 302: 1-18. 10.1016/j.aquaculture.2010.02.007.

123. Boutin S, Bernatchez L, Audet C, Derôme N: Antagonistic effect of indigenous skin bacteria of brook charr (*Salvelinus fontinalis*) against *Flavobacterium columnare* and *F. psychrophilum*. Vet Microbiol. 2012, 155: 355-361. 10.1016/j.vetmic.2011.09.002.

124. Arias CR, Cai W, Peatman E, Bullard SA: Catfish hybrid *Ictalurus punctatus* × *I. furcatus* exhibits higher resistance to columnaris disease than the parental species. Dis Aquat Organ. 2012, 100: 77-81. 10.3354/dao02492.

125. LaFrentz BR, Shoemaker CA, Booth NJ, Peterson BC, Ourth DD: Spleen index and mannose-binding lectin levels in four channel catfish families exhibiting different susceptibilities to *Flavobacterium columnare* and *Edwardsiella ictaluri*. J Aquatic Anim Health. 2012, 24: 141-147. 10.1080/08997659.2012.675936.

126. Snieszko SF: Columnaris disease of fishes. USFWS Fish Leafl. 1958, 46: 1-3.

127. Amend DF, Ross AJ: Experimental control of columnaris disease with a new nitrofuran drug, P-7138. Prog Fish-Cult. 1970, 32: 19-25. 10.1577/1548-8640(1970)32[19:ECOCDW]2.0.CO;2.

128. Ross AJ: *In vitro* studies with nifurpirinol (P-7138) and bacterial fish pathogens. Prog Fish-Cult. 1972, 34: 18-20. 10.1577/1548-8640(1972)34[18:IVSWNP]2.0.CO;2.

129. Shiraki K, Miyamoto F, Sato T, Sonezaki I, Yano K: Studies on a new chemotherapeutic agent nifurprazine (HB-115) against fish infectious diseases. Fish Pathol. 1970, 4: 130-137. 10.3147/jsfp.4.130.

130. Deufel J: Prophylactic measures against bacterial diseases of salmonid fry (Prophylaktische Massnahmen gegen bakterielle Erkrankungen der Salmonidenbrut). Osterr Fisch. 1974, 27: 1-5.

131. Endo T, Ogishima K, Hayasaka H, Kaneko S, Ohshima S: Application of oxolinic acid as a chemotherapeutic agent against infectious diseases in fishes-I. Antibacterial activity, chemotherapeutic effects and pharmacokinetics of oxolinic acid in fishes. Bull Jpn Soc Sci Fish. 1973, 39: 165-171. 10.2331/suisan.39.165.

132. Soltani M, Shanker S, Munday BL: Chemotherapy of *Cytophaga/Flexibacter*-like bacteria (CFLB) infections in fish: studies validating clinical efficacies of selected antimicrobials. J Fish Dis. 1995, 18: 555-565. 10.1111/j.1365-2761.1995.tb00360.x.

133. Wood JW: Treatment and drugs. Diseases of Pacific Salmon: their prevention and treatment. 1974, State of Washington: Department of Fisheries, Hatchery Division, 7-16. 2

134. Koski P, Hirvelä-Koski V, Bernardet JF: *Flexibacter columnaris* infection in arctic char (*Salvelinus alpinus* (L.)). First isolation in Finland. Bull Eur Ass Fish Pathol. 1993, 31: 66-69.

135. Wolf K, Snieszko SF: Uses of antibiotics and other antimicrobials in therapy of diseases of fishes. Antimicrob Agents Chemother. 1963, 161: 597-603.

136. Gaunt PS, Gao D, Sun F, Endris R: Efficacy of florfenicol for control of mortality caused by *Flavobacterium columnare*infection in channel catfish. J Aquat Anim Health. 2010, 22: 115-122. 10.1577/H09-057.1.

137. Serrano PH: Responsible use of antibiotics in aquaculture. FAO Fish Tech Pap. 2005, 469: 1-97.

138. Declercq AM, Boyen F, Van den Broeck W, Bossier P, Karsi A, Haesebrouck F, Decostere A: Antimicrobial susceptibility pattern of *Flavobacterium columnare* isolates collected worldwide from 17 fish species. J Fish Dis. 2013, 36: 45-55. 10.1111/j.1365-2761.2012.01410.x.

139. Darwish AM, Mitchell AJ: Evaluation of diquat against an acute experimental infection of *Flavobacterium columnare* in channel catfish, *Ictalurus punctatus* (Rafinesque). J Fish Dis. 2009, 32: 401-408. 10.1111/j.1365-2761.2009.01024.x.

140. Marchand P-A, Phan T-M, Straus DL, Farmer BD, Stüber A, Meinelt T: Reduction of *in vitro* growth in *Flavobacterium columnare* and *Saprolegnia parasitica* by products containing peracetic acid. Aquac Res. 2012, 43: 1861-1866. 10.1111/j.1365-2109.2011.02995.x.

141. Laanto E, Sundberg L-R, Bamford JKH: Phage specificity of the freshwater fish pathogen *Flavobacterium columnare*. Appl Environ Microbiol. 2011, 77: 7868-7872. 10.1128/AEM.05574-11.

142. Prasad Y, Arpana , Kumar D, Sharma AK: Lytic bacteriophages specific to *Flavobacterium columnare* rescue catfish, *Clarias batrachus* (Linn.) from columnaris disease. J Environ Biol. 2011, 32: 161-168.

CHAPTER 3

Anaesthesia and Handling Stress Effects on Pigmentation and Monoamines in Arctic Charr

Tobias Backström Martina Heynen Eva Brännäs

Department of Wildlife, Fish, and Environmental Studies, Swedish University of Agricultural Sciences, Umeå, Sweden

Department of Ecology and Environmental Science, Umeå University, Umeå, Sweden

Department of Neuroscience, Uppsala University, Uppsala, Sweden

ABSTRACT

Stress responsiveness differs between individuals and is often categorized into different stress coping styles. Using these stress coping styles for selection in fish farming could be beneficial, since stress is one main factor affecting welfare. In Arctic charr (*Salvelinus alpinus*) carotenoid pigmentation is associated with stress responsiveness and stress coping styles. Thus this could be an important tool to use for selection of stress resilient charr. However, anaesthetics seem to affect carotenoid pigmentation, and it would be better if the method for selection could be implemented during normal maintenance, which usually includes anaesthetics. Therefore, this study investigated how the use of anaesthetics affected carotenoid pigmentation, i.e. number of spots, over time compared to no-anaesthetic treatment. Additionally, the stress indicators monoamines and glucocorticoids were investigated. The results indicate that the anaesthetic MS-222 affects number of spots on the right side. This anaesthetic also increased dopaminergic activity in the telencephalon. Both brain dopaminergic and serotonergic activity was associated with spottiness. Further, behaviour during anaesthetization was associated with spots on the left side, but not the right side. Repetition of the same treatment seemed to affect spot numbers on the right side. In conclusion, this study shows that inducing stress in charr affects the carotenoid spots. Thus, it is possible to use anaesthetics when evaluating spottiness although careful planning is needed.

Keywords: Anaesthesia Carotenoid pigmentation Lateralization Monoamine Salmonid Stress

INTRODUCTION

Intraspecific divergence in stress response profiles is common among various animal taxa. This divergence in stress responses is often related to specific behavioural traits and has been referred to as stress coping styles. Typically, there are two extremes, namely proactive and reactive coping style (Koolhaas et al. 1999; Koolhaas et al. 2007). Proactive individuals are more aggressive, have a higher general activity, forming behavioural routines more easily, and respond to stress with a lower hypothalamic-pituitary-adrenal (HPA) axis reactivity as compared to reactive individuals (Koolhaas et al. 1999; Koolhaas et al. 2007). In recent years, a plethora of studies have shown that stress coping styles are evident in teleost fish (Øverli et al. 2005; Schjolden and Winberg 2007; Koolhaas et al. 2010; Conrad et al. 2011; Castanheira et al. 2015).

To reduce within-group variance and hence number of animals used would be beneficial, for example, in growth studies (Erhard et al. 1997). This could be achieved by selection of animals based on stress coping style. Further, if personality is a heritable trait, it could be beneficial to include this trait in the selection of farmed animals. Fish are relatively new in farming industry compared to terrestrial animals and plants, and intensive fish farming just started during the last century (Duarte et al. 2007). Welfare is an important issue within aquaculture and relates to stress coping styles (Huntingford et al. 2006). For instance, selection of fast-growing fish might lead to more aggressive and risk-taking fish (i.e. proactive copers), which could impair welfare by increased fighting (Huntingford and Adams 2005). Thus, selection using specific stress coping styles could be beneficial for aquaculture.

Stress coping styles are usually evaluated using tests such as confinement, dominance, and resident-intruder (Koolhaas et al. 1999). However, these tests are very time consuming and hence not practical if large numbers of individuals would need to be screened. Therefore there is a need for better and faster methods to test for stress coping styles. Interestingly, animal's visual appearance often signals condition or social status (Hill 2011). In vertebrates the two pigment groups, carotenoids and melanins, cause a colour variation that sometimes reflect physiological condition. More melanised animals seem to be more stress-resilient (Ducrest et al. 2008). This has been reported in teleost fish, such as Atlantic salmon (*Salmo salar*) and rainbow trout (*Oncorhynchus mykiss*), in which fish with more melanin-based skin spots have a lower physiological stress response than the fish with few spots (Kittilsen et al. 2009). However, social subordination induces skin darkening in salmonids (O'Connor et al. 1999; O'Connor et al. 2000; Eaton and Sloman 2011) including the Arctic charr (*Salvelinus alpinus*) (Höglund et al. 2000; Höglund et al. 2002). Thus, in Arctic charr darker animals do not seem to be more stress resilient. Instead, a recent study reported that the carotenoid pigmentation in Arctic charr could be used as an indicator of physiological stress responses (Backström et al. 2014). Additionally, socially subordinate charr have more carotenoid spots compared to dominant individuals (Backström et al. 2015a). Therefore, carotenoid pigmentation, rather than melanin-based pigmentation, seems to be signalling stress in Arctic charr. We predict that carotenoid-based pigmentation in Arctic charr could be usable as a tool for predicting stress responsiveness. Such an assessment could be made during other, routine evaluation of brood stock, such as weighing and measuring. These procedures, however, usually involve anaesthetics; however, anaesthetics may increase the number of carotenoid spots (Backström et al. 2015b). There is also an association between pigmentation and monoaminergic activity (Backström et al. 2015c). The monoaminergic activity of serotonin (5-hydroxytryptamine; 5-HT), dopamine (DA), and norepinephrine (NE) are all physiological stress indicators (Blanchard et al. 2001) and are also evident in teleost fish (Summers and Winberg 2006).

Based on these earlier studies, we investigated how carotenoid spots were affected by the photographing procedure on two consecutive days using or not using anaesthetics in all possible combinations (without anaesthetics during first, second, or both days). We monitored the behaviour during anaesthesia and the physiological stress responses of plasma cortisol and brain monoamines on the final day to get a better understanding of the specific stress responses. Specifically, two hypotheses were tested: (1) spots were affected by consecutive use of anaesthetic compared to no-anaesthetic, and (2) spots were correlated to stress responsiveness independent of treatment.

MATERIAL AND METHODS

Experimental animals and location

This experiment was done at the Umeå Marine Research Station (UMF) using 1 year old juvenile Arctic charr from the 7th generation of the Swedish Arctic charr breeding programme [Arctic superior, for details on the programme see Nilsson et al. (2010)]. The fish were transported to UMF several months before the experiment, and were kept in tanks supplied with running brackish water (3–4‰) from the Bothnian Bay (temperature range 5 to 10 °C, photoperiod 12 h light/12 h dark). The fish were fed commercial pellets (BioMar; 4 mm, www.biomar.com) at 1.0–1.5% of body mass (MB) per day depending on temperature. The experiment was done in May-June 2012 with a methodology approved by the Umeå Animal Research Ethical Committee.

Experiment setup

On day 1, fish (body mass: 141.2 ± 21.4 g, fork length (L_F): 23.9 ± 1.1 cm, mean ± SD, N = 32) were netted from the stock, photographed on both sides as described by Backström et al. (2015a) using a Canon EOS 500D digital camera and a box with a transparent lid. Immediately after photographing, fish were anaesthetized with Tricaine methane sulfonate (MS-222, 0.15 g/L), weighed and measured. The fish were not anaesthetized prior to photographing based upon the apparent spot inducing effects by anaesthetics in Arctic charr (Backström et al. 2015b). Isolation of individual fish was created by separating experimental aquaria (170 L, 95 x 41 x 44 cm) into four equally-sized 42.5 L compartments using removable dark PVC walls. The fish were allowed to acclimate for one week during which they were hand-fed commercial pellets. Each aquarium was continuously supplied with running water from the Bothnian Bay (5–10 °C), and the full spectrum daylight light was turned on at 06:00 and off at 18:00 h (12 h light/12 h dark).

After the isolation, fish were divided into four treatment groups aimed at deducing how different handling procedures during photographing would affect carotenoid pigmentation and subsequent stress responses. Earlier results show that anaesthesia increases number of spots (Backström et al. 2015b), and therefore this procedure was aimed at deducing if this effect was sustained over time and/or exacerbated by repeated treatment. MS-222 was used as anaesthetic because it is one of two anaesthetics allowed in the European Union (EEC Council regulation 2377/90) and is the one used in the Swedish Arctic charr aquaculture. The four treatment groups were: 1) being photographed without anaesthesia (N for non-anaesthetic) on day 8 and without anaesthesia (N) on day 9 (group NN, $N = 8$), 2) being photographed without anaesthesia on day 8 and with anaesthesia (A for anaesthetic) on day 9 (group NA, $N = 8$), 3) being photographed with anaesthesia on day 8 and without anaesthesia on day 9 (group AN, $N = 8$), and 4) being photographed with anaesthesia on day 8 and with anaesthesia on day 9 (group AA, $N = 8$).

During anaesthesia, the fish were transferred to a 17-L bucket (Ø 36 cm) filled with 4 L of a low dose of MS-222 (0.15 g/L). The behaviour during the anaesthetisation process was recorded via a camcorder for further analysis [see below and Backström et al. (2015b)]. Fish were kept in the solution until they lost equilibrium and had no response to tactile stimuli. Immediately following this stage, the fish were photographed (see description above). The fish without anaesthesia were taken immediately from the aquaria to be photographed.

After photographing on day 9, the fish were sacrificed in a high dose of MS-222 (0.30 g/L). Blood was immediately sampled via a heparinized syringe from the caudal vein, then the fish was decapitated and the brain dissected out. Sex was determined by visual inspection of gonads. The blood was subsequently centrifuged, the plasma collected and stored at −20 °C until further analysis. The brains were divided into the easily identified telencephalon, cerebellum, optic tectum and brainstem (including hypothalamus) and were stored at −80 °C until further analysis.

Image analysis

Photographs were analysed as described and validated earlier (see Fig. 1) (Backström et al. 2015a). Briefly, carotenoid-based spots were counted systematically in a rectangle (2 cm x 10 cm) on each side of the fish using ImageJ (Rasband, W.S., ImageJ, U. S. National Institutes of Health, Bethesda, Maryland, USA, http://imagej.nih.gov/ij/, 1997–2012.).

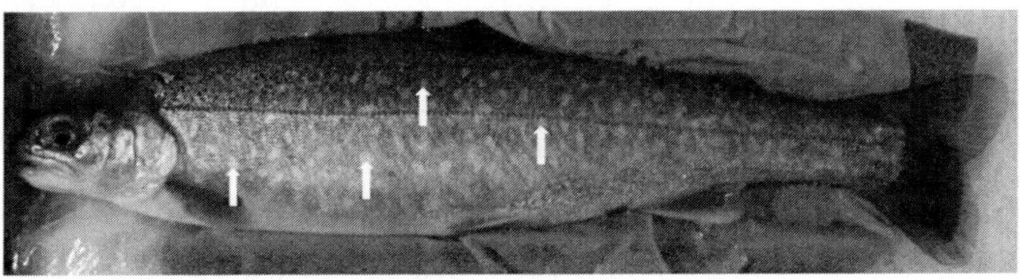

Figure. 1 Typical photograph from the experiment. White arrows indicate typical spots that were counted

Assays

Plasma was analysed for cortisol using a commercial enzyme linked immunosorbent assay (ELISA) kit (product # 402710, Neogen Corporation, Lexington, USA). Each sample was run in duplicate during a single assay with an intra-assay coefficient of variation of 1.30%.

Tissue levels of 5-HT and the 5-HT metabolite, 5-hydroxyindoleacetic acid (5-HIAA), DA and the DA metabolite 3,4-dihydroxyphenylacetic acid (DOPAC), and NE were analysed in telencephalon, cerebellum, optic tectum and brainstem samples using high performance liquid chromatography with electrochemical detection (HPLC-EC) as described by Höglund et al. (2000). The concentrations were standardized against the weight of the brain tissues. The ratios of 5-HIAA to 5-HT concentration and DOPAC to DA concentration were calculated and used as an index of serotonergic and dopaminergic activity, respectively (Shannon et al. 1986).

Behavioural observations

In the analysis of the behaviour during treatment with anaesthetics we recorded: (1) activity until loss of equilibrium [anaesthesia stage III plane 1 (Zahl et al. 2012)] monitored by dividing the anaesthetization bucket into quadrants via a cross and counting number of crossed lines per second, (2) latency to loss of equilibrium, and (3) latency to anaesthesia [anaesthesia stage III plane 2 (Zahl et al. 2012)] defined as lost equilibrium and no response to three repeated tactile stimuli. This was done by one observer (MH).

Statistical analyses

Normally distributed data were analysed using *t*-test, ANOVA or repeated measures ANOVA with LSD post hoc tests. Further, correlations between behaviour, physiology and pigmentation were investigated using either Pearson correlation test if data were normally distributed or Spearman rank correlation test if they were not. All statistics were performed in IBM SPSS Statistics 20 (IBM Corporation, New York, USA) and data are presented as mean ± SEM if not stated otherwise.

RESULTS

Carotenoid pigmentation

The number of carotenoid spots on the right side differed between treatment groups on day 9 (Table 1, Fig. 2). The charr in group AA had more spots than the groups AN and NN, but not more spots than the NA group. However, on day 9 no difference was seen in spots on the left side (Table 1, Fig. 1). The number of carotenoid spots did not differ between treatment groups on either side on day 1, or on day 8 (Table 1). When considering the photographing procedures separately, that is day 8 and 9, there were some differences between anaesthetized and non-anaesthetized charr. Number of spots on the right side were higher in anaesthetized compared to non-anaesthetized charr on day 9 (*t*-test, $P = 0.005$), and a similar trend was seen on day 8 (*t*-test, $P = 0.059$) (see Fig. 3). On the left side, a similar trend could be noted on day 9 (*t*-test, $P = 0.069$) but not on day 8 (*t*-test, $P = 0.262$) (see Fig. 2). All data are available in the Supplemental table 1.

Table 1 ANOVA showing the effect of treatment (AA, AN, NA, NN, A is for anaesthetic, N for non-anaesthetic, and procedure for counting spots was done on two consecutive days for all individuals) on the number of spots on left and right side, calculated from photographs taken on day 1, 8, and 9

	Right side			**Left side**		
	$F_{3,28}$	**P**	**Post-hoc**	**$F_{3,28}$**	**P**	**Post-hoc**
Day 1	1.128	0.355	–	1.084	0.372	–
Day 8	2.771	0.060	–	1.759	0.178	–
Day 9	4.417	**0.012**	AA > AN, NN	2.060	0.128	–

Post-hoc column show significant differences (LSD, $P < 0.05$) between treatments. Days were analysed separately

Significant results are shown in bold font

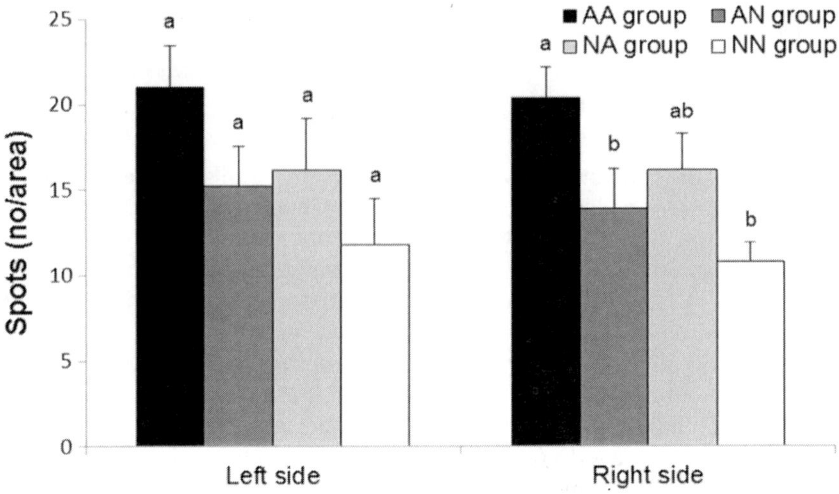

Figure. 2 Number of spots on day 9 on left and right side of each treatment (A is for anaesthetic, N for non-anaesthetic, and procedure for counting spots was done on two consecutive days for all individuals). *Different superscript capitals* indicates significant differences (LSD post hoc, $P < 0.05$). Values are mean ± S.E.M and $N = 8$ for all treatments

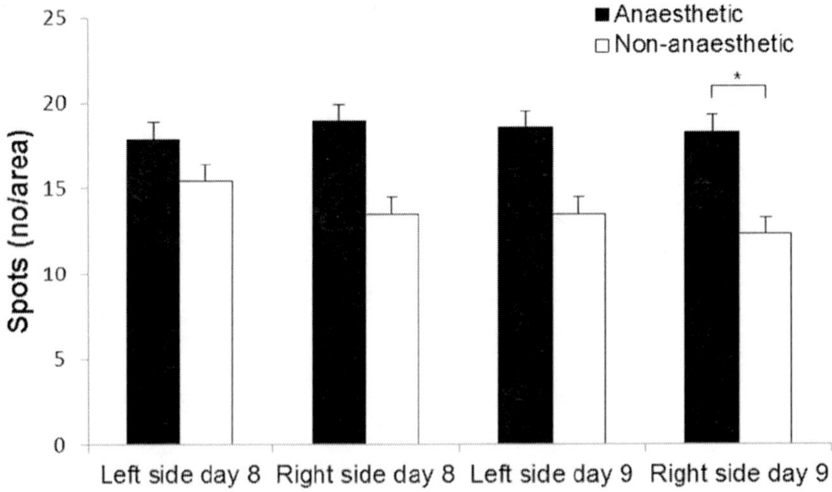

Figure. 3 Number of spots on day 8 and 9 on left and right side of anaesthetised (A) and non-anaesthetised (N) individuals. * indicates significant differences (*t*-test, $P < 0.05$). Values are mean ± S.E.M and $N = 16$ for all treatments

Considering a change over time in the number of spots, both the AA group and the NN group differed in number of spots on the right side (Table 2). In the AA group there were more spots on the right side on day 8 compared to day 1, whereas in the NN group there were less spots on the right side on day 9 compared to day 1 (Table 2). On the right side, no difference was evident in the AN group or in the NA group. Additionally, no difference could be seen on the left side in any group (Table 2).

Table 2 Repeated measures ANOVA on number of spots on left and right side, from photographs taken on day 1, 8, and 9

	Right side			**Left side**		
	$F_{2,14}$	**P**	**Post-hoc**	**$F_{2,14}$**	**P**	**Post-hoc**
AA	4.341	0.034	8 > 1	0.005	0.995	–
AN	0.349	0.712	–	0.180	0.837	–
NA	1.627	0.232	–	0.041	0.960	–
NN	3.848	0.047	9 < 1	1.363	0.288	–

Post-hoc column show significant differences between days (LSD, $P < 0.05$). Treatments (AA, AN, NA, NN, A is for anaesthetic, N for non-anaesthetic, and procedure for counting spots was done on two consecutive days for all individuals) were analysed separately

There were several correlations between spots on separate days and sides (see Table 3). For instance, on the same day left and right side spots were positively correlated with spots on the other side on day 1, 8 and 9. The spots on the right side were also positively correlated across all days (day 1, 8 and 9). Additionally, on day 8 and 9, left and right side spots were all positively correlated with each other. Finally, spots on the left side on day 1 were positively correlated with spots on the right side on day 9.

Table 3 The Pearson correlation coefficients and P-value for number of carotenoid spots of Arctic charr after handling stress

	Day 1, Left side	Day 1, Right side	Day 8, Left side	Day 8, Right side	Day 9, Left side	Day 9, Right side	Mean ± S.D. (N)
Day 1, Left side	1.00						16.84 ± 8.44 (32)
Day 1, Right side	**0.478, P = 0.006**	1.00					14.97 ± 6.39 (32)
Day 8, Left side	0.318, P = 0.077	0.276, P = 0.126	1.00				16.66 ± 6.06 (32)
Day 8, Right side	0.248, P = 0.170	**0. 507, P = 0.003**	**0.509, P = 0.003**	1.00			16.22 ± 8.19 (32)
Day 9, Left side	0.303, P = 0.092	0.347, P = 0.052	**0.574, P = 0.001**	**0.588, P < 0.001**	1.00		16.03 ± 7.89 (32)
Day 9, Right side	**0.377, P = 0.034**	**0.416, P = 0.018**	**0.541, P = 0.001**	**0.689, P < 0.001**	**0.743, P < 0.001**	1.00	15.28 ± 6.29 (32)

Significant results are shown in bold font

Physiological parameters

Concerning monoamines, DOPAC concentration in telencephalon differed between treatments (ANOVA, $F_{3,27} = 3.941$, $P = 0.019$; see Fig. 4), with the NA group having higher concentration of DOPAC than group AN (LSD post hoc, $P = 0.004$) and NN (LSD post hoc, $P = 0.011$), but not AA (LSD post hoc, $P = 0.064$). However, no other differences in monoamines between treatments were evident (see Table 4). Similarly, there were no differences in plasma cortisol between the groups (ANOVA, $F_{3,28} = 2.407$, $P = 0.088$; see Fig. 5).

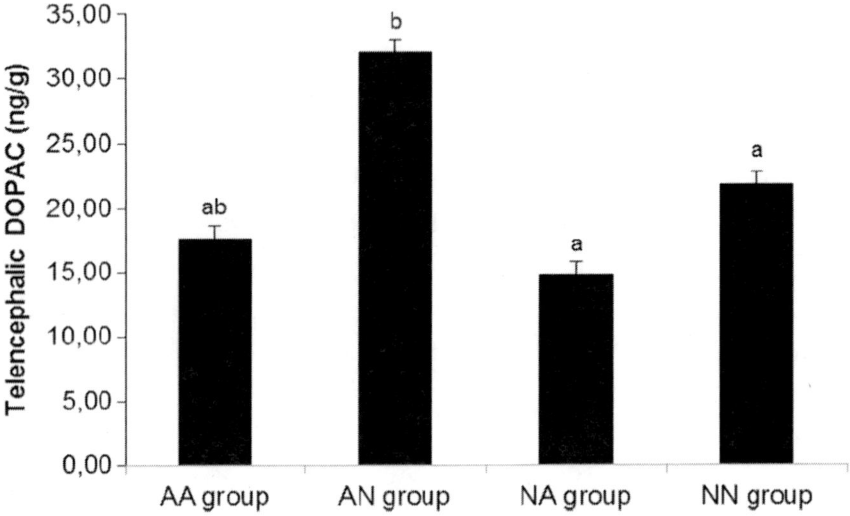

Figure. 4 DOPAC levels in telencephalon after treatments (A is for anaesthetic and N for non-anaesthetic). Different superscript capitals indicates significant differences (LSD post hoc, $P < 0.05$). Values are mean ± S.E.M and $N = 8$ for all treatments except for group NA ($N = 7$)

Table 4 Concentrations of NE, DOPAC, DA, 5-HIAA and 5-HT, and DOPAC/DA and 5-HIAA/5-HT ratios in the brainstem, cerebellum, optic tectum, and telencephalon in all of the Arctic charr combined after handling stress

Brain part	NE (ng/g)	DOPAC (ng/g)	DA (ng/g)	DOPAC/DA*10^{-3}	5-HIAA (ng/g)	5-HT (ng/g)	5-HIAA/5-HT*10^{-3}
Brain stem	63 ± 11 (32)	14 ± 9 (32)	59 ± 37 (32)	264 ± 113 (32)	4 ± 1 (31)	36 ± 9 (32)	100 ± 30 (31)
Cerebellum	18 ± 5 (32)	32 ± 25 (32)	n.d.	n.d.	n.d.	n.d.	n.d.
Optic tectum	50 ± 11 (32)	25 ± 18 (32)	12 ± 18 (32)	3129 ± 1432 (32)	3 ± 1 (32)	20 ± 7 (32)	163 ± 42 (32)
Telencephalon	96 ± 32 (32)	22 ± 12 (31)	48 ± 37 (32)	673 ± 546 (32)	6 ± 2 (32)	39 ± 15 (32)	168 ± 71 (31)

Values are mean ± S.D. (N). n.d. indicates that monoamines were not detected or too few to be analysed further

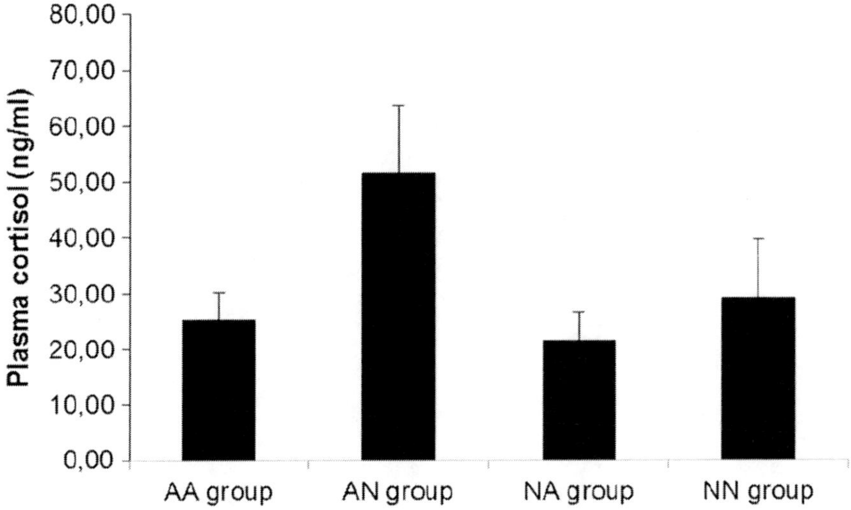

Figure. 5 Plasma cortisol levels after treatments (A is for anaesthetic and N for non-anaesthetic). Values are mean ± S.E.M and $N = 8$ for all treatments

Several associations between monoamines and the carotenoid pigmentation on day 9 were noted in the treatment groups. For instance, in the AA group spots on the left side on day 9 were positively correlated to 5-HIAA in the optic tectum (Pearson correlation, r = 0.859, $P = 0.006$, $N = 8$). In the NA group spots on the right side on day 9 were positively correlated to DOPAC in the telencephalon (Pearson correlation, r = 0.817, $P = 0.013$, $N = 8$), and negatively correlated to 5-HT in the telencephalon (Pearson correlation, r = -0.739, $P = 0.036$, $N = 8$). In the NN group spots on the right side on day 9 were negatively correlated with DA in the brainstem (Pearson correlation, r = -0.772, $P = 0.025$, $N = 8$). However, no associations were found in the AN group.

Behaviour

During the acclimation all fish started eating. There were no behavioural differences between the groups during anaesthetisation (see Table 5). However, behaviour during anaesthetisation was associated with carotenoid pigmentation. Spots on the left side on day 1 (Pearson correlation, r = 0.539, $P = 0.031$, $N = 16$) and on day 8 (Pearson correlation, r = 0.591, $P = 0.016$, $N = 16$) were positively correlated to activity until loss of equilibrium on day 8. Additionally, there were some associations within the treatment groups. For instance, in the NA group spots on the left side on day 1 were positively correlated with activity until loss of equilibrium (Pearson correlation, r = 0.837, $P = 0.009$, $N = 8$), and with time until anaesthesia (Pearson correlation, r = 0.757, $P = 0.030$, $N = 8$) on day 9. In the AA group, spots on the left side on day 8 were positively correlated with activity until loss of equilibrium on day 8 (Pearson correlation, r = 0.857, $P = 0.007$, $N = 8$), and time until loss of equilibrium (Pearson correlation, r = 0.751, $P = 0.032$, $N = 8$) on day 9.

Table 5 The behaviour during anaesthesia in different treatment groups (A is for anaesthetic, N for non-anaesthetic, and procedure for counting spots was done on two consecutive days for all individuals) of juvenile Arctic charr after handling stress

Treatment	AA group	AN group	NA group	NN group
Day 8				
Activity until loss of equilibrium, stage III plane 1 (no/s)	0.17 ± 0.10 (8)	0.16 ± 0.06 (8)	–	–
Time until loss of equilibrium, stage III plane 1 (s)	42.0 ± 9.9 (8)	49.9 ± 9.7 (8)	–	–
Time until anaesthesia, stage III plane 2 (s)	113.3 ± 57.7 (8)	107.4 ± 32.8 (8)	–	–
Day 9				
Activity until loss of equilibrium, stage III plane 1 (no/s)	0.15 ± 0.11 (8)	–	0.09 ± 0.05 (8)	–
Time until loss of equilibrium, stage III plane 1 (s)	39.1 ± 10.7 (8)	–	47.8 ± 8.2 (8)	–
Time until anaesthesia, stage III plane 2 (s)	79.1 ± 10.7 (8)	–	106.5 ± 34.6 (8)	–

Loss of equilibrium, stage III plane 1, and anaesthesia, stage III plane 2 are based upon definitions from Zahl et al. (2012), see text for further information. Values are mean ± S.D. (N)

DISCUSSION

The main purpose of this study was to investigate the possibility to use carotenoid pigmentation in Arctic charr as a tool for evaluating stress responsiveness, especially under normal maintenance, i.e. with the use of the anaesthetic MS-222. Some of the results indicate that anaesthetization affects number of spots. For instance, the group that was anaesthetized twice (AA) had more spots on the right side than the groups not anaesthetized the last day (AN and NN). This was also evident when considering spots on the right side in anaesthetized compared to not anaesthetized on day 9, when anaesthetized charr had more spots, and a similar trend was found on day 8. Further this trend was to some extent evident in spots on the left side on day 9 but not on day 8. This is in line with the report that anaesthesia induced carotenoid spots in Arctic charr (Backström et al. 2015b), as well as the reported stress-inducing effect of the anaesthetic MS-222 in teleost fish (Zahl et al. 2010; Readman et al. 2013). Thus the anaesthetic treatment before photographing could arguably be the more stressful handling. This is further corroborated by the higher dopaminergic activity in telencephalon of anaesthetised individuals the last day in our study, which could be indicative for higher stress responsiveness (Fuchs and Flugge 2003; Höglund et al. 2005; Backström et al. 2011). Thus there are indications of both spots and physiological stress being induced by anaesthetics. However, we could not see any differences in the plasma cortisol levels which was expected from an earlier study (Backström et al. 2015b). Plasma for measuring cortisol was taken quickly (range of time until anaesthesia 60-185 s and an approximate additional 30 s for photographing) after netting fish from isolation aquaria, and is therefore not expected to show a full stress response (Wendelaar Bonga 1997; Cockrem 2013; Gesto et al. 2013; López-Patiño et al. 2014). Since both treatments are stressful small differences were expected. Further, in this study there were three similar stressful events consecutively and thus stress could be predictable. Predictable stress has been shown to reduce the cortisol response (Galhardo et al. 2011), and a comparison to an earlier study (Backström et al. 2015b) indicates that there is a lower elevation of cortisol in our study. However, the number of repeated stimuli is low so it might not be a plausible explanation.

Some of the monoaminergic activities were associated with carotenoid pigmentation. Serotonergic activity in the optic tectum was positively correlated with left side spots in the AA group, supposedly the most stressed group (see above). During stress the serotonergic activity increases (Winberg and Nilsson 1993; Summers and Winberg 2006), and thus this follows predictions and supports earlier results on telencephalic serotonergic activity association with left side spots in social dominants (Backström et al. 2015c). Dopaminergic activity in telencephalon was positively correlated to right side spots in the PA group. This could be indicative for higher stress responsiveness (Fuchs and Flugge 2003; Höglund et al. 2005; Backström et al. 2011), and is similar to an earlier study of social subordinate Arctic charr concerning dopaminergic activity in optic tectum and its association with right side spots (Backström et al. 2015c), although there is a temporal difference in sampling time. The same group also had a positive correlation between left side spots and activity during anaesthesia. Thus, the increased activity was reflected in both DA concentrations and spots. However, all individuals were also tested the day before and therefore might have a lingering stress effects.

Behaviour during stress was associated to carotenoid pigmentation in this study. During the anaesthetization, activity until loss of equilibrium was higher in individuals with more spots on the left side. Similarly, in some groups (NA and AA) individuals with more left side spots took longer time until loss of equilibrium. This is the opposite result compared to an earlier report that showed a negative correlation between spots and both activity and time until loss of equilibrium during anaesthetization comparing the effect of several anaesthetics (Backström et al. 2015b). This was however not observed for MS-222 (Backström et al. 2015b), that was used in this study. This difference could be the result of predictable stress (see above). Further, the protocols were slightly different with two more photographing sessions (in total three) in this study, which could influence the behaviour.

Thus, several of the results from this study indicate that MS-222 affect the spots and therefore would not be recommended to be used when evaluating stress responsiveness. However, carotenoid spots were highly correlated between days and especially on the last days, during the treatments. Further, there also seems to be a lateralized effect of the carotenoid pigmentation. It has been suggested that the left side is associated to stress responsiveness whereas the right side is associated to aggressiveness (Backström et al. 2015a; Backström et al. 2015c). The left side association to stress responsiveness seems to be corroborated in this study. For instance, activity during anaesthesia, which indicates more stress (see above), was positively correlated to left side spots. Additionally, serotonergic activity in the optic tectum of the group AA was positively associated with the left side spots. Since increased serotonergic activity is associated with stress response (see above), this follows predictions. However, 5-HT in the optic tectum was negatively correlated to the right side spots of group NA. In the groups NA and NN, dopaminergic activity in the optic tectum was positively correlated to right side spots. Thus there are some connections between right side spots and the monoaminergic systems as well.

The mechanism controlling size and number of carotenoid spots is not known at present. However, since it has been shown earlier that carotenoid spot patterns can change rapidly (Backström et al. 2015b), it is probably an effect of physiological colour change (Mills and Patterson 2009; Nilsson Sköld et al. 2013). One of the main factors controlling physiological colour change is α-melanocyte-stimulating hormone (α-MSH) (Mills and Patterson 2009; Nilsson Sköld et al. 2013), and it has been shown to be involved in skin darkening of Arctic charr during social interactions (Höglund et al. 2000; Höglund et al. 2002). Thus it seems likely that this process can also be involved in and have an effect on the carotenoid spots. The contrast between the spots and the background coloration of the skin could be an effect of this skin darkening, but in an earlier study we found it unlikely (Backström et al. 2015a). Further, since there also seems to be a link between carotenoid spots counted the day before and the stress responsiveness the following day (Backström et al. 2014), the physiological colour change is not the only factor involved in the association between carotenoid spots and stress responsiveness. Therefore the mechanism of the change as well as the origin of the spots needs further studies.

In conclusion, this study adds further support to the hypothesis that carotenoid spots, at least on the left side, can predict stress responsiveness in Arctic charr. This is evident in behavioural responses as well as monoaminergic responses. Finally, it seems plausible that using anaesthetics for photographing is possible without confounding the results, however careful planning is needed and especial attention concerning which side to use.

ACKNOWLEDGMENTS

The methodology of this study was approved by the Swedish University of Agricultural Sciences Ethical Committee. Financing was granted by the Swedish Research Council Formas (Dnr 2011-888).

REFERENCES

1. Backström T, Brännäs E, Nilsson J, Magnhagen C (2014) Behaviour, physiology and carotenoid pigmentation in Arctic charr Salvelinus alpinus. J Fish Biol 84(1):1–9. doi:10.1111/jfb.12240
2. Backström T, Heynen M, Brännäs E, Nilsson J, Magnhagen C (2015a) Dominance and stress signalling of carotenoid pigmentation in Arctic charr (Salvelinus alpinus): lateralization effects? Physiol Behav 138:52–57. doi:10.1016/j.physbeh.2014.10.003
3. Backström T, Heynen M, Brännäs E, Nilsson J, Magnhagen C (2015b) The effect of anesthetics on carotenoid pigmentation and behavior in Arctic char (Salvelinus alpinus). J Vet Behav: Clin Appl Res 10(2):179–184. doi:10.1016/j.jveb.2014.11.007
4. Backström T, Heynen M, Brännäs E, Nilsson J, Winberg S, Magnhagen C (2015c) Social stress effects on pigmentation and monoamines in Arctic charr. Behav Brain Res 291:103–107. doi:10.1016/j.bbr.2015.05.011
5. Backström T, Schjolden J, Øverli Ø, Thörnqvist PO, Winberg S (2011) Stress effects on AVT and CRF systems in two strains of rainbow trout (*Oncorhynchus mykiss*) divergent in stress responsiveness. Horm Behav 59(1):180–186. doi:10.1016/j.yhbeh.2010.11.008
6. Blanchard RJ, McKittrick CR, Blanchard DC (2001) Animal models of social stress: effects on behavior and brain neurochemical systems. Physiol Behav 73(3):261–271. doi:10.1016/S0031-9384(01)00449-8
7. Castanheira MF, et al. (2015) Coping styles in farmed fish: consequences for aquaculture. Rev Aquacult:n/a-n/a doi:10.1111/raq.12100
8. Cockrem JF (2013) Individual variation in glucocorticoid stress responses in animals. Gen Comp Endocrinol 181:45–58. doi:10.1016/j.ygcen.2012.11.025
9. Conrad JL, Weinersmith KL, Brodin T, Saltz JB, Sih A (2011) Behavioural syndromes in fishes: a review with implications for ecology and fisheries management. J Fish Biol 78(2):395–435
10. Duarte CM, Marbá N, Holmer M (2007) Rapid domestication of marine species. Science 316(5823):382–383. doi:10.1126/science.1138042

11. Ducrest A-L, Keller L, Roulin A (2008) Pleiotropy in the melanocortin system, coloration and behavioural syndromes. Trends Ecol Evol 23(9):502–510. doi:10.1016/j.tree.2008.06.001
12. Eaton L, Sloman KA (2011) Subordinate brown trout exaggerate social signalling in turbid conditions. Anim Behav 81(3):603–608. doi:10.1016/j.anbehav.2010.12.005
13. Erhard HW, Mendl M, Ashley DD (1997) Individual aggressiveness of pigs can be measured and used to reduce aggression after mixing. Appl Anim Behav Sci 54(2–3):137–151. doi:10.1016/S0168-1591(97)00068-3
14. Fuchs E, Flugge G (2003) Chronic social stress: effects on limbic brain structures. physiology & behavior. a tribute to Paul MacLean. Neurobiol Relev Soc Behav 79(3):417–427. doi:10.1016/S0031-9384(03)00161-6
15. Galhardo L, Vital J, Oliveira RF (2011) The role of predictability in the stress response of a cichlid fish. Physiol Behav 102(3–4):367–372. doi:10.1016/j.physbeh.2010.11.035
16. Gesto M, López-Patiño MA, Hernández J, Soengas JL, Míguez JM (2013) The response of brain serotonergic and dopaminergic systems to an acute stressor in rainbow trout: a time course study. J Exp Biol 216(23):4435–4442. doi:10.1242/jeb.091751
17. Hill GE (2011) Condition-dependent traits as signals of the functionality of vital cellular processes. Ecol Lett 14(7):625–634. doi:10.1111/j.1461-0248.2011.01622.x
18. Huntingford F, Adams C (2005) Behavioural syndromes in farmed fish: implications for production and welfare. Behaviour 142(9-10):1207–1221. doi:10.1163/156853905774539382
19. Huntingford FA et al (2006) Current issues in fish welfare. J Fish Biol 68(2):332–372. doi:10.1111/ j.0022-1112.2006.001046.x
20. Höglund E, Balm P, Winberg S (2000) Skin darkening, a potential social signal in subordinate arctic charr (*Salvelinus alpinus*): the regulatory role of brain monoamines and pro-opiomelanocortin-derived peptides. J Exp Biol 203(11):1711–1721
21. Höglund E, Balm PHM, Winberg S (2002) Behavioural and neuroendocrine effects of environmental background colour and social interaction in Arctic charr (*Salvelinus alpinus*). J Exp Biol 205(16):2535–2543
22. Höglund E, Weltzien F-A, Schjolden J, Winberg S, Ursin H, Døving KB (2005) Avoidance behavior and brain monoamines in fish. Brain Res 1032(1–2):104–110. doi:10.1016/j.brainres.2004.10.050
23. Kittilsen S et al (2009) Melanin-based skin spots reflect stress responsiveness in salmonid fish. Horm Behav 56(3):292–298. doi:10.1016/j.yhbeh.2009.06.006
24. Koolhaas JM, de Boer SF, Buwalda B, van Reenen K (2007) Individual variation in coping with stress: a multidimensional approach of ultimate and proximate mechanisms. Brain Behav Evol 70(4):218–226
25. Koolhaas JM, de Boer SF, Coppens CM, Buwalda B (2010) Neuroendocrinology of coping styles: Towards understanding the biology of individual variation. Front Neuroendocrinol 31(3):307–321. doi:10.1016/j.yfrne.2010.04.001
26. Koolhaas JM et al (1999) Coping styles in animals: current status in behavior and stress-physiology. Neurosci Biobehav Rev 23(7):925–935
27. López-Patiño MA, Hernández-Pérez J, Gesto M, Librán-Pérez M, Míguez JM, Soengas JL (2014) Short-term time course of liver metabolic response to acute handling stress in rainbow trout, Oncorhynchus mykiss. Comp Biochem Physiol A Mol Integr Physiol 168:40–49. doi:10.1016/j.cbpa.2013.10.027
28. Mills MG, Patterson LB (2009) Not just black and white: pigment pattern development and evolution in vertebrates. Semin Cell Dev Biol 20(1):72–81. doi:10.1016/j.semcdb.2008.11.012
29. Nilsson J, Brännäs E, Eriksson LO (2010) The Swedish Arctic charr breeding programme. Hydrobiologia 650(1):275–282. doi:10.1007/s10750-010-0232-9
30. Nilsson Sköld H, Aspengren S, Wallin M (2013) Rapid color change in fish and amphibians—function, regulation, and emerging applications. Pigment Cell Melanoma Res 26(1):29–38. doi:10.1111/pcmr.12040

31. Øverli Ø, Winberg S, Pottinger TG (2005) Behavioral and neuroendocrine correlates of selection for stress responsiveness in rainbow trout - a review. Integr Comp Biol 45(3):463–474
32. O'Connor KI, Metcalfe NB, Taylor AC (1999) Does darkening signal submission in territorial contests between juvenile Atlantic salmon, Salmo salar ? Anim Behav 58(6):1269–1276. doi:10.1006/anbe.1999.1260
33. O'Connor KI, Metcalfe NB, Taylor AC (2000) Familiarity influences body darkening in territorial disputes between juvenile salmon. Anim Behav 59(6):1095–1101. doi:10.1006/anbe.2000.1401
34. Readman GD, Owen SF, Murrell JC, Knowles TG (2013) Do fish perceive anaesthetics as aversive? PLoS One 8(9):e73773. doi:10.1371/journal.pone.0073773
35. Schjolden J, Winberg S (2007) Genetically determined variation in stress responsiveness in rainbow trout: behavior and neurobiology. Brain Behav Evol 70(4):227–238
36. Shannon NJ, Gunnet JW, Moore KE (1986) A comparison of biochemical indices of 5-hydroxytryptaminergic neuronal activity following electrical stimulation of the dorsal Raphe nucleus. J Neurochem 47(3):958–965. doi:10.1111/j.1471-4159.1986.tb00704.x
37. Summers CH, Winberg S (2006) Interactions between the neural regulation of stress and aggression. J Exp Biol 209(23):4581–4589
38. Wendelaar Bonga SE (1997) The stress response in fish. Physiol Rev 77(3):591–625
39. Winberg S, Nilsson GE (1993) Roles of brain monoamine neurotransmitters in agonistic behaviour and stress reactions, with particular reference to fish. ComparaT Biochem Physiol Part C: Comparat Pharmacol Toxicol 106(3):597–614
40. Zahl I, Kiessling A, Samuelsen O, Olsen R (2010) Anesthesia induces stress in Atlantic salmon (Salmo salar), Atlantic cod (Gadus morhua) and Atlantic halibut (Hippoglossus hippoglossus). Fish Physiol Biochem 36(3):719–730. doi:10.1007/s10695-009-9346-2
41. Zahl I, Samuelsen O, Kiessling A (2012) Anaesthesia of farmed fish: implications for welfare. Fish Physiol Biochem 38(1):201–218. doi:10.1007/s10695-011-9565-1

CHAPTER 4

Vibrio Diseases of Marine Fish Populations

R. R. Colwell & D. J. Grimes

Department of Microbiology, University of Maryland; College Park, MD 20742 USA

ABSTRACT

Several *Vibrio* spp. cause disease in marine fish populations, both wild and cultured. The most common disease, vibriosis, is caused by *V. anguillarum.* However, increase in the intensity of mariculture, combined with continuing improvements in bacterial systematics, expands the list of *Vibrio* spp. that cause fish disease. The bacterial pathogens, species of fish affected, virulence mechanisms, and disease treatment and prevention are included as topics of emphasis in this review

INTRODUCTION

It is most appropriate that the introductory paper at this Symposium on Diseases of Marine Organisms should be about vibrios. The disease syndrome termed vibriosis was one of the first diseases of marine fish to be described (Sindermann, 1970; Home, 1982). Called "red sore", "red pest", "red spot" and "red disease" because of characteristic hemorrhagic skin lesions, this disease was recognized and described as early as 1718 in Italy, with numerous epizootics being documented throughout the 19th century (Crosa et al., 1977; Sindermann, 1970). Today, while "red sore" is, without doubt, the best understood of the marine bacterial fish diseases, new diseases have been added to the list of those caused by *Vibrio* spp.

The more recently published work describing *Vibrio* disease, primarily that appearing in the literature since 1977, will be the focus of this paper. Bacterial pathogens, the species of fish affected, virulence mechanisms, and disease treatment and prevention are included as topics of emphasis. For discussions of earlier work, the reader is referred to Sindermann (1970) and Roberts (1982).

VIBRIOS AS BACTERIA PATHOGENIC FOR FISH

All members of the genus *Vibrio* are Gram-negative rods ranging in size and morphology from coccobacilli to definite rod-shape cells that may exhibit some degree of curvature, i.e., vibrioid. Under conditions of nutrient depletion or in the natural environment, including estuaries and the ocean, where oligotrophic conditions occur, vibrios will occur as small coccoid forms, representing a strategy for survival (Singleton et al., 1982; Xu et al., 1982; Novitsky & Morita, 1976). *Vibrio* spp. are motile by means of a sheathed, polar flagellum, one or more at one pole of the cell. Some species also form non-sheathed lateral flagella, when grown on solid media. Lateral flagella play a role in attachment to surfaces and also appear to be involved in swarming (Belas & Colwell, 1982a; 1982b).

Biochemical features of the genus include fermentation of glucose, usually with out the production of gas; sensitivity to 150 btg vibriostat 0/129 (2, 4 - diamino - 6, 7 - diisopro -pylpteridine)per ml of medium; positive oxidase reaction, except for V. *Metschnikovii* and V. *gazogenes;* and an absolute requirement for NaCI (West & Colwell, 1984). The over - all DNA base composition , i.e., percent guanine plus cytosine, for the genus ranges from 38 to 51%, with the type species, V. *cholerae,* having an average % G+C of 47. Table 1 lists salient features of the genus, useful in identifying the fish pathogens.

The natural habitat of *Vibrio* spp. is estuarine and marine water and sediment. Vibrios often occur in association with in vertebrate and vertebrate animal surfaces and internal organs (Huq e t al., 1983; Grimes etal , 1984b}, Stress can compromise the animal host, resulting in a commensal species switching to pathogenicity as the mode of interaction(Grimes et al., 1984c).

Table 1. Characteristics useful in identifying those members of the genus *Vibrio* that have been shown to be pathogenic for fish*

Characteristic	*Vibrio*								
	alginolyticus	*anguillarum*	*carchariae*	*cholerae*	*damsela*	*ordalii*	*parahaemolyticus*	*vulnificus* 1	*vulnificus* 2
0/129									
10 µg	R	S	R	S	S	S	R	S	S
150 µg	S	S	S	S	S	S	S	S	S
Swarming	+	–	+	–	–	–	V	–	–
Amino acids**									
arginine	–	+	–	–	+	–	–	–	–
lysine	+	–	+	+	–	–	+	+	+
ornithine	+	–	+	+	–	–	+	+	–
Fermentation									
glucose	A	A	A	A	AG	A	A	A	A
lactose	–	–	–	–	–	–	–	A	A
mannose	A	A	A	A	AG	–	A	A	A
salicin	–	–	A	–	–	–	–	A	A
sucrose	A	A	A	A	–	A	–	–	–
Urease	–	–	+	–	+	–	V	V	–
Indole	+	+	+	+	–	–	+	+	–
Voges-Proskauer	+	+	–	+	+	–	–	–	–
Growth at % NaCl									
0	–	V	–	V	–	–	–	–	–
3	+	+	+	+	+	+	+	+	+
6	+	+	+	V	V	+	+	+	–
8	+	V	+	–	–	–	+	–	–
10	+	–	–	–	–	–	–	–	–
Growth at 42 °C	+	–	–	+	–	–	+	+	–

* + = 90 % of all strains give a positive reaction, (+) = weak and/or delayed positive, – = 90 % of all strains are negative, V = variable reactions, A = acid, AG = acid + gas, R = resistant, S = sensitive; test procedures are given by West & Colwell (1984).

** arginine dihydrolase, lysine and ornithine decarboxylase

Vibrio alginolyticus

Formerly classified as V. *parahaemolyticus* biotype 2, V. *alginolyticus* often exhibits pronounced swarming on complex media, produces a positive Voges-Proskauer reaction, tolerates 10 % NaC1, and grows at temperatures up to 42 °C. It ferments sucrose and decarboxylates lysine and omithine (Table 1).

V. alginolyticus is frequently isolated from fin fish, shellfish, seawater, and sediment (Gjerde & Boe, 1981; Joseph et al., 1982). It is not widely recognized as a fish pathogen, even though it has been associated with ulcer disease (Akazawa, 1968) and mortality of sea-bream (Colorni et al., 1981; Iwata et al., 1978) (see Table 2). Recently, Burke &

Table 2. *Vibrio* species known to cause disease in marine fish

Organism	Disease
V. alginolyticus	ulcer disease
V. anguillarum	vibriosis
V. carchariae	vasculitis
V. cholerae non-01	petechial hemorrhage
V. damsela	skin lesions
*V. ordalii**	vibriosis
V. parahaemolyticus	ulcer disease
V. vulnificus 2	skin lesions

* Formerly *V. anguillarum* biovar II

Rodgers (1981) found V. *alginolyticus* to be a secondary invader of "red spot", a disease caused by *V. "anguillarum.* Gauthier & Clement (1979) observed a streptomycin-resistant strain of V. *alginolyticus* to be present not only in sediment but also water, when the water temperature exceeded 16 °C. However, food chain transmission was not observed when reconstructed marine food chains were studied in the laboratory (Gauthier & Clement, 1979).

It is well established that *V. alginolyticus* can colonize the human gut (Sakazaki et al., 1963). In fact, Hiratsuka et al. (1980) observed the cause of acute enterocolitis in a patient who consumed trout roe contaminated with the same bacteria to be *V. alginolyticus.* More frequently, *V. alginolyticus* is associated with skin infections (Blake et al., 1980), such as conjunctivitis and stump ulcer cases described by Schmidt et al. (1979). Even though Blake et al. (1980) reported that little work has been done on virulence mechanisms, it is probable, based on available evidence, that the collagenase and several extracellular proteases produced by *V, alginolyticus* (Long et al., 1981) participate in the invasion of fish and human skin.

Vibrio anguillarum

This vibrio, the first recognized to be a fish pathogen, includes strains capable of utilizing arginine and tolerant of 6 % NaC1; some strains tolerate 8 % salt and others are capable of growth in the presence of trace amounts of salt, i.e., they are reported to grow in 0 % salt, but in fact, require the concentration of NaC1 present in peptone used to prepare the medium (Singleton et al., 1982). 1/. *anguillarum* is sensitive to 10 ~g 0/129, gives a positive Voges-Proskauer reaction and ferments sucrose (Table 1). Recently, strains referred to as V. *anguillarum* biovar II were described as a new species, V. *Ordalii* sp. nov. (Schiewe et al., 1981) vide infra.

Vibriosis of V. *anguillarum* etiology has been described in over 42 species of fish, including ayu, eel, cod, pike, brown trout, flounder, striped bass and salmon. Vibriosis has been especially devastating to salmon reared in Pacific coastal waters {Crosa et al., 1977). Vibriosis is a typical Gram-negative septicemia, exhibiting extensive haemorrhaging and necrosis in the internal organs and musculature. It is a rapidly fulminating disease, affecting both cultured and wild marine fish. Gross clinical signs of the disease include hemorrhaging of the fins, eyes, and ventral surface.

It is interesting to note the progression of discovery of V. *anguillanlm* throughout the world. It was first described in Europe, as the agent of "red spot" in eels (Bergman, 1909; Horne, 1982; Sindermann, 1970). Its recovery from diseased fish in North America did not occur until 1953 (Crosa et al., 1977), and Muroga et al. {1976) suggest that the first isolation of *V. anguillarum* in Japan in 1975 may have resulted from contaminated eels imported from Prance. Whether or not there was a true geographic spread (natural or anthropogenic) of *V. anguillarum,* or merely a circumstantial sequence of discovery is open to speculation.

Two of the more recent developments with regard to *1/. anguillarum* are very exciting and should lead to effective control of this fish pathogen. Virulence factors, and their genetic control, are receiving much attention in several laboratories. Also promising are several approaches to vaccination being investigated.

An interesting phenomenon was recently reported for V.. *anguillarum* isolates from ayu cultured in Japan. Muroga et al. (1979a) found that 27 of 52 isolates were not sensitive to vibriostatic agent 0/129, previously thought to be a significant attribute of *Vibrio* spp. These isolates were obtained from ayu that had been raised in ponds that contained prophylactic trimethoprim-sulfadoxine (TS). Similar findings were reported by Aoki et al. (1981), after examination of 132 strains of V. *anguillarum* for resistance to 0/129 and trimethoprim. Transferable R plasmids were suspected, but could not be demonstrated.

Vibrio carchanae

The original isolate of V. *carchariae* was obtained from a dead sandbar (brown) shark *(Carcharhinusplumbeus)* (Grimes et al., 1984a). Subsequently, it has been isolated from lemon sharks *(Negaprion brevirostris)* and from trematodes infesting the skin of lemon sharks (Grimes et al., 1984c). V. *carchariae* is a swarming vibrio that exhibits mixed flagellation. Morphologically, and in terms of several biochemical attributes, V. *carchariae* is practically indistinguishable from *V. alginolyticus* and the sucrose-positive, urease-positive strains of *Vibrio parahaemolyticus.*

However, unlike these two vibrios, V. *carchariae* can not grow at 42°C; other differences include growth in 10 % NaC1 and Voges-Proskauer (Table 1), as well as DNA homology.

V. carchariae is described in detail elsewhere in this Symposium volume (GAmes et al., 1984b). The host range of *V. carchariae* has not yet been thoroughly investigated, since the type strain was isolated only in 1982. With regard to sharks, 1/. *Carchariae* causes a vasculitis, especially in organs of the reticuloendothelial system. It also causes, or is frequently isolated from, dermal lesions and cysts in sharks. Recently, Grimes et al. {1984c} hypothesized that shark trematodes *(Dermophthirius nigrellii)* may serve as a vector for V. *carchariae.*

Vibrio cholerae

Serovars of V. *cholerae,* other than serovar 01, are occasionally isolated from fish. These V. *cholerae* serovars have been referred to as non-agglutinable (NAG) vibrios and as non-cholera vibrios (NCV), a practice that has created much confusion and is not acceptable nomenclature [West & Colwell, 1984). The non-01 V. *cholerae* strains do not react with the 0 antiserum originally designated as 01.

In a study of the Danube River, Strusiewicz et aL (1980) isolated V. *cholerae* non-01 from water, fish, and sewage outfalls. The highest incidence of positive samples occurred during September (the warmest month) and 81.9 % of the isolates were enterotoxigenic, measured by rabbit ileal loop assay. Enterotoxigenic strains were recovered more frequently from fish than from water samples (Strusiewicz & Serban, 1980). In a similar study, Szeness et al. (1979) isolated V. *cholerae* non-01 from fish, frogs, and birds. Neither of these studies, however, commented on the health status of the fish from which the V. *cholerae* isolates were obtained.

Muroga et al. {1979b) isolated a *Vibrio* sp. from diseased ayu and demonstrated that the isolate was the causative agent of an ayu epizootic. Briefly, the disease was characterized by petechiae on the body surface and by internal organ congestion. The isolates were morphologically and biochemically very similar to *V. cholerae,* but did not agglutinate in either Ogawa or Inaba antisera, i.e., they were non-01 serovar. Muroga et al. (1979b) also determined the % G+C base ratio for their isolates (47 to 47.7 %), and demonstrated 86 % DNA homology between their isolates and If. *cholerae* NIH35A3. There is little doubt that the cause of the ayu epizootic was the human pathogen, V. *cholerae.*

In a later study, Yamanoi et al. (1980) demonstrated that one of the ayu V. *Cholera* isolates (strain PS-7701) survived for 320 days at 25°C in freshwater, 0.85 % NaC1 solution, Ringer solution, seawater, and in diluted seawater. They further noted that survival in water was greatly reduced at 2 °C, a finding similar to that of Singleton et al. {1982). Related to this observation was the fact that healthy ayu could not be reinfected with V. *cholerae,* when the ayu were held at 16°C. However, when infected by the immersion method (ca. 1.26 × 104 cells/ml) and held at 21 °C and 26°C, most of the ayu died within 4 days post infection. The *V. cholerae* strain was also pathogenic for eels *(Anguilla japonica]* and for mice.

Vibrio damsela

First isolated from damselfish off the coast of southern California [Love et al., 1981), *V. damsela* is now known to infect humans (Love et al, 1981) and sharks (Grimes et al., 1984a). Distinguishing attributes of V. *damsela* include its ability to metabolize arginine and to produce gas during fermentation of glucose and other selected sugars (Table 1).

V. damsela causes characteristic skin ulcers in the blacksmith, a temperate-water damselfish *(Chromis punctipinnis),* and in garibaldi *(Hypsypops rubicunda),* and was thought by Love et al. (1981) to have a narrow host range. However, we isolated V. *damsela,* along with *V. carchariae,* from a dead sandbar shark *(C. plumbeus},* and subsequently demonstrated *V. damsela* to be virulent for the spiny dogfish shark *(Squalus acanthias)* (Grimes et al., 1984a). Although we have not been successful in producing disease in lemon sharks *(N. brevirostris)* experimentally infected with V. *damsela, V. damsela* was cultured from the stomach of a wild lemon shark (Grimes et al., unpubl, data).

Vibrio ordalii

V. ordalii was recently proposed by Schiewe et al. (1981) to include all strains previously designated as *V. anguillarum* biovar II. It is differentiated from *V. Anguillarum* by giving negative reactions for: Voges-Proskauer; arginine dihydrolase; citrate; starch; ONPG; lipase; growth at 37 °C; and acid from cellobiose, glycerol, sorbitol, and trehalose (Table 1). Detailed discussion of *V. ordalil* is provided in the recent review by Schiewe (1983).

V. ordalii causes vibriosis in salmonid fish in the Pacific Northwest and in Japan. This species appears to have a more host-dependent mode of existence than does V. *anguillarum,* because it is rarely isolated from water, sediment, or other abiotic marine samples; V. *ordalii* is recovered only from moribund fish (Schiewe, 1983). V. *Ordalii* preferentially colonizes skeletal muscle, cardiac muscle, gills, and the gastrointestinal tract of salmonids, in contrast to V. *anguillarum,* which presents disseminated infections, with highest numbers in blood, loose connective tissue, kidney, spleen, gills, and in the posterior gastrointestinal tract. Several investigators have observed a marked leukopenia in moribund fish, suggesting the presence of a powerful leukocytolytic factor in V. *ordalii* (Schiewe, 1983).

Vibrio parahaemolyticus

First recognized as the etiologic agent of human seafood-borne disease in Japan in the 1950s, *V. parahaemolyticus* is the most thoroughly studied *Vibrio* sp., next to V. *cholerae.* Notable features of *V. parahaemolyticus* include swarming, growth at 42 °C, negative arginine reaction, and resistance to 10 t~g of 0/129. *V. parahaemol~icus* is considered to be non-sucrose fermenting. However, sucrose positive variants do exist, and are usually overlooked durin 9 routine clinical isolation (West & Colwell, 1984).

V. parahaemolyticus causes gastroenteritis in humans and has also been implicated in extraintestinal infections (Blake et al., 1980; Joseph et al., 1982). An enterotoxin appears to be involved, but has yet to be conclusively demonstrated (Joseph et at., 1982). *V. parahaemolyticus* was so named because of its ability to produce beta-hemolysis on high-salt blood agar, a reaction that has been designated the Kanagawa phenomenon. For reasons as yet unknown, only human clinical isolates seem to give a positive Kanagawa reaction; environmental isolates of V. *parahaemolyticus,* including fish isolates, are rarely Kanagawa positive (Joseph et al., 1982).

The role of *V. parahaemolyticus* in fish disease is more obscure, and fish may simply function as a passive carrier. Like V. *alginolyticus, V. parahaemolyticus* is frequently isolated from both healthy and diseased fish throughout the world (Lall et al., 1979; Lhuillier, 1977; Franca et al., 1980; Qadri & Zuberi, 1977; Nair et al., 1980). Important habitats for V. *parahaemolyticus* include sediment and shellfish (both oysters and crustaceans), with movement into the water column occurring during warmer months (Kaneko & Colwell, 1978; Joseph et al., 1982; Gjerde & Boe, 1981; Natarajan et al., 1980). *V. parahaemolyticus* is occasionally isolated from skin ulcers of fish, but usually in association with other vibrios (Yasunaga & Yamamoto, 1977). Kusuda et al. (1979) has isolated 1I. *parahaemolyticus* from vibriosis of red sea-bream cultured in Japan, and Gilmour {1977) recovered the species from fish farm tank water. Ortiz & Decker (1976) detected anti-V, *parahaemolyticus* antibody in the sera of 13 of 26 flounder examined; Robohm et al. (1979), on the other hand, did not detect significant titers of anti-V. *parahaemolyticus* antibody in the fish they examined. The status of V. *Parahaemolyticus* as a fish pathogen needs to be more completely documented.

Vibrio vulnificus

V. vulnificus is a lactose-fermenting vibrio that, in many respects resembles V. *alginolyticus* and *V. parahaemolyticus* (Table 1). Originally referred to as a lactosepositive vibrio, V. *vulnificus* is recognized as a highly virulent, opportunistic human pathogen with a low incidence of disease (Oliver et al., 1983; Joseph et al., 1982; Blake et al., 1980). Oliver et al. (1983) studied the distribution of sucrose-negative vibrios in seawater, sediment, plankton, and animal samples collected at 80 sites from Miami, Florida to Portland, Maine. They concluded that V. *vulnificus* was ubiquitous, both geographically and in terms of presence in a variety of environmental samples. However, absolute numbers of V. *vulnificus* were always found to be low, since Oliver et al. (1983) demonstrated that only 33 isolates (0.83 %) of 3,887 sucrose-negative vibrios examined fermented lactose.

Tison et al. (1982) reported on a new subgroup, biogroup 2, of V. *vulnificus* that is pathogenic for eels. Previously, the names V. *anguillarum* type B and V. *anguillicida* had been used to describe these vibrios (Nishibuchi et al., 1979). Briefly, the eel disease caused by *V. vulnificus* biogroup 2 is a vibriosis characterized by a red patch or swollen lesion on the trunk or tail. In early stages of the disease, the skin and lateral musculature is primarily involved; advanced stages involve histopathological changes in the spleen, liver, kidney, heart, gills, and intestine (Miyazaki et al., 1977).

V. vulnificus biogroup 2 differs from V. *vulnificus* biogroup I in several characteristics (Table 1), and the two biogroups do not share common surface antigens (Tison et al., 1982). However, because of the strong reassociation (> 90 %), under stringent conditions, of DNA from clinical *V. vulnificus* strains (i.e., biogroup 1) with DNA from eel isolates, Tison et al. (1982) proposed that the eel isolates, and strains phenotypically resembling the eel isolates, be classified as *V. vulnificus* biogroup 2.

Luminous *Vibrio* spp.

Although not known to be pathogenic for fish, the luminous vibrios deserve brief mention. These interesting species are symbionts of fish, livin 9 in the luminous organ (Ruby et al., 1980; Yetinson & Shilo, 1979; Hastings & Nealson, 1977). The principal species are *V. fischeri* and *1,I. harvey1,* although occasional strains of V. *cholerae* are also luminous. Light organs, located variously depending on the species of fish, usually contain approximately l0 g cells/ml, and the organ emits constant light (Haygood & Nealson, 1984}. Because of this high cell density, Haygood & Nealson {1984} believe that free-living luminous vibrios recovered from seawater are "over-flow" from light organs of fish. Presently, the precise reason for this unusual species interaction is unknown.

FISH POPULATIONS AFFECTED BY *VIBRIO*

Vibrio spp. infect, and often cause disease in, a variety of estuarine and marine fish, some of which are listed in Table 3. Generally speaking, disease is attributable to some form of stress that renders the fish more susceptible. The disease that follows is typically a systemic infection known as vibriosis or some form of skin ulcer (Table 3).

Table 3. Selected marine fish known to be susceptible to *Vibrio* infection and disease

Fish	Disease
Ayu	vibriosis, hemorrhagic disease
Akame	skin ulcers
Sharks	skin ulcers, vasculitis
Cod	skin ulcers, fin rot
Eels	vibriosis
Flounder	vibriosis
Gray mullet	fin rot
Salmon	vibriosis
Sea-bream	acute septicemia, vibriosis
Sole	red spot
Striped bass	vibriosis
Turbot	acute edema

Ayu. Ayu (Plecoglossus altivelis), a fish cultured in Japan, is susceptible to a variety of infections, including streptococcal and vibrio disease. *Vibrio* disease is usually a classic vibriosis caused by V. *anguillarum* (Jo & Muroga, 1977; Kusuda et al., 1981; Aoki & Kitao, 1978). However, *V. cholerae* serovar non-01 has also been implicated in ayu disease (Muroga et al., 1979b; Yamanoi et al., 1980}.

Akame. Wild akame *(Liza akame)* were observed by Muroga {1979} to have skin ulcers caused by a parasitic copepod, *Caligus orlentalls. The* copepod was detected on 35 of 36 akame examined, with an infestation rate of 27 copepods per fish. Several species of bacteria were isolated from the lesions, including *Vibrlo* spp.

Cod. Atlantic cod *(Gadus morhua)* are susceptible to vibriosis, especially when exposed to polluted waters. Larsen et al. (1978) observed that cod living in Danish coastal waters suffered from a specific ulcer syndrome, but only when exposed to water contaminated with municipal sewage or effluent from sugar and cellulose plants. Larsen et al. (1978} were not able to elucidate etiology of the primary lesion, but could demonstrate that V. *anguillarum* infected the ulcer thereby establishing a generalized vibriosis. In a later study, it was shown that V. *anguflarum* could not elicit the disease, even when the organism was inoculated into the animal by a variety of routes (Jenson & Larsen, 1982). Interestingly, an icosahedral virus isolated from the diseased cod was shown to produce the ulcer syndrome. Thus, it was hypothesized that the icosahedral virus was necessary to initiate pathogenesis, with V. *anguillarum* playing a role in progression of the disease. Most interesting was the relationship between the incidence of *V. anguillarum* and wastewater discharge. The frequency of occurrence of *V. Anguillarum* was highest in areas of known contamination, and the numbers of V. *Anguillarum* in water and sediment immediately increased in direct response to effluent discharge from a sugar plant (Larsen et al., 1978). Robohm et al. (1979) also studied the effect of pollution on fish disease, and were able to demonstrate significantly higher antibody titers to V. *anguillarum* in flounder collected from the New York Bight apex than in flounder from unpolluted coastal waters.

In a study of captive Atlantic cod, it was observed that 320 of 621 cod died from fin rot, over an 8-year period. Most deaths occurred within 2 months after capture, and death was probably due to excessive blood loss associated with fin erosion. Three genera of bacteria were isolated from the infected tissues, mainly *Pseudomonas* spp., but also *Aeromonas* and *Vibrio* spp. (Khan et al., 1981).

Ee1s. Cultivated eels are susceptible to a variety of bacterial infections, producing several clinical symptoms. Correct diagnosis is, therefore, dependent on isolation and identification of the pathogen (Nishibuchi et al., 1980). In addition to diseases caused by *Aeromonas* and *Pseudomonas* spp., discussed above, eels are susceptible to infection with *V. anguillarum* and 11. *vulnificus* biogroup 2 (Nishibuchi et al., 1980; Tison et al., 1982). Rodsaether et al. (1977) concluded that V. *anguillarum* is part of the normal flora of eels *(A. anguilla},* and that copper contamination changes the commensal association between fish and the bacterium to pathogenicity. In contrast, is the finding of Kanai et al. (1977) who reported they could not recover *Vibrio* spp. from healthy eels *(A. japonica},* except during the month of October. However the results of Kanai et al. (1977) must be seriously questioned, with regard to *Vibrio* recovery, since the recovery medium used was nutrient agar without salt added. Vibrios, in general, require NaC1 for growth (Singleton et al., 1982).

F1ounder. Both winter flounder *(Pseudopleuronectes americanus}* and smooth flounder *(Liopsetta putnami}* are susceptible to infection and disease caused by V. *anguillarum* (Hoornbeek et al., 1982; Watkins et al., 19811 Robohm et al., 1979). Myxobacter and protozoan parasites have also been shown to be associated with flounder mortality observed by Hoombeek et al. (1982), but it was not stated whether there were mutualistic, mixed infection, or vector relationships with V. *anguillarum.* Watkins et al. (1981) studied laboratory infections of winter flounder, and found that the LDs0 was lower (by 2-3 logs) for challenge by injection of *V. anguillarum,* compared with oral challenge. In addition, the mean LDs0 for fish collected during the winter was about 3 logs less than that for specimens collected in the summer.

Graymu11et. Healthy gray mullet *(Mugil cephalus}* were shown by Hamid et al. (1978) to carry *Enterobacter, Bacillus,* and *Micrococcus* spp. as the principal intestinal bacteria. When transferred to seawater, the dominant genera recovered from the fish included *Vibrio, Pseudomonas,* and *Aeromonas. In* subsequent studies, it was shown that the intestinal *Vibrio* and *Enterobacter* spp. were strongly proteolytic and also capable of producing amylase, chitinase, and lecithinase (Hamid et al., 1979). In a later study, Kakimoto & Mowlah (1980) showed these gray mullet isolates to be V. *anguillarum* and *E. aerogenes*

Minchew & Yarbrough (1977) demonstrated that experimental crude oil spills significantly increased the incidence of fin rot in mullet, from 6 % in clean, control ponds to 96 % in ponds contaminated with 4-5 ~l/1 (ppm) crude oil. A *Vibrio* sp. was considered to be the primary pathogen involved in the oil-associated fin erosion. Interestingly, Hada & Sizemore (1981) showed that the incidence of *Vibrio* spp. in the water column near a Gulf of Mexico oil field production platform was much greater than incidence in a noncontaminated control area. Furthermore, these vibrios were more likely to contain one or more plasmids.

Salmonids. Vibriosis of salmonid fish, both wild and cultured, is a serious disease of worldwide distribution and importance. As discussed above, both *V. Anffuillarum* and V. *ordafii* can cause vibriosis in salmon (Schiewe, 1983; Crosa et al., 1977; Harrell et al., 1976) although in specific cases other *Vibrio* sl0p. have been isolated {Tajima et al., 1981).

Yoshimizu et al. (1976a, 1976b, 1976c) published a series of studies on the intestinal microflora of salmonids, Masu salmon reared in freshwater were found to contain *Aeromonas* as the predominant bacterial genus in the intestine, When the fish were moved to seawater, a succession in the bacterial flora composition occurred, going from *Aeromonas* to *Pseudomonas* to *Vibrio* dominance (Yoshimizu et al., 1976a). Yoshimizu et al. (1976b) also studied the intestinal microflora of salmonids living in the Bering Sea, as well as bacteria associated with seawater and zooplankton samples collected from the Bering Sea. *Vibrio* spp. were found to be dominant in the intestine of salmonids. The microbial flora of the animals differed from that isolated from seawater and zooplankton. Finally,

Yoshimizu et al. (1976c) compared the intestinal microflora of adult pink salmon *(Oncorhynchus gorbuscha)* with that of anadromous chum salmon *(0. keta)* and cultured masu salmon *(0. masu).* Predominant bacterial genera in the anadromous pink and chum salmon were *Vibrio, Pseudomonas,* and *Aeromonas.* In the masu salmon, *Aeromohas* dominated. It would appear, therefore, that *Vibrio* spp. comprise the normal intestinal flora of salmonids, with exposure to seawater serving as a selective pressure.

Sawyer et al. (1979) studied the comparative susceptibility of atlantic salmon *(Salmo salar)* and coho salmon *(Oncorhynchus kisutch)* to V. *anguillarum* administered via water. Exposure to ~ 105 cells/ml for 1 h at 10 °C and at 15 °C resulted in the death of 80 to 100 % of both salmonid species. It was concluded that newly released smolts could encounter lethal concentrations of V. *anguiltarum* if the animals were released in Maine (USA) estuaries.

Finally, with regard to salmon, MacMillan et al. (1980) studied infection of chum salmon *(0. keta)* suffering from erythrocytic necrosis virus (ENV). Fish infected with ENV were less tolerant of dissolved oxygen depletion, exhibited characteristic cytopathogenic effects in erythroid cells, and were at least 3 times more susceptible to vibriosis.

Sea-bream. Acute septicemia was observed by Colorni et al. (1981) to be a frequent cause of mortality among gilthead sea-bream *(Sparus aurata)* raised in a laboratory at Elat, Israel. The most frequently isolated bacterium was 1/. *alginolyticus,* although 6 distinct strains were identified. Yasunaga & Yamamoto (1977) studied a similar outbreak among cultured red sea-bream *(Pagrus major)* and isolated several cultures of *Vibrio* that were able to be classified as 3 distinct groups. Iwata et al. (1978), in a study of the virulence of 1/. *alginolyticus* for red sea-bream, discovered that red seabream mortality increased if rotifers were present in the seawater along with the V. *alginolyticus,* If *Chlorella* was mixed with V. *alginolyticus,* mortality was less than if V. *alginolyticus* was present alone. When all three microorganisms, *V. alginotyticus, Chlorella,* and rotifers, were mixed in an aquarium, the number of V. *Alginolyticus* increased as the rotifers grazed and consumed the *Chlorella.*

Sharks. Diseases of sharks have been described recently (Grimes et al., 1984a, b, c), with *Vibrio* spp. definitely involved in certain of the shark infections reported. Genera other than *Vibrio* have been isolated from shark specimens examined in our laboratory, and these bacteria are now being characterized. Cheung et al. (1982) isolated and described a vibrio associated along with a trematode *(Dermophthirius nigrellii)* (Cheung & Ruggieri, 1983) which induced ulcers in captive lemon sharks. Unfortunately, the few biochemical characteristics reported by Cheung et al. (1982) are insufficient to classify their isolate as a *Vibrio* sp.; the characteristics reported also fit the description for *Aeromonas hydrophila* and *A. salmonicida. Vibrio* spp. have been isolated from the great white shark *(Carcharodon carcharias),* but not in association with specific disease (J. Buck, pets. comm).

So1e. Vibrio spp. can cause disease in sole. Williams & Caldwell (1978) reported that laboratory-reared English sole *[Parophyrs vetulus]* experienced vibrio mortality, and Fluechter (1979) described a red spot disease of the common sole *(Solea solea)* caused by a *Vibrio* sp.

S t r i p e d b a s s. Substantial losses of striped bass *(Morone saxatilis)* have occurred in the Chesapeake Bay during the last decade. Eight distinct biochemical groups of *Vibrio* spp. have been isolated from moribund striped bass (Toranzo et al., 1983a), two groups of which were identified as V. *anguillarum* (Toranzo et al., 1983b). Hetrick et al. (1984) found that the virulence of *V. anguillarum* for striped bass was directly influenced by age of the fish and water temperature. Exposure of the striped bass to sub-lethal chlorine levels (0.05-0,20 rag/l, total residual chlorine) did not affect susceptibility to infection. Virulence (LDs0), cytotoxicity, hemolytic activity, and adhesiveness of V. *anguillarum* have been studied (Toranzo et al., 1983b), as well as plasmid content and envelope protein profiles (Toranzo et al., 1983a). Details of this work are discussed in the section on mechanisms of virulence (vide infra).

Turbot. A vibriosis described by Home et al. (1977) as an acute edematous disease and characterized by orbital and abdominal swelling and high mortality (20 %), occurred in turbot *(Scophthalmus maximus]* in Scotland. V. *anguillarum* was isolated, in pure culture, from all organs of all of the fish examined. Antibiotic therapy was not successful, but reduction of the water temperature to less than 10 °C reduced mortality from 30 % to 5 %. Austin (1982) studied the surface bacterial flora of healthy turbot, as well as specimens collected from turbot suspected of having vibriosis, and failed to isolate *V. anguillarum* from any of the samples.

VIRULENCE MECHANISMS

The most recent work on *Vibrio* diseases of marine fish has dealt with elucidation of virulence mechanisms. Much of this exciting work involves molecular genetics and the bulk of the studies have been directed toward V. *anguillarum* and V. *ordalii* because of the extensive involvement of these pathogens in fish diseases throughout the world.

In 1977, Crosa, Schiewe, and Falkow (Crosa et al., 1977) discovered that high virulence strains of V. *anguitlarum,* i.e., strains that cause death of infected salmonids, possess a large plasmid with a mass of ca 47 megadaltons (Mdal). This plasmid, pJM1, *was found in aH high virulence strains examined, judged by plasmid homology experi*ments (Crosa et al., *1977).* Furthermore, cured strains lost their virulence (Crosa et al., 1977, 1980). Crosa (1979, 1980) subsequently demonstrated that the plasmid pJM1 codes for production of a highly efficient iron sequestering system which allows V. *Angufllarum* to obtain iron necessary for metabolism, even in the presence of host defense factors that have high affinity for unbound iron (i.e., transferrin and lactoferrin, iron chelating proteins present in body fluids and secretions of vertebrates). Briefly, the iron sequestering system consists of a low molecular weight siderophore and two outer membrane proteins, OM2 and OM3, which have molecular weights of 86,000 and 79,000 daltons, respectively (Crosa et al., 1983). Both the siderophore and OM2 are coded by pJM1; OM3 is chromosome product of unknown involvement in the system. The siderophore diffuses into the environment and complexes with iron, *even* iron that has been bound by vertebrate transferrin (Crosa et al., 1983). The siderophore-iron complex is attached to outer membrane protein OM2, where iron is then presumably transported into the cell (Crosa et al., 1983).

Trust et al. (1981) studied the ability of marine vibrios, including V. *anguillarum,* to agglutinate erythrocytes and to resist the bactericidal effect of normal (nonimmune) serum. High virulence strains of V. *anguillarum* exhibited resistance to normal trout *(Salmo gairdneri)* serum and agglutinated trout erythrocytes, even in the presence of mannose. However, neither serum resistance nor mannose-resistant hemagglutination was coded by virulence plasmid pJM1.

Recently, Schiewe & Crosa (1981) demonstrated the presence of a 20 Mdal plasmid, pMJ101, in *V. ordalii.* Homology was not observed between pJM1 and pMJ101, and the function of pMJ101 was concluded to be cryptic. Interestingly, the presence of this 20-Mdal plasmid is universal in all strains of *V. ordalii* examined thus far, and it is the only plasmid present.

Toranzo et al. (1983a, b, c) extended the molecular genetic and virulence studies of *V. anguillarum* to strains isolated from moribund striped bass (M. *saxatilis)* captured in Chesapeake Bay. In a study of molecular factors associated with virulence, Toranzo et ak (1983b) found that, of 8 distinct groups of *Vibrio* spp. isolated from striped bass and Chesapeake Bay water samples, two vibrio groups were pathogenic for striped bass. These two groups were identified as V. *anguillarum,* based on biochemical tests and DNA homology with reference strains. None of the *V. anguillanzm* strains contained detectable plasmids, although all strains were capable of growing under conditions of iron limitation and in normal fish serum. When grown under iron limitation, new outer membrane protein bands appeared in SDS-polyacrylamide gels. Furthermore, the striped bass *V. anguillarum* strains were shown to synthesize a sideophore, and preliminary experiments revealed the presence of chromosomal DNA sequences that hybridized with pJM1 (Toranzo et al., 1983b). In a related study, Toranzo et al. (1983c) showed that the striped bass 1I. *anguitlarum* strains were not cytotoxic for a chinook salmon embryo cell culture, although all strains were hemolytic for striped bass, trout, and sheep erythrocytes. Munn (1978) has also investigated hemolysin production by V. *anguillarum,* and he found the toxin to be a thermolabile protein produced in stationary growth phase. In addition, the V. *anguillarum* strains studied by Toranzo et al. (1983c) also produced strong hemagglutinins for trout, striped bass, and other erythrocytes. Adhesion of V. *anguillarum* to rainbow trout intestinal tissue sections was recently examined by Home & Baxendale (1983). They found that maximum attachment occurred within 100 min, and that all regions of the gut were subject to heavy (ca. 103 cells per cm 2) colonization.

Kreger (1984) studied the virulence of V. *damsela,* and detected the production of a cytolytic toxin. The toxin has a molecular weight of 57,000 daltons and is antigenically distinct from toxins produced by V. *cholerae, V. parahaemolyticus,* and *V. vulnificus.* Using mouse erythrocytes as the cytolytic assay system, Kreger found that partially purified cytolysin from an unnamed fish isolate contained ca. 3 x 105 hemolytic units per mg of protein. Mice injected subcutaneously with 1 LDs0 (1,000 hemolytic units) of the cytotysin became lethargic, had ruffled fur, developed encrustations around the eyelids, and exhibited severe local edema at the site of inoculation; mice so injected usually died within 24 h. To date, there have been no reported virulence studies conducted on fish.

Both V. *vulnificus* (Smith & Merkel, 1982) and V. *anginolyticus* (Long et al., 1981) produce collagenase which presumably aids in the cutaneous infections caused by these two vibrios. Other *Vibrio* spp. have also been shown to be capable of elaborating collagenase, but the role of the proteolytic enzyme in pathogenesis has not been reported (Merkel & Dreisbach, 1978}.

TREATMENT AND PREVENTION

Antimicrobialagents

Attempts to control fish disease have been reported almost since development of the first clinically successful antimicrobics in the late 1940's (Snieszko et al., 1952). Treatment of freshwater fish species, wild or cultured, is relatively easy, since one is usually dealing with a small body of water. Saltwater species on the other hand, unless they are cultured or confined, present obvious difficulties because of the immense size of the habitat. Generally speaking, the most widely used antimicrobials have been the sulfonamides, the nitrofurans, the tetracyclines, and, more recently, substituted quinolines. Regardless of what antimicrobic is used, bacteria can develop resistance by aquisition of R plasmids (Aoki et al., 1980; Kimura et al., 1983; Toranzo et al., 1983a).

One of the more useful antimicrobials has been the combination of two bacteriostatic chemicals, sulfonamides and trimethoprim. In concert, these two chemicals have a synergistic effect, and can be used effectively to control resistant organisms (Kimura et al., 1983; Sako & Kusuda, 1978; Sako et al., 1979). Recently, Keck et al. (1980) reported that increased oxygen tension (ca 2.2 atm.) markedly potentiated a bacteriostatic effect (i.e., lowered the minimum inhibitory concentration, MIC) of folic acid inhibitors, such as sulfonamides and trimethoprim-sulfonamides.

Some of the newer agents include piromidic acid (Tashiro et al., 1979), furanace (Egidius & Andersen, 1979), and substituted quinolines, such as halquinol (Austin et al., 1981, 1982). Also, some degree of success in preventing transmission of several Gramnegative fish pathogens, including V. *anguillarum,* has been achieved by ultraviolet irradiation of the fish culture water (Bullock & Stuckey, 1977). In Table 4 are listed some of the more effective antimicrobials.

Table 4. Recommended rates of application of fish antimicrobics

Antimicrobic	Dosage[b]	Route	Target	Reference
Piromidic acid	10–40 mg/kg	oral	salmonids & eels	Tashiro et al. (1979)
Sulfadoxine-TMP[a]	30 mg/kg	oral	ayu	Sako et al. (1979)
Sulfamethoxazole-TMP	25 mg/l	bath	turbot	Austin et al. (1981)
Furanace	2 mg/l	bath	rainbow trout	Egidius & Andersen (1979)
Halquinol	25 mg/l	bath	turbot	Austin et al. (1982)
	75 mg/kg	oral	turbot	Austin et al. (1982)

[a] TMP-trimethoprim
[b] wt/kg indicates amount per kg fish body wt.; wt/l indicate amount per l of immersion water

Immunization

As with antimicrobial therapy, immunization has also been practiced for many years (Snieszko & Friddle, 1949; Krantz et al., 1963; Hayashi et al., 1964). Early attempts at immunization involved crude antigen preparations, if not simply whole cells, and was labor intensive. Recent approaches to immunization are novel and much more efficacious than the early procedures. A wide variety of antigen preparations, admin istered in different ways, have been t r i e d and are presented in Table 5.

Table 5. Summary of approaches to immunization against *Vibrio anffuillarum*

Delivery route	Average dose	Protection: Site	Antibody (titer)	Duration	Reference
Oral	0.4 g/kg	body surface	agglutinin (1:64)	>4 wk	Kawai et al. (1981)
Hyper-osmotic	0.2 g/kg	serum	agglutinin (1:8)	>100 wk	Antipa & Amend (1977) Nakajima & Chikahata (1979)
Direct immersion	5×10^5 cells/ml	NA*	NA	>200 da	Gould et al. (1979) Johnson et al. (1928b)
Spray	100 mg/g	serum	agglutinin (1:147)	>16 wk	Gould et al. (1978) Itami & Kusuda (1980b)
Intraperi-toneal inj.	150 mg/kg	serum	agglutinin (1:64)	>6 mo	Antipa (1976) Antipa & Amend (1977) Horne et al. (1984) Agius et al. (1983)

* NA-data not available

The simplest delivery system appears to be the administration of the vaccine by the oral route, usually with food. Kusuda et al. (1978) studied four different oral vaccines prepared from formalin-killed V. *anguillarum:* (1)formalin-killed cells, (2) formalinkilled, ultrasonicated cells, (3) formalin-killed, heated cells, and (4) formalin-killed cells mixed with Freund's incomplete adjuvant. The four bacterins were administered to cultured ayu *(P. attivelis)* at a rate of 0.4 g (wet wt) of cells per kg body wt. per day. Formalin-killed and formalin-killed, ultrasonicated bacterins were the most effective, lowering mortality from 90 % in controls to approximately 10 %. Interestingly, circulating antibody could not be detected. In a follow-up study, Kawai et al. (1981) compared orally immunized ayu with non-immunized controls to determine the site of immunity. Immunized fish were highly resistant to challenge with *V. anguillarum* by immersion, but were equally susceptible, as were the controls, when the challenge was by intramuscular injection. It was determined that the skin of immunized ayu was free from V. *anguillarum,* when challenged by immersion, and serological examination of the body surface mucus revealed an agglutinin titer of 1 : 64 (Kawai et al., 1981). Neither the serum nor the intestinal mucus of immunized fish contained increased agglutinin titers. It was concluced that the mechanism of immune resistance involved prevention of bacterial attachment by agglutinins secreted in the surface mucus (Kawai & Kusuda, 1981).

Vaccination by direct immersion of fish has been practiced in the commercial sector by Antipa & Amend (1977). They dipped the salmon (O. *kisutch* and O. *tshawytscha)* in a hyperosmotic salt solution, a step that was hypothesized to enhance antigen uptake. The salmon were next dipped in killed V. *anguillarum* antigen preparations, and immunized animals were compared with non-immunized controls and intraperitoneally-injected immune controls. Antipa & Amend (1977) concluded that the hyperosmotic infiltration method of vaccine delivery afforded the best protection and yielded the most rapid rise in antibody titer. Croy & Amend (1977) studied the hyperosmotic infiltration method of immunizing sockeye salmon *(0. nerka)* against vibriosis, using killed V. *anguillarum* as antigen. They found that pre-immersion of the fish in Hanks' balanced salt solution with 8 % NaC1 was markedly superior to immunization without hyperosmotic treatment. In subsequent studies on immunizing against *V. anguillarum,* Lannan (1978) applied the hyperosmotic approach for chum salmon *(0. keta)* and Nakajima & Chikahata (1979) tried it for ayu *(P. altivelus).* Most recently, Aoki et al. (1984) modified the hyperosmotic approach by removing undesirable toxic components of *V. anguillarum,* Specifically, they formalized a culture and sedimented the cells by centrifugation to remove soluble toxic factors in the supernatant The precipitated ceils were used to immunize ayu previously immersed in a 5.32 % NaC1 solution. Controls consisted of non-immunized and intraperitoneally immunized ayu. Removal of the supernatant increased the survival of immunized ayu. The conclusion was that an exotoxin was present in the formalized cultures. Upon intraperitoneal challenge with viable V. *anguillarum,* survival was found to be greatest (near 100 %) among ayu immersed in precipitated cells, next highest (60-70 %) among ayu injected intraperitoneally, and least (ca. 20 %) among nonimmunized controls. One month after immunization, no agglutinating titers were detected in the serum or mucus of fish immunized by immersion or in non-immunized controls. Intraperitoneally immunized ayu had elevated serum titers, but no anti- *V. anguillarum* activity was detected in the mucus (Aoki et al., 1984).

Later studies revealed prior treatment with hyperosmotic salt immersion was unnecessary. Gould et al. (1979) obtained long-lasting immunity in sockeye salmon *(0. nerka)* by immersing the fish in formalin-killed V. *anguillarum* (ca 5 × 105 cells per ml of immersion solution), types I and II (i.e., *V. anguillamm* and *V. ordalii,* respectively). Mortality was reduced to 2 %, with little cross protection observed by immunization with the two different *Vibrio* bacterins. Bivalent immunization, i.e., simultaneous immersion in both bacterins, was as effective as type I and II bacterins individually applied (Gould et al., 1979). Johnson et al. (1982a, b) examined the direct immersion method as a means of immunizing the fry of different salmonids. In general, the best results were obtained with fry 1.0 to 2.5 g, total body wt, and the immersion time required was ca 5 seconds. Direct immersion as a means of immunizing against vibriosis has also been applied to cultured ayu (Kusuda et al., 1980; Takeji et al., 1981) and to striped bass (Roberson et al., 1982). Roberson et al. (1982) used a 2-min immersion in *I/. anguillamm* 0-antigen (acetone-ethanol extract), and enhanced antibody levels were detected in serum collected from the striped bass.

Spray, or shower, vaccination is basically a modification of direct immersion, in that fish are immunized by a high pressure spray directed at netted fish or at fish in shallow channels. This approach has been used successfully to immunize both ayu and salmonids against V. *anguillarum* (Gould et al., 1978; Itami & Kusuda, 1978, 1980a, b; Johnson et al., 1982b), In general, immunity was comparable to that obtained with direct immersion, and better than that by oral (Gould et al., 1978).

Intraperitoneal injection is perhaps the most effective, in terms of producing a high degree of immunity, but is both expensive and labor intensive. It also subjects the fish to stress associated with handling. Single injections of V. *anguillarum* bacterins (heatkilled, formalin-killed, or both) gave significant protection in salmonids for over 7 months post-inoculation (Antipa, 1976). Sawyer & Strout (1977) compared immunization with oxytetracycline treatment as methods for controlling vibriosis in coho salmon, and found intraperitoneal immunization significantly superior to antibiotic prophylaxis. Finally, Home et al. (1982) compared intraperitoneal, direct immersion, and oral administration of formalized V. *anguillarum* to rainbow trout *(Salmo gairdneri)using* a vaccine containing 35 mg (wet wt.) of bacterin per ml of vaccine and 2.5 % (w/v) potassium aluminum sulfate adjuvant. The intraperitoneal route employed 100 ~1. Fish were immersed for 30 min in a 10-fold dilution for direct immersion. For oral route of entry, 3.5 mg of bacteria were given with food over a 32-day period, After 85 days, all fish, including non-immunized controls, were challenged by intraperitoneal injection of 2 x l0 T viable cells of V, *anguillarum.* Survival rates at 21 days after challenge were 93 % for intraperitoneal delivery of the vaccine; 47 % for direct immersion, 6 % for oral, and 0 % for controls. In a follow-up study, Agius et al. (1983) used a perchloric acid extract as the antigen preparation, and compared it to formalin-killed V. *Anguillarum* cells as an immunizing agent. The antigen extract was concluded to be superior when intraperitoneal delivery was employed, but formalized cells were better in delivery by the oral route. Alum appears to offer little additional protection, leading Home et al. (1984) to question the use of adjuvants.

In addition to improvements in methods of immunization, improvements have also been made in serological techniques. Roberson (1981) combined the methods of thin layer immunoassay and enzyme-linked immunosorbent assay (ELISA) to detect anti- *V. anguillarum* antibody in striped bass immunized by intraperitoneal injection with 0- antigen. The sensitivity was found to be comparable to passive hemagglutination. Gosting et al. (1981) used an immunocytoadherence assay to monitor progression of the immune response in sockeye salmon immunized against V. *anguillarum* by immersion and intraperitoneal injection. Both delivery methods initiated an elevated immune adherence in less than 1 day, and the response persisted 1 week longer in injected than in immersed fish.

CONCLUSION

The understanding of *Vibrio* diseases of marine fish populations has increased significantly since the first description of V. *anguillarum* appeared in 1893 (Canestrini, 1893). We now r e c o g n i z e several *Vibrio* spp. as b e i n g c a p a b l e of c a u s i n g d i s e a s e in fish. The number of species of *Vibrio* involved will no doubt i n c r e a s e as mariculture activities are extended. However, modern immunization t e c h n i q u e s , more effective antimicrobials, and better understanding of the ecology of the genus *Vibrio* offer means for improved control of *Vibrio* disease. In the near future, it will b e p o s s i b l e to prevent vibriosis by specific g e n e t i c manipulations of the pathogens (Colwell, 1983, 1984).

LITERATURE CITED

1. Agius, C., Horne, M. T. & Ward, P. D., 1983. Immunization of rainbow trout, *Salmo gairdneri* Richardson, against vibriosis: comparison of an extract antigen with whole cell bacterins by oral and intraperitoneal routes. - J. Fish Dis. 6, 129-134.
2. Akazawa, H., 1968. Bacterial disease of marine fishes. - Bull. Jap. Soc. scient. Fish. *34,* 271-272.
3. Antipa, R., 1976. Field testing of injected *Vibrio anguillarum* bacterins in pen reared Pacific salmon. - J. Fish. Res. Bd Can. *33,* 1291-1296.
4. Antipa, R. & Amend, D. F., 1977. Immunization of Pacific salmon: comparison of intraperitoneal injection and hyperosmotic infiltration of *Vibrio anguillarum* and *Aeromonas salmonicida* bacterins. - J. Fish. Res. Bd Can. *34,* 203-208.
5. Aoki, T. & Kitao, T., 1978. Vibriosis of ayu. - Fish Pathol. *13,* 19-24.
6. Aoki, T., Jo, Y. & Egusa, S., 1980. Frequent occurrence of drug resistant bacteria in ayu *(Plecoglossus altivelis)* culture. - Fish Pathol. *15,* 1-6.
7. Aoki, T., Kitao, T. & Kawano, K., 1981. Changes in drug resistance of *Vibrio anguillarum* in cultured ayu, *Plecofflossus altivelis* Temminck and Schlegel, in Japan. - J. Fish Dis. 4, 223-230.
8. Aoki, T., Kitao, T., Fukudome, M., Takahashi, S. & Egusa, S., 1984. Modification of the hyperosmatic infiltration method of vaccination against vibriosis in cultured ayu, *Plecoglossus altivelis* Temminck and Schlegel. - J. Fish Dis. 7, 149-156.
9. Austin, B., 1982. Taxonomy of bacteria isolated from a coastal marine fish rearing unit. - J. appl. Bact. *53,* 253-268.
10. Austin, B., Morgan, D. A. & Alderman, D. J., 1981. Comparison of antimicrobial agents for control of vibriosis in marine fish. - Aquaculture *26,* 1-12.
11. Austin, B., Johnson, C. & Alderman, D. J., 1982. Evaluation of substituted quinolines for the control of vibriosis in turbot *(Scophthalmus maximus}.* - Aquaculture *29,* 227-239.

12. Belas, M. R. & Colwell, R. R, 1982a. Scanning and electron microscope observation of the swarming phenomenon in *Vibrio parahaemolyticus.* - J. Bact. *150,* 956-959.
13. Belas, M. R. & Colwell, R. R., 1982b. Adsorption kinetics of laterally flagellated *Vibrio.* - J. Bact. *15l,* 1568-1580.
14. Bergman, A. M., 1909. Die rote Beulenkrankheit des Aals. - Ber. K. bayer, biol. Vers.-Stn 2, 10-54.
15. Blake, P. A., Weaver, R. E. & Hollis, D. G., 1980. Diseases of humans (other than cholera) caused by vibrios. - A. Rev. Microbiol. *34,* 341-367.
16. Bullock, G. L. & Stuckey, H. M., 1977. UV treatment of water for destruction of 5 gram-negative bacteria pathogenic to fishes. - J. Fish. Res. Bd Can. *34,* 1244-1249.
17. Burke, J. & Rodgers, L., 1981. Identification of pathogenic bacteria associated with the occurrence of "red spot" in sea mullet, *Mugil cephalus* L., in s o u t h - e a s t e r n Queensland. - J. Fish Dis. 4, 153-159.
18. Canestrini, G, 1893. La malattia dominante delle anguille. - Atti Ist. veneto Sci. 7, 809-814,
19. Cheung, P. J. & Ruggieri, G. D., 1983. *Dermophthirius nigrellii* n. sp. (Monogenea: Microbothriidae), an ectoparasite from the skin of the lemon shark, *Neffaprion brevirostris.* - Trans. Am. microsc. Soc. *102,* 129-134.
20. Cheung, P. J., Nigrelli, R. F., Ruggieri, G. D. & Cilia, A., 1982. Treatment of skin lesions in captive lemon sharks, *Neffaprion brevirostris* (Poey), caused by monogeneans *(Dermophthirius* sp.). -J. Fish Dis. 5, 167-170.
21. Colorni, A., Paperna, I. & Gordin, H., 1981. Bacterial infections in gilthead sea-bream *(Sparus aurata)* cultured at Elat, Israel. - Aquaculture *23,* 257-268.
22. Colwell, R. R., 1983. Biotechnology in the marine sciences. - Science, N. Y. *222,* 19-24.
23. Colwell, R. R., 1984. The industrial potential of marine biotechnology. - Oceanus *27"* 3-12.
24. Crosa, J. H., 1979. *Vibrio anguillarum: An* iron u p t a k e mechanism as a factor of virulence. – Fish Health News *8,* 7-10.
25. Crosa, d. H., 1980. A plasmid associated with virulence in the marine fish pathogen *Vibrio*
26. *anguillarum* specifies an iron sequestering system. - Nature, Lond. *284,* 566-568.
27. Crosa, d. H., Schiewe, M. H. & Falkow, S., 1977. Evidence for plasmid contribution to t h e virulence of the fish pathogen *Vibrio anguillarum.* - Infect. Immun. 18, 509-513.
28. Crosa, d. H., Hodges, L. L. & Schiewe, M. H., 1980. Curing of a plasmid is correlated with an attenuation of virulence in the marine fish pathogen *Vibrio anguillarum.* - Infect. Immun. *27,* 897-902.
29. Crosa, J. H., Walter, M. A. & Potter, S. A., 1983. The genetics of plasmid-mediated virulence in the marine fish pathogen *Vibrio anguitlarum.* In: Bacterial and viral diseases of fish, molecular studies. Ed. by J. H. Crosa. Univ. of Washington, Seattle, 21-30.
30. Croy, T. R. & Amend, D. F., 1977. Immunization of sockeye salmon *(Oncorhynchus nerka)* against vibriosis using the hyperosmotic infiltration technique. - Aquaculture *12,* 317-326.
31. Egidius, E. & Andersen, K., 1979. The u s e of Furanace against vibriosis in rainbow trout *Salmo galrdnerl* Richardson in salt water. - J. Fish Dis. 2, 79-80.
32. Fluechter, J., 1979. Identification and t r e a t m e n t of diseases in the common sole *(Solea solea).* - Aquaculture *16,* 271-274.
33. Franca, S. M. C., Gibbs, D. L., Samuels, P. & Johnson, W. D. Jr., 1980. *Vibrioparahaemolyticusin* Brazilian coastal waters. - J. Am. med. Ass. *244,* 587-588.
34. Gauthier, M. J. & Clement, R., 1979. Experimental s t u d y of the transfer of *Vibrioparahaemolyticus* biotype 2 from water and sediments to organisms in benthic marine food chains. - Can. J. MicrobioL *25,* 499-507.
35. Gilmore, A., 1977. Characteristics of marine vibrios isolated from fish farm tanks.- Aquaculture I 1, 51-62.

36. Gjerde, J. & Boe, B., 1981. Isolation and characterization of *Vibrio alginolyticus* and *Vibrio parahaemolyticus* from the Norwegian coastal environment. - Acta vet. scand. *22,* 331-343.
37. Gosting, L., Mirando, D. M. & Gould, R. W., 1981. Antigen b i n d i n g cells in the peripheral blood of sockeye salmon *(Oncorhynchus nerka)* induced by immersion or intraperitoneal injection of *Vibrio anguillarum* bacteria. - J. Fish Biol. *19,* 83-86.
38. Gould, R. W., O'Leary, P. J., Garrison, R. L., Rohovec, J. S. & Fryer, J. L., 1978. Spray vaccination: A method for the immunization of fish. - Fish Pathol. *13,* 63-68.
39. Gould, R. W., Anfipa, R. & Amend, D. F., 1979. Immersion vaccination of sockeye salmon *(Oncorhynchus nerka)* with two pathogenic strains of *Vibrio anguillarum. - J.* Fish. Res. Bd Can. *36,* 220-225.
40. Grimes, D. J., Stemmler, J., Hada, H., May, E. B., Maneval, D., Hetrick, F. M., Jones, R. T., Stoskopf, M. & Colwell, R. R., 1984a. *Vibrio* species associated with mortality of sharks held in captivity. - Microb. Ecol. *10* (In press).
41. Grimes, D. J., Colwell, R. R, Stemmler, J., Hada, H., Maneval, D., Hetrick, F. M., May, E. B., Jones, R. T. & Stoskopf, M., 1984b. *Vibrio* species as agents of elasmobranch disease. – HelgoRinder Meeresunters. 37, 309-315.
42. Grimes, D. J., Gruber, S. H. & May, E. B., 1984c. Experimental infection of lemon sharks, *Negaprion brevirostris,* with *Vibrio* species. - J. Fish Dis. (Submitted).
43. Hada, H. S. & Sizemore, R. K., 1981. Incidence of plasmids in marine *Vibrio* spp. isolated from an oil field in the Northwestern Gulf of Mexico. - Appl. environ. Microbiol. *41,* 199-202.
44. Hamid, A., Sakata, T. & Kakimoto, D., 1978. Microflora in the alimentary tract of gray mullet. Part 2. A comparison of the mullet intestinal microflora in fresh water and s e a water. - Bull. Jap. Soc. scient. Fish. *44, 53-58.*
45. I-lamid, A., Sakata, T. & Kakimoto, D., 1979. Microflora in the alimentary tract of gray mullet. Part 4. Estimation of enzymic activities of the intestinal bacteria. - Bull. Jap. Soc. scient. Fish. *45,* 99-106.
46. Harrell, L. W., Novotny, A. J., Schiewe, M. H. & Hodgins, H. O., 1976. Isolation and description of two vibrios pathogenic to Pacific salmon in Puget Sound, Washington. - Fish. Bull. U. S. *74,* 447-449.
47. Hastings, J. W. & Nealson, K. H., 1977. Bacterial bioluminescence. - A. Rev. Microbiol. *31,549-595.*
48. Hayashi, K., Kobayashi, S., Kamata, T. & Ozaki, H., 1964, Studies on the vibrio disease of rainbow trout *(Salmo gairdneri irideus)* II. Prophylactic vaccination against the vibrio disease. - J. Fac. Fish., pref. Univ. Mie. 6, 181-191.
49. Haygood, M. G. & Nealson, K. H., 1984. Effects of iron on bacterial growth and luminescence: Ecological implications. In: Current perspectives in microbial ecology. Ed. by M. J. Klug & A. Reddy. Amer. Soc. Microbiol., Washington, D.C., 56-61.
50. Hetrick, F. M., Hall, L. W. Jr., Wolski, S., Graves, W. C., Roberson, B. S. & Burton, D. T., 1984. Effect of exposure of striped bass *(h4orone saxatilis)* to chlorine on their susceptibility to *Vibrio anguillarnm.* - Can. J. Fish. aquat. Sci, (In press}.
51. Hiratsuka, M., Saitoh, Y. & Yamane, N,, 1980. Isolation of *Vibrio alginolyticus* from a patient with acute enterocolitis. - Tohoku J. exp. Med. *132,* 469-472.
52. Hoornbeek, F. K., Sawyer, P. J. & Sawyer, E. S., 1982, Growth of winter flounder *Pseudopleuronectes americanus* and smooth flounder *I~'opsetta putnami* in h e a t e d and unheated water. - Aquaculture *28,* 363-374.
53. Home, M. T., 1982. The pathogenicity of *Vibrio anguillarum* (Bergman}. In: Microbial disease of fish. Ed. by R. J. Roberts. Acad. Press, New York, 171-187.
54. Horne, M. T. & Baxendale, A., 1983. The adhesion of *Vibrio anguillarum* to host tissues and its role in pathogenesis. - J. Fish Dis. 6, 461-471.
55. Home, M. T., Richards, R. H., Roberts, R. J. & Smith, P. C., 1977. Per acute vibriosis in juvenile turbot *(Scophthalmus* max/runs). - J. Fish Biol. *11,* 355-362.
56. Hornme, M. T., Tatner, M., McDerment, S., Agius, C. & Ward, P., 1982. Vaccination of rainbow trout, *Salmo gairdneri* Richardson, at low t e m p e r a t u r e s and the long-term persistence of protection. - J. Fish Dis. 5, 343-345.

57. Home, M. T., Roberts, R. J., Tatner, M. & Ward, P., 1984. The effects of the use of potassium alum adjuvant in vaccines against vibriosis in rainbow trout, *Salmo gairdneri* Richardson. - J. Fish Dis. 7, 91-99.

58. Huq, A., Small, E. B., West, P. A., Huq, M. I., Rahman, R. & Colwell, R. R., 1983. Ecological relationships between *Vibrio choterae* and planktonic crustacean copepods. - Appl. environ. Microbiol. *45,* 2?5-283.

59. Itami, T. & Kusuda, R., 1978. Efficacy of a vaccination by s p r a y administration against vibriosis in cultured ayu. - Bull. Jap. Soc. scient. Fish. *44,* 1413-1414.

60. Itami, T. & Kusuda, R., 1980a. Studies on spray vaccination against vibriosis in cultured ayu *(Plecogtossus altivelis).* 1. Effect of bentonite and pH on vaccination efficacy. - Bull. Jap. Soc. scient. Fish. *46,* 533-536.

61. Itami, T. & Kusuda, R., 1980b, Studies on spray vaccination against vibriosis in cultured ayu *(Plecoglossus altivelis).* 2. Duration of vaccination efficacy and effect of different vaccine preparations. - Bull. Jap. Soc. scient. Fish. *46,* 699-704.

62. Iwata, K., Yanohara, Y. & Ishibashi, O., 1978, Studies on factors related to mortality of young red sea-bream *(Pagrus major)* in the artifical seed production. - Fish Pathol. *13,* 97-102.

63. Jensen, N. J. & Larsen, J. L., 1982. The ulcus syndrome in cod *Gadus* morhua. 4. Transmission experiments with 2 viruses isolated from cod and *Vibrio anguillarum.* - Nord. VetMed. *34,* 136-142.

64. Jo, Y. & Muroga, K., 1977. Studies on vibriosis in ayu. Part 1. Virulence of a culture of *Vibrio anguillarum.* - Fish Pathol. *12,* 151-156,

65. Johnson, K. A., Flynn, J. K. & Amend, D. F., 1982a. Onset of immunity in salmonid fry vaccinated by direct immersion in *Vibrio anguillarum* and *Yersinia ruckeri* bacterins. - J. Fish Dis. 5, 197-206.

66. Johnson, K. A., Flynn, J. K. & Amend, D. F., 1982b. Duration of immunity in salmonids vaccinated by direct immersion with *Yersinia ruckeri* and *Vibrio anguillarum* bacterins. - J. Fish Dis. 5, 207-214.

67. Joseph, S. W., Colwell, R. R. & Kaper, J. B., 1982. *Vibrio parahaemolyticus* and related halophilic vibrios. - CRC crit. Rev. Microbiol. *10,* 77-124.

68. Kakimoto, D. & Mowlah, A. H., 1980. Microflora in the alimentary tract of gray mullet *(h4ugil cephalus).* 8. Utilization of amino acids by *Vibrio* and *Enterobacter* isolates. - Mem. Fac. Fish. Kagoshima Univ. *29,* 349-354.

69. Kanai, K., Wakabayashi, H. & Egusa, S., 1977. Comparison of intestinal microflora between healthy and diseased pond cultured eels. - Fish Pathol. *12,* 199-204.

70. Kaneko, T. & Colwell, R. R., 1978. The a n n u a l cycle of *Vibrioparahaemolyticusin* Chesapeake Bay. -Microb. Ecol. 4, 135-155.

71. Kawai, K. & Kusuda, R., 1981. Mechanisms of protection in ayu *(Plecoglossus altivelis)* orally vaccinated for vibriosis. 1. Protection activity in body surface mucus. - Fish Pathol. *16, 1-8.*

72. Kawai, K., Kusuda, R. & Itami, T., 1981. Mechanisms of protection in ayu *(Plecoglossus altivetis)* orally vaccinated for vibriosis. - Fish Pathol. *15,* 257-262.

73. Keck, P. E., Gottlieb, S. F. & Conley, J., 1980. Interaction of increased pressures of oxygen and sulfona ides on t h e *in vitro* growth of pathogenic bacteria. - Undersea biomed. Res. 7, 95-106.

74. Khan, R. A., Campbell, J. & Lear, H., 1981. Mortality in captive Atlantic cod *Gadus morhua* associated with fin rot disease. -J. Wildl. Dis. *17,* 521-528.

75. Kimura, T., Yoshimizu, M. & Wada, M., 1983. *In vitro* antibacterial activity of the combination of sulphadiazine and trimethoprim on bacterial fish pathogens. - J. Fish Dis. 6, 525-532.

76. Krantz, G. E., Reddecliff, J, M. & Heist, C. E., 1963, Development of antibodies against *Aeromonas salmonicida* in trout. - J. Immun. *91,* 757-760.

77. Kreger, A. S., 1984. Cytolytic activity and virulence of *Vibrio damsela.* - Infect. Immun. *44,* 326-331.

78. Kusuda, R., Kawai, K., Jo, Y., Akizuki, T., Fukunaga, M. & Kotake, N., 1978. Efficacy of oral vaccination for vibriosis in cultured ayu. - Bull. Jap. Soc. scient. Fish. *44,* 21-26.

79. Kusuda, R., Sake, H. & Kawai, K., 1979. Classification of vibrios isolated from diseased fishes. Part 1. On the morphological and biochemical properties. - Fish Pathol. *13,* 123-138.
80. Kusuda, R., Kawai, K. & Itami, T., 1980. Efficacy of bath immunization against vibriosis in cultured ayu *(Plecoglossus altivelis).* - Bull. Jap. Soc. scient. Fish. *46,* 1053.
81. Kusuda, R., Sugiyama, A., Kawai, K., Inada, Y. & Yoneda, M., 1981. Pathogenicity of *Streptococcus* sp. and *Vibrio anguillarum* in cultured ayu. - Bull. Jap. Soc. scient. Fish. *47,* 993-998.
82. LalI, R., Sen, D., Saha, M. R., Bose, A. K., De, S. P., Palchowdhury, N. C. & Pal, S. C., 1979. Prevalence of *Vlbrio parahaemolyticus* in Port-Blair, India. - Indian J. reed. Res. *69,* 217-221.
83. Lannan, J. E., 1978. Vibriosis vaccination of chum salmon by hyperosmotic infiltration. – Progve Fish Cult. *40,* 43-45.
84. Larsen, J. L., Jensen, N. J. & Christensen, N. D., 1978. Water pollution and the ulcer syndrome in the cod *Gadus morhua.* - Vet. Sci. Commun. 2, 207-216.
85. Lhuillier, M., 1977. *Vibrio parahaemolyticus* and *Vibrio alginolyticus* isolation, diagnosis and epidemiology: 74 strains isolated in Madagascar. - Archs Inst. Pasteur Madagascar *45,* 27-50.
86. Long, S., Mothibeli, M. A., Robb, F. T. & Woods, D. R., 1981. Regulation of extracellular alkaline protease activity by histidine in a collagenolytic *Vibrio alginolyticus* strain. - J. gen. Microbiol. *127,* 193-199.
87. Love, M., Teebken-Fisher, D., Hose, J. E., Farmer III, J. J,, Hickman, F. W. & Fanning, G. R., 1981. *Vibrlo damsela,* a marine bacterium, causes skin ulcers on t h e damselfish *Chromispunctlpinnis.* - Science, N. Y. *214,* 1139-1140.
88. MacMillan, J. R., Mulchay, D. & Landolt, M,, 1980. Viral erythrocytic necrosis. Some physiological consequences of infection in chum salmon *(Oncorhynchus keta).* - Can. J. Fish. aquat. Sci. *37,* 799-804.
89. Merkel, J. R. & Dreisbach, J. H., 1978. Purification and characterization of a marine bacterial collagenase. - Biochemistry *17,* 2857-2863.
90. Minchew, C. D. & Yarbrough, J. D., 1977. The occurrence of fin rot in mullet *(Mugil cephalus)* associated with crude oil contamination of an estuarine pond ecosystem. - J, Fish Biol. *10,* 319-323,
91. Miyazaki, T., Jo, Y., Kubota, S. S. & Egusa, S., 1977. Histopathological studies on vibriosis of the Japanese eel *Anffuilla japonica.* Part 1. Natural infection. - Fish Pathol. 12, 163-170.
92. Munn, C. B., 1978. Haemolysin production by *Vibrio anguillarum.* - FEMS MicrobioL Lett. 3, 265-268.
93. Muroga, K., 1979. Ulcer disease of akame (Mugilidae) in the estuary of the river Ashida, Japan. - Fish Pathol. *13,* 163-168.
94. Muroga, K., Jo, Y. & Nishibuchi, M., 1976. *Vibrio anguillarum* isolated from the European eel *Anguilla anguilla* cultured in Japan. - J. Fac. Fish Anim. Husb. Hiroshima Univ. *15,* 29-34.
95. Muroga, K., Yoneyama, N. & Jo, Y., 1979a. Vibriostatic agent resistant *Vibrio anguillarum* isolated from ayu. - Fish Pathol. *13,* 159-162.
96. Muroga, K., Takahashi, S., Yamanoi, H. & Nishibuchi, M., 1979b. Noncholera vibrio *Vibrio anguillarum* isolated from d i s e a s e d ayu *Plecoglossus altivelis.* - Bull. Jap. Soc. scient. Fish. *45,* 829-834.
97. Nair, G. B., Abraham, M. & Natarajan, R., 1980. Distribution of *Vibrioparahaemolyticusin* finfish harvested from Porto Novo (S. India) environs: a seasonal study. - Can. J. MicrobioL *26,* 1264-1269.
98. Nakajima, M. & Chikahata, H., 1979. Efficiency of oral and hyperosmotic vaccinations for vibriosis in ayu *(Plecoglossus altivelis).* - Fish Pathol. *14,* 9-14.
99. Natarajan, R., Abraham, M. & Nair, G. B., 1980. Distribution of *Vibrio parahaemolyticus* in the Porto-Novo, India environment. - Indian J. med. Res. *71,* 679-687.

100. Nishibuchi, M., Muroga, K., Seidler, R. J. & Fryer, J. L., 1979. Pathogenic *Vibrio* isolated from cultured eels. IV. Deoxyribonucleic acid studies, - Bull. Jap. Soc. scient. Fish. *45,* 1469-1473.

101. Nishibuchi, M., Muroga, K. & Jo, Y., 1980. Pathogenic *Vibrio* isolated from cultured eels. 6. Diagnostic tests for the disease due to present bacterium. - Fish Pathol. *14,* 125-132.

102. Novitsky, J. A. & Morita, R. Y., 1976. Morphological characterization of small cells resulting from nutrient starvation of a psychrophilic marine *Vibrio.* - AppL environ. Microbiol. *32,* 617-622.

103. Oliver, J. D., Wamer, R. A. & Cleland, D. R., 1983. Distribution of *Vibrio vulnificus* and other lactose-fermenting vibrios in the marine environment. - Appl. environ. Microbiol. *45,* 985-998.

104. Ortiz, J. S. & Decker, D. L., 1976. Demonstration of agglutinating antibodies against *Vibrio parahaemolFticus* in the yellowtail flounder, *Limanda ferruginea.* - Health Lab. Sci. *13,* 197-202.

105. Qadri, R. B. & Zuberi, R., 1977. Survey on the occurrence of *Vibrio parahaemolyticus* and *Vibrio alginolyticus* in fish and shellfish from the Karachi, Pakistan coastal waters. - Pakist. J. scient. ind. Res. *20,* 183-187.

106. Roberson, B. S., 1981, Detection of fish antibody by thin-layer ELISA. - Devs biol. Stand. *49,* 113-118.

107. Roberson, B. S., Wolski, S. & Hetrick, F. M., 1982. *Anti-Vibrio anguillarum* antibody among striped bass *(Morone saxatilis)* following injection with viable bacteria, or ambient exposure to isolated 0-antigen or environmental vibrios. - Abstracts of the Annual Meeting of the American Society for Microbiology *82, 33* (B91).

108. Roberts, R. J. (Ed.)., 1982. Microbial diseases of fish. Acad. Press, New York, 305 pp. Rohohm, R. A., Brown, C. & Murchelano, R. A., 1979. Comparison of antibodies in marine fish from clean and polluted waters of the New York Bight: Relative levels against 36 bacteria. - Appl. environ. Microbiol. *38,* 248-257.

109. Rodsaether, M. C., Olafsen, J., Rea, J., Myhre, K. & Steen, J. B., 1977. Copper as an initiating factor of vibriosis *(Vibrio anguillarum)* in eel *(Anguilla anguilla).* - J. Fish Biol. *10,* 17-21.

110. Ruby, E. G., Greenberg, E. P. & Hastings, J. W., 1980. Planktonic marine luminous bacteria: species distribution in the water column. - Appl. environ. Microbiol. *39,* 302-306.

111. Sakazaki, R., Iwanami, S. & Fukumi, H., 1963. Studies on the enteropathogenic facultatively halophilic bacteria, *Vibrio parahaemolFticus.* - Jap. J. reed. Sci. Biol. *15,* 161-168.

112. Sako, H. & Kusuda, R., 1978. Chemotherapeutical studies on trimethoprim against vibriosis of pond cultured ayu. Part 1. Microbiological evaluation of trimethoprim and sulfonamides on the causative agent, *Vibrio anguillarum.* - Fish Pathol. *13,* 91-96.

113. Sako, H., Kusuda, R. & Tabata, T., 1979. Chemotherapeutical studies on trimethoprim against vibriosis of pond cultured ayu *(Plecoglossus altivelis).* 2. Tissue levels after administration of trimethoprim-sulfadoxine mixture. - Fish Pathol. *I4,* 65-70.

114. Sawyer, E. S. & Strout, R. G., 1977. Survival and growth of vaccinated, medicated and untreated coho salmon *(Oncorhynchus klsutch]* exposed to *Vlbrio anguillarum.* - Aquaculture I0, 311-315.

115. Sawyer, E. S., Strout, R. G. & Coutermarsh, B. A., 1979. Comparative susceptibility of Atlantic *(Salmo salar)* and coho *(Oncorhynchus kisutch)* salmon to 3 strains of *Vibrio anguillarum* from the Maine-New Hampshire coast. - J. Fish. Res. Bd Can. *36,* 280-282.

116. Schmidt0 U., Chmel, H. & Cobbs, C., 1979. *Vibrio alginolyticus* infections in humans. - **J. clin.** Microbiol. *10,* 666--668.

117. Schiewe, M. H., 1983. *Vibrio ordalii* as a cause of vibriosis in salmonid fish. In: Bacterial and viral diseases of fish, molecular studies. Ed. by J. H. Crosa. Univ. of Washington, Seattle, 31-40.

118. Schiewe, M. H. & Crosa, J, H., 1981. Molecular characterization of *Vibrio anguillarum* biotype 2. - Can. J. Microbiol. *27,* 1011-1018.

119. Schiewe, M. H., Trust, T. J. & Crosa, J. H., 1981. *Vlbrlo ordalii* sp. nov.: a causative agent of vibriosis in fish. - Curt. Microbiol. 6, 343-348.

120. Sindermann, C. J., 1970. Principal diseases of marine fish and shellfish. Acad. Press, New York, 369 pp.

121. Singleton, F. L., Attwell, R., Jangi, S. & Colwell, R. R., 1982. Effects of temperature and salinity on *Vlbrio cholerae* growth. - Appl. environ. Microbiol. *44,* 1047-1058.

122. Smith, G. C. & Merkel, J. R., 1982. Collagenolytic activity of *Vibrio vulnificus:* Potential contribution to its invasiveness. - Infect. Immun. *35,* 1155-1156.

123. Snieszko, S. F. & Friddle, S. B,, 1949. Prophylaxis of furunculosis in brook trout *(Salvelinus fontinalis)* by oral immunization and sulfamerazine. - Progve Fish Cult. *I1,* 161-168.

124. Snieszko, S. F., Griffin, P. J. & Friddle, S. B., 1952. Antibiotic treatment of ulcer disease and furunculosis in trout. -Trans. N. Am. Wildl. Conf. *17,* 197-213.

125. Strusiewicz, B. & Serban, D., 1980. The enterotoxicity of nonagglutinating vibrios. - Rev. Ig. Bacteriol. VirusoL Parazitol. Epidemiol. Pneumoftiziol. (Ser, Bacteriol. Virusol. Parazitol. Epidermiol.) *28,* 215-218.

126. Strusiewicz, B., Cvasniuc, D., Florian, V., Cvasniuc, T. & Busca, M., 1980. Nonagglutinating vibrios, etiologic agents of diarrheic disease: Their presence and dissemination in the external environment. - Rev. Ig. Bacteriol. Virusol. Parazitol. Epidemiol. PneumoftizioL (Ser. Ig.) *29,* 135-140.

127. Szeness, L, Sey, L. & Szeness, A., 1979. Bacteriological studies of the intestinal content of aquatic birds, fishes, and frogs with special reference to the presence of noncholera vibrios. - Zbl. Bakt. ParasitKde Infektionskr. Hyg. (Abt. 1, Orig., R. A.: Med. Mikrobiol. Parasitol.) *245,* 89-95.

128. Tajima, K., Yoshimizu, M., Ezura, Y. & Kimura, T., 1981. Causative organisms of vibriosis among pen cultured coho salmon *(Oncorhynchus kisutch)* in Japan. - Bull. Jap. Soc. scient. Fish. *47,* 35-42.

129. Takeji, S., Susumu, I., Shiteru, M. & Akira, G., 1981. Immersion and oral bacterins used in combination to immunize ayu *(Plecoglossus altivelis)* against vibriosis. - Devs biol. Stand. *49,* 461-462.

130. Tashiro, F., Morikawa, S., Motonishi, A., Sanjo, L, Kimura, N., Inoue, K., Nomura, T., Ushiyama, M. & Jo, Y., 1979. Chemotherapy of fish diseases with piromidic acid. 2. Its clinical studies against bacterial infections in cultured salmonids and eels *(Anguilla japonica).* - Fish PathoL *14,* 93-101.

131. Tison, D. L., Nishibuchi, M., Greenwood, J. D. & Seidler, R. J., 1982. *Vibrio vulnificus* biogroup 2: new biogroup pathogenic for eels. - Appt. environ. Microbiol. *44,* 640--646.

132. Toranzo, A. E., Barja, J. L, Colwell, R. R. & Hetrick, F. M., 1983a. Characterization of plasmids in bacterial fish pathogens. - Infect. Immun. *39,* 184-192.

133. Toranzo, A. E., Barja, J. L, Potter, S. A., Colwell, R. R., Hetrick, F. M. & Crosa, J. H., 1983b. Molecular factors associated with virulence of marine vibrios isolated from striped bass in Chesapeake Bay. - Infect. Immun. *39,* 1220--1227.

134. Toranzo, A. E., Barja, J. L., Colwell, R. R., Hetrick, F. M. & Crosa, J. H., 1983c. Hemagglutinating, haemolytic and cytotoxic activities of *Vibrio anguiflarum* and related vibrios isolated from striped bass on the Atlantic Coast. - FEMS Microbiol. Lett. *18,* 257-262.

135. Trust, T. J., Courtice, I. D., Khouri, A. G., Crosa, J. H. & Schiewe, M. H., 1981. Serum resistence and hemagglutination ability of marine vibrios pathogenic for fish. - Infect. Imrnun. *34,* Y02-707.

136. Watkins, W. D., Wolke, R. E. & Cabelli, V. J., 1981. Pathogenicity of *Vibrio anguillarum* for juvenile winter flounder, *Pseudopleuronectes amencanus.* - Can. J. Fish. aquat. Sci. *38,* 1045-1051.

137. West, P. A. & Colwell, R. R., 1984. Identification and classification of Vibrionaceae - an overview. In: Vibrios in the environment. Ed. by R. R. Colwell. Wiley, New York, 285-363.

138. Williams, S. F. & Caldwell, R. S., 1978. Growth, food conversion, and survival of 0 group English sole *{Parophrys vetulus]* at 5 temperatures and 5 rations. - Aquaculture *15,* 129--140.

139. Xu, H.-S., Roberts, N., Singleton, F. L., Attwell, R. W., Grimes, D. J. & Colwell, R. R., 1982. Survival and viability of nonculturable *Escherichia coli* and *Vibrio cholerae* in the estuarine and marine environment. - Microb. Ecol. *8,* 313-323.

140. Yamanoi, H., Muroga, K. & Takahashi, S., 1980. Physiological characteristics and pathogenicity of NAG vibrio isolated from diseased ayu. - Fish Pathol. *15,* 69-73.

141. Yasunaga, N. & Yamamoto, N., 1977. Characteristics of bacterial strains isolated from so-called vibriosis of cultured red sea-bream in the winter of 1977. - Fish Pathol. *12,* 209-214.

142. Yetinson, T. & Shilo, M., 1979. Seasonal and geographic distribution of luminous bacteria in the eastern Mediterranean Sea and the Gulf of Elat - Appl. environ. Microbiol. *37,* 1230-1238.

143. Yoshimizu, M., Kimura, T. & Sakai, M., 1976a. Studies on the intestinal microflora of salmonids. Part 2. Effects of artificial transplanting from fresh water into sea water on the intestinal microflora of feeding and nonfeeding fish. - Bull. Jap. Soc. scient Fish. *42,* 863--873,

144. Yoshimizu, M., Kimura, T. & Sakai, M., 1976b. Studies on the intestinal microflora of salmonids. Part 3. The intestinal microflora of salmon living in the open sea. - Bull. Jap. Soc. scient. Fish. *42,* 875-884.

145. Yoshimizu, M., Kimura, T. & Sakai, M., 1976c. Studies on the intestinal microflora of salmonids. Part 5. The intestinal microflora of the anadromous salmon. - Bull. Jap. Soc. scient Fish. *42,* 1291-1298.

CHAPTER 5

Black Spot Infection in Juvenile Steelhead Trout Increases with Stream Temperature in Northern California

Cody J. Schaaf Suzanne J. Kelson Sébastien C. Nusslé Stephanie M. Carlson

Department of Environmental Science, Policy, and Management, University of California, 130 Mulford Hall #3114, Berkeley, CA 94720-3114, USA

ABSTRACT

Climate change will increase water temperature in rivers and streams that provide critical habitat for imperiled species. Warmer water temperatures will influence the intensity and nature of biotic interactions, including parasitism. To better understand the factors influencing a neascus-type parasitic infection known as black spot disease, we examined the relationship between infection rate in juvenile steelhead trout (*Oncorhynchus mykiss*), abundance of another intermediate host (ramshorn snail, *Planorbella trivolvis*), and water temperature. We quantified infection patterns of trout at seven sites within the South Fork Eel River in northern California, visiting each site on three different occasions across the summer, and recording water temperature at each site. We also quantified infection patterns in trout captured from two tributaries to the South Fork Eel River. Overall, trout infection rates were highest in sites with the warmest temperatures. The abundance of ramshorn snails was positively related to both water temperature and black spot infection rates in juvenile trout.

Both snail abundance and infection rates increased rapidly above a 23 °C daily maximum, suggesting a threshold effect at this temperature. We suggest that warmer temperatures are associated with environmental and biotic conditions that increase black spot disease prevalence in threatened steelhead trout. A comparison of our results with similar data collected from a more northern latitude suggests that salmonids in California may be warm-adapted in terms of their parasite susceptibility.

Keywords: Trematode Digenea Parasitism Neascus Salmonid *Oncorhynchus Planorbella trivolvis* Cold-water refuge

INTRODUCTION

Climate change is expected to increase aridity, temperature, and drought frequency in many parts of the world over the coming decades (Dai 2011, 2013; Trenberth et al. 2014), including western North America (Cook et al. 2004; Seager et al. 2007; Cook et al. 2015). In fact, the recent drought in California may have been the most severe drought experienced in the region in the past 1200 years (Griffin and Anchukaitis 2014), and several studies predict that drought severity and frequency will continue to increase in this region (Diffenbaugh et al. 2015; Shukla et al. 2015). Decreased stream flow coupled with increased air temperature during periods of drought has the potential to dramatically increase water temperatures in rivers (Mantua et al. 2010; van Vliet et al. 2011), with many potential consequences for riverine taxa and ecosystems.

Beyond the direct effects of temperature on individual performance, elevated stream temperatures can alter and, in some cases, intensify ecological interactions. Increased parasitism and disease with warmer temperatures has been observed across a range of freshwater taxa from invertebrates (e.g., *Biomphalaria glabrata*, Marcogliese 2008) to amphibians (e.g., *Rana boylii*, Kupferberg et al. 2009) to fishes (e.g., *Micropterus salmoides*,Marcogliese 2001). For example, proliferative kidney disease, which is caused by a myxosporean (Hedrick et al. 1993), can increase mortality of brown trout (*Salmo trutta*) in Swiss rivers by 70% when water temperatures exceed 15 °C for over two weeks (Borsuk et al. 2006).

Cold-water taxa including salmonid fishes may be particularly vulnerable to the effects of increased water temperature. Warm temperatures can have direct physiological effects on salmonids with negative fitness consequences. For example, warm water temperatures reduce the ability of Chinook salmon (*Oncorhynchus tshawytscha*) to successfully undergo smoltification (Marine and Cech 2004) and increase chronic hormonal stress in juvenile coho salmon (*O. kisutch*, Thomas et al. 1986). Moreover, salmonids that are physiologically stressed by warm water temperatures may be unable to handle further stressors like pollutants or pathogens (Richter and Kolmes 2005). For example, the dual effects of warmer temperatures – including reduced physiological performance and increased parasite outbreaks – may result in increased mortality (Udey et al. 1975; Ray et al. 2012). In 2002, for example, a bacterial pathogen that proliferated in low flows and elevated water temperature in the Klamath River induced a significant die-off of Chinook salmon (Lynch and Risley 2003).

Trematode flukes are freshwater parasitic flatworms that proliferate in warm temperatures (Poulin 2006; Paull and Johnson 2011). Several digenetic trematode species (e.g., *Uvulifer ambplopitis*, *Crassiphilia bulboglossa*) have metacercarial stages that induce a condition known as black spot disease in fish (Hoffman and Putz 1965; Berra and Au 1978; Harrison and Hadley 1982). These black spot-causing trematodes are often referred to as neascus-type or neascus-like in recent work (Ferguson et al. 2010; Markle et al. 2014). Many families of fish are vulnerable to neascus-induced black spot disease including Cyprinidae (Evans and Mackiewicz 1958; Mahon et al. 1979; Wisenden et al. 2012), Gasterosteidae (Fitzgerald et al. 1994; Karvonen et al. 2013), and Salmonidae (Cairns et al. 2005; Rodnick et al. 2008; Ferguson et al. 2010, 2012), among others. Neascus-type trematodes have a complex life cycle with multiple hosts and life stages. The first intermediate host is often a snail (e.g., *Planorbella* (aka *Helisoma*) *trivolvis,* in California streams, Turgeon et al. 1998). After emerging into the water column from the snail, infectious cercariae penetrate the skin of a fish, the second intermediate host, and encyst as metacercariae that produce a melanin-induced black spot around the raised, fibrous cyst (Berra and Au 1978; Teixeira-De Mello and Eguren 2008; Ferguson et al. 2010). The life cycle of the parasite is completed as infected fish are eaten by a belted kingfisher (*Megaceryle alcyon*), the definitive host, and the trematode carries out reproduction in the bird's intestine (Berra and Au 1978; Lane and Morris 2000). The trematode's eggs are shed with bird feces into the river where larvae hatch and infect a snail, continuing the cycle of infection.

Parasites and pathogens, in combination with increased temperatures, have been considered a potential factor in the decline of wild Pacific salmon populations (Miller et al. 2014). Moreover, previous studies have noted that parasites, including those that cause black spot disease, are more prevalent in coho salmon in warmer stream reaches of the Pacific Northwest (Cairns et al. 2005; Rodnick et al. 2008; Ferguson et al. 2010, 2012). However, the relationship between temperature and neascus-type infection prevalence has yet to be explored in salmonids at their more southern range in California, where salmonids are already on the brink of extinction (Katz et al. 2013).

To better understand the factors influencing black spot prevalence in salmonids near the southern end of their range, we examined the relationship between water temperature and black spot disease infection rates in juvenile steelhead trout (*O. mykiss*) in the South Fork Eel River in northern California. We also explored other variables that may influence infection rates, including intermediate host (i.e., snail) abundance and fish size. We predicted that black spot prevalence in *O. mykiss* would increase with water temperature, leading to spatial variation in infection rates between sites with different temperature histories. We also predicted that infection rates would increase as the summer progressed and water temperature and exposure history increased. We predicted that sites with high infection rates would have high abundance of snails (the first intermediate host). Finally, we predicted that infection rates would be higher in older/larger fish, which have had a longer period of exposure than younger/smaller fish.

MATERIALS AND METHODS

Study system

Our sampling sites were distributed along the mainstem of the South Fork Eel River (SFER) and two of its tributaries in Mendocino County, California, USA (Fig. 1), including sites in and around the UC-Berkeley Angelo Coast Range Reserve (39.733°N, 123.65°W). The SFER experiences Mediterranean seasonality characterized by wet winters and dry, hot summers (with daily summer air temperatures averaging 29 °C in the reserve, Mast and Clow 2000). The dry season extends from May through October and is associated with low river flows and warm water temperatures. The fish community in this region includes our focal organism, steelhead trout (*O. mykiss*), as well as threespine stickleback (*Gasterosteus aculeatus*), Sacramento sucker (*Catostomus occidentalis*), California roach (*Hesperoleucus symmetricus*), and non-native Sacramento pikeminnow (*Ptychocheilus grandis*). Steelhead trout from this region are part of the Northern California Coast Distinct Population Segment (DPS), which is currently listed as threatened under the U.S. Endangered Species Act.

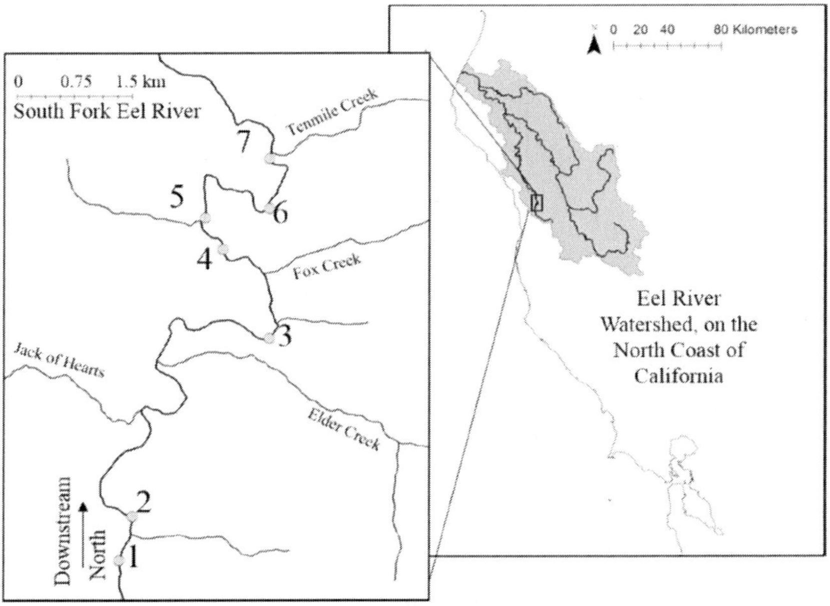

Figure. 1 Map of our study reach on the South Fork Eel River (SFER) in northern California. Numbered circles represent sample sites distributed along the mainstem SFER Our tributary sample sites on Elder and Fox creeks are labelled . The arrow pointing downstream/north represents the direction of water flow

Fish and snail sampling

From June 5 to August 21 2014, we sampled juvenile trout from a 9 km section of the SFER mainstem. We sampled seven sites, spaced from 1 to 2.8 km apart, chosen to represent a gradient of increasing water temperature from upstream to downstream (Fig. 1).

To assess individual fish for black spot infection, we captured young of year (YOY) trout (≤ 80 mm fork length) using a combination of seine and hand nets. At capture, each fish was weighed (±0.01 g), measured (fork length, mm), and assessed for black spot disease infection, always by the same author (CJS). Infection was confirmed via a tactile examination of the black spot to determine whether it was a natural marking or a metacercarial infection (infection spots are raised off the skin, Vaughn 1962; Ferguson et al. 2010). While the vast majority of metacercarial cysts were black, raised nodules that were yellow or brown (Fig. 2) were also considered infectious, since these likely represented new infections that had not yet produced a melanin pigment reaction (Ferguson et al. 2010; Markle et al. 2014).

Figure. 2 Photos showing large (panel **a**) and small (panel **b**) trout with evidence of black spot infection. Cysts on infected trout were small, fibrous, and rounded black spots that were raised off the skin. A few of the heavily infected fish had parasitic cysts that were yellow and brown due to developing infections (e.g., top photo)

We sampled fish at each of the seven mainstem sites on three different dates across the summer, resulting in a total capture of 556 steelhead. To explore infection rate in fish of different size classes, we used a backpack electrofisher (Smith-Root LR-24) to capture 36 additional larger steelhead (i.e., > 80 mm fork length) from our seven mainstem sampling sites during the last sampling effort. Due to length between sample sites (> 1 km), it is highly unlikely that any fish moved between sites during the study period (Edmundson et al. 1968; Keely 2000).

To compare infection rates in the SFER with its reliably cooler tributaries, we report black spot infection rates of 1426 trout that were captured in Elder and Fox creeks (Fig. 1). These fish were captured as part of another ongoing study via a backpack electrofisher between July 20 and Aug 17, 2014 in Elder Creek and July 10–11, 2014 in Fox Creek. Observations from these sites were not included in any statistical models.

To determine the abundance of intermediate snail hosts (*Planorbella trivolvis*), we surveyed snails once (on July 14, 2014) at all seven mainstem sites. We qualitatively observed that snail abundance was highest near the stream bank, where the snails graze algae. Consequently, we surveyed one random site of dimensions 3 m × 2 m at each site, with the long axis paralleling the bank. Within this area, we counted and collected all *P. trivolvis* encountered within a 15-min period, including those visible to the naked eye and those that were found via touch within algae-covered substrates.

Water temperature

At each study site, and in each tributary, we placed one thermistor (HOBO Pendant Temperature/Light Data Logger 64 K - UA-002-64) 10 cm from the substrate, which recorded water temperature at 10-min intervals between the first and last sampling events (6/5/2014–8/21/2014). We use a two-pronged approach to explore temperature variation through space and time.

First, we calculated the daily average temperature at each site to visualize continuous trends in temperature throughout the summer at each sampling site.

Next, we summarized water temperature data for each site and each sampling effort by averaging the maximum daily water temperature that was recorded during a seven-day timeframe including the sampling date (details below). We chose to use the average daily maximum (ADM) water temperature as a metric because it has been used previously in other studies exploring black spot infection rates (Cairns et al. 2005). This resulted in three ADM temperature metrics for each sample site, i.e., one for each of three sampling events that were used in analyses.

For our first sampling effort, we deployed thermistors on the date of the initial sampling. Thus, for the early summer ADM for each site, the seven-day average includes the initial sampling date and the six following days. During our mid-summer sampling event, the seven-day ADM was centered on our sampling date. Because temperature loggers were removed on our last sampling event, the late summer ADM included the six days leading up to and including the final sampling date.

ADM temperature at one site (site three on figures and table) was excluded from analyses because the data logger was exposed to unrealistically warm in-stream conditions due to severe water level drop at that site.

Statistical analyses

To determine the effect of water temperature (measured as the seven-day ADM) on black spot infection rate in YOY trout from our seven mainstem sampling sites, we used a mixed modeling approach, with infection status (infected/not infected) as the response variable (logistic regression). We chose to use mixed models so that temperature and sample date could be included as fixed effects, and site could be included as a random effect on the intercept to avoid pseudo-sampling (Lindstrom and Bates 1990). We tested the fit of three candidate models to determine the effects of temperature and sample date on the proportion of infected fish: 1) temperature as a fixed effect, 2) sample date as a fixed effect, and 3) both as fixed effects. Sample site was included as a random effect for all models. We used Akaike Information Criteria (AIC) (Akaike 1974) to choose between the three candidate models (Sakamoto et al. 1986). All analyses were performed in R (R Core Team 2016). Mixed models were fitted with the *glmer()* function from the package *lme4* (Bates et al. 2015).

To determine the relationship between snail count and water temperature, we performed exponential regressions on snail abundance, i.e., abundance as the response variable, and *e* to the power of temperature as the explanatory variable, this time using the 30-day ADM for each site. We used a longer temperature history for this analysis since there was only a single sampling event for the snails. The 30-day ADM was calculated by averaging the maximum temperature at each site for a 30-day period centered on the date of the snail abundance survey (July 14, 2014). To test the hypothesis that trout infection rates increase with snail abundance, we used a logistic regression with snail count as the independent variable and the proportion of infected steelhead at each site as the response variable, focusing on YOY trout and combining data from all seven mainstem sites from all three sampling events.

Finally, we tested whether infected trout were on average larger than non-infected trout using a Welch two-sample t-test accounting for heteroscedasticity. For this analysis, we only used fish captured in the final sampling event, to include the age-1 and older fish captured during this event.

RESULTS

Water temperature varied among sites and across the summer (Table 1, Fig. 3). In general, downstream sites were consistently warmer than upstream sites (Table 1, Fig. 3), and temperatures increased rapidly from June to July, after which temperature remain elevated and decreased slowly through our final sampling event in mid-August (Table 1, Fig. 3). Water temperatures in the two tributaries were considerably lower than all mainstem sites, ranging from 13 to 18 °C over the course of the summer. The seven-day ADM water temperature in the mainstem ranged from a minimum of 19.0 °C to a maximum 25.3 °C from June 5 to August 21, 2014 (Table 1).

Table 1 Summary of the 7-day average daily maximum temperature (ADM, °C) and black spot infection rates (%) in young of the year steelhead trout at each site and each sampling event (n = sample size). Sites are listed in order from upstream (Site 1) to downstream (Site 7)

Site	Date	ADM	% Infected	n
1	6/5/2014	20.9	0	18
	6/30/2014	20.5	8.6	35
	8/21/2014	21.5	33.3	18
2	6/18/2014	19.0	0	44
	7/7/2014	23.0	7.6	66
	8/21/2014	21.7	36.4	11
3	6/18/2014	20.4	0	28
	7/7/2014	25.2	52	25
	8/21/2014	22.4	100	19
4	6/18/2014	21.5	0	45
	6/30/2014	23.2	11.5	26
	8/21/2014	NA	92.9	28
5	6/23/2014	21.7	6.4	16
	7/13/2014	24.7	100	22
	8/21/2014	24.7	100	42
6	6/25/2014	24.1	76.5	17
	7/13/2014	25.3	100	13
	8/21/2014	24.8	100	11
7	6/30/2014	23.1	96.2	26
	7/13/2014	24.3	87.0	25
	8/21/2014	23.7	100	21

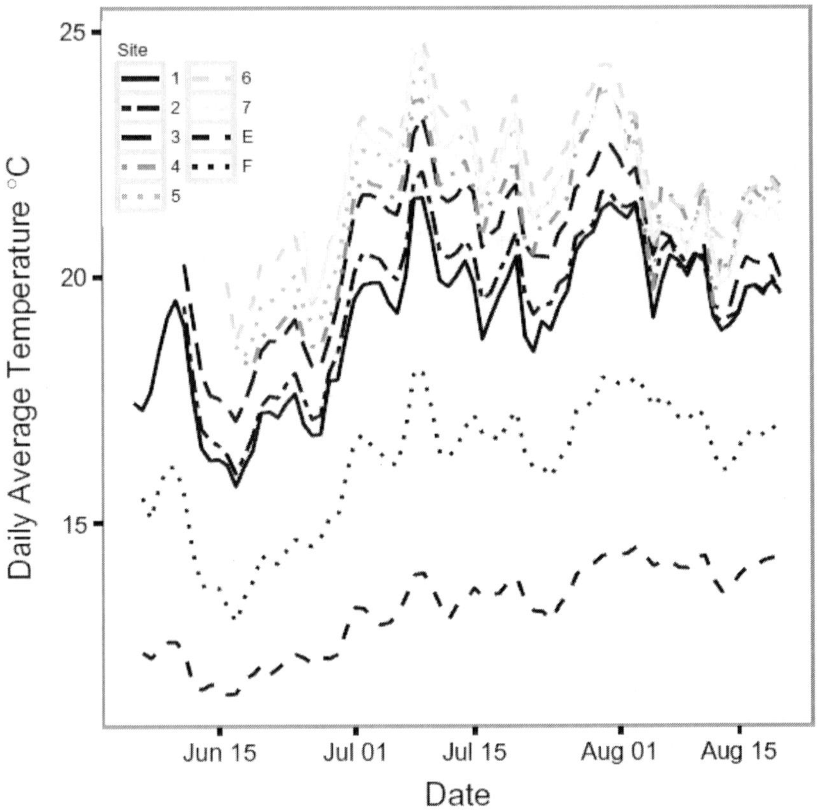

Figure. 3 Average daily water temperature at each site over the course of the summer. E and F refer to Elder Creek and Fox Creek, respectively. Refer to Fig. 1 for the location of sites 1–7 along the mainstem South Fork Eel River

Nearly half (249 of 556, 44.8%) of the juvenile trout that were sampled from the seven mainstem SFER sites showed evidence of black spot infection. In contrast, we observed minimal black spot infection (2 of 1426, < 0.1%) among trout sampled from the cool tributary sites in Elder and Fox creeks.

The mixed models of the seven mainstem sampling sites revealed that both temperature and sample date are important covariates in determining the proportion of infected steelhead at each site. Temperature was significant in the first (i.e., temperature [fixed] and site [random]) model ($z = 6.86, p < 0.01$, AIC = 178.4). Sample date was significant in the second (i.e., sample date [fixed] and site [random]) model ($z = 9.88, p < 0.01$,

AIC = 124.6). In the final model (temperature [fixed], sample date [fixed], and site), both fixed effects were statistically significant, and this model was associated the lowest AIC value (temperature: $z = 4.35, p < 0.01$, sample date: $z = 7.22, p < 0.01$, AIC = 98.0). From these comparisons, we base our interpretations on the final model, including both sample date and temperature as fixed effects, and site as a random effect.

Overall, the top-supported model suggests that temperature and sample date both influence observed infection rates, with infections accumulating over the course of the summer, and infection rates increasing with temperature. The model suggests that at 23.3 °C, 50% of fish will be infected with black spot disease. In general, trout infection rates tended to be higher at downstream sites (Figs. 4 and 5), reaching infection rates of 95–100%, where water temperatures were the highest (Fig. 3), in comparison to cooler sites, where infection rates never exceeded 40% (Table 1, Fig. 5). In addition, at every site, infection rates increased over the course of the study, peaking at the last sample event on August 21, 2014 (Table 1, Fig. 5).

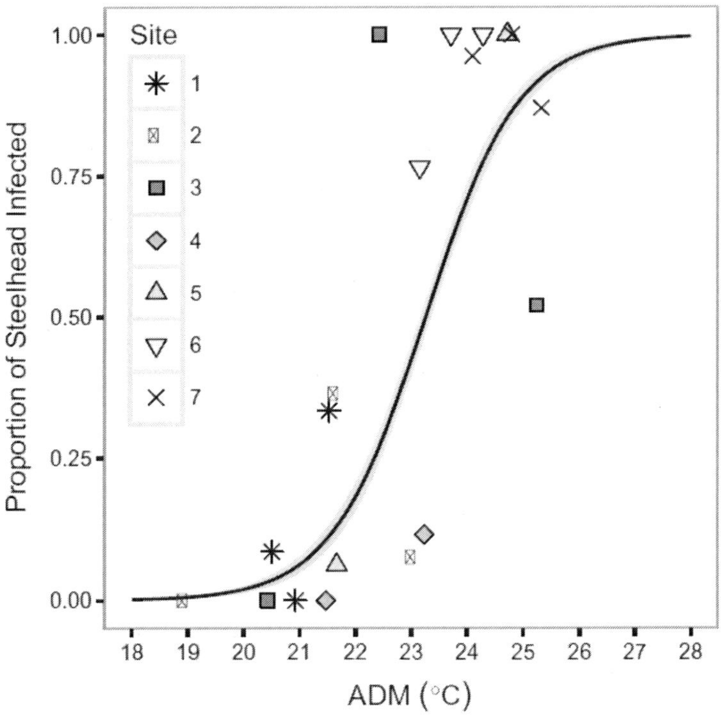

Figure. 4 Relationship between trout infection rates and average daily maximum (ADM) water temperature calculated over 7-day windows

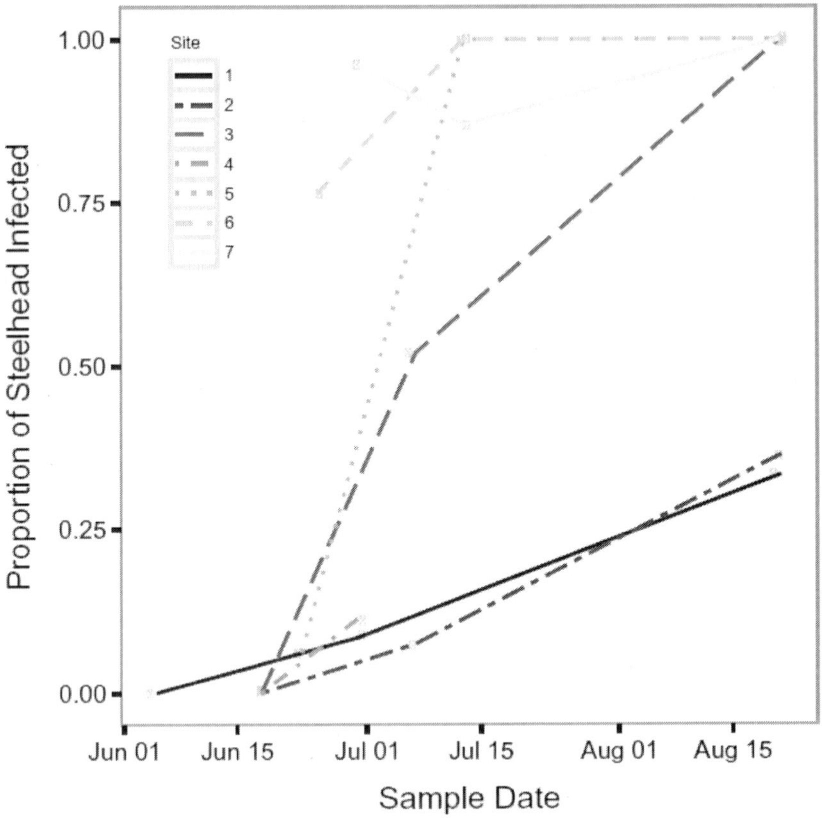

Figure. 5 Temporal trend in trout infection rates at each site over the study period, highlighting differences among early, middle, and late sampling events at each site. Refer to Fig. 1 for the location of sites 1–7 along the mainstem South Fork Eel River

The relationship between snail abundance at each site and water temperature (as measured by the 30-day ADM) was best described by an exponential relationship ($F_{1,5} = 51.6$, $p < 0.001$, r-squared =0.91). We found <10 snails per plot until water temperatures exceeded 23 °C (Fig. 6a). Snail counts were lowest at upstream (cool) mainstem sites ($n = 0$) and increased at the warmer downstream sites (Fig. 6a). The highest number of snails encountered in any given survey ($n = 41$) occurred at the furthest downstream site (i.e., site 7). The relationship between snail abundance and proportion of infected fish was best described by a logistical regression relationship ($z = 12.15$, $p < 0.001$). When there were 11 snails per plot, at least 50% of the steelhead at that site were infected with black spot disease.

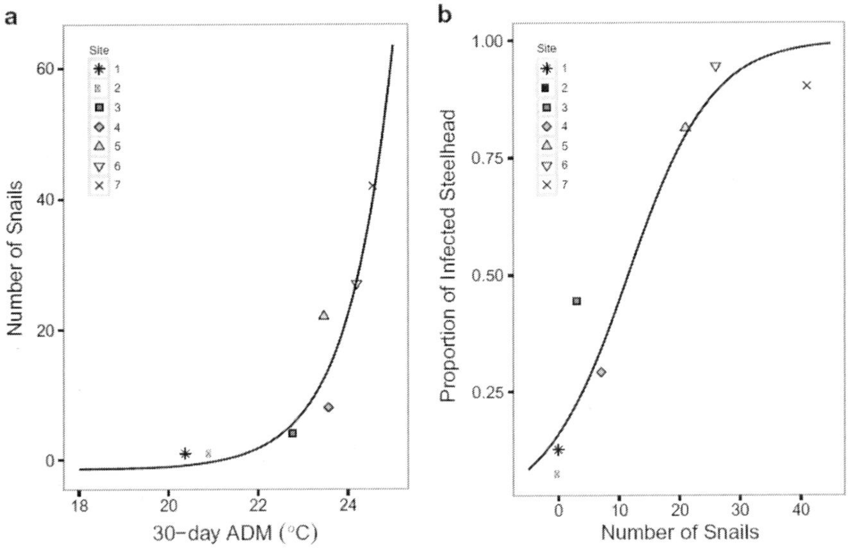

Figure. 6 Relationship between snail counts and average daily maximum temperature (ADM) (panel **a**) calculated over 7-day windows, and between trout infection rate and snail counts (panel **b**)

Finally, we found that the mean size of infected fish was significantly larger than the mean size of not-infected fish (infected: 72.8 ± 23.3 mm, not-infected: 64.6 ± 13.4 mm, $t_{43} = 2.24$, $p = 0.03$).

DISCUSSION

Black spot infection rates in juvenile steelhead trout increased with maximum water temperature in the South Fork Eel River watershed of northern California (Fig. 4). Moreover, spatiotemporal variation in water temperature was associated with spatiotemporal variation in blackspot infection rates. Black spot infection rates were highest in juvenile steelhead trout occupying warmer downstream sites than cooler upstream sites, and infection rates increased over the course of the summer season at every site, possibly due to increased exposure or increased cercarial releases.

We found that infected fish tended to be larger, a pattern that is commonly seen in other parasitized salmonids (e.g., *Salvelinus fontinalis*, Poulin et al. 1991; *O. kisutch*, Barndt and Stone 2003). One possible reason is that larger fish are usually older and, thus, have had a longer cumulative period of exposure to parasites compared to smaller, younger fish (Berra and Au 1978). In fact, metacercarial cysts can persist for up to 4.5 years in some fish species (Hoffman and Putz 1965), suggesting that some of the black spots detected on age-1 and older fish may reflect infections from previous years.

Spatial and temporal variation in infection rates

At the South Fork Eel River, trout captured from downstream sites had the highest infection rates throughout the summer (Table 1, Fig. 5). These same sites were associated with the highest abundance of intermediate-host ramshorn snails (Fig. 6). Since infectious trematode cercariae survive for less than 24 h (Pechenik and Fried 1995), they would likely not influence infection rates at downstream sites by being carried downstream in low flows. Thus, snail host abundance is the most obvious factor that would influence local infection. Increased abundance of intermediate host snails within certain stream reaches may encourage growth of the trematode population, increasing local black spot infection in fish. The proliferation of snails in downstream sites may be linked to features of these sites. Previous work has shown that the abundance of aquatic grazers like snails increases steadily with an increasing biomass of epiphytes and macroalgae (Cattaneo 1983). In the South Fork Eel River, increased solar radiation in wide, downstream channels can cause proliferations of macroalage (Power et al. 2008, 2013), which may explain increased snail abundance at downstream sites.

Infection rates in cool tributaries were considerably lower than in mainstem sites (<1% versus 44.8%). A similar pattern - of lower infection rates in cooler waters - was also observed in other salmonid species (Cairns et al. 2005; Ferguson et al. 2010, 2012) and a cyprinid (Mahon et al. 1979). The tributary sites are shaded and cooler due to groundwater inputs, and lower productivity may limit algal growth (Power et al. 2013) and snail establishment, both of which would impede parasite transmission to trout residing in tributary habitat. Shaded tributary habitats may represent critical refugia for salmonids at the southern end of their range, both because of the cooler temperatures but also by providing escape from warm-water parasites and predators, both of which are more common in the warmer mainstem waters of the South Fork Eel River. Moreover, models predict that such tributary sites are less likely to warm as a result of climate change (Daigle et al. 2015) possibly due to their tendency to be shaded and fed by cool groundwater, which further highlights their importance for conserving salmonid fishes in a warming world.

Black spot infection rates increased with sample date at our South Fork Eel River sampling sites, suggesting that infection increases with exposure history. That both time and temperature are significant in our best-supported model suggests that both are important. However, the large variation in infection rates among sites early in the summer suggests that other (unmeasured) variables could also be important for explaining spatiotemporal variation in infection patterns – e.g., water flow, parasite dispersal, and abundance of other fish hosts.

Water temperature and trematode activity

We found that black spot infection rates increased with water temperature (Fig. 4) in the mainstem South Fork Eel River, as did black spot infection in coho salmon from an Oregon watershed (Cairns et al. 2005; Rodnick et al. 2008). Previous studies have revealed that warm temperatures trigger the growth and release of trematode cercariae from snails (Ataev 1991; Poulin 2006). In one trematode (*Ribeiroia ondatrae*), development was four times faster at 26 °C than 17 °C (Paull and Johnson 2011). In a trematode that commonly infects *O. mykiss*(*Diplostomum spathaceum*), cercariae were shed from snails more than five times faster at 20 °C than 10 °C (Lyholt and Buchmann 1996). We postulate that infection rates increased over the summer because rising temperatures in June and July accelerated trematode development and cercarial releases. Another possibility is that the cumulative period of exposure increases over the summer.

Fitness and health effects of black spot disease

Previous studies on black spot disease have revealed assorted information about the extent to which the disease can negatively impact infected fish. Research by Ferguson et al. (2010) revealed lower overwinter survival in coho salmon heavily infected with *Apophallus sp.* and neascus-type metacercariae. Other freshwater fish species show increased mortality when infected with neascus-type metacercariae, possibly due to increased predation susceptibility (*Esox lucius*, Harrison and Hadley 1982; *Chasmistes brevirostris*, Markle et al. 2014). Reported sub-lethal impacts of neascus-type infection vary greatly from no effect on salmonid metabolism and growth (Ferguson et al. 2012) to significant weight loss at high levels of parasite load in Centrarchids (e.g., *Micropterus dolomieu*, Hunter and Hamilton 1941; *Lepomis macrochirus*, Lemly and Esch 1984). In our study, heavily parasitized steelhead felt rough to the touch (CJS, personal observation), suggesting a possible loss of streamlining and increased swimming burden. Increased swimming burden due to ectoparasites has been observed previously for marine fish (Binning et al. 2012). Future studies on the fitness effects of the parasitic condition in steelhead are warranted given the potential sub-lethal and lethal effects of black spot disease reported in other species. However, Ferguson et al. (2012) stress the importance of examining more than one type of parasite when assessing fitness impacts and mortality.

Temperature thresholds

As in our study, black spot infection rates also increased with stream temperature in coho salmon in the West Fork Smith River, Oregon, USA (Cairns et al. 2005; Rodnick et al. 2008). However, one important difference emerged between the two studies: the threshold temperature at which infection rates increased was *lower* for the Oregon (West Fork Smith River) population compared to the California (South Fork Eel River) population. Specifically, Cairns et al. report over 50% infection in coho salmon from Oregon at 21 °C ADM, while we found that the 50% threshold was not reached until 23.3 °C ADM for steelhead trout in California (Fig. 4).

There are several non-mutually exclusive explanations for the observed geographical differences in temperature thresholds for black spot infection rates in salmonid fishes. One possibility is that the threshold for infection differs for different species of salmonid fishes, due to differences in their physiology or habitat use. Another intriguing possibility is local adaptation to warmer conditions at more southern latitudes. Overall, the South Fork Eel River, California experiences warmer temperatures than the West Fork Smith River, Oregon (average seven-day maximum of 22.8 °C in the SFER (this study) vs 19.3 °C (Rodnick et al. 2008)), suggesting that the more southern salmonids experience warmer temperatures overall may have adjusted to their thermal habitat through local adaptation and increased resistance to parasites. Such local adaptation in salmonids to warm waters is not without precedent (e.g., Verhile et al. 2016). This possibility of local adaptation to thermal regime could provide another reason to conserve salmonids at the southern end of their range as a valuable genetic reservoir for persisting in a changing world (Katz et al. 2013).

Uncertainty regarding the neascus-type trematode in our study

Although we did not collect and taxonomically identify the black spot-causing trematode in our study from any host (fish or snail), we believe that infection was likely caused by a neascus- type trematode. The parasitic cysts were small, fibrous and raised off the skin upon tactile examination (Fig. 2), characteristics of neascus-type trematodes, rather than other black spot-inducing trematodes (e.g., *Apophallus sp.*) (Vaughn 1962; Ferguson et al. 2010; V. Tkach, pers. comm.; J. Ferguson pers. comm.). Two possible trematodes include *Uvulifer amblopitis* and *Crassiphilia bulboglossa*, both of which use *P. trivolvis*, the snail that is very abundant in our study sites, as an intermediate host (Hoffman and Putz 1965; Berra and Au 1978; V. Tkach, pers. comm.).

However, it is possible that non-neascus-type trematodes were responsible for some of the observed infections, since observation of metacercarial cysts under a microscope is the only true way to identify trematode-specific infections (Ferguson et al. 2010). Further investigation into the exact trematode species and corresponding intermediate snail host involved in the infection cycle is needed to confirm linkages between snail and infection data, and would greatly improve our understanding of the causes of black spot infection in fishes, including imperiled salmonid fishes.

Broader implications and conclusions

In summary, we found that the prevalence of black spot disease in juvenile steelhead trout can be high in a large northern California river, but that disease prevalence varies through space and time. Importantly, we found that infection rates increased rapidly once water temperatures exceeded 23 °C ADM, a threshold most likely to be exceeded in sunny, warm mainstem reaches. As climate changes and stream temperatures warm, it is possible that other populations inhabiting sunny river reaches may experience increased vulnerability to black spot disease, highlighting the importance of shaded and cool tributary habitats for sustaining salmonids in a changing climate.

ACKNOWLEDGEMENTS

We thank the National Science Foundation for the Eel River Critical Zone Observatory financial support (CZP EAR-1331940), including an REU that supported CJZ (EAR-1443164), and a Graduate Research Fellowship to SJK (NSF 201517964). Additionally, we thank Tina Mendez, Kurt Spreyer, and Mary Power for feedback on an earlier version of the manuscript, and Katie Kobayashi for assistance in the field. We also thank Vasyl Tkach and Jayde Ferguson for providing expertise in parasite identification. The research activities presented in this manuscript were carried out with approval from the University of California Berkeley Animal Care and Use Committee.

REFERENCES

1. Akaike H (1974) A new look at the statistical model identification. IEEE Trans Autom Control 19:716–723. doi:10.1109/TAC.1974.1100705
2. Ataev G (1991) Temperature influence on the development and biology of rediae and cercariae of *Philophthalmus rhionica* (Trematoda). Parazitologiya 25:349–359
3. Barndt S, Stone J (2003) Infestation of *Salmincola californiensis* (Copepoda: Lernaeopodidae) in wild coho salmon, steelhead, and coastal cutthroat trout juveniles in a small Columbia River tributary. Trans Am Fish Soc 132:1027–1032. doi:10.1577/T02-122
4. Bates D, Maechler M, Bolker B, Walker S (2015) Fitting linear mixed-effect models using lme4. J Stat Softw 67:1–48. doi:10.18637/jss.v067.i01
5. Berra TM, Au R-J (1978) Incidence of black spot disease in fishes in Cedar Fork Creek, Ohio. Ohio J Sci 78:318–322
6. Binning SA, Roche DG, Layton C (2012) Ectoparasites increase swimming costs in a coral reef fish. Biol Lett 9:20120927. doi:10.1098/rsbl.2012.0927
7. Borsuk ME, Reichert P, Peter A et al (2006) Assessing the decline of brown trout (*Salmo trutta*) in Swiss rivers using a Bayesian probability network. Ecol Model 192:224–244. doi:10.1016/j.ecolmodel.2005.07.006
8. Cairns MA, Ebersole JL, Baker JP et al (2005) Influence of summer stream temperatures on black spot infestation of juvenile coho salmon in the Oregon Coast range. Trans Am Fish Soc 134:1471–1479. doi:10.1577/T04-151.1
9. Cattaneo A (1983) Grazing on epiphytes. Limnol Oceanogr 28:124–132. doi:10.4319/lo.1983.28.1.0124
10. Cook ER, Woodhouse CA, Eakin CM et al (2004) Long-term aridity changes in the western United States. Science 306:1015–1018
11. Cook BI, Ault TR, Smerdon JE (2015) Unprecedented 21st century drought risk in the American Southwest and Plains. Sci Adv 1:e1400082
12. Dai A (2011) Drought under global warming: a review. Wiley Interdiscip Rev Clim Chang 2:45–65. doi:10.1002/wcc.81
13. Dai A (2013) Increasing drought under global warming in observations and models. Nat Clim Chang 3:52–58
14. Daigle A, Jeong D Il, Lapointe MF (2015) Climate change and resilience of tributary thermal refugia for salmonids in eastern Canadian rivers. Hydrol Sci J 60:1044–1063. doi: 10.1080/02626667.2014.898121
15. Diffenbaugh NS, Swain DL, Touma D (2015) Anthropogenic warming has increased drought risk in California. Proc Natl Acad Sci 112:3931–3936. doi:10.1073 /pnas.1422385112
16. Edmundson E, Everest FE, Chapman DW (1968) Permanence of station in juvenile Chinook salmon and steelhead trout. J Fish Res Bd Can 25:1453–1464
17. Evans HE, Mackiewicz JS (1958) The incidence and location of metacercarial cysts (Trematoda: Strigeida) on 35 species of New York fishes. J Parasitol 44:231–235. doi:10.2307/3274708
18. Ferguson JA, Schreck CB, Chitwood R, Kent ML (2010) Persistence of infection by metacercariae of *Apophallus sp.*, *Neascus sp.*, and *Nanophyetus salmincola* plus two myxozoans (*Myxobolus insidiosus* and *Myxobolus fryeri*) in coho salmon *Oncorhynchus kisutch*. J Parasitol 96:340–347. doi:10.1645/GE-2289.1
19. Ferguson JA, Romer J, Sifneos JC et al (2012) Impacts of multispecies parasitism on juvenile coho salmon (*Oncorhynchus kisutch*) in Oregon. Aquaculture 362–363:184–192. doi:10.1016/j.aquaculture.2011.07.003
20. Fitzgerald GJ, Fournier M, Morrissette J (1994) Sexual selection in an anadromous population of threespine sticklebacks --- no role for parasites. Evol Ecol 8:348–356. doi:10.1007/BF01238187

21. Griffin D, Anchukaitis KJ (2014) How unusual is the 2012–2014 California drought? Geophys Res Lett 41:9017–9023. doi:10.1002/2014GL062433.1
22. Harrison EJ, Hadley WF (1982) Possible effects of black-spot disease on northern pike. Trans Am Fish Soc 111:106–109. doi:10.1577/1548-8659(1982)111<106:PEOBDO>2.0.CO;2
23. Hedrick RP, MacConnell E, de Kinkelin P (1993) Proliferative kidney disease of salmonid fish. Annu Rev Fish Dis 3:277–290. doi:10.1016/0959-8030(93)90039-E
24. Hoffman GL, Putz RE (1965) The black-spot (*Uvulifer ambloplitis*: Trematoda: Strigeoidea) of centrarchid fishes. Trans Am Fish Soc 94:143–151. doi:10.1577/1548-8659(1965)94[143:TBUASO]2.0.CO;2
25. Hunter GW, Hamilton JM (1941) Studies on host-parasite reactions to larval parasites. IV. The cyst of *Uvulifer ambloplitis*. Trans Am Microsc Soc 60:498–507
26. Karvonen A, Kristjansson BK, Skulason S et al (2013) Water temperature, not fish morph, determines parasite infections of sympatric Icelandic threespine sticklebacks (*Gasterosteus aculeatus*). Ecol Evol 3:1507–1517. doi:10.1002/ece3.568
27. Katz J, Moyle PB, Quiñones RM et al (2013) Impending extinction of salmon, steelhead, and trout (Salmonidae) in California. Environ Biol Fish 96:1169–1186. doi:10.1007/s10641-012-9974-8
28. Keely ER (2000) An experimental analysis of territory size in juvenile steelhead trout. Anim Behav 59:477–490
29. Kupferberg SJ, Catenazzi A, Lunde K et al (2009) Parasitic copepod (*Lernaea cyprinacea*) outbreaks in foothill yellow-legged frogs (*Rana boylii*) linked to unusually warm summers and amphibian malformations in northern California. Copeia 2009:529–537. doi:10.1643/CH-08-011
30. Lane RL, Morris JE (2000) Biology, prevention, and effects of common grubs (digenetic trematodes) in freshwater fish. NCRAC Tech Bull Series 115
31. Lemly AD, Esch GW (1984) Effects of the trematode *Uvulifer ambloplitis* on juvenile bluegill sunfish, *Lepomis macrochirus*: ecological implications. J Parasitol 70:475–492. doi:10.2307/3281395
32. Lindstrom MJ, Bates D (1990) Nonlinear mixed effects models for repeated measures data. Biometrics 46:673–687
33. Lyholt HCK, Buchmann K (1996) *Diplostomum spathaceum*: effects of temperature and light on cercarial shedding and infection of rainbow trout. Dis Aquat Org 25:169–173
34. Lynch DD, Risley JC (2003) Klamath River Basin hydrologic conditions prior to the September 2002 die-off of salmon and steelhead: U.S. Geological Survey Water-Resources Investigations Report 03–4099
35. Mahon R, Balon EK, Noakles DLG (1979) Distribution, community structure and production of fishes in the upper speed river, Ontario: a preimpoundment study. Environ Biol Fish 4:219–244. doi:10.1007/BF00005480
36. Mantua N, Tohver I, Hamlet A (2010) Climate change impacts on streamflow extremes and summertime stream temperature and their possible consequences for freshwater salmon habitat in Washington State. Clim Chang 102:187–223. doi:10.1007/s10584-010-9845-2
37. Marcogliese DJ (2001) Implications of climate change for parasitism of animals in the aquatic environment. Can J Zool 79:1331–1352. doi:10.1139/z01-067
38. Marcogliese DJ (2008) The impact of climate change on the parasites and infectious diseases of aquatic animals. Rev Sci Tech 27:467–484
39. Marine KR, Cech JJ (2004) Effects of high water temperature on growth, smoltification, and predator avoidance in juvenile Sacramento River chinook salmon. North Am J Fish Manag 24:198–210. doi:10.1577/M02-142
40. Markle DF, Terwilliger MR, Simon DC (2014) Estimates of daily mortality from a neascus trematode in age-0 shortnose sucker (*Chasmistes brevirostris*) and the potential impact of avian predation. Environ Biol Fish 97:197–207. doi:10.1007/s10641-013-0141-7

41. Mast MA, Clow DW (2000) Environmental characteristics and water quality of hydrologic benchmark network stations in the Western United States, 1963–95. U.S. Geological Survey Circular 1173-D.
42. Miller KM, Teffer A, Tucker S et al (2014) Infectious disease, shifting climates, and opportunistic predators: cumulative factors potentially impacting wild salmon declines. Evol Appl 7:812–855. doi:10.1111/eva.12164
43. Paull SH, Johnson PTJ (2011) High temperature enhances host pathology in a snail–trematode system: possible consequences of climate change for the emergence of disease. Freshw Biol 56:767–778. doi:10.1111/j.1365-2427.2010.02547.x
44. Pechenik JA, Fried B (1995) Effect of temperature on survival and infectivity of *Echinostoma trivolvis*: a test of the energy limitation hypothesis. Parasitology 111:373–378
45. Poulin R (2006) Global warming and temperature-mediated increases in cercarial emergence in trematode parasites. Parasitology 132:143–151. doi:10.1017 /S0031182005008693
46. Poulin R, Curtis MA, Rau ME (1991) Size, behaviour, and acquisition of ectoparasitic copepods by brook trout, *Salvelinus fontinalis*. Oikos 2:169–174. doi:10.2307/3545334
47. Power ME, Parker MS, Dietrich WE (2008) Seasonal reassembly of a river food web: floods, droughts, and impacts of fish. Ecol Monogr 78:263–282. doi:10.1890/06-0902.1
48. Power ME, Holomuzki JR, Lowe RL (2013) Food webs in Mediterranean rivers. Hydrobiologia 719:119–136. doi:10.1007/s10750-013-1510-0
49. R Core Team (2016) R: A language and environment for statistical computing
50. Ray RA, Holt RA, Bartholomew JL (2012) Relationship between temperature and *Ceratomyxa shasta* -induced mortality in Klamath River salmonids. J Parasitol 98:520–526. doi:10.1645/JP-GE-2737.1
51. Richter A, Kolmes SA (2005) Maximum temperature limits for chinook, coho, and chum salmon, and steelhead trout in the Pacific Northwest. Rev Fish Sci 13:23–49. doi:10.1080/10641260590885861
52. Rodnick KJ, St.-Hilaire S, Battiprolu PK et al (2008) Habitat selection influences sex distribution, morphology, tissue biochemistry, and parasite load of juvenile coho salmon in the West Fork Smith River, Oregon. Trans Am Fish Soc 137:1571–1590. doi:10.1577/T07-138.1
53. Sakamoto Y, Ishiguro M, Kitagawa G (1986) Akaike information criterion statistics. Reidel, Dordrecht
54. Seager R, Ting M, Held I et al (2007) Model projections of an imminent transition to a more arid climate in southwestern North America. Science 316:1181–1184. doi:10.1126/science.1139601
55. Shukla S, Safeeq M, AghaKouchak A et al (2015) Temperature impacts on the water year 2014 drought in California. Geophys Res Lett 42:4384–4393. doi:10.1002 /2015GL063666
56. Teixeira-De Mello F, Eguren G (2008) Prevalence and intensity of black-spot disease in fish community from a subtropical stream (Santa Lucía river basin, Uruguay). Limnetica 27:251–258
57. Thomas RE, Gharrett JA, Carls MG et al (1986) Effects of fluctuating temperature on mortality, stress, and energy reserves of juvenile coho salmon. Trans Am Fish Soc 115:52–59. doi:10.1577/1548-8659(1986)115<52:EOFTOM>2.0.CO;2
58. Trenberth KE, Dai A, van der Schrier G et al (2014) Global warming and changes in drought. Nat Clim Chang 4:17–22
59. Turgeon DD, Quinn Jr JF, Bogan AE, et al (1998) Common and scientific names of aquatic invertebrates from the United States and Canada: Mollusks Second edition Am Fish Soc Spec Publ 26
60. Udey LR, Fryer JL, Pilcher KS (1975) Relation of water temperature to ceratomyxosis in rainbow trout (*Salmo gairdneri*) and coho salmon (*Oncorhynchus kisutch*). J Fish Res Board Can 32:1545–1551. doi:10.1139/f75-181

61. van Vliet MTH, Ludwig F, Zwolsman JJG et al (2011) Global river temperatures and sensitivity to atmospheric warming and changes in river flow. Water Resour Res 47:W02544. doi:10.1029/2010WR009198

62. Vaughn LR (1962) The incidence and location of metacercarial cysts (*Neascus spp.*, in "black spot") on species of fishes in Clay County, Missouri. Beta Beta Beta Biol Soc 33:216–220

63. Verhile CE, English KK, Cocherell DE, Farrell AP, Fangue NA (2016) High thermal tolerance of a rainbow trout populaiton near its southern range limit suggests local thermal adjustment. Conserv Physiol 4:1–12. doi:10.1093/conphys/cow057

64. Wisenden BD, Martinez-Marquez JY, Gracia ES, McEwen DC (2012) High intensity and prevalence of two species of trematode metacercariae in the fathead minnow (*Pimephales promelas*) with no compromise of minnow anti-predator competence. J Parasitol 98:722–727. doi:10.1645/GE-2454.1

CHAPTER 6

Surveillance of Fish Diseases in the Nordic Countries

By T. Håstein[1], A. Hellstrøm[2], G. Jonsson[3], N. J. Olesen[4], E. R.-Pärnänen[5]

[1]National Veterinary Institute, P.O.Box 8156 DEP, 0033 Oslo, Norway, [2]National Veterinary Institute, P.O.Box 7073, SE-75007 Uppsala, Sweden, [3]Veterinary Officer for Fish Diseases, Keldur, Vesturlandsveg, IS-112, Reykjavik, Iceland, [4]Danish Veterinary Laboratory, Aarhus, Denmark, [5]National Veterinary and Food Research Institute (EELA), P. O. Box 368, Fin-00231 Helsinki, Finland.

ABSTRACT

Due to the increasing importance of disease problems in the fish farming industry and the impact disease may have on both feral and farmed fish in the Nordic countries, monitoring and surveillance on diseases have for many years been considered to be of socioeconomic importance. All the Nordic countries have a national legislation as basis for their surveillance and disease control in aquatic animals and regulations listing notifiable diseases of concern to the countries. The list of diseases vary between the countries. In addition, Denmark, Finland and Sweden are ruled by [3] as regards placing on the market of aquaculture animals and products. The surveillance for viral diseases in all the Nordic countries has mainly been based on the testing procedures given in the EU Commission [2].

The list of diseases to be monitored or surveyed for varies between the Nordic countries. All the Nordic countries run a surveillance programme for Viral Haemorrhagic Septicaemia (VHS) and Infectious Haematopoietic Necrosis (IHN) based on EU regulations and a monitoring programme for Infectious Pancreatic Necrosis (IPN) on a national level. Iceland and Norway have also established a surveillance programme for Viral Nervous Necrosis (VNN) on a national level, while Finland and Sweden has established a surveillance programme for Spring Viraemia of carp (SVC) on a national level in order to obtain additional guarantees within the EU for that disease. Norway is the only country that has established a control programme for *Herpesvirus scophthalmi* in turbot, for *Anguillicola* spp in eel and for pasteurellosis in sea bass. These programmes are due to start year 2000. Monitoring of Bacterial Kidney Disease (BKD) takes place in all the Nordic countries; in some of the countries it is being compulsory and in others it is carried out on a voluntary basis. Due to the importance of the haptorworm *Gyrodactylus salaris* for wild stocks of Atlantic salmon, a surveillance programme for this parasite is under consideration in Norway, while in Finland regular monitoring for the parasite takes place in certain regions recognised to be free of the parasite in order to obtain additional guarantees within the EU. Similarly, Finland, Iceland and Sweden run a national control programme for furunculosis for the same reason and Finland has asked for additional guarantees for certain areas in which the disease do not occur.

The measures taken when disease occur vary from restrictions on movements to stamping out procedures as well as disinfection procedures and other means to establish control of a disease. The ultimate goal is to eradicate or keep the level of disease to a minimum.

INTRODUCTION

On a global scale, fish and fishery products are the main food supply for human beings and these products constitutes more than 80% of the total amount of seafood consumed. It is widely known that the supplies of fish from traditional fisheries are more or less constant [4] and that the shortage in fish and fish products has to be met by aquaculture. The potential for the aquaculture industry to meet the challenge as regards food security has been clearly demonstrated by the rapid expansion of the fish farming industry world wide [14]. Fig. 1 shows the development of aquaculture compared to global fisheries in the period 1987–1997 (FAO). The figures show that the total capture in fisheries is relatively constant at some 90 millions metric tonnes while there has been an increase in aquaculture production from 10 millions metric tonnes to 28 millions metric tonnes in the same period.

The way fish farming is conducted and the species farmed vary considerably in the world depending on geographical conditions, water resources, temperature etc.

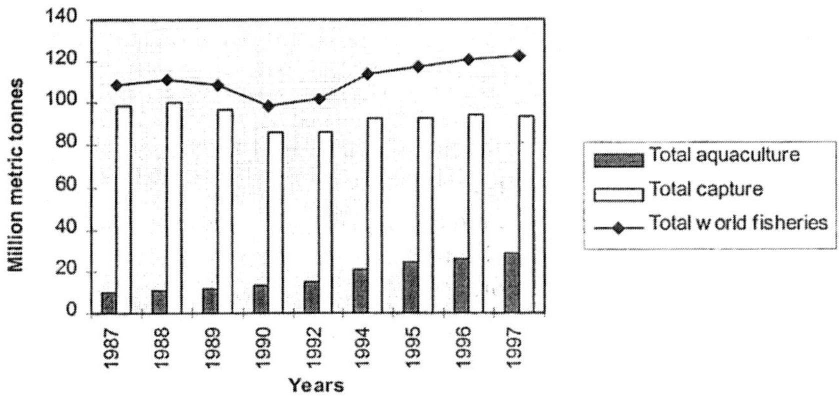

Figure 1 Aquaculture and global fisheries production (FAO information).

FISH FARMING IN THE NORDIC COUNTRIES

In the Nordic countries, the fish farming industry is mainly based on salmonid species, Atlantic salmon (*Salmo salar*) and rainbow trout (*Oncorhynchus mykiss*), but in the later years several different marine species such as halibut (*Hippoglossus hippoglossus*), turbot (*Scophthalmus maximus*) and sea bass (*Dicentrarchus labrax*) have gained increased importance.

Table 1 gives an overview over the fish species farmed in the Nordic countries as well as the size and production of these species.

The rational for disease surveillance

Over the last decades, several emerging or serious diseases in fish have been diagnosed in farmed and feral populations, creating large problems in the fish farming industry and thus being the subject of surveillance and monitoring programmes in many countries [6]. The differences between infectious diseases in fish and those of terrestrial animals means that the approach to the problems and the eradication efforts differ as the diseases may spread effectively through flowing water [7].

Following is the described basis for and the results of monitoring and surveillance programmes as well as established measures on some of the diseases considered to be of major concern in the Nordic countries.

Table 1 Fish farming in the Nordic countries; No. of farms, species and production

Country	No. of farms	Species kept under farming conditions	Total production all species Metric tons
Denmark	478	Eel, Rainbow trout	46.000
Finland	333	Arctic char, Baltic salmon rainbow trout, whitefish	16.597
Iceland	40	Atlantic salmon, Arctic char, rainbow trout, halibut, Sea bass	3.669
Norway	1169 (salmonids) 381 (others)	Atlantic salmon, Arctic char, brown trout, Cod, eel, halibut, turbot, sea bass	390.764
Sweden	177	Arctic char, Baltic salmon, rainbow trout, grass carp	6.200

Basis for the monitoring and surveillance in the Nordic countries

All the Nordic countries have a national legislation as basis for their surveillance and disease control in aquatic animals, as well as regulations listing notifiable diseases of concern to the countries. The list of notifiable diseases vary from country to country. In addition to their national legislation, Denmark, Finland and Sweden are ruled by [3] as regards placing on the market of aquaculture animals and products, but even Iceland and Norway have accepted the principles laid down in the Directive.

The monitoring and surveillance for viral diseases in all the Nordic countries has mainly been based on the testing procedures given in the EU Commission [2] while for the bacterial and parasitic diseases, standard diagnostic procedures for such diseases has been used for screening purposes.

The list of diseases under monitoring and surveillance varies between the Nordic countries. All the countries runs a surveillance programme for Viral Haemorrhagic Septicaemia (VHS) and Infectious Haematopoietic Necrosis (IHN) based on EU regulations [1] and a monitoring programme for Infectious Pancreatic Necrosis (IPN) on a national level. The Nordic countries have also a clinical monitoring for infectious salmon anaemia (ISA) as no valididated surveillance methods so far have been established. Iceland and Norway has also established a surveillance programme for Viral Nervous Necrosis (VNN) on a national level, while Finland and Sweden have established a surveillance programme for Spring Viraemia in carp (SVC) on a national level in order to obtain additional guarantees within the EU for that disease. Norway is the only country that have established a control programme for *Herpesvirus scophthalmi* in turbot, for Anguillicola spp in eel and for pasteurellosis in sea bass which will start year 2000.

Monitoring for Bacterial Kidney Disease (BKD) takes place in all the Nordic countries; in some of the countries being compulsory and in others carried out on a voluntary level.

Due to the importance of the haptorworm *Gyrodactylus salaris* for wild stocks of Atlantic salmon, a surveillance programme for this parasite is under consideration in Norway, while in Finland regular monitoring for the parasite takes place in certain regions recognised to be free of the parasite to obtain additional guarantees within the EU. Similarly Finland, Iceland and Sweden runs a national control programme as regards furunculosis for the same reason and Finland has asked for additional guarantees for certain areas in which the disease due not occur.

Aim of surveillance

The aim of the monitoring and surveillance programmes for fish diseases in question in the Nordic countries is either to document and maintain freedom of disease, to eradicate a disease or to keep a disease under control within certain bonds. For some of the diseases that may be present on a low level such as IPN in Finland and Sweden, the ultimate aim will probably be eradication.

Surveillance programmes and target populations

The main target population for the monitoring and surveillance programmes is salmonids for most of the diseases under surveillance.

The size of the target populations is given in Table 1. All types of farms are included in the survey such as hatcheries, brood stock farms as well as grow out farms, table 2 shows the diseases under surveillance and the target species for the disease in question.

Denmark, Norway and Scotland run a surveillance programme with the aim of deriving information on VHS virus and other viral infections in wild marine populations is carried out in European coastal waters from the English channel in the south to the Spitsbergen area as well as in the Baltic sea [9, 12]. Similar examinations were carried out by Sweden in the Baltic sea and Kattegat in 1998 and the beginning of 1999.

Table 2 Diseases and target species under surveillance in the Nordic countries

Disease	Target species	Country
Viral Diseases		
Infectious salmon anaemia (ISA)	Salmonids	Denmark, Finland, Iceland, Norway, Sweden
Infectious pancreatic necrosis (IPN)	Salmonids	Denmark, Finland, Iceland, Norway, Sweden
Infectious haematopoietic necrosis (IHN)	Salmonids	Denmark, Finland, Iceland, Norway, Sweden
Viral haemorrhagic septicaemia (VHS)	Salmonids	Denmark, Finland, Iceland, Norway, Sweden
Spring viremia in Carp (SVC)	Carp	Finland, Sweden
Viral nervous necrosis (VNN)	Halibut	Iceland, Norway,
Herpesvirus scophthalmi infection	Turbot	Norway
Bacterial diseases		
Bacterial kidney disease (BKD)	Salmonids	Denmark, Finland, Iceland, Norway, Sweden
Furunculosis	Salmonids	Finland, Iceland, Sweden
Pasteurellosis	Sea bass	Norway
Parasitic diseases		
Gyrodactylus salaris	Salmonids	Finland, Norway
Anguillicola spp	Eel	Norway

Organisation

The basis for the different surveillance and monitoring programmes in formal terms is partly based on EU regulations, OIE criteria or criteria derived from national legislation. For some of the diseases such as IHN and VHS, the participation is compulsory in all the Nordic countries as regards approval and maintenance of disease free status for the two diseases either for the whole country, for zones or for farms in non approved zones. In Denmark a control program for VHS started as early as in 1965, while in Finland, Iceland, Norway and Sweden an official surveillance program started in 1995, 1993, 1994 and 1993 respectively. For other diseases such as furunculosis the participation is compulsory in some countries such as Sweden, while it is voluntary or even non existing on a regular basis in other countries such as Norway.

In Finland, the surveillance programme for *G. salaris* has been compulsory for the last three years.

Sampling

The sampling of fish for surveillance vary between the Nordic countries depending on the infrastructure of the health control in the different countries. In Sweden the National Fish Health Service (Fiskhälsan FH AB) are responsible for all sampling, while in Norway, the responsible Authorithy is the district veterinary officers (DVO) or local fish health services operating under instruction of the DVO. In Finland the municipal veterinarians do the necessary sampling when inspecting the farms which is done biannually.

In Iceland, the responsible authority as regards sampling is under the supervision of the Veterinary Officer for Fish Diseases.

In Denmark, the Danish Veterinary and Food Administration do the necessary sampling and control.

Analyses of the diseases under surveillance are carried out at official veterinary laboratories i.e. National Veterinary Institute in Norway and Sweden, the Veterinary Officer for Fish Diseases, Keldur, in Iceland, Danish Veterinary Laboratory in Denmark and National Veterinary and Food Research Institute (EELA) in Finland.

Test and testing procedures

The surveillance for viral diseases in all the Nordic countries has mainly been based on the testing procedures given in the EU Commission [2]. According to the decision, fish farms are inspected clinically biannually and samples for virological examinations are collected by rotation in 50% of the fish farms each year to document freedom for IHN and VHS. In Denmark, Finland, Iceland and Sweden, all or some of the same samples are used for IPN testing. In Denmark, the control of IPN is based on the monitoring of IPN free broodstock farms and ongrowing farms as IPN is considered to be endemic in non controlled farms. In Norway testing for IPN virus is done in a specific monitoring programme in brood stocks of Atlantic salmon and rainbow trout as IPN virus neutralisation is used in the examinations for IHN/VHS virus.

The examination procedures given in the [10] are the basis for examinations as regards BKD, *Gyrodactylus salaris* and VNN while for furunculosis, pasteurellosis and *Anguillicola* spp, standard bacteriological or parasitological procedures are used, respectively. The frequency in the testing of these diseases is varying between the countries.

Measures taken when positive diagnosis occur

In all the Nordic countries, it is the Veterinary Authorities that is responsible for the implementation of measures that will be used in order to control a given notifiable disease. The implementation involves both central and regional veterinary officers.

In all the Nordic countries except non-approved zones in Denmark, stamping out procedures followed by cleaning, disinfection and fallowing will be carried out if VHS is diagnosed [11, 13]. In non-approved zones in

Denmark, a stamping out program will be started if 1/3 of the farms in a river system agrees to an eradication program to get rid of the disease.

Since IHN and SVC is considered to be absent in all the Nordic countries, stamping out procedures will be applied to these diseases. As regards IPN, the detection of the disease or the virus as such, will be subject to restrictions followed by a sanitation programme laid down by the Board of Agriculture in Sweden, while in Norway the mere detection of IPN virus will not bring about any restrictions. Clinical IPN will, however, lead to restrictions on movement of live fish in Norway as long as mortality and signs of the disease occur. In Finland, isolation of IPN virus will lead to restrictions on movement of live fish in continental zones, while clinical IPN probably will result in more strong measures. These have so far not been determined. Prevention may be achieved by avoiding introduction of disease free eggs and/or fish into disease free farms as well as using protected water supply (e.g. spring-, borehole water).

In Iceland, detection of furunculosis is also followed by stamping out because the disease is recognised as a List A disease, while in Sweden active measures are taken only if the disease occur in freshwater, but not in sea water where the disease is considered to be enzootic. In Norway, restrictions are laid upon farms if the furunculosis is diagnosed. Hatcheries are not allowed to sell live fish while affected fish farms in sea water may keep the fish until reaching marketable size for slaughtering.

In all the Nordic countries an infected farm may restock after fallowing if no signs of infection appear after a sanitation programme has been carried out.

Economic compensation

In Sweden, all expenses will be paid by the Board of Agriculture for IHN, VHS, SVC and IPN as these diseases are included in the epizootic act. Similarly, in Finland, the government will pay for all expenses if IHN and VHS are detected, probably also for SVC. In Norway affected farm(s) will have to pay themselves for any measures imposed by the authorities for the time being, because no compensation is granted.

In Denmark expenses will probably be paid in case of IHN and ISA outbreaks and in case of a VHS outbreak in approved zones free of VHS, expenses might be covered as well.

Certification

Health certificates and/or transportation documents is needed in connection with deliveries of live fish for stocking into grow out farms and restocking into rivers. Additionally, in Norway it is not allowed to move fish prior to slaughter, when they are stocked in a sea water site. This is to prevent spread of disease if disease occur.

Record keeping

In all the Nordic countries, the record of findings are kept by the responsible authorities, both regionally and centrally. The diagnostic laboratories also keep the necessary documentation on the examinations carried out. Furthermore all farms have to keep records on events in the farms that can be requested by Competent Responsible Authorithy.

Current status

While still a few cases of VHS is diagnosed in Denmark each year, Finland, Iceland, Norway and Sweden are considered to be free from VHS. However, VHS was diagnosed in one farm in Norway and Sweden in 1998. All Nordic countries have status as free from IHN and SVC although only Finland and Sweden is carrying out a specific surveillance programme for SVC.

IPN is considered to be the main viral disease problem in Norway. The disease is endemic in Denmark but not considered as a mayor problem. The disease level of IPN is low in Finland and Sweden and so far not reported in Iceland. Iceland is considered to be free from classical furunculosis, while the disease appears to be endemic in Denmark, Finland, Norway and Sweden. However, some areas in Finland are reported free of furunculosis and the authorities has established so-called protective zones to which live fish and fish eggs only can be transported if special permission is granted.

BKD has been reported from all the Nordic countries but the level appear to be relatively low [5, 8]. In Iceland. the disease has occasionally been detected in wild Atlantic salmon returning to spawning grounds while it has not been detected in farms for the last 4 years. BKD was for the first time detected in wild salmon in the Baltic sea in 1999, so it is quite apparent that a reservoir of the disease exists in the wild.

Gyrodactylus salaris has been described from Denmark, Finland, Norway and Sweden, but it is only in Norway that disease caused by this parasite has given high mortality in feral fish populations in Norway. The parasite has now totally been detected in 40 rivers and 37 farms/hatcheries since its first detection.

Anguillicola crassus has so far been reported from Denmark, Norway and Sweden, but the distribution in the wild is not known.

CONCLUSION

Although some differences occur, all the Nordic countries have established appropriate surveillance and monitoring for fish diseases of concern to the fish farming industry. Due to these systems and good management practises, the fish disease situation in the Nordic countries is generally good compared to other countries in the world.

REFERENCES

1. *Ariel E, Helgason S, Mortensen HF, Olesen NJ* (1999). Report of the Third annual Meeting of EU National Reference Laboratories for Fish Diseases, Aarhus, Denmark 9.th-11th June 1999. 31 pp.
2. *Decision 96/240/EC:* Laying down the sampling plans and diagnostic methods for the detection and confirmation of certain fish diseases.
3. *Directive 91/67/EEC:* Placing on the market of aquaculture animals and products.
4. FAO Aquaculture production and statistics 1988-1997, Fishery Information, Data and Statistics Unit, FAO Fisheries Department, 203pp.
5. *Heuer OE, Lorenzen E, Korsholm H, Hansen RG, Olesen NJ* (1999). BKD i Danmark: Status efter to sæsoner. Ferskvandsfiskeribladet, 97, 5, p 116-118.
6. *Håstein T.* (1995). Disease problems, use of drugs, resistance problems and preventive measures in fish farming world wide. In Sustainable Fish Farming, eds. Reinertsen H & Haaland H., 183-194.
7. *Håstein T, Hill BJ, Winton JR* (1999). Succesful aquatic animal disease emergency programmes. Rev. sci. tech. Off. int. Epiz., 18 (1), 214-227.
8. *Lorenzen E, Korsholm H, Olesen NJ, Heuer OE* (1997). Påvisning af bakteriel nyresyge (BKD) hos regnbueørred i Danmark. Ferskvandsfiskeribladet, 95, 7, p 160-168.
9. *Mortensen HF, Heuer OE, Lorenzen N, Otte L, Olesen NJ* (1999). Isolation of Viral Haemorrhagic Septicaemia Virus (VHSV) from Wild Marine Fish Species in the Baltic Sea, Kattegat, Skagerrak and the North Sea. Virus Research, *63,* 95-106.
10. *OIE Diagnostic Manual for Aquatic Animal Diseases* (1997), 251pp.
11. *Olesen NJ, Korsholm H* (1997). Control measures for viral diseases in aquaculture: Eradication of VHS and IHN. Bull. Eur. Ass. Fish Pathol.17, 229-233.
12. *Olesen NJ* (1997). Isolering og karakterisering af VHS virus fra saltvandsfisk . Ferskvandsfiskeribladet, 95, 6, p 129-130.
13. *Olesen NJ* (1998). Sanitation of Viral Haemorrhagic Septicaemia (VHS). J. Appl. Ichthyol. 14, 173-177.
14. *Roberts RJ, Muir JF* (1995). 25 years of world aquaculture: Sustainability, a global problem. In Sustainable Fish Farming, eds. Reinertsen H & Haaland H., 167-181.

CHAPTER 7

The Application of Epidemiology in Aquatic Animal Health -Opportunities and Challenges

Edmund J Peeler* and Nicholas GH Taylor

Centre for Environment, Fisheries and Aquaculture Science (Cefas), Barrack Road, Weymouth, Dorset, DT4 8UB, UK

ABSTRACT

Over recent years the growth in aquaculture, accompanied by the emergence of new and transboundary diseases, has stimulated epidemiological studies of aquatic animal diseases. Great potential exists for both observational and theoretical approaches to investigate the processes driving emergence but, to date, compared to terrestrial systems, relatively few studies exist in aquatic animals. Research using risk methods has assessed routes of introduction of aquatic animal pathogens to facilitate safe trade (e.g. import risk analyses) and support biosecurity. Epidemiological studies of risk factors for disease in aquaculture (most notably Atlantic salmon farming) have effectively supported control measures. Methods developed for terrestrial livestock diseases (e.g. risk-based surveillance) could improve the capacity of aquatic animal surveillance systems to detect disease incursions and emergence. The study of disease in wild populations presents many challenges and the judicious use of theoretical models offers some solutions. Models, parameterised from observational studies of host pathogen interactions, have been used to extrapolate estimates of impacts on the individual to the population level. These have proved effective in estimating the likely impact of parasite infections on wild salmonid populations in Switzerland and Canada (where the importance of farmed salmon as a reservoir of infection was investigated). A lack of data is often the key constraint in the application of new approaches to surveillance and modelling. The need for epidemiological approaches to protect aquatic animal health will inevitably increase in the face of the combined challenges of climate change, increasing anthropogenic pressures, limited water sources and the growth in aquaculture.

1 INTRODUCTION

Epidemiology was originally considered as the study of pathogen lifecycles, but is now more comprehensively defined in the Oxford English dictionary as "*the branch of medicine which deals with the incidence, distribution, and possible control of diseases and other factors relating to health*". The terms "epizootic" and "epizootiology" have been used in the veterinary literature but in keeping with current practice we refer to "epidemics" and "epidemiology" in this paper [1]. The birth of modern veterinary epidemiology can be traced back to work by Schwabe [2], who identified that traditional approaches to disease control were failing to resolve animal health problems occurring in intensive livestock production systems. From its inception, modern epidemiology was primarily an applied discipline, practised in the field, with the purpose of preventing and resolving animal health problems. The requirement for inter-disciplinary approaches was central to modern epidemiology, and it encompasses aspects of biostatistics, animal health economics, risk analysis and theoretical modelling studies [3]. A founding principle of modern epidemiology is that disease is caused by multiple interacting factors, which was formulated in the early 1970s as the causal triad of pathogen, host and environment [4]. However, it has since been argued that pathogens should be considered as a component of the environment [5].

Until relatively recently, epidemiological research focused on human and terrestrial animal systems with comparatively little effort directed towards aquatic animal health. However, with 71% of the Earth's surface being covered in water, we live on a wet planet. Life began in water, and the diversity of aquatic organisms is greater than on land. For example, viruses are the most abundant life form in the oceans. There are estimated to be millions of virus-like particles in every millilitre of seawater [6] and that every second 10^{23} viral infections occur in the marine environment [7].

Humans use the aquatic environment for a range of services: potable water, wild capture fisheries, aquaculture, irrigation, travel, recreation and waste disposal. The interaction between humans and aquatic environment means that the study of pathogens in aquatic animals is important to protect a valuable food and recreational resource, and becaise diseases of humans may originate in the aquatic environment [8]. Globally aquaculture and fisheries supplied 110 million tonnes of food in 2006, of which 47% was accounted for by aquaculture [9], compared with only 3.9% in 1970 [10]. The majority of wild capture fisheries are over-exploited [9], growth in consumer demand and new technologies has led to a 10% per annum growth in aquaculture over between 1970 to 2008 [9]. Over 230 aquatic animal species are now farmed [10]. The development of aquaculture has, to a significant extent, depended on the use of non-native species [11] (e.g. marine culture of Altantic salmon in Chile and tilapia and sea-bass in land-based systems in the UK). The diversity and number of species being produced has also resulted in new culture systems and practices, which themselves could influence the emergence and spread of pathogens and occurrence of disease [12]. For example, polyculture (keeping several species on the same site) provides the opportunity for pathogen transmission between species [13]. Farmed fish may be exposed to wild aquatic animals and their pathogens, to which the they would not normally exposed. Similarly, aquaculture may expose wild fish to novel pathogens. Both cases are illustrated by tuna ranching in southern Australia. Juvenile tuna are caught in the open ocean and moved inshores, where they come in to contact with parasites to which they would normally have little or no exposure, resulting in disease emergence [14, 15, 16]. Additionally, the tuna are fed imported frozen fish which was very likely to have been the route of introduction of pilchard herpes virus, which has caused epidemics in wild populations [17].

Fish are one of the most commonly kept companion animals (in the UK there are estimate to be 135 million [18]) and as a consequence the global trade in ornamental aquatic animals is massive. Over 1 billion ornamental fish comprising more than 4000 freshwater and 1400 marine species are traded internationally each year [19, 20]. Trade in live fish (for aquaculture, food and as ornamental animals) and changing climates has expanded the geographic range of many aquatic animal species (and their pathogens), facilitating disease emergence through host-switching [21, 22, 23].

The uses of aquatic animals produce interactions between industry sectors that are not obvious and may drive disease emergence and pathogen spread. For example, within the UK cyprinid fish (e.g. carp and goldfish) industry there are two main interacting sectors: the ornamental and coarse (sport) fish (Figure 1). In summary, cyprinid fish may be imported from both European and non-European countries, the majority of which will be intended for the ornamental (fish keeping) sector. In addition cyprinid farm production with the UK supplies both sectors. Fish are also moved between fisheries either directly or via dealers. Data on fish movement in this network is patchy. It is believed that many movements between fisheries are non-consented and therefore not officially recorded [24], and additionally hobbyists are known to stock fish into lakes and rivers [19, 25, 26, 27]. The cross-over that occurs between the ornamental and coarse fishing sectors has the potential to contribute substantially to pathogen spread in fisheries, and is believed to have been one of the main drivers in the early stages of the koi herpesvirus epidemic in the UK [24].

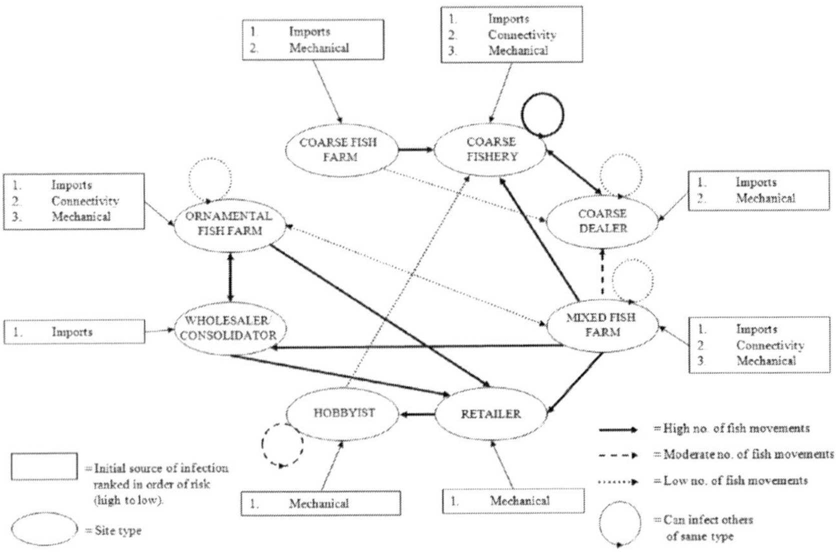

Figure 1 Contact network structure occurring between businesses in the UK cyprinid fish sector, demonstrating the potential routes of pathogen introduction and spread (connectivity refers to sites being connected by the river network, and mechanical refers to transmission via vectors and fomites).

The diversity and interactions of aquatic systems presents a series of unique challenges and complexities that requires the holistic approach which is fundamental to modern epidemiology. This review aims to summarise the application of aquatic animal epidemiology in three key areas of aquatic animal health: i) transboundary and emerging aquatic animal diseases, ii) disease monitoring, surveillance and surveys, and iii) spread and impact of aquatic animal pathogens.

In addressing these questions, epidemiological approaches needed for diseases of livestock are compared with those required by studies of aquatic animal disease. Examples of where these approaches have been applied, their success and limitations, are discussed. Unique challenges and constraints that arise in epidemiological studies of aquatic systems are identified. The underlying theme running through this paper is the role of epidemiology in improving aquatic animal health.

2 THE DEVELOPMENT OF AQUATIC EPIDEMIOLOGY

It has been argued that fish health research has focused on isolating and characterising pathogens, at the expense of host and environmental factors [28]. Only relatively recently have epidemiological studies investigated causes of aquatic animal diseases under field conditions. Inevitably the development of epidemiological methodology has focused firstly human, and secondly terrestrial livestock health issues. As a consequence the development of aquatic epidemiology has lagged behind its terrestrial counterparts, however in recent years the field has developed rapidly. Epidemiological studies of aquatic animals systems are becoming more common place and the techniques used have increased in complexity.

The beginning of aquatic animal epidemiology mirrored the evolution of modern epidemiology in terrestrial livestock. The first risk factor study was published by Thorburn [29], who applied approaches previously used in terrestrial systems to investigate mortality due to vibriosis in farmed rainbow trout. Intensive Atlantic salmon (the first large scale commercially produced fish) production was pioneered in Norway in the 1970s. Inevitably, disease problems emerged, most notably infectious salmon anaemia (ISA) [30]. ISA and other diseases stimulated epidemiological research to support cost-effective control strategies in farmed Atlantic salmon [31, 32, 33]. Subsequently, other important diseases have emerged in aquaculture, some with serious economic and environmental consequences. *Gyrodacytlus salaris* emerged in Norway in the early 1970s causing serious declines in many wild Atlantic salmon populations [34]. Farmed Atlantic salmon may act as a reservoir for infection, thus contributing to populations declines in wild salmon [35]. A number of viral diseases, notably white spot syndrome virus (WSSV), have emerged in the nascent paenid shrimp farming sector, with devastating economic consequences in some countries [36]. Cold water strawberry disease emerged in 2006 in the UK, causing financial losses to trout producers [37]. Disease emergence has clearly provided an important spur to the application of epidemiology.

The growth of aquatic animal epidemiology is illustrated by an analysis of the number and type of studies presented at the International Society of Veterinary Epidemiology and Economics (ISVEE) conferences, held every three years since 1976 (Table 1). The number of aquatic abstracts and the species being studied have increased, the questions addressed and the techniques used have become more diverse (the location of the conference may influence the number of aquatic animal submissions). Shellfish (molluscs and crustacea) production accounts for over a quarter of the aquatic animals consumed annually [9], but have been the subject of comparatively few studies (which are predominantly of cultured shrimp). This may be explained by the limited options for disease control.

Table 1 Analysis of oral presentations at meetings of the International Society for Veterinary Epidemiology and Economics (ISVEE), 1976-2009

Year	Total papers presented	Aquatic papers presented	Species (number of presentations)	Subject area (number of presentations)
1976 - 1982	246	0 (0%)		
1985	117	1 (0.85%)	Fish (1)	Economics/Control (1)
1988	164	1 (0.61%)	Fish (1)	Diagnostic tools (1)
1991	207	2 (0.97%)	Fish (2)	Control (1) Pathogen dynamics (1)
1994	228	0 (0%)	NA	NA
1997	423	9 (1.83%)	Fish (7) Shellfish (2)	Baseline/monitoring/investigation (6) Control (1) Risk analysis (1) Risk factors (1)
2000	603	15 (2.49%)	Fish (13) Shellfish (2)	Baseline/monitoring/investigation (6) Risk factors (4) Control (2) Diagnostic testing x2 Risk analysis (1)

2003	691	19 (2.75%)	Fish (16) Shellfish (3)	Baseline/monitoring/investigation (7) Risk factors (4) Risk analysis (2) Zoonosis (2) Epidemiological tools (2) Disease modelling (1) Control (1)
2006	911	67 (7.35%)	Fish (41) Shellfish (12) Amphibians (4) General (10)	Baseline/monitoring/investigation (25) Risk factors (13) Epidemiological tools (9 Control (6) Risk analysis (5) Diagnostic testing (4) Pathogen dynamics (2) Disease modelling (1) Zoonosis (1) Economic (1)
2009	755	26 (3.44%)	Fish (24) Shellfish (1) General (1)	Baseline/monitoring/investigation (11) Risk factors (6) Epidemiological tools (2) Control (2) Risk analysis (2) Diagnostic testing (1) Pathogen dynamics (1) Disease modelling (1)

3 TRANSBOUNDARY AND EMERGING DISEASES

A key role for epidemiology is to identify and assess exotic and emerging disease threats. Transboundary diseases are a major threat to both farmed and wild terrestrial and aquatic animal populations [21]. In recent years, a number of aquatic animal diseases have emerged and spread rapidly across international political boundaries, causing serious financial and ecological impacts. WSSV emerged in shrimp (*Paenus monodon*) culture in China (Fuijan Province) in 1992 and subsequently spread to Taiwan and Japan and then became panzootic [36]. The main route of transmission was the movement of live animals [36]. *G. salaris* emerged in Norway in the early 1970s, and is hypothesised to have been introduced from Sweden with the introduction of a Baltic strain of Atlantic salmon [34]. A new variant of oyster herpes virus recently emerged in European oyster production [38] and has spread rapidly through live animals movements. Koi herpesvirus achieved widespread distribution across the globe within a few years of emerging through movements of carp, mainly destined for the ornamental trade industry [39]. The movements of bullfrogs (as pets, food and for laboratory use) has been implicated in the spread of the fungal disease, chytridiomycosis in wild amphibia [40]. Movements of animals drive disease emergence not only through extending the geographic range of known diseases but by moving species outside of their natural ranges, thus providing the opportunity for putative pathogens to find new hosts [22]. The scale and diversity of aquatic animal movements, mainly for the ornamental trade but also for food and aquaculture, far outweighs international terrestrial movements, and presents important challenges to epidemiologists.

3.1 Import risk analysis (IRA)

IRA assesses the risk of disease spread through trade in animals or their products. In aquatic animal health they have been undertaken for a variety of reasons (reviewed by Peeler et al., [41]), but mainly to support animal health policy making with respect to trade and biosecurity. Epidemiological approaches are used within the import risk analysis framework [42]. Consistent, defensible and scientifically sound assessments of disease threats and disease freedom are required to fulfil the requirement of the Agreement on the Application of Sanitary and Phytosanitary (the SPS agreement) of the World Trade Organisation (WTO) [43] (which was enacted to allow Members to adopt measures to protect human, animal or plant life or health, whilst preventing restrictions on international trade disguised as sanitary measures). Thus this application of epidemiology, for both terrestrial and aquatic animal diseases, have developed in response to both requirements of international trade agreements and the real threat of transboundary diseases. One of first aquatic animal IRA was Biosecurity Australia's assessment of the risk of disease introduction with the import of wild caught salmon carcasses from Canada [44]. IRA requires information on the epidemiology of the diseases in the exporting country, biophysical characteristics of the pathogen and product information to be integrated. As part of the consequence assessment epidemiological and economic modelling are needed to assess disease spread and impact on local industries and wild populations. More broadly, an assessment of routes of introduction (through trade, movement of shipping, lorries etc.) is the necessary basis on which to review a country's biosecurity measures. However, there are constraints to IRA, as currently practised. Frequently, relevant information is not available. IRA cannot account for unidentified hazards; this is particularly relevant for aquatic animal health in general [45] and the trade in ornamental aquatic animals in particular [19].

3.2 Aquaculture and disease emergence

The same risk framework which has been used for examining disease spread through trade in live animals and animal products have also been applied to disease emergence. Murray and Peeler [12] and Bridges et al. [46] used risk models to investigate drivers for the emergence of disease in aquaculture. However, disease emergence in terrestrial systems has been studied in greater depth [47, 48, 49], in particular the evolution of virulence [50, 51, 52], the importance of animal reservoirs [53] and multiple hosts [54, 55, 56]. These studies have in large part been driven by the emergence of zoonotic pathogens [57, 58]. Epidemiological studies need to more fully explore the conditions under which putative pathogens spread between wild and farmed aquatic animals. The introduction of non-native species has been fundamental to the expansion of aquaculture. Rainbow trout, a native of North America, is the most important freshwater finfish species in Europe and is cultured across the globe [59]. *Paenus monodon* and other species have been shipped internationally to establish shrimp aquaculture. A recent review demonstrated that the introduction of non-native aquatic animal species underpinned many instances of disease emergence in Europe [22] and this is likely to be true for other parts of the world. It is likely that the culture of new species, the expansion of aquaculture into new areas and the global trade in both live aquatic animals and commodity will continue to increase as the demand for animal protein increases with the growth in human population and the decline of wild capture fisheries [10]. Diseases which emerge in farmed populations invariably originate in wild populations. Thus interaction between wild and farmed populations facilitated by the open design of many aquaculture systems [60, 61] and the current fish farming practices [62] is fundamental to disease emergence. Additionally, restrictions in the use of chemicals or the development of resistance could lead to the re-emergence of old diseases that had previously be controlled. Applying the epidemiological modelling approaches which have been used to study disease emergence in terrestrial systems may provide insights into the processes driving disease emergence in the aquatic environment, and inform the development of improved mitigation measures.

3.3 Climate change and disease emergence

Climate change will influence the development of aquaculture and the epidemiology of aquatic animal diseases. Many important aquatic animal pathogens have temperature thresholds, above or below which disease or infection does not occur. A risk assessment model to screen pathogens, to identify those whose threat will increase under established climate change scenarios has been developed for freshwater fish diseases and applied to the UK [23], which identified *Lactococcus garvieae* and *Anphanomyces invadans* as the exotic diseases whose threat will increase most with rising water temperatures. Epidemiological modelling and the application of geographic information systems (GIS) have been used to good effect to assess the likely impact of climate change scenarios on terrestrial livestock diseases (especially vector borne diseases such as blue tongue virus and trypanosomiasis) [63], and huge scope exists to apply these approaches to aquatic animal diseases.

3.4 Outbreak investigations

Outbreak investigation is a core activity for epidemiologists and uses key methods of modern epidemiology: measurement of disease frequency in time and space and in different populations. More recently molecular analyses have been incorporated into outbreak investigations in order to aid source tracking. Outbreak investigations of transboundary diseases are generally undertaken by the veterinary services and generally remain unpublished. However, research papers investigating the origin of transboundary diseases have been published. Investigations of infectious haematopoietic necrosis virus (IHNV) in Europe have attributed its introduction to imported rainbow trout eggs from the US [64]. This hypothesis has been supported by molecular epidemiology studies showing that all European IHNV isolates belong to one of the four N. American genogroups [65]. Conversely, genotyping demonstrated that viral haemorrhagic septicaemia virus (VHSV) in N. America strain had emerged separately from a marine reservoir with no epidemiological link to Europe [66, 67] (where the disease first emerged). Similarly genotyping of ISAV demonstrated emergence from the marine reservoir on separate occasions in Norway, Scotland and N. America [68, 69]. However, genetic analysis of the isolate from the recent outbreak in Chile is identical to Norwegian isolates suggesting spread via imported eggs in 1996 [70, 71].

Despite its growing use and popularity, aquatic molecular epidemiology is yet to reach the level of sophistication applied to terrestrial systems. Examples of the potential power of molecular epidemiology can be seen in studies such as Cottam et al. [72] who, in studying the transmission of foot and mouth disease virus (FMDV) used small mutations found in the entire viral sequence to identify possible previously unidentified contacts in the 2001 epidemic. The availability of technologies that allow for rapid, large-scale sequencing, such techniques provide exciting potential to develop aquatic epidemiology, and may provide a powerful tool to enhance our ability to accurately trace the source and course of infections.

4 SURVEILLANCE AND SURVEYS

4.1 Investigation of disease prevalence

Epidemiological textbooks [73, 74] generally define disease monitoring as ongoing efforts to assess health and disease in a specified population. The term surveillance is reserved for a more active system and implies action will be taken based on the results. Surveys collect information for a specific aim or to test a hypothesis. Monitoring and surveillance for aquatic animal disease is undertaken by governments to demonstrate disease freedom or progress in disease control, and generally remains unpublished (the exception being work by East et al. [75] demonstrating freedom in Australia from WSSV). In the scientific literature there are a number of reports of surveys of disease in farmed and wild fish. Studies of gyrodactylid infections in farmed rainbow trout in Denmark [76, 77] showed considerable between farm and seasonal variation. Investigations of the prevalence of renal myxosporidiosis in wild [78, 79, 80, 81] and farmed salmonids [82] were undertaken to assess the impact of these diseases. Studies of *Renibacterium salmoninarum* (Rs) revealed a low prevalence in wild salmonid fish populations [83] and farms [84], and considerable variation between rivers and farms. More recent a sero-survey of koi herpesvirus found that the virus was geographically widespread in fisheries in E&W, but largely absent from farms [25].

Relatively few papers have used more sophisticated spatial and statistical analytical methods However, analysis of data from a large-scale survey of Atlantic salmon in Scotland [85]) modelled the within and between river prevalences to correct the bias that arose from low sample sizes (in some rivers) and pooling [86]. Peeler et al. [81] used multi-level modelling approaches to analyse prevalence data of renal myxosporidiosis and hepatitis in wild brown trout, and concluded that site (compared with factors at the level of the fish or river) level factors exerted most influence.

4.2 Developments in surveillance methodology

In terrestrial animal epidemiology there has been considerable progress in the development of methods and approaches to support the design and analysis of surveillance systems and data [73]. In this section, the application of these methods to aquatic animal diseases is discussed.

4.2.1 Risk-based surveillance and scenario tree modelling

Risk-based surveillance (RBS) and scenario tree modelling are powerful tool for improving the design and analysis of surveillance systems, not least because complex data sources can be used and historic data discounted [87, 88]. To date there are only publications on the use of RBS and scenario tree modelling in terrestrial systems, however, these method have started to be used in aquatic animal production systems. The more sophisticated approaches used in terrestrial animal surveillance have not yet been applied to aquatic systems, in part, because of an inevitable lag that will occur for methods developed for one system to be refined and used elsewhere. The EU aquatic animal health directive 2006/88/EC requires member states to use RBS, and so will drive the application of this approach in aquatic systems.

4.2.2 Spatial and temporal analysis

The use of for example the scan statistic [89] to detect spatial and temporal clusters or the nearest neighbour technique (first developed by Clark and Evans [90]) to detect spatial clusters, helps identify confounding and risk factors for disease, and generate research hypotheses [91, 92]. These techniques could serve the same role in aquatic animal health research, though spatial analysis would need to account for hydrodynamic connections between populations (e.g. for farms in the same river catchment). In the terrestrial field, data from long term data collection exercises have been used. These have often been collected for statutory purposes (e.g. monitoring for bovine spongiform encephalopathy). Arguably, it is the lack of suitable datasets which have held back the adoption of these approaches. The main exception to this are datasets of sealice infections of farmed Atlantic salmon [93, 94, 95] and a single dataset of 14 years of monitoring of two cyprinid populations and their parasites in three rivers in north east England [96], which could be used to assess parasite burdens over time.

4.3 Test validation

Advances in Bayesian statistical methods to determine test characteristics (sensitivity and specificity) in the absence of a gold standard (TAGS) have also contributed to the development of surveillance [97, 98], and are of great importance in aquatic systems as few true gold standard methods currently exist. However, the test characteristics of relatively few diagnostic tests have been validated under field conditions and in the absence of a gold standard, the exception being tests for infectious salmon anaemia [99, 100]. The techniques and expertise exists; once again the main constraint is the lack of resources to test a sufficient number of populations (with different levels of disease). Pooling diagnostic samples occurs more frequently with aquatic compared with terrestrial animals, with many of the tests approved by the World Organisation for Animal health [101] for aquatic animal diseases allow samples from up to 10 animals to be pooled. The effect of pooling on the characteristics of these test have in general not been considered, however, methods to adjust prevalence estimates for pooled samples have been developed [102]. Epidemiologists have an important role to play in designing trials to generate information on test characteristics and to assess the impact of pooling [97].

5 SPREAD, ESTABLISHMENT AND IMPACT OF PATHOGENS

5.1 Identifying routes of spread

Understanding the routes of disease spread is of key importance to containment and control. Transmission mechanisms must first be identified and understood, and then the importance of routes of transmission between areas or sites quantified. Studies of the UK 2001 foot and mouth disease (FMD) epidemic provide examples of techniques which may be of use in the study of aquatic systems: farmed based stochastic modelling [103]; social network analysis [104] and space time interaction [105]. As in terrestrial systems, the main routes of transmission in aquatic systems are the movement of live animals, transmission via vectors and fomites and the contiguous nature of some sites (by connecting water sources). The specific processes involved for each route and their importance can be investigated ex-ante by theoretical probabilistic approaches (e.g. risk assessment), and ex-post observational studies of patterns of disease spread.

5.1.1 *Ex-ante* studies of disease spread

Peeler et al. [41] reviewed the use of risk assessment to study the spread of aquatic animal pathogens between countries, river catchments and farms. A number of risk assessments focused on *G. salaris*. These ex-ante studies tended to focus on one pathway and addressed specific management questions. Paisley et al. [106] used Monte-Carlo simulation methods to study the likelihood of *G. salaris* being introduced to the river Tana in Norway from infected Atlantic salmon smolts in marine cages. The study identified the components of this route that had the most influence in determining the final probability of introduction (salinity at the cage site followed by the time of year fish escaped from the farm). Høgasen et al. [107] used similar methods to assess the risk associated with infected wild Atlantic salmon smolts moving to uninfected catchments and successfully transmitting *G. salaris*. Peeler, Gardiner and Thrush [108] conducted a qualitative risk assessment to establish the most important routes by which *G. salaris* would spread through England and Wales should it be introduced. This study concluded that in England and Wales the most likely transmission routes were the movement of live rainbow trout followed by the movement of other fish species. Such studies provide valuable information for contingency plans to prevent spread should an introduction occur.

Once the contact pathways have been established, their importance in an epidemic and the overall rate of spread can be modelled between sites (or individuals as is often the case with human pathogens). This however relies on an accurate picture of the total population size of susceptible, potential reservoirs, and the degree of contact between them. If the question is simply to determine the rate of pathogen spread under different conditions, simple compartment based epidemic models, in which the host moves from susceptible to infected states [109], may be applied. Murray [110] and Ruane et al. [111] used such models to study the spread and dynamics of infectious pancreatic necrosis virus (IPNV) between salmon farms in Scotland and Ireland, respectively. Both studies suggested that the stocking of infected smolts from freshwater sites drove the epidemic in the marine sites, but Ruane et al. [111] found that endogenous spread between marine sites took over as the dominant transmission route shortly after introduction from freshwater sites. Taylor et al. [112] also found that the role of infection pressures in spreading a pathogen could change as an epidemic progressed: the stocking of imported ornamental fish drove the initial phase of the koi herpesvirus epidemic in England and Wales (E&W), but that live fish movements between lakes became the main driving force after a few years.

Network analysis may be employed to investigate which sites are most likely to be exposed to pathogens. The technique is used to investigate the number and direction of contacts (edges) made by one site to another, and allows several useful measures that may facilitate an understanding of pathogen spread to be derived. In their simplest form they simply model contacts and not transmission. However, they can be combined with epidemic models, and simulations run to determine how a disease might spread through a population. Such approaches are powerful, but rely on a full knowledge of the network structure, which is also assumed to be stable over time, i.e. sites always contact the same sites. In aquatic systems, network approaches have received relatively little attention, as the required contact data is often not available. However, some systems do lend themselves to this approach, and two good examples of this are demonstrated by Thrush and Peeler [113], who looked at live trout movements in E&W, and Green [114] who studied live fish movements between Scottish trout and salmon sites. In the UK, data on live fish movements from farms exists farmers are obliged to keep these records. Both studies used these data to construct contact networks. Thrush and Peeler developed a stochastic model to examine how rapidly a pathogen may spread on the network. Green et al. [115] used a network model to identify sites where movement controls should be targeted to most effectively slow the spread of disease. Sharkey et al. [116] further developed the work of Thrush and Peeler [113] by adding other pathways such as connectivity via the river network to the analysis. Hydrographical connections between farm sites have also been studied in the marine environment using particle tracking, advection decay models. These have been used to good effect in the study of the dispersal of sea lice [117], ISAV [118], pancreas disease virus [119] and IPNV [120]. If network and hyodrographical models were accurately parameterised, for a particular pathogen, they could be used during an outbreak to estimate the likely spread of the pathogen. In addition they can be used to develop outbreak scenarios which to test contingency plans and specifically competing control measures. Such studies are, however, often limited by a lack of available data on the contact network and key epidemiological parameters such as transmission rate, latent periods, duration of shedding and immunity.

5.1.2 *Ex-post* observational studies

Retrospective observational studies (e.g. case control studies) of spread are a natural extension of outbreak investigations. They have been used extensively in terrestrial epidemiology to identify routes of spread, e.g. analysis of data from the FMD 2001 outbreak [121]. Again *G. salaris* provides an example of a case-control study [122] which used logistic regression to assess routes of spread (whilst accounting for confounding variables); and concluded that the amount of freshwater inflow into fjords and the distance travelled by infected fish were the most important factors. A number of retrospective studies of ISA have provided important insights into routes of spread. Early work in Norway demonstrated that spread from neighbouring infected farms and slaughter houses were important [32, 123]. Mechanical routes of transmission (e.g. divers moving between sites) were also significant [123]. Investigations of the ISA outbreak in Scotland [124] and Canada [125] identified the use of well boats, for moving and harvesting fish, as an important factor in the spread of the disease.

The limitation of studies such as case-control are that it is often not known if the pathogen has been introduced to a control site but has not be detected as conditions at the site were not conducive to disease expression or establishment. Serological testing may provide insight into whether this may have been the case, and was used by Taylor et al. [24] to identify carp fishery sites that may have been exposed to koi herpesvirus. Subsequent contact tracing studies also using such tests subsequently suggested that live fish movements were the main route of transmission [126]. Unfortunately, at present few validated serological test methods exist for aquatic animal diseases.

5.2 Identifying risk factors for disease establishment

Classical epidemiological studies such as cross-sectional, case-control and cohort are commonly applied to identify risk factors associated with pathogen establishment and disease. In terrestrial systems individual animals were often the unit of study. In aquatic systems, although these study types are now commonly used by aquatic epidemiologists, they tend to be conducted at the level of the site (i.e. farm or fishery) or batch/cohort of fish due to the logistics of tracking individual fish.

There are a growing number of observational epidemiological studies that have identified risk factors for aquatic animal diseases. Most commonly these studies have undertaken to support the development of management advice; examples include diseases of farmed salmon [31, 32, 33], including sealice [127], and white spot syndrome virus (WSSV) infection in paenid shrimp [128]. The same approaches have also been applied to investigate risk factors for non-infectious causes of disease, for example fin erosion in rainbow trout [129], cataracts [130] and skin lesions [131] in Atlantic salmon. To date, few studies have focused on ornamental fish or fishery populations. With some exceptions these studies have investigated diseases in managed stillwater fisheries (e.g. *Argulus* spp. [132, 133] and KHV [25]). Most of these studies use standard generalised linear modelling techniques such as logistic regression to determine risk factors, but due to the difficulties in accurately measuring many factors of interest, latent variable modelling techniques may provide an approach that should be considered in future studies. The lack of studies in ornamental fish and fishery systems is likely a reflection of difficulties in collecting data from these systems. However, with a growing focus on conservation of wild stocks, their study and interactions with other sectors is likely to be of increased interest and should be the focus of future research efforts.

Most of the risk factor studies have been cross-sectional, which have the advantage over case-control studies that they provide an estimate of disease prevalence, but are not suitable for studying factors which change over time. Few prospective longitudinal studies of disease frequency in aquatic animals have been published, possibly because these studies face a number of constraints not encountered in terrestrial systems. Animals are not individually identified, and on farms groups of fish are split and mixed during the production cycle. These factors make repeated observations over time problematic (possible solutions were discussed by Thorburn et al. [134]). However, successful prospective studies of WSSV in shrimp [128] and a cohort studies in Atlantic salmon [131, 135, 136] have been completed.

5.3 Assessing impact at the population level

It is important to have accurate assessments of the impact of disease at the population level to justify and prioritise expenditure on aquatic animal health. However, a range of impacts need to be considered. Most classically one thinks of mortality and the economic impact attributed to it; however there may also be indirect financial impacts. The key question about the impact of pathogens on wild populations is whether mortality, reduced fecundity, recruitment etc. actually reduce the overall population size, or whether their influence is mitigated by other density-dependent processes, i.e. can impacts at the level of the individual be scaled to the population.

5.3.1 Recording mortality

Although mortality is a commonly used measure of impact, in the aquatic environment even this may not be easy to measure. In the fish farm environment dead and moribund fish are relatively easy to observe compared to wild fish or fishery populations, and there is generally a legal obligation on farmers to record mortalities. However due to the large number of fish that may die (especially in the case of juvenile fish), accurate recording may be a problem. Additionally many farms experience relatively high and variable levels of "natural" background mortality compared to terrestrial farms, and multiple infections may occur at the same time. Consequently, attributing a cause of death to each individual and identifying when a disease first occurs is not always easy (although with a time-series of mortality data simple statistical techniques such as cumulative sum plots are useful). Fish farmers will often record the most obvious potential cause (often an ectoparasite, as they are easy to see and identify), to which all mortalities are attributed. The quality and value of diagnoses by farmers is likely to be highly variable, and this must be taken into consideration when using it for the purpose of epidemiological studies.

Assessing mortality attributed to a disease in wild or even recreational fisheries is more difficult, and as a consequence has been little studied. Even small recreational fisheries are relatively extensive and deep compared to farm systems. The likelihood of detecting mortality or capturing moribund fish for the purpose of diagnosis is therefore low, especially in the early stages of and epidemic. Methods for assessing impact wild populations is discussed in section 5.3.3.

5.3.2 Farm health and production records

Some farmers use production software that records mortalities and their causes. Such information is potentially of great use to the epidemiologist, but commercial sensitivities may constrain its use. Wheatley et al. [137] used production records from Irish salmon production to assess the impact of endemic diseases. Attempts have been made to create a national database to store production and mortality information for the UK trout industry (Development of a scheme for monitoring sentinel farms in the UK trout industry, Scottish Aquaculture Research Forum, project number 028). This database was designed to compile a national picture that would allow farmers to compare their sites performance with their sectors baseline performance (a common practice in cattle production where extensive health and production databases exist). Currently this system is of limited use to the research scientist, but a similar database, MonAqua which holds Norwegian salmon production data has successfully been used by epidemiologists to assess health related impacts [138]. The Faroe Islands have also successfully implemented such a database created following outbreaks of infectious salmon anaemia between 2000 and 2005, which could be used for similar purposes. This is a government initiative which requires by law that farmers report production figures, fish movements, mortalities and their causes on a regular basis for each cage. Where such systems are available, care must be taken to note the purpose of the database when compiling and analysing data, as data collected by the industry for their own monitoring purposes is likely to be subject to a different set of biases to that collected to comply with government regulations.

Of the few databases used to compile disease information in aquatic systems the most extensive are of sea lice infections in farmed Atlantic salmon. This is possibly due to the serious impact of this pathogen across the industry, and the fact that the majority of farms are owned by only a few companies that want to tightly monitor their lossess and the efficacy of treatments across sites. These data have been extensively used to build statistical and mathematical models of the epidemiology of *Lepeophtheirius salmonis* [139], *Caligus elongatus* [140] and other sea lice [141]. The studies include the identification of risk factors [127], time series analysis for trends [142] the impact of treatment timing [143], and economic impacts [144].

5.3.3 Assessing the impact of disease in wild populations

Although the incidence and dynamics of disease in wild aquatic animal populations has been studied, relatively few have adequately assessed impact at the population level. A few observational studies have provided assessments of pathogen induced. Johnsen and Jensen [145] compared Atlantic salmon catch rates and parr densities in Norwegian rivers where *G. salaris* was known to be present and absent. Although no formal statistical associations were provided, the data clearly demonstrated a decline in infected compared with uninfected salmon populations. Several longitudinal studies have been conducted to establish statistical associations between pathogen prevalence/abundance and impact. Taylor et al. [146] for example, studied the impact of *Argulus foliaceus* abundance and other risk factors on fish capture rates in several recreational trout fisheries and demonstrated a statistically significant reduction in capture rates when the parasite burden was high but it was not possible to quantify host mortality. Murray et al. [147] used mark recapture methods combined with non-lethal pathogen sampling methods in a longitudinal study to assess the impact of chytridiomycosis, demonstrating a clear decline in frog survival attributable to the infection.

The effort required to collect the data sets required to conduct observational studies is costly and labour intensive, especially if the host under study is not of interest in terms of commercial fishing, and a dedicated survey is therefore needed. Due to their extensive nature, such surveys are difficult in freshwater systems, but exceptionally demanding in marine systems. Consequently, theoretical modelling approaches that specifically model the host-pathogen interaction may be more appropriate (given knowledge of host-pathogen life cycles and processes), since they provide a method of extrapolating data on individual host level impacts to the population level. The application of such techniques is reflected in the scientific literature. The majority of these modelling studies are based on deterministic differential equation based models, but the model structure can vary greatly in nature. Patterson [148] extended a fisheries capture model to incorporate different mortality rates for infected and uninfected fish. This was used to study the influence of *Ichthophonus hoferi* infections in herring and demonstrated a 21% reduction in stock size. The study highlighted one of the short comings of standard fisheries models, in that they normally do not account for dynamic natural mortality processes such as disease.

The impact of pathogens reproducing on fish farms and effecting wild stocks is a major area of concern, and a promising subject for theoretical studies. Krkošek et al. [35] modelled the impact of sea lice (*L. salmonis*) emerging from Canadian fish farms on the survival of two species of migrating wild Pacific salmon. Deterministic differential equation based models were combined with several different model types. They used an advection-diffusion-decay model to track parasite dispersal and transmission, and coupled this to a survival model for the two host species. The study showed that the sea lice substantially reduced host survival, and that fish farms produced the number of parasite required to induce large scale population effects.

Other approaches to modelling have also been used to good effect in studying the impact of pathogens in wild aquatic animal populations. Borsuk et al. [149] used a Bayesian probabilistic network to study declining brown trout populations in Swiss rivers. This network incorporated many factors thought to be influential in the life-cycle of this fish. The parasite *T. bryosalmonae* (the aetiological agent of proliferative kidney disease) was one of the factors included, and the analysis demonstrated that this was one of the most influential factors in driving population declines, especially at higher temperatures. The approach provides evidence to support recommendations for risk mitigation and the identification of sites that are likely to be at high risk, thus allowing monitoring efforts to be targeted.

One of the major challenges in modelling the impact of epidemics and their impact in wild aquatic animal populations is they can occur over vast areas in many distinct populations and subpopulations. Murray and O'Callagan [150] addressed some of these issues in their review of the different approaches used to study the pilchard herpesvirus epidemic in Australian stocks in 1998 to 1999. The authors demonstrated the need to apply several different model types to understand the processes occurring at different scales; from the epidemic within a shoal, to between shoal spread, and the large scale transmission around the southern Australian coastline.

In summary, epidemiological studies to study pathogen impact in wild aquatic animal populations are sparse compared to farmed species. This is in part likely to be due to the challenges of data collection, the scales over which these processes occur, and the many interacting factors. Adjusting for the effect of density-dependent processes may pose a significant challenge in accurately assessing impact, however, in exploited populations such as wild capture fisheries it may be argued that such processes are of less concern. Sub-lethal effects of pathogens (such as reduced growth or fecundity) that are often observed at the individual level are also difficult to assess at the population level, and modelling approaches provide and important tool in their assessment.

A lack of research into the impact of pathogens on wild fish populations may be because it is difficult to influence events, and some fisheries scientists may argue that pathogen-induced mortality is of little consequence in the presence of high fishing mortality. However, with increasing pressures on fish stocks, understanding the influence of these combined pressures on a population may be of great importance when trying to identify tipping points and set catch limits, in order to prevent population collapses. Many of the challenges described here for the study of impacts in fisheries and wild aquatic animals are similar to those faced by researchers studying wildlife epidemiology and population ecology. These disciples may provide further tools helpful to the study of aquatic systems (e.g research on the impact of parasites on gamebirds [120, 151] and rabies in wildlife populations [20]).

6 CONCLUSIONS

Georgiadis et al. [152] reviewed how epidemiology was and could be used in aquatic animal health. Peeler et al. [41] reviewed the application of risk analysis in aquatic animal health management. In this review we have tried to demonstrate the wide breadth of epidemiological studies and the range of tools, approaches currently being applied to study disease aquatic systems, and the constraints and challenges facing epidemiologists. The single unifying actor is the purpose of epidemiology, namely the improvement of animal health. Although epidemiological studies in aquatic systems have, and continue to lag behind those applied to terrestrial systems in terms of scope and complexity, rapid development in the application of epidemiology to aquatic systems over a relatively short period has occurred. Consequently, epidemiological studies have made a significant contribution to the health of both wild and farmed aquatic animals, through improved biosecurity and surveillance of exotic diseases and control of endemic diseases.

The role for epidemiology in aquatic animal health became clear as pathogens emerged and rapidly spread in intensive aquaculture systems and disease interactions between wild and farmed populations became increasingly problematic. These complex aquatic animal health issues required inter-disciplinary approaches and epidemiologists are best placed to integrate and analyse diverse data sources and lead teams of pathogens specialists, economists, modellers and statisticians. Despite its success to date many challenges remain. Difficulties in obtaining data on host populations, especially in extensive systems require epidemiologists to work with fishery scientists and ecologists to better integrate wild fish demographics into epidemiological studies and models. Economic, population and disease model needed to be better integrated to improve assessments of disease spread to support aquatic animal health policy making. Surveillance methods developed for terrestrial systems need to be adapted for aquatic animals, and the lack of diagnostic test methods suited to surveillance (as opposed to disease confirmation) is currently a major limitation. More sophisticated models of the processes underlying disease emergence that take into account, *inter-alia*, the role of multi-host pathogens and reservoir populations and a greater understanding of the way pathogens are transmitted in aquatic systems are needed to support measures to minimise the number and impact of new diseases.

The volume of both legal and illegal trade in live aquatic animals far surpasses that of terrestrial animals, and is likely to be the biggest challenge in trying to prevent the spread of emerging aquatic diseases. Additionally, over the next few decades, not only is such trade set to increase, but aquaculture is likely to further grow, and pressures of limited water resources and climate change effects are likely to combine to increase the rate at which diseases emerge and spread. In the face of these combined challenges the demand for epidemiological approaches to support measures to protect aquatic animal health will undoubtedly increase.

9 ACKNOWLEDGEMENTS

This work was supported by research funding (FC 1185 and FB001) from the Department for Environment, Food and Rural Affairs (Defra, UK).

REFERENCES

1. Dohoo IR, Morris RS, Martin SW, Perry BD, Bernardo T, Erb H, Thrusfleld M, Smith R, Welte VR: Epidemiology. Nature. 1994, 368: 284-
2. Schwabe C: The current epidemiological revolution in veterinary medicine. Part 1. Prev Vet Med. 1982, 1: 5-15. 10.1016/0167-5877(82)90003-4.
3. Thrusfield M: Veterinary Epidemiology. 2005, Blackwell Science Ltd, 3
4. MacHohan B, Pugh TF: Epidemiology: Principles and Methods. 1970, Boston: Little Brown
5. Martin SW, Meek AH, Willeberg P: Veterinary Epidemiology - principles and methods. 1987, Iowa: Iowa Statue University Press/Ames, 1
6. Bergh O, Borsheim KY, Bratbak G, Heldal M: High abundance of viruses found in aquatic environments. Nature. 1989, 340: 467-468. 10.1038/340467a0.
7. Suttle CA: Marine viruses [mdash] major players in the global ecosystem. Nat Rev Micro. 2007, 5: 801-812. 10.1038/nrmicro1750.

8. Taylor NGH, Verner-Jeffreys DW, Baker-Austin C: Aquatic systems: maintaining, mixing and mobilising antimicrobial resistance?. Trends Ecol Evol. 2011, 26: 278-284. 10.1016/j.tree.2011.03.004.
9. F.A.O: State of the World Fisheries and Aquaculture. Book State of the World Fisheries and Aquaculture. 2008, City: Food and Agriculture Organisation Fisheries and Aquaculture Department, 196-
10. Subasinghe RP: Epidemiological approach to aquatic animal health management: opportunities and challenges for developing countries to increase aquatic production through aquaculture. Prev Vet Med. 2005, 67: 117-124. 10.1016/j.prevetmed.2004.11.004.
11. Gozlan RE: Introduction of non-native freshwater fish: is it all bad?. Fish Fish. 2008, 9: 106-115.
12. Murray AG, Peeler EJ: A framework for understanding the potential for emerging diseases in aquaculture. Prev Vet Med. 2005, 67: 223-235. 10.1016/j.prevetmed.2004.10.012.
13. Mclnerney J: Old economics for new problems - livestock disease: presidential address. J Agric Econ. 1996, 47: 295-314. 10.1111/j.1477-9552.1996.tb00695.x.
14. Aiken HM, Hayward CJ, Nowak BF: An epizootic and its decline of a blood fluke, Cardicola forsteri, in farmed southern bluefin tuna, Thunnus maccoyii. Aquaculture. 2006, 254: 40-45. 10.1016/j.aquaculture.2005.10.013.
15. Deveney MR, Bayly TJ, Johnston CJ, Nowak BF: A parasite survey of farmed Southern bluefin tuna, *Thunnus maccoyii*(Castelnau). J Fish Dis. 2005, 28: 279-284. 10.1111/j.1365-2761.2005.00629.x.
16. Nowak BF: Parasitic diseases in marine cage culture - An example of experimental evolution of parasites?. Int J Parasitol. 2007, 37: 581-588. 10.1016/j.ijpara.2007.01.003.
17. Whittington RJ, Jones JB, Hine PM, Hyatt AD: Epizootic mortality in the pilchard *Sardinops sagax neopilchardus* in Australia and New Zealand in 1995. 1. Pathology and epizootiology. Dis Aquat Organ. 1997, 28: 1-16.
18. Ornamental fish.
19. Whittington RJ, Chong R: Global trade in ornamental fish from an Australian perspective: the case for revised import risk analysis and management strategies. Prev Vet Med. 2007, 81: 92-116. 10.1016/j.prevetmed.2007.04.007.
20. Lembo T, Hampson K, Haydon DT, Craft M, Dobson A, Dushoff J, Ernest E, Hoare R, Kaare M, Mlengeya T, Mentzel C, Cleaveland S: Exploring reservoir dynamics: a case study of rabies in the Serengeti ecosystem. J Appl Ecol. 2008, 45: 1246-1257. 10.1111/j.1365-2664.2008.01468.x.
21. Hedrick RP: Movements of pathogens with the international trade of live fish: problems and solutions. Rev Sci Tech. 1996, 15: 523-531.
22. Peeler EJ, Oidtmann BC, Midtlyng PM, Miossec L, Gozlan RE: Non-native aquatic animals introductions have driven disease emergence in Europe. Biol Invasions.
23. Marcos-López M, Gale P, Oidtmann BC, Peeler EJ: Assessing the impact of climate change on disease emergence in freshwater fish in the United Kingdom. Transbound Emerg Dis. 2010, 57: 293-304. 10.1111/j.1865-1682.2010.01150.x.
24. Taylor NGH, Dixon PF, Jeffery KR, Peeler EJ, Denham KL, Way K: Koi herpesvirus: distribution and prospects for control in England and Wales. J Fish Dis. 2010, 33: 221-230. 10.1111/j.1365-2761.2009.01111.x.
25. Schikorski D, Faury N, Pepin JF, Saulnier D, Tourbiez D, Renault T: Experimental ostreid herpesvirus 1 infection of the Pacific oyster Crassostrea gigas: Kinetics of virus DNA detection by q-PCR in seawater and in oyster samples. Virus Res. 2011, 155: 28-34. 10.1016/j.virusres.2010.07.031.
26. Copp G, Villizi L, Gozlan R: The demography of introduction pathways, propagule pressure and non-native freshwater fishes occurrence in England. Aquatic Conserv: Mar Freshw Ecosyst. 2010, 20: 595-601. 10.1002/aqc.1129.
27. Copp GH, Templeton M, Gozlan RE: Propagule pressure and the invasion risks of non-native freshwater fishes: a case study in England. J Fish Biol. 2007, 71: 148-159.

28. Smith P: An overview of current fish disease research; Hyper-reality in a post-modern world?. Bull Eur Assoc Fish Pathol. 1999, 19: 310-312.
29. Thorburn MA: Factors influencing seasonal vibriosis mortality rates in Swedish pen reared rainbow trout. Aquaculture. 1987, 67: 79-85. 10.1016/0044-8486(87)90010-X.
30. Thorud K, Djupvik H: Infectious salmon anaemia in Atlantic salmon (*Salmo salar* L.). Bull Eur Assoc Fish Pathol. 1988, 8: 109-111.
31. Jarp J, Gjevre AG, Olsen AB, Bruheim T: Risk factors for furunculosis, infectious pancreatic necrosis and mortality in post-smolt Atlantic salmon, *Salmo salar* L. J Fish Dis. 1994, 18: 67-78.
32. Jarp J, Karlsen E: Infectious salmon anaemia (ISA) risk factors in sea-cultured Atlantic salmon Salmo salar. Dis Aquat Organ. 1997, 28: 79-86.
33. Jarp J, Tangen K, Willumsen FV, Dyupvik HO, Tveit AM: Risk factors for infection with Aeromonas salmonicida subsp. salmonicida in Norwegian freshwater hatcheries. Dis Aquat Organ. 1993, 17: 81-86.
34. Johnsen BO, Jensen AJ: The *Gyrodactylus* story in Norway. Aquaculture. 1991, 98: 289-302. 10.1016/0044-8486(91)90393-L.
35. Krkosek M, Ford JS, Morton A, Lele S, Myers RA, Lewis MA: Declining wild salmon populations in relation to parasites from farm salmon. Science. 2007, 318: 1772-1775. 10.1126/science.1148744.
36. Walker PJ, Mohan CV: Viral disease emergence in shrimp aquaculture: origins, impacts and the effectiveness of health management strategies. Rev Aquac. 2009, 1: 125-154. 10.1111/j.1753-5131.2009.01007.x.
37. Verner-Jeffreys DW, Pond MJ, Peeler EJ, Rimmer GSE, Oidtmann B, Way K, Mewett J, Jeffrey K, Bateman K, Reese RA, Feist SW: Emergence of cold water strawberry disease of rainbow trout Oncorynchus mykiss in England and Wales: Outbreak investigations and transmission studies. Dis Aquat Organ. 2008, 79: 207-218.
38. Segarra A, Pépin JF, Arzul I, Morga B, Faury N, Renault T: Detection and description of a particular Ostreid herpesvirus 1 genotype associated with massive mortality outbreaks of Pacific oysters, Crassostrea gigas, in France in 2008. Virus Res. 2010, 153: 92-99. 10.1016/j.virusres.2010.07.011.
39. Haenen OLM, Way K, Bergmann SM, Ariel E: The emergence of koi herpesvirus and its significance to European aquaculture. Bull Eur Ass Fish Pathol. 2004, 24: 293-307.
40. Daszak P, Strieby A, Cunningham AA, Longcore JE, Brown CC, Porter D: Experimental evidence that the bullfrog (Rana catesbeiana) is a potential carrier of chytridiomycosis, an emerging fungal disease of amphibians. Herpetolog J. 2004, 14: 201-207.
41. Peeler EJ, Murray AG, Thebault A, Brun E, Giovaninni A, Thrush MA: The application of risk analysis in aquatic animal health management. Prev Vet Med. 2007, 81: 3-20. 10.1016/j.prevetmed.2007.04.012.
42. O.I.E: Section 2.2. Risk analysis. Aquatic Animal Health Code. 2009, Paris: World Animal Health Organisation, 1: 10
43. WTO: Agreement on the Application of Sanitary and Phytosanitary Measures. Book Agreement on the Application of Sanitary and Phytosanitary Measures. 1995, City: World Trade Organisation, 21-
44. Kahn SA, Beers PT, Findlay VL, Peebles IR, Durham PJ, Wilson DW, Gerrity SE: Import Risk Analysis on Non-viable Salmonids and Non-salmonids Marine Finfish. Book Import Risk Analysis on Non-viable Salmonids and Non-salmonids Marine Finfish. 1999, City: Australian Quarantine and Inspection Service, 409-
45. Gaughan DJ: Disease-translocation across geographic boundaries must be recognized as a risk even in the absence of disease identification: the case with Australian Sardinops. Rev Fish Biol Fish. 2002, 11: 113-123.
46. Bridges VE, Akkina J, Grannis J, Johnson C, Johnson R, Tuszynski C: A qualitative assessment tool for the potential of infectious disease emergence and spread. Prev Vet Med. 2007, 81: 80-91. 10.1016/j.prevetmed.2007.04.008.

47. Woolhouse MEJ: Population biology of emerging and re-emerging pathogens. Trends Microbiol. 2002, 10: 3-7. 10.1016/S0966-842X(02)02428-9.
48. Woolhouse MEJ: Where do emerging pathogens come from?. Microbe. 2006, 1: 511-515.
49. Woolhouse MEJ: Epidemiology: emerging diseases go global. Nature. 2008, 451: 898-899. 10.1038/451898a.
50. Brown SP, Hochberg ME, Grenfell BT: Does multiple infection select for virulence?. Trends Microbiol. 2002, 10: 401-405. 10.1016/S0966-842X(02)02413-7.
51. Davies CM, Webster JP, Woolhouse MEJ: Trade-offs in the evolution of virulence in an indirectly transmitted macroparasite. Proc Biol Sci. 2001, 268: 251-257. 10.1098/rspb.2000.1367.
52. Frank SA: Models of parasite virulence. Q Rev Biol. 1996, 71: 37-78. 10.1086/419267.
53. Haydon DT, Cleaveland S, Taylor LH, Laurenson MK: Identifying reservoirs of infection: A conceptual and practical challenge. Emerg Infect Dis. 2002, 8: 1468-1473.
54. Woolhouse MEJ, Gowtage-Sequeria S: Host range and emerging and reemerging pathogens. Emerg Infect Dis. 2005, 11: 1842-1847. 10.3201/eid1112.050997.
55. Woolhouse MEJ, Haydon DT, Antia R: Emerging pathogens: the epidemiology and evolution of species jumps. Trends Ecol Evol. 2005, 20: 238-244. 10.1016/j.tree.2005.02.009.
56. Woolhouse MEJ, Taylor LH, Haydon DT: Population biology of multihost pathogens. Science. 2001, 292: 1109-1112. 10.1126/science.1059026.
57. Taylor LH, Latham SM, Woolhouse MEJ: Risk factors for human disease emergence. Philos Trans R Soc Lond B Biol Sci. 2001, 356: 983-989. 10.1098/rstb.2001.0888.
58. Woolhouse M, Gaunt E: Ecological origins of novel human pathogens. Crit Rev Microbiol. 2007, 33: 231-242. 10.1080/10408410701647560.
59. Maccrimmon HR: World distribution of rainbow trout (*Salmon gairdneri*). J Fish Res Board Can. 1971, 28: 663-704. 10.1139/f71-098.
60. Amos K, Thomas J: Disease interactions between wild and cultured fish: observation and lessons learned in the Pacific Northwest. Bull Eur Ass Fish Pathol. 2002, 22: 95-102.
61. Coutant CC: What is normative for fish pathogens? A perspective on the controversy over interactions between wild and cultured fish. J Aquat Anim Health. 1998, 10: 101-106. 10.1577/1548-8667(1998)010<0101:WINFFP>2.0.CO;2.
62. Mennerat A, Nilsen F, Ebert D, Skorping A: Intensive Farming: Evolutionary Implications for Parasites and Pathogens. Evol Biol. 2010, 37: 59-67. 10.1007/s11692-010-9089-0.
63. Rogers DJ, Randolph SE: Studying the global distribution of infectious diseases using GIS and RS. Nat Rev Microbiol. 2003, 1: 231-237. 10.1038/nrmicro776.
64. Bovo G, Giorgetti G, Jørgensen PEV, Olesen NJ: Infectious hematopoietic necrosis: first detection in Italy. Bull Eur Ass Fish Pathol. 1987, 7: 124-
65. Enzmann PJ, Kurath G, Fichtner D, Bergmann SM: Infectious hematopoietic necrosis virus: monophyletic origin of European IHNV isolates from North-American genogroup M. Dis Aquat Organ. 2005, 66: 187:195-
66. Basurco B, Benmansour A: Distant strains of the fish rhabdovirus VHSV maintain a sixth functional cistron which codes for a nonstructural protein of unknown function. Virology. 1995, 212: 741-745. 10.1006/viro.1995.1534.
67. Basurco B, Vende P, Monnier AF, Winton JR, de Kinkelin P, Benmansour A: Genetic diversity and phylogenetic classification of viral hemorrhagic septicemia virus (VHSV). Vet Res. 1995, 26: 460-463.
68. Cunningham CO, Snow M: Genetic analysis of infectious salmon anaemia virus (ISAV) from Scotland. Dis Aquat Organ. 2000, 41: 1-8.
69. Kibenge FSB, Lyaku JR, Rainnie D, Hammell KL: Growth of infectious salmon anaemia virus in CHSE-214 cells and evidence for phenotypic differences between virus strains. J Gen Virol. 2000, 81: 143-150.

70. Kibenge FS, Godoy MG, Wang Y, Kibenge MJ, Gherardelli V, Mansilla S, Lisperger A, Jarpa M, Larroquete G, Avendão F, Lara M, Gallardo A: Infectious salmon anaemia virus (ISAV) isolated from the ISA disease outbreaks in Chile diverged from ISAV isolates from Norway around 1996 and was disseminated around 2005, based on surface glycoprotein gene sequences. Virol J. 2009, 6: 88-10.1186/1743-422X-6-88.
71. Godoy MG, Aedo A, Kibenge MJT, Groman DB, Yason CV, Grothusen H, Lisperguer A, Calbucura M, Avendaño F, Imilán M, Jarpa M, Kibenge FS: First detection, isolation and molecular characterization of infectious salmon anaemia virus associated with clinical disease in farmed Atlantic salmon (Salmo salar) in Chile. BMC Vet Res. 2008, 4: 28-10.1186/1746-6148-4-28.
72. Cottam EM, Thébaud G, Wadsworth J, Gloster J, Mansley L, Paton DJ, King DP, Haydon DT: Integrating genetic and epidemiological data to determine transmission pathways of foot-and-mouth disease virus. Proc Biol Sci. 2008, 275: 887-895. 10.1098/rspb.2007.1442.
73. Salman MD: Animal disease surveillance and survey systems - methods and applications. 2003, Iowa: Iowa State Press, 1
74. Cameron A: Survey toolbox for aquatic animal diseases. A practical Manual and Software Package. Book Survey toolbox for aquatic animal diseases. A practical Manual and Software Package. 2002, City: ACIAR, 375-
75. East IJ, Black PF, McColl KA, Hodgson RAJ, Bernoth EM: Survey for the presence of White Spot Syndrome virus in Australian crustaceans. Aust Vet J. 2004, 82: 236-240. 10.1111/j.1751-0813.2004.tb12688.x.
76. Lindenstroem T, Nielsen B, Buchmann K: Gyrodactylids on salmonids from Danish streams. Systematics and Phylogeny of Platyhelminthes; St Petesburg, Russia. 1999, 62-64.
77. Nielsen CV, Buchmann K: Occurrence of *Gyrodactylus* parasites in Danish fish farms. Bull Eur Assoc Fish Pathol. 2001, 21: 19-25.
78. Wahli T, Bernet D, Steiner PA, Schmidt-Posthaus H: Geographic distribution of *Tetracapsuloides bryosalmonae* infected fish in Swiss rivers: an update. Aquat Sci. 2007, 69: 3-10. 10.1007/s00027-006-0843-4.
79. Wahli T, Knuesel R, Bernet D, Segner H, Pugovkin D, Burkhardt-Holm P, Escher M, Schmidt-Posthaus H: Proliferative kidney disease in Switzerland: current state of knowledge. J Fish Dis. 2002, 25: 491-500. 10.1046/j.1365-2761.2002.00401.x.
80. Feist SW, Peeler EJ, Gardiner R, Smith E, Longshaw M: Proliferative kidney disease and renal myxosporidiosis in juvenile salmonids from rivers in England and Wales. J Fish Dis. 2002, 25: 451-458. 10.1046/j.1365-2761.2002.00361.x.
81. Peeler EJ, Feist SW, Longshaw M, Thrush MA, St-Hilaire S: An assessment of the variation in the prevalence of renal myxosporidiosis and hepatitis in wild brown trout, *Salmo trutta* L., within and between rivers in South-West England. J Fish Dis. 2008, 31: 719-728. 10.1111/j.1365-2761.2008.00942.x.
82. Seagrave CP, Bucke D, Hudson EB, McGregor D: A survey of the prevalence and distribution of proliferative kidney disease (PKD) in England and Wales. J Fish Dis. 1981, 4: 437-439. 10.1111/j.1365-2761.1981.tb01155.x.
83. Chambers E, Gardiner R, Peeler EJ: An investigation into the prevalence of *Renibacterium salmoninarum* in farmed rainbow trout (*Onchorhynchus mykiss*) and wild fish populatins in selected river catchments in England and Wales between 1998 and 2000. J Fish Dis. 2008, 31: 89-96. 10.1111/j.1365-2761.2007.00868.x.
84. Austin B, Rayment JN: Epizootiology of Renibacterium salmoninarum, the causal agent of bacterial kidney disease in salmonid fish. J Fish Dis. 1985, 8: 505-509. 10.1111/j.1365-2761.1985.tb00965.x.
85. Rodger HD, Turnbull T, Muir F, Millar S, Richards RH: Infectious salmon anemia (ISA) in the United Kingdom. Bull Eur Assoc Fish Pathol. 1998, 18: 115-116.
86. Raynard RS, Murray AG, Gregory A: Infectious salmon anaemia virus in wild fish from Scotland. Dis Aquat Organ. 2001, 46: 93-100.

87. Martin PAJ: Current value of historical and ongoing surveillance for disease freedom: Surveillance for bovine Johne's disease in Western Australia. Prev Vet Med. 2008, 84: 291-309. 10.1016/j.prevetmed.2007.12.002.
88. Martin PAJ, Cameron AR, Barfod K, Sergeant ESG, Greiner M: Demonstrating freedom from disease using multiple complex data sources. 2: Case study-Classical swine fever in Denmark. Prev Vet Med. 2007, 79: 98-115. 10.1016/j.prevetmed.2006.09.007.
89. Weinstock MA: A generalised scan statistic test for the detection of clusters. Int J Epidemiol. 1981, 10: 289-293. 10.1093/ije/10.3.289.
90. Clark PJ, Evans FC: Distance to nearest neighbor as a measure of special relationship in populations. Ecology. 1954, 159-163.
91. Carpenter TE: Methods to investigate spatial and temporal clustering in veterinary epidemiology. Prev Vet Med. 2001, 48: 303-320. 10.1016/S0167-5877(00)00199-9.
92. Ward MP, Carpenter TE: Techniques for analysis of disease clustering in space and in time in veterinary epidemiology. Prev Vet Med. 2000, 45: 257-284. 10.1016/S0167-5877(00)00133-1.
93. Heuch PA, Revie CW, Gettinby G: A comparison of epidemiological patterns of salmon lice, Lepeophtheirus salmonis, infections on farmed Atlantic salmon, Salmo salar L., in Norway and Scotland. J Fish Dis. 2003, 26: 539-551. 10.1046/j.1365-2761.2003.00490.x.
94. Revie CW, Hollinger E, Gettinby G, Lees F, Heuch PA: Clustering of parasites within cages on Scottish and Norwegian salmon farms: Alternative sampling strategies illustrated using simulation. Prev Vet Med. 2007, 81: 135-147. 10.1016/j.prevetmed.2007.04.004.
95. Lees F, Gettinby G, Revie CW: Changes in epidemiological patterns of sea lice infestation on farmed Atlantic salmon, Salmo salar L., in Scotland between 1996 and 2006. J Fish Dis. 2008, 31: 259-268. 10.1111/j.1365-2761.2007.00897.x.
96. Longshaw M, Frear PA, Nunn AD, Cowx IG, Feist SW: The influence of parasitism on fish population success. Fish Manag Ecol. 2010, 17: 426-434. 10.1111/j.1365-2400.2010.00741.x.
97. Georgiadis MP, Johnson WO, Gardner IA: Sample size determination for estimation of the accuracy of two conditionally independent tests in the absence of a gold standard. Prev Vet Med. 2005, 71: 1-10. 10.1016/j.prevetmed.2005.04.004.
98. Georgiadis MP, Johnson WO, Gardner IA, Singh R: Correlation-adjusted estimation of sensitivity and specificity of two diagnostic tests. Appl Stat. 2003, 52: 63-76. 10.1111/1467-9876.00389.
99. Nérette P, Stryhn H, Dohoo I, Hammell L: Using pseudogold standards and latent-class analysis in combination to evaluate the accuracy of three diagnostic tests. Prev Vet Med. 2008, 85: 207-225. 10.1016/j.prevetmed.2008.01.011.
100. Nérette P, Dohoo I, Hammell L: Estimation of specificity and sensitivity of three diagnostic tests for infectious salmon anaemia virus in the absence of a gold standard. J Fish Dis. 2005, 28: 89-99. 10.1111/j.1365-2761.2005.00612.x.
101. O.I.E: Manual of Diagnostic Tests for Aquatic Animals. 2009, Paris: Office International des Epizooties, 5
102. Williams CJ, Moffitt CM: Estimation of fish and wildlife disease prevalence from imperfect diagnostic tests on pooled samples with varying pool sizes. Ecol Inform. 2010, 5: 273-280. 10.1016/j.ecoinf.2010.04.003.
103. Keeling MJ, Woolhouse MEJ, May RM, Davies G, Grenfell BT: Modelling vaccination strategies against foot-and-mouth disease. Nature. 2003, 421: 136-142. 10.1038/nature01343.
104. Ortiz-Pelaez A, Pfeiffer DU, Soares-Magalhaes RJ, Guitian FJ: Use of social network analysis to characterize the pattern of animal movements in the initial phases of the 2001 foot and mouth disease (FMD) epidemic in the UK. Prev Vet Med. 2006, 76: 40-55. 10.1016/j.prevetmed.2006.04.007.

105. Picado A, Guitian FJ, Pfeiffer DU: Space-time interaction as an indicator of local spread during the 2001 FMD outbreak in the UK. Prev Vet Med. 2007, 79: 3-19. 10.1016/j.prevetmed.2006.11.009.
106. Paisley LG, Karlsen E, Jarp J, Mo TA: A Monte Carlo simulation model for assessing the risk of introduction of *Gyrodactylus salaris* to the Tana river, Norway. Dis Aquat Organ. 1999, 37: 145-152.
107. Høgåsen HR, Brun E: Risk of inter-river transmission of *Gyrodactylus salaris* by migrating Atlantic salmon smolts, estimated by Monte Carlo simulation. Dis Aquat Organ. 2003, 57: 247-254.
108. Peeler EJ, Gardiner R, Thrush MA: Qualitative risk assessment of routes of transmission of the exotic fish parasite *Gyrodactylus salaris* between river catchments in England and Wales. Prev Vet Med. 2004, 64: 175-189. 10.1016/j.prevetmed.2004.05.005.
109. Anderson RM, May RM: Population biology of infectious diseases: Part 1. Nature. 1979, 280: 361-367. 10.1038/280361a0.
110. Murray AG: A model of the emergence of infectious pancreatic necrosis virus in Scottish salmon farms 1996-2003. Ecol Modell. 2006, 199: 64-72. 10.1016/j.ecolmodel.2006.06.010.
111. Ruane NM, Murray AG, Geoghegan F, Raynard RS: Modelling the spread of Infectious Pancreatic Necrosis Virus (IPNV) in the Irish salmon farming industry: the role of inputs. Ecol Modell. 2009, 220: 1369-1374. 10.1016/j.ecolmodel.2009.02.016.
112. Taylor NGH, Norman RA, Way K, Peeler EJ: Modelling the koi herpesvirus (KHV) epidemic highlights the importance of active surveillance within a national control policy. J Appl Ecol. 2011, 48: 348-355. 10.1111/j.1365-2664.2010.01926.x.
113. Thrush M, Peeler E: Stochastic simulation of live salmonid movement in England and Wales to predict potential spread of exotic pathogens. Dis Aquat Organ. 2006, 72: 115-123.
114. Green DM: A strategic model for epidemic control in aquaculture. Prev Vet Med. 2010, 94: 119-127. 10.1016/j.prevetmed.2009.12.004.
115. Green DM, Gregory A, Munro LA: Small- and large-scale network structure of live fish movements in Scotland. Prev Vet Med. 2009, 91: 261-269. 10.1016/j.prevetmed.2009.05.031.
116. Sharkey KJ, Fernandez C, Morgan KL, Peeler E, Thrush M, Turnbull JF, Bowers RG: Pair-level approximations to the spatio-temporal dynamics of epidemics on asymmetric contact networks. J Math Biol. 2006, 53: 61-85. 10.1007/s00285-006-0377-3.
117. Murray AG, Gillibrand PA: Modelling dispersal of salmon lice in Loch Torridon, Scotland. Mar Pollut Bull. 2006, 53: 128-135. 10.1016/j.marpolbul.2005.09.013.
118. Murray AG, Amundrud TL, Gillibrand PA: Models of hydrodynamic pathogen dispersal affecting Scottish salmon production: modelling shows how Scotland eradicated ISA, but not IPN. Bull Aquacult Assoc Can. 2005, 105: 79-86.
119. Viljugrein H, Staalstrøm A, Molvær J, Urke H, Jansen P: Integration of hydrodynamics into a statistical model on the spread of pancreas disease (PD) in salmon farming. Dis Aquat Organ. 2009, 88: 35-44.
120. Murray AG, Amundrud TL, Gillibrand PA: Models of hydrodynamic pathogen dispersal affecting Scottish salmon production: modelling shows how Scotland eradicated ISA, but not IPN. Bull Aquacult Assoc Can. 2005, 105: 80-87.
121. Keeling MJ, Woolhouse MEJ, Shaw DJ, Mattthews L, Chase-Topping M, Haydon DT, Cornell SJ, Kappey J, Wilesmith JW, Grenfell BT: Dynamics of the 2001 foot and mouth epidemic: stochastic dispersal and a hetergenous landscape. Science. 2001, 294: 812-817. 10.1126/science.1064627.
122. Jansen PA, Matthews L, Toft N: Geographic risk factors for inter-river dispersal of Gyrodactylus salaris in fjord-systems in Norway. Dis Aquat Organ. 2007, 74: 139-149.

123. Vågsholm I, Djupvik HO, Willumsen FV, Tveit AM, Tangen K: Infectious salmon anaemia (ISA) epidemiology in Norway. Prev Vet Med. 1994, 19: 277-290. 10.1016/0167-5877(94)90095-7.
124. Murray AG, Smith RJ, Stagg RM: Shipping and the spread of infectious salmon anemia in Scottish aquaculture. Emerg Infect Dis. 2002, 8: 1-5. 10.3201/eid0801.010144.
125. McClure CA, Hammell KL, Dohoo IR: Risk factors for outbreaks of infectious salmon anemia in farmed Atlantic salmon, Salmo salar. Prev Vet Med. 2005, 72: 263-280. 10.1016/j.prevetmed.2005.07.010.
126. Taylor NGH, Way K, Jeffery KR, Peeler EJ: The role of live fish movements in spreading koi herpesvirus throughout England and Wales. J Fish Dis. 2010, 33: 1005-1007. 10.1111/j.1365-2761.2010.01198.x.
127. Revie CW, Gettinby G, Treasurer JW, Wallace C: Identifying epidemiological factors affecting sea lice *Lepeophtheirus salmonis* abundance on Scottish salmon farms using general linear models. Dis Aquat Organ. 2003, 57: 85-89.
128. Corsin F, Turnbull JF, Hao NV, Mohan CV, Phi TT, Phuoc LH, Tinh NTN, Morgan KL: Risk factors associated with white spot syndrome virus infection in a Vietnamese rice-shrimp farming system. Dis Aquat Organ. 2001, 47: 1-12.
129. St-Hilaire S, Ellis T, Cooke A, North BP, Turnbull JF, Knowles T, Kestin S: Fin erosion on rainbow trout on commercial trout farms in the United Kingdom. Vet Rec. 2006, 159: 446-451. 10.1136/vr.159.14.446.
130. Ersdal C, Midtlyng PJ, Jarp J: An epidemiological study of cataracts in seawater farmed Atlantic salmon Salmo salar. Dis Aquat Organ. 2001, 45: 229-236.
131. Vaagsholm I, Djupvik HO: Risk factors for skin lesions in Atlantic salmon, Salmo salar L. J Fish Dis. 1998, 21: 449-453. 10.1046/j.1365-2761.1998.00123.x.
132. Thrush MA, Murray AG, Brun E, Wallace S, Peeler EJ: The application of risk and disease modelling to emerging freshwater diseases in wild aquatic animals. Freshwat Biol. 2011, 56: 658-675. 10.1111/j.1365-2427.2010.02549.x.
133. Taylor NGH, Sommerville C, Wootten R: The epidemiology of Argulus spp. (Crustacea: Branchiura) infections in stillwater trout fisheries. J Fish Dis. 2006, 29: 193-200. 10.1111/j.1365-2761.2006.00704.x.
134. Thorburn MA: Variables associated with the use of chemotherapeutic agents on trout farms in Ontario, Canada. Prev Vet Med. 1995, 24: 1-14. 10.1016/0167-5877(95)00472-9.
135. Vågsholm I, Djupvik HO: Risk factors for abdominal adhesions in Atlantic salmon, *Salmo salar* L. J Fish Dis. 1999, 22: 53-58. 10.1046/j.1365-2761.1999.00137.x.
136. Vågsholm I, Djupvik HO: Risk factors for spinal deformities in Atlantic salmon, *Salmo salar* L. J Fish Dis. 1998, 21: 47-53. 10.1046/j.1365-2761.1998.00069.x.
137. Wheatley SB, McLoughlin MF, Menzies FD, Goodall EA: Site management factors influencing mortality rates in Atlantic salmon (*Salmo salar* L.) during marine production. Aquaculture. 1995, 136: 195-207. 10.1016/0044-8486(95)01058-0.
138. Aunsmo A, Valle PS, Sandberg M, Midtlyng PJ, Bruheim T: Stochastic modelling of direct costs of pancreas disease (PD) in Norwegian farmed Atlantic salmon (Salmo salar L.). Prev Vet Med. 2010, 93: 233-241. 10.1016/j.prevetmed.2009.10.001.
139. Revie CW, Gettinby G, Treasurer JW, Rae GH, Clark N: Temporal, environmental and management factors influencing the epidemiological patterns of sea lice (*Lepeophtheirus salmonis*) infestations on farmed Atlantic salmon (*Salmo salar*) in Scotland. Pest Manag Sci. 2002, 58: 576-584. 10.1002/ps.476.
140. Revie CW, Gettinby G, Treasurer JW, Rae GH: The epidemiology of the sea lice, *Caligus elongatus* Nordmann, in marine aquaculture of Atlantic salmon, *Salmo salar* L. in Scotland. J Fish Dis. 2002, 25: 391-400. 10.1046/j.1365-2761.2002.00388.x.
141. Zagmutt-Vergara FJ, Carpenter TE, Farver TB, Hedrick RP: Spatial and temporal variations in sea lice (Copeda: Caligidae) infestations of three salmonid species farmed in net pens in southern Chile. Dis Aquat Organ. 2005, 64: 163-173.

142. McKenzie E, Gettinby G, McCart K, Revie CW: Time-series models of sea lice Caligus elongatus (Nordmann) abundance on Atlantic salmon Salmo salar L. in Loch Sunart, Scotland. Aquacult Res. 2004, 35: 764-772. 10.1111/j.1365-2109.2004.01099.x.

143. Tucker CS, Norman R, Shinn AP, Bron JE, Sommerville C, Wooten R: A single cohort time delay model of the life-cycle of the salmon louse *Lepeophtheirus salmonis* on Atlantic salmon *Salmo salar*. Fish Pathol. 2002, 37: 107-118. 10.3147/jsfp.37.107.

144. Pike AW, Wadsworth SL: Sea lice on salmonids: their biology and control. Adv Parasitol. 2000, 44: 231-337.

145. Johnsen BO, Jensen AJ: Infestations of Atlantic salmon, *Salmo salar*, by *Gyrodactylus salaris* in Norwegian rivers. J Fish Biol. 1986, 29: 233-241. 10.1111/j.1095-8649.1986.tb04941.x.

146. Taylor NGH, Wootten R, Sommerville C: The influence of risk factors on the abundance, egg laying habits and impact of Argulus foliaceus in stillwater trout fisheries. J Fish Dis. 2009, 32: 509-519. 10.1111/j.1365-2761.2009.01007.x.

147. Murphy PJ, St-Hilaire S, Bruer S, Corn PS, Peterson CR: Distribution and pathogenicity of Batrachochytrium dendrobatidis in boreal toads from the Grand Teton area of western Wyoming. Ecohealth. 2009, 6: 109-120. 10.1007/s10393-009-0230-4.

148. Patterson KR: Modelling the impact of disease-induced mortality in an exploited population of the fungal parasite Icthyophonus hoferi in the North Sea herring (*Clupea harengus*). Can J Fish Aquat Sci. 1996, 53: 2870-2877.

149. Borsuk ME, Reichert P, Peter A, Schager E, Burkhardt-Holm P: Assessing the decline of brown trout (*Salmo trutta*) in Swiss rivers using a Bayesian probability network. Ecol Model. 2006, 192: 224-244. 10.1016/j.ecolmodel.2005.07.006.

150. Murray AG, O'Callaghan M: The numerical and programming methods used to implement models of the spread and impact of a major epidemic disease: pilchard herpesvirus, Australia 1995 and 1998/1999. Environ Model Software. 2005, 20: 575-585. 10.1016/j.envsoft.2004.03.011.

151. Barber I, Scharsack JP: The three-spined stickleback-*Schistocephalus solidus* system: an experimental model for investigating host-parasite interactions in fish. Parasitology. 2010, 137: 411-424. 10.1017/S0031182009991466.

152. Georgiadis MP, Gardner IA, Hedrick RP: The role of epidemiology in the prevention, diagnosis and control of infectious disease of fish. Prev Vet Med. 2001, 48: 287-302. 10.1016/S0167-5877(00)00202-6.

CHAPTER 8

Multilocus Variable-Number Tandem-Repeat Genotyping of Renibacterium Salmoninarum, a Bacterium Causing Bacterial Kidney Disease in Salmonid Fish

Iveta Matejusova[1*], Nicola Bain[1], Duncan J Colquhoun[2], Edward J Feil[3], Una McCarthy[1], Darry[l] McLennan[1], Michael Snow[4], David Verner-Jeffreys[5], I Stuart Wallace[1], Sarah J Weir[1] and Malcolm Hall[1]

[1]*Marine Scotland Science, Victoria Road, P.O. Box 101, Aberdeen AB11 9DB, UK*

ABSTRACT

Background: Bacterial kidney disease (BKD), caused by *Renibacterium salmoninarum*, is a bacterial disease of fish, which is both geographically widespread and difficult to control. Previously, application of various molecular typing methods has failed to reliably discriminate between *R. salmoninarum* isolates originating from different host species and geographic areas. The current study aimed to utilize multilocus variable number tandem repeats (VNTR) to investigate inter-strain variation of *R. salmoninarum* to establish whether host-specific populations exist in Atlantic salmon and rainbow trout respectively. Such information would be valuable in risk assessment of transmission of *R. salmoninarum* in a multispecies aquaculture environment.

Results: The present analysis utilizing sixteen VNTRs distinguished 17 different haplotypes amongst 41 *R. salmoninarum* isolates originating from Atlantic salmon and rainbow trout in Scotland, Norway and the US. The VNTR typing system revealed two well supported groups of *R. salmoninarum* haplotypes. The first group included *R. salmoninarum* isolates originating from both Atlantic salmon and rainbow trout circulating in Scottish and Norwegian aquaculture, in addition to the type strain ATCC33209^T originating from Chinook salmon in North America. The second group comprised isolates found exclusively in Atlantic salmon, of mainly wild origin, including isolates NCIB1114 and NCIB1116 associated with the original Dee disease in Scotland.

Conclusions: The present study confirmed that VNTR analysis can be successfully applied to discriminate *R. salmoninarum* strains. There was no clear distinction between isolates originating from Atlantic salmon and rainbow trout as several haplotypes in group 1 clustered together *R. salmoninarum* isolates from both species. These findings indicate a potential exchange of pathogens between Atlantic salmon and rainbow trout in Scottish and Norwegian aquaculture during the last 20 years. In a scenario of expansion of rainbow trout farming into the marine environment, appropriate biosecurity measures to minimize disease occurrence are advised. The present results also suggest that *R. salmoninarum* isolates circulating in European aquaculture over the last 20 years are genetically distant to the wild strains originally causing BKD in the rivers Dee and Spey.

Keywords: Bacterial kidney disease VNTR Genotyping *Renibacterium* Salmonids

BACKGROUND

Renibacterium salmoninarum[1] is a Gram-positive bacterium, belonging to the Micrococcus-Arthrobacter subgroup of the actinomycetes [2, 3, 4] and the causative agent of bacterial kidney disease (BKD), a chronic systemic disease of salmonid fish in both marine and freshwater environments [5]. Bacterial kidney disease was first reported in wild Atlantic salmon (*Salmo salar*) in the Rivers Dee and Spey (Scotland, United Kingdom) in 1930 [6, 7] and similar disease signs were reported from North America in 1935 in brook trout (*Salvelinus fontinalis*), brown trout (*Salmo trutta*) and rainbow trout (*Oncorhynchus mykiss*) [8, 9]. *Renibacterium salmoninarum* has an intracellular lifecycle and transmission, both horizontally through contact with infected fish/water or vertically inside fish ova, has been confirmed in many salmonid species [10, 11, 12, 13, 14]. Recent epidemiological studies have identified an association between the spread of BKD and anthropogenic activities [15, 16].

Bacterial kidney disease is geographically widespread and has been reported from most countries where salmonid fish are cultured or naturally occurring. The disease is known to have the potential to cause high mortalities [17, 18] and represents one of the most difficult bacterial diseases of fish to control due to its slow progression and lack of effective treatment. In Scotland, farmed Atlantic salmon and rainbow trout may be infected in both seawater and freshwater environments [19], although the contribution of wild fish to infection transmission is considered low [16].

Sensitive *R. salmoninarum* typing tools are required to improve BKD control through identification of sources of infection and transmission routes. Molecular typing methods, such as direct sequencing of the intergenic spacer (ITS1 and ITS) of ribosomal RNA and randomly amplified polymorphic DNA (RAPD) analysis [20, 21, 22, 23], have shown a limited power to distinguish *R. salmoninarum* strains. Amplification of length polymorphisms in the tRNA intergenic spacer (tDNA-ILPs) has, however, offered improved discriminatory power with some potential for identification of *R. salmoninarum* isolates known to come from the same hatchery [23].

In Scotland, BKD and infection with *R. salmoninarum* are regulated under The Aquatic Animal Health (Scotland) Regulation 2009. From the available farm data, it appears that BKD persists longer on rainbow trout farms [24], compared with Atlantic salmon farms [16, 19]. To date, all typing systems have failed to distinguish between *R. salmoninarum* strains originating from Atlantic salmon and rainbow trout [20, 22, 23], suggesting that individual isolates may represent a risk to both host species. Confirmation of this, applying a more sensitive typing tool, would be beneficial, for example, in a scenario of an expansion of rainbow trout sea water aquaculture. Application of appropriate biosecurity measures could then be applied to minimise risk of pathogen transmission.

In recent years, multilocus variable number tandem repeat analysis, based on amplification of short repetitive DNA sequences, has been found to be a rapid and simple typing technique that enables differentiation of bacterial strains displaying otherwise low genomic variation. The method has been used to discriminate between closely related strains of various human pathogenic microorganisms such as *Clostridium difficile*[25], *Bartonella henselae*[26], or *Streptococcus agalactiae*[27] as well as fish pathogenic species such as *Francisella noatunensis*[28]. The primary purpose of this study was therefore to investigate the genetic variation in *R. salmoninarum* isolated from Atlantic salmon and rainbow trout farms in Scotland using multilocus variable number tandem repeat analysis (VNTR).

Additional samples from other countries were also included in the present study to put any observed variation into context and identify whether the present VNTR typing scheme can distinguish between *R. salmoninarum* collected from different geographic areas.

RESULTS

Characterization of tandem repeat loci

In total, 32 tandem repeat loci were identified using either the Microorganisms Tandem Repeat Database or Tandem Repeats Finder (Additional file 1: Table S1). All loci were successfully amplified in 41 *R. salmoninarum* isolates (Additional file 2: Table S2) and sequences were analyzed for polymorphism (differences in number of tandem repeat units) (Accession numbers KF903677-KF904322). Sixteen of 32 studied loci were polymorphic (Table 1). The 16 monomorphic loci were excluded from the VNTR genotyping scheme.

Table 1 Number of alleles and variation in repeat span per polymorphic locus

Marker locus name*	Number of alleles	Repeat number/span (bp)
BKD23	4	3.7–6.7/33–60
BKD92	2	2.5–5.5/27–63
BKD143	5	9–14/37–57
BKD305	5	2.2–8.2/15–51
BKD396	2	2.6–4.6/16–32
BKD494	2	1.5–2.5/42–72
BKD526	2	1.5–2.5/42–72
BKD584	3	2.1–4.1/48–95
BKD694	3	1.9–2.9/19–27
BKD1023	2	1–2/37–71
BKD1506	3	5–8/34–56
BKD1850	2	4–5/43–55

BKD1935	3	1–3/52–154
BKD2126	8	5.3–11.3/37–79
BKD2770	4	2.8–4.8/26–44
BKD3038	2	2.6–4.6/46–81

*locus name reflects its position within genome of *R. salmoninarum* reference isolate ATCC33209[T] (Accession number NC_010168). Loci in italics represent a minimum combined loci required to sufficiently recognized 17 *R. salmoninarum*haplotypes with the HGDI of value 0.81.

The allelic diversity ranged from two (BKD 92, 396, 494, 526, 1023, 1850 and 3038) to eight different alleles (BKD 2126) per locus. The largest observed variation in allele size was found in locus BKD2126 which varied between five to eleven repeats (Table 1). The VNTR typing system has a discriminatory power value of 0.81 and seventeen different haplotypes of *R. salmoninarum* were distinguished using 16 combined polymorphic VNTRs (Table 1, Table 2). A VNTR typing system relying on only six combined loci (BKD23, BKD305, BKD694, BKD1506, BKD1935, BKD2126) also sufficiently recognized 17 *R. salmoninarum*haplotypes, with the same discriminatory power value of 0.81.

Table 2 *Renibacterium salmoninarum* isolates haplotype identified using multilocus tandem repeat sequencing

Haplotype	Isolate name	Country of origin	Host species	Environment (wild/farmed fish)	Data of isolation
A	MT1470, MT1511[b], MT2119[c], MT2622[c]	Scotland	RT	FW, SW (F)	1994–2002
B	MT452[a], MT839, MT1351, MT1363, MT1880, MT2979, MT3277[a], MT3314, MT3315[b],	Scotland, Norway	AS, RT	FW, SW (F)	1988–2009

	MT3402, N4245, N6642, N6552[d], N6553[d], N6694, N6765, N6863[e], N6864[e]				
C	MT2943, MT3320	Scotland	AS	SW (F)	2005–2008
D	MT3482, MT3483	Scotland	AS, RT	SW (F)	2009
E	N3769, N6695	Norway	AS, RT	FW, SW (F)	1997–2008
F	N5298	Norway	AS	SW (F)	2005
G	MT3106, MT3479, TERV	Scotland	AS, RT	FW, SW (F)	2006–2009
H	MT861	Scotland	AS	FW (F)	1990
I	MT1262	Scotland	AS	FW (F)	1992
J	ATCC33209	N. America	Chinook salmon	SW (F)	1974
K	MT3313	Scotland	RT	FW (F)	2008
L	MT444	Scotland	AS	SW (F)	1988
M	N5223	Norway	AS	SW (F)	2005
N	N6975	Norway	AS	SW (F)	2009
O	NCIMB1116	Scotland	AS	FW (W)	1960
P	NCIMB1114	Scotland	AS	FW (W)	1960
Q	N7443	Norway	AS	FW (W)	1985

[a,b,c,d,e]represent *R. salmoninarum* isolates from different disease outbreaks occurring on the same aquaculture site. RT – rainbow trout, AS – Atlantic salmon, FW – freshwater, SW – seawater, FA – farmed fish, W – wild wish.

Phylogenetic relationships among *R. salmoninarum*isolates inferred from VNTRs

The phylogenetic relationships among the *R. salmoninarum* strains inferred from 16 polymorphic VNTRs are illustrated in Figure 1. Two distinct groups comprising haplotypes A-L (group 1) and M-Q (group 2) were supported with a high bootstrap value (92%). Group 1 comprised *R. salmoninarum* from both Atlantic salmon and rainbow trout farmed in Scotland and Norway, recovered over a period of more than 40 years. This group also includes the type strain of *R. salmoninarum* ATCC33209^{T}, recovered from a Chinook salmon (*Oncorhynchus tshawytscha*) in Oregon (USA) in the early 1970s. Strain B represented the most common haplotype, comprising 18 *R. salmoninarum* isolates from Atlantic salmon and rainbow trout farmed in Scotland and Norway over a period of more than 20 years. Strain B was one of five closely related strains (A, B, C, D, E) differing from each other at a single locus.

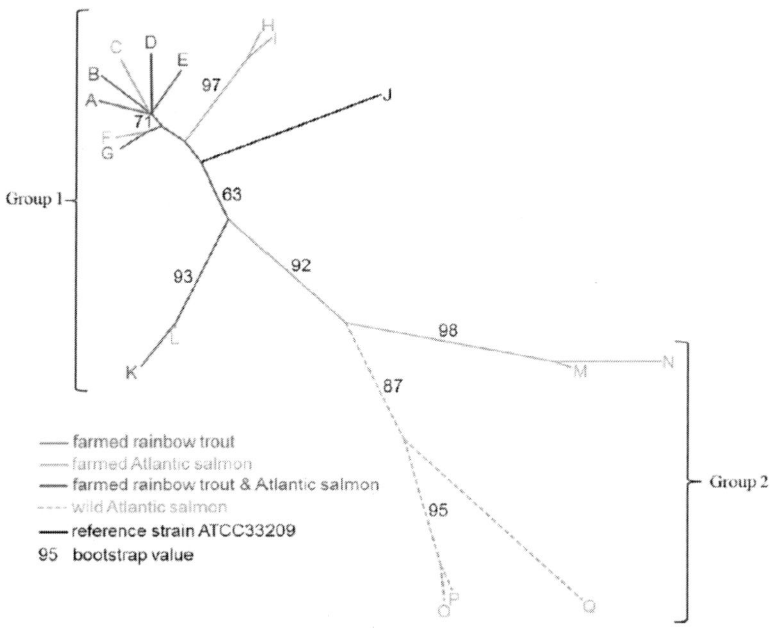

Figure 1 Relationship among the observed haplotypes described in Table2. Group 1 includes haplotypes A to L and Group 2 includes haplotypes M to Q. Bootstrap values indicate the level of support for clusters if higher than 50%.

Group 2 represents *R. salmoninarum* isolates obtained uniquely from Atlantic salmon originating from Scotland and Norway. These isolates differed from group 1 at loci BKD396 and BKD1935. A moderately supported cluster within group 2, comprising strains O-Q, represented isolates exclusively from wild Atlantic salmon, including the Dee disease isolates NCIMB 1114 and 1116 associated with first occurrence of BKD in Scotland.

Similar clustering of *R. salmoninarum* isolates into two main groups was achieved using the eBURST algorithm based on either 16 or 6 polymorphic loci (Figure 2A,B). Using 16 polymorphic loci, a large radial cluster of 7 closely related haplotypes (A-G) was defined. Haplotype B was assigned as the most parsimonious "founder" of this group. Group 2 haplotypes occurred as a single pair O/P representing the Dee disease isolates and three singletons (L,M,N). Using eBURST, a loss in resolving power of the genotyping system was observed when the number of polymorphic loci included was reduced to 6 (Figure 2B).

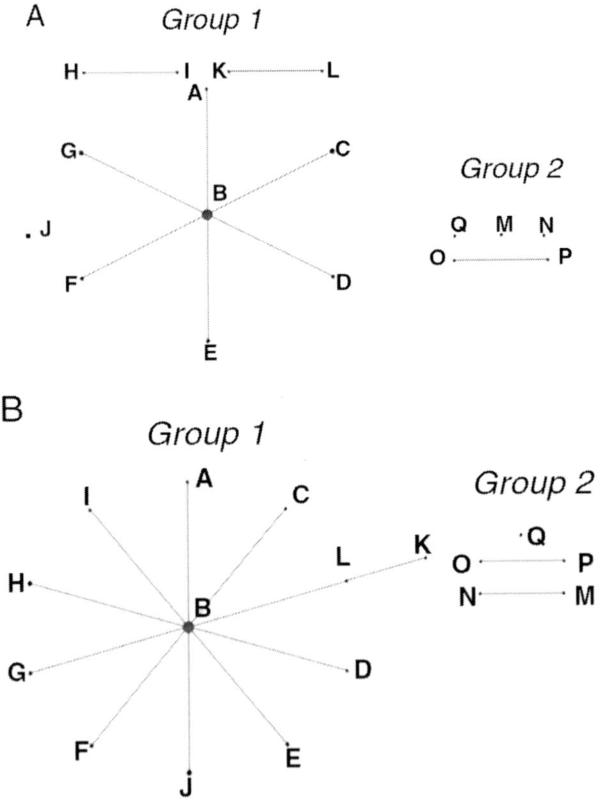

Figure 2 eBURST diagram of *R. salmoninarum* population derived from the allelic variation in (A) 16 polymorphic loci or (B) 6 polymorphic loci.

The present VNTR typing scheme was also applied to investigate whether Scottish *R. salmoninarum* isolates can be distinguished from isolates originating from Norway. Within group 1, some association with country of origin was observed for haplotypes A, C, and G, uniquely obtained from Scottish aquaculture, while haplotype E represented *R. salmoninarum* from Norway. On the contrary, the most common haplotype B contained isolates obtained from aquaculture establishments in both countries.

DISCUSSION

The present study describes development and application of a VNTR typing system for *R. salmoninarum*, a bacterium affecting salmonid aquaculture worldwide and discusses the potential implications for disease management. In comparison to other genotyping methods used to study *R. salmoninarum* such as RAPD, tDNA-ILPs [20, 21, 22, 23], multilocus VNTR typing offers a considerable improvement. Using a combination of sixteen VNTRs, 17 different haplotypes can be identified among 41 *R. salmoninarum* isolates. The discriminatory power of the present combined VNTR scheme was high, characterized by HGDI index of 0.81, indicating that two unrelated isolates will on 81% of occasions fall into different haplotypes. In contrast, no more than nine different tDNA-ILP profiles could be identified from a much larger collection of *R. salmoninarum* isolates [23]. In addition, VNTR represents a more reproducible typing system in comparison to techniques relying on random amplification under low-stringency parameters and accurate data from individual isolates can readily be shared between different laboratories. Although the discriminatory power of VNTR when applied to *R. salmoninarum* is lower than has been achieved with some human pathogenic bacteria such as *Bartonella* or *Streptococcus*[26, 27], these later studies are based on significantly larger data sets usually gathered from wider geographic areas. If a larger *R. salmoninarum* data set becomes available in future, the VNTRs described in the present study should be applied to test its ability to trace disease outbreaks and connect individual infected farms with a source of infection.

The developed VNTR typing system separated the studied isolates into two well-supported groups. Group 1 clustered together 12 out of 17 *R. salmoninarum* haplotypes, including a wide range of isolates from Scotland, Norway and North America, from three different species of salmonid fish, spanning the period between 1974 and 2009. Several haplotypes of group 1 (B, D, E and G) comprised multiple isolates causing disease in both Atlantic salmon and rainbow trout, suggesting a relatively common historical transfer of the pathogen between these fish species. On the other hand, some association was found between rainbow trout and *R. salmoninarum* haplotype A and between Atlantic salmon and *R. salmoninarum* haplotypes C, F, H, I and L-Q. However, with the exception of haplotypes A and C, these haplotypes were represented by single isolations.

The present study concludes that using a data set of 41 isolates representing bacterium circulating in Scotland over a period of more than 20 years, there was no consistent division of *R. salmoninarum* isolates into two host specific populations. This result is consistent with the possibility that individual *R. salmoninarum* strains can infect both host species in environments where both species co-occur. The transfer of *R. salmoninarum* free stock to the marine environment could in theory eliminate disease transmission. However, the possibility that a carrier would be not detected, as a consequence of a potentially low infection prevalence and low diagnostic sensitivity of tests for asymptomatic stock, have to be considered [29]. The spatial separation of marine rainbow trout and Atlantic salmon farms into separate disease management areas in marine environment, as described in [16], can further reduce the risk of pathogen transfer between host species.

All previous *R. salmoninarum* typing systems have failed to reliably discriminate between European and US isolates [20, 22, 23]. This study identified the type strain ATCC33209^T originating from Oregon (US) as a unique strain (haplotype J) and its position within group 1 supports previous studies suggesting intercontinental spread of this pathogen related to anthropogenic activities such as movement of fish during expansion of the rainbow trout industry in Europe [30]. The present genotyping system could, however, clearly separate *R. salmoninarum* strains M-Q (group 2), indicating an origin not associated with ATCC33209^T. The majority of these strains were of wild origin and have not been reported in wild or farmed fish since their original description in the 1930s. The locus BKD1935 was previously described by [22] referred to as the Exact Tandem Repeat A (ETR-A). It was demonstrated that the ETR-A can successfully separate the wild-fish isolates such as NCIMB1114 and NCIMB1116 (tandem repeat 1) from the farmed isolates such as MT452 and MT1363 (tandem repeat 2). Further investigation on a larger data set, focusing on loci BKD396 and BKD1935, which are solely responsible for differentiation between groups 1 and 2, might bring more insight into a relationship between farmed and wild *R. salmoninarum* strains and confirm the origin of *R. salmoninarum* in Scottish aquaculture.

CONCLUSIONS

Cross-species infectivity of *R. salmoninarum* strains also has wider implications for marine ecosystems; including possible transfer of *R. salmoninarum* from farmed to wild fish or *vice versa*. In Scotland, recent studies provided evidence of a relatively low prevalence of *R. salmoninarum* in wild fish captured in close proximity to farms, suggesting that the transmission of this pathogen between wild and farmed fish is limited [16, 31]. However, this scenario might not apply for other regions or countries such as England or Norway [32] and the described VNTR typing system can be utilized to identify and understand farmed and wild fish interactions in terms of *R. salmoninarum* transmission if a larger data set should become available.

METHODS

Preparation of Renibacterium isolates and DNA extraction

Twenty-five *R. salmoninarum* isolates from confirmed disease outbreaks on Scottish farms were selected for this study. Number and selection of Scottish *R. salmoninarum* isolates represents the geographic range, habitat, frequency of disease outbreaks in the salmonid aquaculture sector, supply of fish stock and takes into account difficulties of bacteria culturing from asymptomatic fish and resuscitation of archived material. In addition, 14 Norwegian isolates and two isolates derived from the first successful cultivation of *R. salmoninarum* from the River Dee [7] were included. Isolate details including country of origin, date of isolation, host species and environment are summarized in Additional file 2: Table S2.

For Scottish strains, lyophilised cultures were resuscitated onto Mueller-Hinton L-cysteine agar (MHCA) containing polymyxin-B-sulphate, D-cycloserine, oxolinic acid and cycloheximide and incubated at 15°C for several weeks to allow growth. Suspensions of culture in 0.85% sterile saline were made to a turbidity of McFarland 1, serially diluted to 10^{-4} in sterile saline and 0.1 ml volume of the dilutions spread onto MHCA to achieve single colony purification. The Gram appearance and purity of individual colonies was confirmed before sub-culturing onto fresh MHCA and additional checks for purity and identity (API ZYM, ELISA using the Bios Chile kit) were carried out. At 4–6 weeks, each agar culture was scraped from the plate and suspended in sterile saline before pelleting at 2,400 × g. Genomic DNA was extracted from bacterial cultures using the MagAttract DNA mini M48 kit (Qiagen) and quantified using a ND-1000

Nanodrop Spectrophotometer (NanoDrop Technologies). For Norwegian strains, cryo-preserved isolates (−80°C) were resuscitated on kidney disease (KD) medium [33] followed by KD broth culture to an approximate turbidity of McFarland 1 prior to extraction of genomic DNA using the Gentra Puregene cell kit (Qiagen).

Tandem repeat identification and amplification

The complete genome sequence of *R. salmoninarum* reference strain ATCC33209^T[4] (Accession number NC_010168) was utilized to identify the repetitive DNA sequence regions using the Microorganisms Tandem Repeat Database (http://minisatellites.u-psud.fr) [34] and Tandem Repeats Finder (TRF version 4.03) (http://tandem.bu.edu) [35]. Tandem repeats with at least two repeat units per locus and a repeat unit length of between 4 and 80 bp were selected for further analysis.

Primers for amplification of each locus were designed using OligoPerfect™ Designer (http://tool.invitrogen.com) and their specificity tested using BLAST (blastn) searches. Loci were amplified using the primer pairs listed in Additional file 1: Table S1. Each reaction consisted of 1 × PCR buffer (Bioline), 1.5 mM $MgCl_2$, 200 μM dNTPs, 10 μM of each primer, 1 U BioTaq (Bioline) in a final volume of 20 μl. The cycling conditions were 35 cycles of: 95°C for 1 min, 50 or 55°C (see Additional file 1: Table S1) for 1 min, 72°C for 1 min, followed by a final elongation step of 72°C for 5 min. Amplified products were visualized on a 1% ethidium bromide-stained agarose gel (Invitrogen) and purified using ExoSAP IT or ExoStar 1-Step (GE Healthcare). Approximately 15 ng of purified PCR product was sequenced, utilising the same primers as in the amplification reaction using the GenomeLab DTCS Quick Start kit (Beckman Coulter) and the automated CEQ8800 DNA Sequencer (Beckman Coulter).

Tandem repeat analysis

Each type (size) of repeat, identified by sequencing, at each locus was assigned a unique allele identifier. Data were imported from a Microsoft Office Excel 2003 generated comma-separated-value data file and analysed using version 2.14.0 of the R statistical computing environment [36]. The permutations of alleles across 16 polymorphic loci were used to define distinct haplotypes.

Gross differences between haplotypes (identified using 16 polymorphic loci) were measured with the Hamming distance [37] and used to construct an unrooted neighbor-joining tree using version 1.6-0 of the supplementary R package phangorn [38]. To simplify interpretation of results, haplotypes were named A-Q on the basis of their respective position in the phylogenetic tree. Support for clusters was evaluated using the bootstrap test of phylogeny (1000 repeats) and clusters with values of less than 50% collapsed [39]. The clustering of very closely related haplotypes, defined as those differing at only one locus, was examined using eBURST v 3.0 [40]. Homoplasy and extent of recombination events were investigated using Splits Decomposition, as implemented in Splitstree v 4 [41], by depicting conflicting signals in the data caused by recombination events. The resulting network was consistent with the phylogenetic analysis, and no reticulation was evident, indicating that the evolutionary relationships have not been affected by recombination or homoplasy (data not shown).

The discriminatory power of a typing system was estimated using the Hunter-Gaston discriminatory index HGDI [42]. The index provides a probability that two randomly sampled unrelated isolates will be placed into different typing groups/haplotypes. The minimum number of loci required to distinguish all the strains was determined.

ACKNOWLEDGEMENTS

The authors would like to thank Drs. Sandy G. Murray and David Bruno (Marine Scotland Science, Aberdeen, United Kingdom) for valuable comments which greatly improved the manuscript draft.

REFERENCES

1. Sanders JE, Fryer JL: *Renibacterium salmoninarum* gen. nov., sp. nov., the causative agent of bacterial kidney disease in salmonid fishes. Int Syst Bacteriol. 1980, 30: 496-502. 10.1099/00207713-30-2-496.
2. Gutenberger SK, Giovannoni SJ, Field KG, Fryer JL, Rohovec JS: A phylogenetic comparison of the 16S rRNA sequence of the fish pathogen, *Renibacterium salmoninarum*, to gram-positive bacteria. FEMS Microbiol Lett. 1991, 77: 151-156. 10.1111/j.1574-6968.1991.tb04339.x.
3. Koch CF, Rainey FA, Stackebrandt E: 16S rDNA studies on members of Arthrobacter and Micrococus: and aid for their future taxonomic restructuring. FEMS Microbiol Lett. 1994, 123: 167-172. 10.1111/j.1574-6968.1994.tb07217.x.
4. Wiens GD, Rockey DD, Wu Z, Chang J, Levy R, Crane S, Chen DS, Capri GR, Burnett JR, Sudheesh PS, Shipma MJ, Burd H, Bhattacharyya A, Rhodes LD, Kaul R, Strom MS: Genome sequence of the fish pathogen *Renibacterium salmoninarum* suggests reductive evolution away from an environmental Arthrobacter ancestor. J Bacteriol. 2008, 190: 6970-6982. 10.1128/JB.00721-08.
5. Evelyn TPT: Bacterial kidney disease – BKD. Bacterial Diseases of Fish. Edited by: Inglis V, Roberts RJ, Bromage NR. 1993, Oxford, United Kingdom: Blackwell Scientific Publications, 177-195.

6. Mackie TJ, Arkwright JA, Pryce-Tannatt TE, Mottram JC, Douglas Johnston WD, Menzies WJM, Martin W: Second Interim Report of the Furunculosis Committee. 1933, London: Ministry of Agriculture and Fisheries
7. Smith IW: Freshwater and Salmon Fisheries Research. The Occurrence and Pathology of Dee Disease. 1964, Edinburgh: Her Majesty's Stationery Office, 34, 34
8. Belding DL, Merrill B: A preliminary report upon a hatchery disease of the Salmonidae. Trans Am Fish Soc. 1935, 65: 76-84. 10.1577/1548-8659(1935)65[76:APRUAH]2.0.CO;2.
9. Fryer JL: Bacterial kidney disease of salmonid fish. Ann Rev Microbiol. 1981, 35: 273-298. 10.1146/annurev.mi.35.100181.001421.
10. Mitchum DL, Sherman LE: Transmission of bacterial kidney disease from wild to stocked hatchery trout. Can J Fish Aquat Sci. 1981, 38: 547-551. 10.1139/f81-077.
11. Bruno D, Munro ALS: Observation on *Renibacterium salmoninarum* and the salmonid egg. Dis Aquat Org. 1986, 1: 83-87.
12. Evelyn TPT, Prosperi-Porta L, Ketcheson JE: Experimental intra-ovum infection of salmonid eggs with *Renibacterium salmoninarum* and vertical transmission of the pathogen with such eggs despite their treatment with erythromycin. Dis Aquat Org. 1986, 1: 197-202.
13. Balfry SK, Albright LJ, Evelyn TPT: Horizontal transfer of *Renibacterium salmoninarum* among farmed salmonids via the faecal-oral route. Dis Aquat Org. 1996, 25: 63-69.
14. McKibben CL, Pascho RJ: Shedding of *Renibacterium salmoninarum* by infected chinook salmon *Oncorhynchus tschawytscha*. Dis Aquat Org. 1999, 38: 75-79.
15. Murray AG, Munro LA, Wallace IS, Peeler EJ, Thrush MA: Bacterial kidney disease: an assessment of risk to Atlantic salmon from infection in trout farms and other sources. Scottish Marine Freshwater Sci. 2011, 2 (3): 1-80.
16. Murray AG, Munro LA, Wallace IS, Allan CET, Peeler EJ, Thrush MA: Epidemiology of *Renibacterium salmoninarum* in Scotland and the potential for compartmentalised management of salmon and trout farming areas. Aquaculture. 2012, 324–325: 1-13.
17. Murray CB, Evelyn TPT, Beacham TD, Barner LW, Ketcheson JE, Prosperi-Porta L: Experimental induction of bacterial kidney disease in Chinook salmon by immersion and cohabitation challenges. Dis Aquat Org. 1992, 12: 91-96.
18. Starliper CE, Smith DR, Shatzer T: Virulence of *Renibacterium salmoninarum* to salmonids. J Aquat Anim Health. 1997, 9: 1-7. 10.1577/1548-8667(1997)009<0001:VORSTS>2.3.CO;2.
19. Bruno D: Prevalence and diagnosis of bacterial kidney disease (BKD) in Scotland between 1990 and 2002. Dis Aquat Org. 2004, 59: 125-130.
20. Grayson TH, Cooper LF, Atienzar FA, Knowles MR, Gilpin ML: Molecular differentiation of *Renibacterium salmoninarum* isolates from worldwide locations. Appl Environ Microbiol. 1999, 65: 961-968.
21. Grayson TH, Alexander SM, Cooper LF, Gilpin ML: *Renibacterium salmoninarum* isolates from different sources possess two highly conserved copies of the rRNA operon. A van Leeuw. 2000, 78: 51-61. 10.1023/A:1002745129625.
22. Grayson TH, Atienzar FA, Alexander SM, Cooper LF, Gilpin ML: Molecular diversity of *Renibacterium salmoninarum*isolates determined by randomly amplified polymorphic DNA analysis. Appl Environ Microbiol. 2000, 66: 435-438. 10.1128/AEM.66.1.435-438.2000.
23. Alexander SM, Grayson TH, Chambers EM, Cooper LF, Barker GA, Gilpin ML: Variation in the spacer regions separating rTNA genes in *Renibacterium salmoninarum* distinguishes recent clinical isolates from the same location. J Clin Microbiol. 2001, 39: 119-128. 10.1128/JCM.39.1.119-128.2001.
24. Murray AG, Hall M, Munro LA, Wallace IS: Modelling management strategies for a disease including undetected sub-clinical infection: Bacterial kidney disease in Scottish salmon and trout farms. Epidemics. 2011, 3: 171-182. 10.1016/j.epidem.2011.10.002.

25. Wei HL, Kao CW, Wei SH, Tzen JTC, Chiou CS: Comparison of PCR ribotyping and multilocus variable-number tandem-repeat analysis (MLVA) for improved detection of *Clostridium difficile*. BMC Microbiol. 2011, 11: 217-10.1186/1471-2180-11-217.
26. Monteil M, Durand B, Bouchouicha R, Petit E, Chomel B, Arvand M, Boulouis H-J, Haddad N: Development of discriminatory multiple-locus variable number tandem repeat analysis for *Bartonella henselae*. Microbiol. 2007, 153: 1141-1148. 10.1099/mic.0.2006/001164-0.
27. Haguenoer E, Baty G, Pourcel C, Lartigue M-F, Domelier A-S, Rosenau A, Quentin R, Mereghetti L, Lanotte P: A multi locus variable number of tandem repeat analysis (MLVA) scheme for *Streptococcus agalactiae* genotyping. BMC Microbiol. 2011, 11: 171-10.1186/1471-2180-11-171.
28. Brevik ØJ, Ottem KF, Nylund A: Multiple-locus, variable number of tandem repeat analysis (MLVA) of the fish-pathogen *Francisella noatunensis*. BMC Vet Res. 2011, 7: 5-10.1186/1746-6148-7-5.
29. Hall LM, Wallace IS, Munro LA, Walker A, Murray AG: Epidemiology informs policy regarding surveillance of a notifiable disease of salmonids. Epidemiol et Santé Anim. 2011, 59–60: 392-394.
30. Munro ALS, Waddell IF: Growth of salmon and trout farming in Scotland. Development in Fisheries Research in Scotland. Edited by: Bailey RS, Parrish BB. 1987, England: Fishing News Books Ltd, 246-263.
31. Wallace IS, Munro LA, Kilburn R, Hall M, Black J, Raynard RS, Murray AG: A report on the effectiveness of cage and farm-level fallowing of the control of bacterial kidney disease and sleeping disease on large cage-based trout farms in Scotland. http://www.scotland.gov.uk/Resource/Doc/356407/0120447.pdf,
32. Chambers E, Gardiner R, Peeler EJ: An investigation into the prevalence of *Renibacterium salmoninarum* in farmed rainbow trout, *Oncorhynchus mykiss* (Walbaum), and wild fish populations in selected river catchments in England and Wales between 1998 and 2000. J Fish Dis. 2008, 31: 89-96. 10.1111/j.1365-2761.2007.00868.x.
33. Ordal EJ, Earp BJ: Cultivation and transmission of etiological agent of kidney disease in salmonid fishes. Proc Soc Eptl Biol Med. 1956, 92: 85-88. 10.3181/00379727-92-22392.
34. Denoeud F, Vergnaud G: Identification of polymorphic tandem repeats by direct comparison of genome sequence from different bacterial strains: a web-based resource. BMC Bioinforma. 2004, 5: 4-10.1186/1471-2105-5-4.
35. Benson G: Tandem repeats finder: a program to analyze DNA sequences. Nucleic Acid Res. 1999, 27: 573-580. 10.1093/nar/27.2.573.
36. Development Core Team R: R: Language and Environment for Statistical Computing. 2011, Vienna: R Foundation for Statistical Computing
37. Hamming RW: Error detecting and error correcting codes. Bell Syst Technic J. 1950, 29: 147-160. 10.1002/j.1538-7305.1950.tb00463.x.
38. Schliep KP: Phangorn: phylogenetic anlaysis in R. Bioinformatics. 2011, 27: 592-593. 10.1093/bioinformatics/btq706.
39. Felsenstein J: Confidence limit on phylogenies: an approach using bootstrap. Evolution. 1985, 39: 783-791. 10.2307/2408678.
40. Feil EJ, Bao CL, Aanensen DM, Hanage WP, Spratt BG: eBURST: inferring patterns of evolutionary descent among clusters of related bacterial genotypes from multilocus sequence typing data. J Bacteriol. 2004, 186: 1518-1530. 10.1128/JB.186.5.1518-1530.2004.
41. Huson DH, Bryant D: Application of phylogenetic networks in evolutionary studies. Mol Biol Evol. 2006, 23: 254-267.
42. Hunter PR, Gaston MA: Numerical index of the discriminatory ability of typing systems: an application of Simpsons's index of diversity. J Clin Microbiol. 1988, 26: 2465-2466.

CHAPTER 9

Taxonomy of Bacterial Fish Pathogens

Brian Austin

Institute of Aquaculture, Pathfoot Building, University of Stirling, Stirling FK9 4LA, Scotland, UK

ABSTRACT

Bacterial taxonomy has progressed from reliance on highly artificial culture-dependent techniques involving the study of phenotype (including morphological, biochemical and physiological data) to the modern applications of molecular biology, most recently 16S rRNA gene sequencing, which gives an insight into evolutionary pathways (= phylogenetics). The latter is applicable to culture-independent approaches, and has led directly to the recognition of new uncultured bacterial groups, i.e. "*Candidatus*", which have been associated as the cause of some fish diseases, including rainbow trout summer enteritic syndrome. One immediate benefit is that 16S rRNA gene sequencing has led to increased confidence in the accuracy of names allocated to bacterial pathogens. This is in marked contrast to the previous dominance of phenotyping, and identifications, which have been subsequently challenged in the light of 16S rRNA gene sequencing. To date, there has been some fluidity over the names of bacterial fish pathogens, with some, for example *Vibrio anguillarum*, being divided into two separate entities *(V. anguillarum* and *V. ordalii)*. Others have been combined, for example *V. carchariae, V. harveyi* and *V. trachuri* as *V. harveyi*. Confusion may result with some organisms recognized by more than one name; *V. anguillarum* was reclassified as *Beneckea* and *Listonella,* with *Vibrio* and *Listonella* persisting in the scientific literature. Notwithstanding, modern methods have permitted real progress in the understanding of the taxonomic relationships of many bacterial fish pathogens.

1. INTRODUCTION

"What's in a name?" (William Shakespeare; Romeo and Juliet)

The Swedish botanist Carl Linnaeus (1707-1778), who was also known as Carolus Linnaeus and Carl von Linné, is undoubtedly the Father of Taxonomy, and was responsible for developing a system for naming and ranking living organisms. His lasting contribution was the development of a simplified naming system in Latin with consistency across all living organisms, i.e. the binomial system, in which each organism has a unique two-word name - incorporating genus and species. A simplistic view is that Linnaeus made order out of chaos. Yet, for Linnaeus and his contemporaries, the process was comparatively easy, and involved only large organisms, which were clearly visible to the naked eye (= macro-organisms) and easily seen morphological characteristics (= a category of phenotypic characters). Thus, these early classifications (= the process of arranging organisms into groups) were based on limited but easily visible data, and the outcomes were largely obvious, for example a dog is notably different from a horse and would therefore belong in separate species.

The founding father of microbiology, the Dutch textile merchant and lens maker, Antonie van Leeuwenhoek (1632-1723), observed small organisms initially from the proximity of his teeth (= bacteria and protozoa?), and these entities were termed "animalcules", which he wrote about in a letter to the Royal Society of London in 1676. His careful illustrations suggested morphological variation between the cells. Yet, another two centuries were to pass before serious attempts at naming and ordering bacteria started. Thus, bacterial taxonomy has progressed from the simplistic approach involving a small number of readily observable characteristics, such as morphology as deduced from observation using light microscopes, to the modern applications of molecular biology. With improvements in knowledge, there have been refinements in taxonomic processes and an increase in reliability. It should be remembered that taxonomy (= the theory of classification, nomenclature and identification) is a man-made process, i.e. the organisms included in any classification have not chosen to be placed in the groups that have been created by human beings. Nevertheless if done properly, taxonomy has value in:

- Understanding biodiversity, namely the range of organisms in a given habitat
- Communication between scientists, thus enabling exchange of information about similar organisms
- Cataloguing information - the name is the key to a catalogue of information about the organism

- Enabling identification, such that new isolates may be readily and reliably identified
- Providing an insight into evolutionary pathways (= phylogenetics).
- To be effective, taxonomy should be
- based on a high information content
- reproducible, and
- stable,

otherwise confusion will surely result.

Since the start of bacterial taxonomic processes in the nineteenth century, there has been a progression in the type of information used in the procedure. It may be argued that early bacteriologists had considerable taxonomic insight judging from the conclusions reached from the comparatively simple data that were available. However, taxonomy is a dynamic science, with new developments/methods being incorporated into processes including the descriptions of bacterial species. Since the 1950 s, bacterial taxonomy has evolved rationally, encompassing numerical methods [1, 2], chemotaxonomy (e.g. [3, 4]), and molecular techniques [5]. Taxonomy has progressed from a highly artificial process involving limited amounts of phenotypic data to the recognition of more natural relationships between organisms, based on comparatively large amounts of varied and reliable data covering multiple aspects of the biology of an organism, and including phenotypic, chemotaxonomic, genotypic and phylogenetic date, i.e. a polyphasic approach. However, the current dominance of 16S rRNA gene sequencing although revolutionising some aspects of bacterial classification needs to be treated cautiously as overreliance on the approach may lead to erroneous conclusions [5]. Nevertheless, it is apparent that sequencing methods are instrumental with the explosion of new species names, which have greeted the arrival of the twenty-first century. Whereas, the information content of many of the new species descriptions is generally high, an unwelcome trend is that many new taxa (= taxonomic groups) are described solely after the study of only single strains. Therefore, the diversity/variability within the new taxon cannot be adequately assessed. Also, it is impossible to determine whether a single strain is effectively an outlier or a median representative of the group (in future years, will it be regarded as typical or atypical of the group?). However, taxonomy is often ignored by many microbiologists in other specialisms, and there may well be concern that basic principles could be forgotten, e.g. is the purity and authenticity of cultures always verified before use? Where culturing is not possible, there is the possibility of analyzing the nucleic acids, determining species composition, and even proposing new taxa, i.e. by the use of culture-independent approaches.

2. BACTERIAL FISH PATHOGENS

There has been a steady increase in the numbers of bacterial species associated with fish diseases, with new pathogens regularly recognised in the scientific literature [6]. However, the names of many bacterial fish pathogens have been subjected to taxonomic change over time, with some species split, for example *Vibrio anguillarum* biotype 2 becoming re-classified as a separate species *V. ordalii*[7, 8]. In other cases, different nomenspecies have been combined, for example *V. carchariae, V. harveyi* and *V. trachuri* into *V. harveyi*, which had precedence because it was the first name to be proposed, albeit as the luminous *Achromobacter harveyi*[9, 10, 11]. The oldest known fish pathogen, *V. anguillarum*, has undergone name changes to *Beneckea*[12] and *Listonella*[13]; neither of which was widely accepted. However, *Listonella*remains a valid name and is mentioned in the current edition of Bergey's Manual of Systematic Bacteriology, and *Beneckea* has been consigned to the history books. A positive aspect of sequencing methods is that there has been a progression towards the Orwellian notion of "Order out of Chaos" even if scientists do not always appreciate the significance of the data.

3. ISOLATION OF FISH PATHOGENS: THE CULTURE-DEPENDENT APPROACH

With the rapid development in molecular biology, it is not always necessary to culture an organism in order to enable its study, including the allocation of a species name. Thus, the concept of culture-independent techniques was developed and refined. Sensitivity and specificity increased, but without culturing there was an inability to carry out associated studies, such as the determination of pathogenicity factors. The attraction of culture-dependent approaches is that a pure culture may be obtained and deposited in culture collections as reference material for use by others. This raises a concern about the usefulness of cultures. An assumption is made that pathological material may be used for the recovery of a pure culture of the aetiological agent. This will depend on using appropriate media and incubation conditions, and assumes that the organism is in a form that may be cultured and that the microbiologist picks the "correct" colony. If mixed growth occurs or if the pathogen is largely overgrown by opportunists/secondary invaders/saprophytes, then there is concern that the actual pathogen will be missed. In addition, infections resulting from two or more organisms working synergistically will undoubtedly be mis-diagnosed if the diagnostician chooses only one culture for study. However, there are only a comparatively few indications of disease resulting from multiple species, such as *Aeromonas hydrophila* with *A. salmonicida*[6]. It is speculative how many diagnoses (if any) are made of contaminants rather than the actual pathogen. Moreover, it is surprising that only two species of anaerobic bacteria, namely *Clostridium botulinum* and *Eubacterium tarantellae*, have been implicated as fish pathogens [6]. Of course, this could reflect the general lack of use of appropriate anaerobic procedures by microbiologists rather than the absence of anaerobic pathogens.

4. APPROACHES TO CHARACTERIZATION

4.1 Phenotype

Traditionally, bacteria were characterized phenotypically, and undoubtedly for some groups, e.g. the Enterobacteriaceae, a wealth of knowledge emerged particularly from the 1950's onwards. Currently, emphasis on phenotype has declined with a concomitant move towards molecular-based approaches. Nevertheless, phenotypic data have a role in polyphasic studies whereby many facets of the biology of an organism are studied [14]. Phenotyping leads the way with diagnoses worldwide; emphasis often being placed on commercial kits and the use of manufacturer's probabilistic databases to achieve an acceptable identification. Although the approach has standardized diagnoses, the weakness is that most identification systems have been developed for medically important bacteria that grow within 24-48 h at 35-37°C. Consequently, the reliability of these kits for use with fish pathogens which need lower incubation temperatures for longer periods needs to be questioned [6, 15]. In particular, the API 20E system includes the use of sugar fermentation reactions, which may be influenced by the presence of plasmids [6]. Moreover, there may be confusion over the interpretation of the profiles. For example, some of the profiles of *A. hydrophila* are similar to those of *A. allosaccharophila* and *A. sobria; Yersinia ruckeri* may be confused with *Hafnia alvei;* moreover *Tenacibaculum maritimum* and *Pseudomonas anguilliseptica* are indistinguishable by API 20E [6]. Problems may result when data from rapid commercial kits are used in conjunction with conventional diagnostic schemes, which have been developed for traditional and often laborious phenotypic characters. Also, some of the traditional tests, e.g. the Voges Proskauer reaction, are not noted for their reproducibility and may introduce errors into the taxonomic process and lead to mis-identification [16].

Undoubtedly, selective media have proved useful for the recovery of some fish pathogens, with an example including selective kidney disease medium (SKDM) for *Renibacterium salmoninarum*[17]. However, selective media are only available for a minority of all fish pathogens, therefore recovery is dependent on more general culturing methods. Specially developed diagnostic procedures have aided identification of some group, e.g. the glucose motility deep cultures have benefitted the recovery and identification of *V. anguillarum*[18].

4.2 Immunological methods

The development and availability of standardized immunological (antibodies and kits) reagents have improved diagnoses considerably [6, 19, 20], and enhanced the reliability of methods for the detection of pathogens, including *Mycobacterium* spp., *Photobacterium damselae* subsp. *piscicida*[21], *Piscirickettsia salmonis*[22], *R. salmoninarum*[23] and *Streptococcus iniae*[24]. Tentative diagnoses, including of asymptomatic infections, may result from use of monospecific polyclonal or monoclonal antibodies in a range of antibody-based procedures, including the indirect fluorescent antibody test (iFAT), whole-cell (slide) agglutination, precipitin reactions, complement fixation, immunodiffusion, antibody-coated latex particles, co-agglutination using antibody-coated staphylococcal cells, passive haemagglutination, immuno-India ink technique (Geck) or enzyme linked immunosorbent assay (ELISA; reviewed by [25]), the latter of which may also be used for serology, i.e. detecting antibodies in the host to specific pathogens [26]. Antibody-based methods are used effectively for detecting exposure to fish viruses, such as Koi herpes virus [27], but bacterial pathogens pose a more complicated picture with cross reactivities likely unless specific known molecules/antigens are used to coat the ELISA plates rather than whole pathogens [19]. Techniques are often sensitive, specific, rapid and reliable, and in some cases may be used in the field [6]. This is in marked contrast to molecular biology, which may be much slower and relies on specialist, well equipped laboratories.

4.3 Chemotaxonomy

Chemotaxonomy involves the investigation of chemical constituents of bacteria, and is particularly useful for the study of Gram-positive bacteria. The molecules studied include fatty acids, polar lipids, lipopolysaccharide (nature of the chain length of the fatty acid and the sugar in the Lipid A moiety; [28]), menaquinones, naphthoquinones, ubiquinones, mycolic acids, peptidoglycan, polyamines, teichoic and teichuronic acids and isoprenoid quinones [29]. Mycolic acids, which are useful taxonomic markers, are present in Gram-positive bacteria with high G+C ratios of their DNA [4], and have been reported for a range of fish pathogens, including *Mycobacterium chelonei* subsp. *piscarium*[3] and *M. shottsii*[30]. The length of the mycolate side chain has been correlated to 16S rRNA gene sequence homology [31]. Specific examples for which reliable chemotaxonomic data exist for Gram-positive bacterial fish pathogens are detailed below:

4.3.1 Lactococccus piscium

The long chain cellular fatty acids of *Lactococcus piscium* were reported to be straight chain saturated, mono-unsaturated and cyclopropane-ring types. The major acids corresponded to hexadecanoic acid, Δ 11-octadecanoic acid and Δ 11-methylenoctadecanoic acid [32].

4.3.2 Mycobacterium neoarum

The cell wall chemotype has been given as IVA, with glycolated muramic acids, mycolic acids and MK-9, as the predominant isoprenoid quinone, being present [33].

4.3.3 Nocardia

Nocardia salmonicida contains LL-diaminopimelic acid (DAP) and glycine but not meso-DAP, arabinose or galactose in the cell wall (i.e. Type I). The major cellular fatty acids are hexdecanoic, octadecanoic, octadecanoic and 10-methyloctadecanoic acid [34].

N. seriolae contains *meso*- diaminopimelic acid, arabinose and galactose, indicative of chemotype IVA. Mycolic acids containing 44-58 carbon atoms are present. The cellular fatty acids are dominated by *n*- $C_{16:0}$, *n*-$C_{16:1}$ and *n*-$C_{18:1}$; 10-methyl-$C_{19:0}$ is also present as a major component in some isolates. The predominant isoprenoid quinone is tetrahydrogenated menaquinone with eight isoprene units [34].

4.3.4 Renibacterium salmoninarum

Chemotaxonomic traits of *R. salmoninarum* have been highlighted in part because of the comparative difficulty with obtaining conventional phenotypic test results. Thus, the cell wall peptidoglycan was deduced to contain D-alanine, D-glutamic acid, glycine and lysine as the diamino acids [35]. The principal cell wall sugar was glucose, although arabinose, mannose and rhamnose were also present; mycolic acids were absent [36]. Methyl-branched fatty acids form > 92% of the total fatty acid component of the cells, with 12-methyltetradecanoic (anteiso-C_{15}), 13-methyldecanoic (iso-C_{15}) and 14-methylhexadecanoic (anteiso-C_{17}) as the major components. Straight chain fatty acids generally account for 1% of the total fatty acids, and unsaturated fatty acids are not detected at all. Over 81% of the total fatty acids are composed of the lower melting point anteiso acids, which may contribute to membrane fluidity at low temperatures. Unsaturated menaquinones with nine isoprene units are present. All strains contain diphosphatydylglycerol, 2 major and 6 or 7 minor glycolipids and two unidentified minor phospholipids [37].

4.4 Molecular/genetic methods

Molecular/genetic methods involving 16S rRNA gene sequencing [38], reverse transcriptase-sequencing [39, 40] and polymerase chain reaction (PCR)-based gene sequencing [41] have been useful additions to the armoury of techniques applicable to bacterial taxonomy [29, 31]. DNA hybridization, which is regarded as the "gold standard" for demonstrating the presence or absence of new species, was introduced into bacterial taxonomy during the 1960s (e.g. [42]). Genotypic classification involving sequencing of the 16S and 23S RNA genes (the latter is less popular) is regarded as the definitive standard for determining phylogenetic relationships of bacteria [29, 38]. In particular, the genes are regarded as having all the attributes of useful, relevant and stable biological markers being present and homologous in all bacteria. Also, they are not prone to the effects of gene transfer [29]. Yet, the exact homology values have a profound effect on interpretation of the outputs. Thus, homology values of ≤ 98.7% (97% according to [31]) indicates membership of different species, and this correlates well with DNA hybridization results. Yet, occasionally higher homology values may be attributed to distinct species groupings [43]. By themselves, 16S rRNA gene sequences are insufficient to describe a new species, but may be used indicatively and in conjunction with DNA:DNA hybridization [31]. However, sequencing has permitted the recognition of new variants. For example, sequencing revealed a new variant among Israeli isolates of *Streptococcus iniae*[44]. Moreover, 16S rRNA cataloguing has been useful in providing information about the position of species in existing classifications. Thus, small-subunit rRNA sequencing and DNA:DNA hybridization revealed that *Pasteurella piscicida* was related to *Photobacterium damsela* leading to the proposal that the pathogen be re-classified as *Ph. damsela(e)* subsp. *piscicida*[45]. Furthermore, *R. salmoninarum* was deduced to be a member of the actinomycete subdivision, being related to *Arthrobacter, Brevibacterium, Cellulomonas, Jonesia, Micrococcus, Promicromonospora, Stomatococcus* and *Terrabacter*[46, 47]. The evolutionary relationship of *R. salmoninarum* to *Arthrobacter* was reinforced as the result of genome sequencing, which suggested that the genome of the former had been reduced significantly since its divergence from a common ancestor [48].

Nucleic acid fingerprinting methods, including amplified fragment-length polymorphism PCR (AFLP), pulsed field gel electrophoresis (PFGE), random amplified polymorphic DNA (RAPD), rep-PCR (repetitive element primed PCR), REP-PCR (repetitive extragenic palindromic-PCR), ERIC-PCR (enterobacterial repetitive intergenic consensus sequences-PCR), BOX-PCR (derived from the boxA element) and ribotyping, provide information at or below the subspecies level [31]. Of these, AFLP and ribotyping are extremely useful and standardized.

It is unfortunate that with the increasing use of molecular methods, the description of bacterial groups has been often met with the use of minimal phenotypic data, which causes problems for diagnostics especially in laboratories, which are not equipped for molecular biology [6]. In these situations where distinguishing phenotypic feature have not been or could not be provided then the species should be referred to as a geno[mo]species. Nevertheless, molecular methods have revolutionized taxonomy, and led to the description of an increasing number of new taxa. The methodologies may be culture-independent, allowing for the study of uncultured organisms but there are issues with genomic fluidity [29]. "*Candidatus*" describes uncultured prokaryotes for which phylogenetic relationships have been determined, and authenticity confirmed by methods such as in situ probing [29].

Sequencing of the 16S rDNA is becoming an accepted procedure for the identification of fish pathogens, for example *V. harveyi*[49] and confirming its synonymy with *V. carchariae*[9, 10], and has been instrumental in the recognition of new pathogens, including *Streptococcus dysgalactiae*[50], *S. parauberis* (previously recognised as *S. uberis* genotype II; [51]) and *Vagococcus salmoninarum*[8, 52], and confirmed the presence of *Lactococcus garvieae* in Taiwan [53].

DNA:DNA and RNA:DNA hybridization, 16S RNA cataloguing, and 5S and 16S rRNA sequencing techniques have been used with increasing regularity and success. A review of PCR with emphasis on validation of the techniques and problems with diagnosis has been published [54]. PCR has been used successfully to identify hard-to-isolate fish pathogens, such as *Mycobacterium* spp. in sea bass (*Dicentrarchus labrax)*[55] and *M. chelonei* in a cichlid oscar *(Astronotus ocellatus)*[56]. Moreover, PCR has been useful with distinguishing different species from within the same genus, such as *Lactococcus garvieae* from *L. lactis*[57], from related genera, i.e. *L. garvieae, S. difficilis, S. iniae* and *S. parauberis*[58], and with an admirable level of specificity [59].

The sensitivity of PCR is clearly a positive attribute particularly with slow growing and/or nutritionally fastidious pathogens that are otherwise difficult to study in the laboratory. Of relevance, a PCR was developed

[60, 61], which detected only 22 cells of *R. salmoninarum;* a sensitivity of 10 cells was reported by others [62]. Similarly, PCR detected only 10^2 colony forming units (CFU) of *N. seriolae* in yellowtail [63].

A recent development is multi locus sequence analysis (MLSA), which permits the genotypic examination of micro-organisms by comparison of the sequences of multiple, i.e. 12 or more, house-keeping genes. The benefit of using multiple genes is that the outputs are more informative and less likely to generate results that are distorted by recombination of single loci [29]. The resulting phylogenetic trees are capable of recognizing deeply branching clusters and permit the delineation of groups within a species or genus [64].

5. NEW SPECIES OF FISH PATHOGENS RECOGNIZED BY 16S RRNA SEQUENCING

16S rRNA sequencing has helped the description of fish pathogens where phenotypic characterization alone does not permit their incorporation in classifications. For example, a new disease of Atlantic salmon (*Salmo salar*) was linked to the *Streptobacillus moniliformis* and the fusobacteria group on the basis of sequence homology; biochemical traits did not permit identification [65]. The newly described cause of a mycobacteriosis in Chesapeake Bay (USA) striped bass *(Morone saxatilis)* was equated to a new species, *M. shottsii*, with confirmation by 16S rRNA sequence homology in which the pathogen was linked most closely to *M. marinum* and *M. ulcerans* (similarity = 99.2%) [30]. In one study, *M. gordonae* was identified by 16S rRNA sequencing [66]. Furthermore, phylogenetic analysis based on 16S rRNA gene sequencing together with partial sequences from the 65 kDa heat-shock protein (hsp65) and the beta-subunit of the bacterial RNA polymerase (*rpoB*) genes and the 16S-23S internal transcribed spacer 1 (ITS 1) region named other novel mycobacteria as *M. stomatepiae* and *M. barombii*[67].

During an examination of 16S rRNA sequences, two isolates of motile aeromonads from diseased elvers in Spain were described as a new species, *Aeromonas allosaccharophila*[68], albeit phenotypically heterogeneous [69]. This heterogenicity has caused problems for reliable phenotypic-based diagnoses.

Francisella became recognized as the cause of a new disease of Atlantic cod (*Gadus morhua*) in Norway in which the affected fish displayed white granuloma in the viscera and skin. Isolates were recovered, and determined to possess the key phenotypic characters of *Francisella*, viz. non-motile, strictly aerobic Gram-negative intracellular coccobacilli which produced H_2S from cysteine-containing media [70]. 16S rRNA sequencing revealed a 99.17% homology to *Francisella philomiragia*[71], although a slightly higher value of 99.3% was published [70] with the proposal for a new subspecies, i.e. *Francisella philomiragia* subsp. *noatunensis*, to accommodate the organisms. There was 92.2-99.0% homology with *Francisella philomiragia*housekeeping genes, *groEL, shdA, rpoB, rpoA, pgm* and *atpA*. A DNA:DNA hybridization of 68% was recorded between the fish pathogen and *Francisella philomiragia*[70].

Pasteurella skyensis was recovered from diseased Atlantic salmon in Scotland, linked to the family Pasteurellaceae by phenotypic analysis, and elevated to a new species largely as a result of 16S rRNA sequencing that identified the closest neighbour as *Pasteurella phocoenarum* (homology = 97.1%; [72]).

Piscirickettsia salmonis was named to accommodate isolates from diseased salmon in Chile, of which LF-89 was studied in detail [73] with 16S rRNA conforming to the gamma subdivision of the Proteobacteria with similarities to the family Rickettsiales, and in particular *Wolbachia persica* (similarity = 86.3%) and *Coxiella burnetii* (similarity = 87.5%) more than to representatives of *Ehrlichia, Rickettsia* or *Rochalimaea* leading to the description of a new genus and species [73]. Other rickettsias not conforming exactly with *Piscirickettia salmonis* have been described. For example, an organism recovered from white sea bass (*Atractascion nobilis*) was reported to have a 96.3-98.7% 16S rDNA homology with *Piscirickettsia salmonis*[74], which was considered by the authors to be too low for a confirmed identity. A Tasmanian isolate from Atlantic salmon was distinct from *Piscirickettsia* in terms of sequence alignment of the 16S rRNA, and for the present regarded as a rickettsial-like organism (RLO; [75]).

Pseudomonas plecoglossicida, the causal agent of bacterial ascites of ayu *(Plecoglossus altivelis)*, was described as a new species as a result of 16S rRNA gene sequence analysis confirming distinctiveness from *P. putida* biovar A. DNA:DNA hybridization confirmed the isolates to be a new centre of variation insofar as < 50% homology was recorded with other pseudomonads, including *P. putida*[76].

Streptococcus phocae was recognized as a cause of systemic disease in Atlantic salmon farmed in Chile. Phenotypic testing linked the pathogen with the streptococci, notably *Gemella*, but analysis of 16S rRNA genes provided a link to *S. phocae*[77].

Tenacibaculum soleae was recovered from diseased sole *(Solea senegalensis)* in Spain, and confirmed as a new species largely on account of 16S rRNA homology values of 94.8-96.7% with other members of the genus [78].

Two groups of bacteria were recovered from Atlantic salmon with winter ulcer disease/syndrome [79], of which one cluster was found to be closest to *Moritella marina* (43% re-association by DNA:DNA hybridization), and was named as *V. viscosus*. By 16S rDNA sequencing, the closest match was with *Moritella*[79] and *M. marina* (99.1% sequence homology) so that the organism was re-classified to *Moritella*, but as a new species, as *M. viscosa*[80], despite the high sequence homology [29]. Separately, 19 Icelandic and one Norwegian isolate from shallow skin lesions on Atlantic salmon, and the type strain of *V. marinus*NCIMB 1144 were identified as *V. marinus* after an examination of phenotypic data and analyses by numerical taxonomy [81]. On the basis of 16S rRNA sequencing, the species was transferred to *Moritella* as *M. marina*[82].

5.1 New and uncultured fish pathogens: "*Candidatus*"

Molecular techniques have permitted the recognition of uncultured pathogens belonging to new groupings for which the name of "*Candidatus*" has been used. "*Candidatus* arthromitus" has been recovered from rainbow trout (*Oncorhynchus mykiss*) with summer enteritic syndrome, which is a gastro-enteritis [83, 84]. The organism was observed in histological preparations to which nested polymerase chain reaction was used, with confirmation by sequencing [85]. "*Candidatus* piscichlamydia salmonis" was detected by RT-DGGE in intracellular inclusions, i.e. epitheliocysts, in Atlantic salmon with proliferative gill inflammation [86]. "*Candidatus* clavochlamydia salmonicola" is an intracellular organism, causing epitheliocystis in Atlantic salmon, which was recognized as novel as a result of 16S rRNA sequencing [87].

6. TAXONOMIC DEVELOPMENTS ASSOCIATED WITH SPECIFIC BACTERIAL FISH PATHOGENS

From the early literature, a question-mark has hung over the reliability of some bacterial names insofar as there was often negligible evidence to support the use of those names. Concern may also be expressed about the value of studies based on only single isolates where concern about the reasons for choice of the culture may be aired. Some of the controversy surrounding specific diseases/pathogens follows:

6.1 Motile aeromonas septicaemia

Aeromonas hydrophila (= *A. formicans* and *A. liquefaciens)* would appear to have worldwide distribution and to be a pathogen, causing motile aeromonas septicaemia, of many species of freshwater fish. Indeed, there are reports of a spread into marine fish, notably ulcer disease of cod [88]. Since its initial recognition in the literature, a wealth of knowledge has been accumulated about many facets of its biology (see [6]). A new variant *A. hydrophila* subsp. *dhakenis*, which was originally recovered from children with diarrhoeae in Bangladesh, was determined to be pathogenic to rainbow trout [89]. However overall, there has been some doubt about the role of *A. hydrophila* as a pathogen, and in some cases it may well be present in fish tissue only as a secondary invader [6]. Moreover with developments in the taxonomy of motile aeromonads [90], the accuracy of some of the early published identifications may be justifiably questioned. Could other motile aeromonads be associated with fish disease and may have been confused previously with *A. hydrophila*[89]?

It is clear that there is phenotypic, serological and genotypic heterogeneity within the descriptions of fish pathogenic *A. hydrophila* [e.g. 91, 92], and other motile aeromonads have been implicated as the aetiological agents of (fish) diseases. Thus, 8 isolates reported as pathogenic to eel in Spain were identified by numerical taxonomy as *A. jandaei*[93, 94]. Certainly, the current approach of allocating species names as a result of the examination of 16S rRNA gene sequences has encompassed fish pathogenic motile aeromonads. For example, isolates from diseased fish which were recovered in *Aeromonas* DNA Hybridization Group 2 (= *A. hydrophila)* were equated with a new group, *A. bestiarum*[95]. Subsequently, *A. sobria* (*A. sobria* biovar sobria and *A. veronii* biovar sobria were reported as pathogenic to rainbow trout [89]. Indeed, *A. sobria* has been previously found to have a role as a fish pathogen, with isolates recovered from wild spawning gizzard shad (*Dorosoma cepedianum)* in Maryland, USA during 1987 [15, 96]. Also, *A. veronii* has been recovered from Siberian sturgeon (*Acipenser baerii*) with identification of the pathogen resulting from phenotyping and 16S rRNA gene sequencing [97].

6.2 Aeromonas salmonicida

Aeromonas salmonicida is one of the oldest described fish pathogens, being isolated initially from diseased hatchery-maintained brown trout (*Salmo trutta*) in Germany, and named as "Bacillus der Forellenseuche" or bacillus of trout contagious disease. The history of the organism reveals a plethora of synonyms including *Bacillus devorans*, *Bacterium salmonica*, *Bacterium salmonicida*, *Bacillus truttae* and *Bacillus salmonicida*[6]. The 7th edition of *Bergey's Manual of Determinative Bacteriology* (1957) placed the pathogen in the genus *Aeromonas* within the family Pseudomonadaceae [98]. Later, there was a transfer to the family Vibrionaceae and subsequently to its own family, i.e. the Aeromonadaceae [99]. Re-classification was based primarily on phenotyping [100]. Thus in 1953, the first detailed description of the pathogen was published, and from an examination of 10 isolates, it was concluded that *Bacterium salmonicida* was homogeneous in cultural and biochemical characteristics [100]. Numerous studies have addressed the homogeneity of the species (e.g. [101]). The basic description is of an organism, which comprises non-motile (motility and *flaA* and *flaB* flagellar genes have been reported [52, 82]), fermentative, Gram-negative rods, which produce a brown water-soluble pigment on tryptone-containing agar, do not grow at 37°C, and produce catalase and oxidase [6]. Cultures have the ability to dissociate into rough, smooth and G-phase (= intermediate) colonies [102]. The pathogen has spread from its dominance in salmonids to cyprinids and marine flatfish [6]. An ongoing issue surrounds the intraspecies structure, i.e. the validity of subspecies *achromogenes, masoucida, pectinolytica, salmonicida* and *smithia*, and the status of so-called atypical isolates.

A. salmonicida subsp. *salmonicida*, isolates of which have been obtained almost exclusively from outbreaks of furunculosis in salmonids, is regarded as homogeneous, and is referred to as "typical" [103]; all other isolates are considered as heterogeneous and "atypical" [6]. So called atypical strains may demonstrate weak, slow or non-pigment production [104, 105], catalase [106] or oxidase-negativity [e.g. 106, 107], nutritional fastidiousness for blood products [108], slow growth, i.e. ≥ 5 days compared with 1-2 days for typical isolates [106, 108], and be pathogenic for fish other than salmonids, e.g. cyprinids (e.g. [106, 108, 109]) and marine flatfish, namely dab (*Limanda limanda*), plaice (*Pleuronectes platessa*), flounder (*Platichthys flesus*) and turbot (*Scophthalmus maximus*) [110, 111, 112], and cause ulceration rather than furunculosis [6, 113]. The deviation in characteristics from the typical to atypical isolates has made typing difficult [114, 115, 116]. Even 16S rDNA sequencing has not helped with the clustering of atypical forms (e.g. [117]).

Smith [118] recognized heterogeneity in the species description of *A. salmonicida*. She examined six isolates of non-pigmented *A. salmonicida*, which were clustered as Group I in her numerical taxonomy study, for which a separate new species name was proposed, i.e. *A. achromogenes*, but the proposal was not adopted widely. A second non-pigmented group was described by Kimura [119], and named as *A. salmonicida* subsp. *masoucida*. Schubert [120] considered these non-pigmented isolates as subspecies, and coined the names of *A. salmonicida* subsp. *achromogenes* and *A. salmonicida* subsp. *masoucida*, respectively. Pigmented strains (= typical) were classified as *A. salmonicida* subsp. *salmonicida*[120]. The precise relationship of the subspecies has been the subject of detailed discussion. In particular, it was contended that subsp. *achromogenes* and *masoucida* were more closely related to *A. hydrophila* than to *A. salmonicida*[121]. Later, it was mooted that subsp. *masoucida* bridged typical *A. salmonicida* and *A. hydrophila*[122]. Yet, *A. salmonicida* subsp. *masoucida* is non-motile, sensitive to *A. salmonicida* bacteriophages, possesses an antigenic profile specific to *A. salmonicida*, and shares a DNA homology of 103% with *A. salmonicida*[92]. By PCR, a combination of *achromogenes* with *masoucida* could be justified, but this was not substantiated by ribotyping and RAPD analyses [114]. Phenotypic data suggest a case for combining subsp. *masoucida* with *salmonicida*, and subsp. *achromogenes* with *Haemophilus piscium*, which is the causal agent of ulcer disease of trout [123]. Examination of the small subunit rRNA gene sequences revealed 99.9% homology of an authentic strain of *H. piscium* with *A. salmonicida* subsp. *salmonicida*[124]. So far, the comparative uniqueness of subsp. *smithia* has been indicated from several studies (e.g. [114]). The complication is with aberrant strains of *A. salmonicida* from fish species other than salmonids.

DNA homology was used to reveal that all isolates of *A. salmonicida* (including *A. salmonicida* subsp. *masoucida)* were highly related, i.e. 96-106% homology, when hybridized against *A. salmonicida* subsp. *salmonicida*[92]. It was opined that *A. salmonicida* subsp. *masoucida* and some atypical isolates did not warrant separate subspecies status, because they were regarded as variants of other well-recognized groups. Also as a result of genotypic analyses, it was reported that typical and atypical isolates of *A. salmonicida* were very closely related, with minimal divergence [125]. Using DNA:DNA re-association, it was concluded that typical *A. salmonicida* were recovered in a homogeneous group, whereas the atypical representatives were more diverse [126]. From numerical taxonomy and DNA:DNA hybridization, similar conclusions resulted regarding the homogeneity of typical isolates of *A. salmonicida*[101]. However using 16S rRNA sequencing techniques, it was reported that subspecies *achromogenes* and *masoucida* were indistinguishable, and only differed from subspecies *salmonicida* by two bases [68].

The relation of *A. salmonicida* to other aeromonads has been discussed. Eddy [127] focused on the inability of *A. salmonicida* to produce 2,3-butanediol from glucose, and the absence of motility, which were in contrast to the genus description [128]. A new genus, i.e. *Necromonas,* was proposed with two species, namely *N. salmonicida* for the typical isolates and *N. achromogenes* for the non-pigmented strains [118]. This proposal was not formally widely accepted, although Cowan [129] used the suggestion in his landmark identification scheme for medically important bacteria. Subsequent serological and bacteriophage sensitivity data supported the relationship between *A. salmonicida* and the motile aeromonads. Common antigens between *A. hydrophila* and *A. salmonicida* subsp. *masoucida* and other isolates of *A. salmonicida*were reported [122, 130]. Furthermore, serological cross-reactions between *A. salmonicida* and motile aeromonads were discussed [131]. Moreover, *A. hydrophila* cultures were found to be sensitive to *A. salmonicida* bacteriophages [132, 133]. The outcome of all the studies is that DNA homology supports the classification of *A. salmonicida* in the genus *Aeromonas* (e.g. [92, 122, 126]).

There are certainly outstanding questions about the validity and taxonomic placing of *Haemophilus piscium*[123], but an authentic reference strain was not deposited any in any recognized culture collection at the time of its first isolation. Later, it was concluded that the organism was not a *bona fide Haemophilus*because of the lack of requirement for haemin or NAD [134]. In particular *H. piscium* differed from the type species of the genus, *H. influenzae*, in the inability to reduce nitrate or alkaline phosphatase and to grow at 37°C, in conjunction with a higher G+C ratio of the DNA. It was commented that there was only a low similarity between *H. piscium* and other *Haemophilus* spp. in a numerical taxonomic study [135]. A link with atypical, achromogenic *A. salmonicida* was made [122]. This link was reinforced by bacteriophage sensitivity, when it was concluded that *H. piscium* is an atypical *A. salmonicida*[136]. Other workers have supported this view (e.g. [114]). However with the absence of an authentic, original type strain, the definitive taxonomic position of *H. piscium* is only speculative.

A lack of congruence has been reported between the results of molecular (PCR, RAPD and ribotyping) and phenotypic methods in taxonomy of aeromonads [114]. Moreover, there are problems of inter-laboratory differences and lack of standardisation in test methods [137]. The outcome is that the definitive classification of *A. salmonicida* has not been achieved, to date.

6.3 Enteric redmouth (ERM)

There has been discussion about the taxonomic position of the aetiological agent of ERM. Strong agglutination with *Salmonella enterica* subsp *arizonae* O group 26, and a weak reaction with O group 29 was reported [138]. In addition, biochemical similarities with enterics, notably *Enterobacter liquefaciens, Serratia marcescens* subsp. *kiliensis* as well as *Salmonella enterica* subsp. *arizonae* were mentioned [138]. Serological cross-reactions were also recorded with *Hafnia alvei*[139]. Nevertheless, a new species, i.e. *Yersinia ruckeri* was described although there was only a 30-31% DNA homology with *Y. enterocolitica* and *Y. pseudotuberculosis* [140]. This compares to DNA homologies of 24-28% and 31% with *Serratia marcescens*and *Serratia liquefaciens*, respectively [141]. Indeed, it has been suggested that the causal agent of ERM should belong in a new genus of the Enterobacteriaceae [142]. A complication developed when a new non-motile form of the pathogen was recovered from rainbow trout. By 16S rRNA sequencing and a homology of 100%, the organisms were linked to *Y. ruckeri* but regarded as a new biogroup [143]. Similar non-motile variants were also recovered from previously vaccinated rainbow trout in Spain [144].

6.4 Vibriosis

The causal agent of "red-pest" in eels was first designated as *Bacterium anguillarum*[6]. Subsequently, an outbreak among eels in Sweden led to the use of the name *Vibrio anguillarum*. Numerous studies have pointed to heterogeneity in *V. anguillarum* initially with the delineation of two sub-groupings/biotypes (e.g. [12]). This increased to 3 [145] and then 4 sub-groups/phena within the species definition [146, 147]. Ribotyping has confirmed the heterogeneity [148], although a single taxon, homogeneous by ribotyping but heterogeneous by LPS profiles, plasmid composition, serogrouping, and BIOLOG-GN fingerprints and API 20E profiles was described [149, 150]. Biotype II became recognized as a separate species, i.e. *V. ordalii*[151], which is homogeneous by plasmid profiling, ribotyping and serogrouping, accommodates two LPS groups, but is heterogeneous by BIOLOG-GN fingerprints and API 20E profiles [150].

Serology has been widely used for diagnosis, but has complicated the understanding of *V. anguillarum*[151], and the establishment of serotypes has to some extent traversed species boundaries. With *V. anguillarum*, serogroup 02 and 05, there are common antigens with *V. ordalii*[152] and *V. harveyi*[149], respectively. Initially, three serotypes were recognised for isolates from salmonids from the northwest USA, Europe, and the Pacific-northwest (USA) [103]. This number increased to 6 [128], and then 10 [153] and upwards [148, 154]. Serogroup 01 dominates the number of isolates available for study and the relative importance to fish pathology [149, 155, 156, 157]. Serogroup 02 has been further subdivided into serogroup 02a and 02b [158].

V. anguillarum was re-classified initially to *Beneckea*[12] and then to a newly proposed genus *Listonella*[13], but the changes were not widely accepted.

7. THE ROLE OF PHYLOGENETICS IN BACTERIAL FISH PATHOLOGY

The techniques described above are relevant for the taxonomy of bacterial fish pathogens. Yet, molecular methods, namely sequencing of the 16S rRNA gene, permit the study of evolutionary relationships, i.e. phylogenetics, which may be viewed as phylogenetic trees, which are interpreted by cladistics and used in defining taxa. The approach is essential in the study of the evolutionary tree of life, but is it strictly necessary for fish pathology and the recognition of species? One concern is the comparative fluidity by which genes may be exchanged, such as by horizontal gene transfer, and the impact of this movement on the outcome of the taxonomic/phylogenetic process.

8. CONCLUSIONS

There has been a resurgence of interest in bacterial taxonomy partially because of the current focus on biodiversity and the development of reliable molecular methods [159], notably 16S rRNA sequencing. Undoubtedly, these molecular approaches have led to greater confidence and accuracy in the reporting of bacterial names. Nevertheless, it is conceded that bacterial taxonomy is a specialist subject, which is not of interest to all fish pathologists. However, it cannot be overstated that there is a real value for good taxonomy as a means of communication. In terms of fish pathology, taxonomy enables the recognition of new pathogens, improvements in the understanding of relationships between taxa, an appreciation of variation within existing nomenspecies including the recognition of new subspecies and biogroups, and facilitates accurate commentary about all aspects of biology from epizootiology to pathogenicity (e.g. [160]), although the position of so-called atypical isolates in taxonomic hierarchies is often difficult to determine [161].

For the future, a range of new techniques, including in situ hybridization, probe hybridization, microarray techniques and restriction enzyme digestion, are entering taxonomic use, and are likely to be used in fish pathology. The impact of these new approaches is difficult to predict, but will undoubtedly be incorporated in some fish bacteriology laboratories.

REFERENCES

1. Sneath PHA: The application of computers to taxonomy. J Gen Microbiol. 1957, 17: 201-226.
2. Sokal RR, Michener CD: Numerical Taxonomy. 1958, WH Freeman, London
3. Arakawa CK, Fryer JL: Isolation and characterization of a new subspecies of *Mycobacterium chelonei* infections for salmonid fish. Helgoländer Meeresuntersuch. 1984, 37: 329-342.
4. Brennan PJ: *Mycobacterium* and other actinomycetes. Microbial Lipids. Edited by: Ratledge C, Wilkinson SG. 1988, Academic Press, London, 1: 203-298.
5. Kämpfer P: The characterization of prokaryote strains for taxonomic purposes. Int J Syst Evol Microbiol. 2010, 60: 7-10.1099/ijs.0.020768-0.
6. Austin B, Austin DA: Bacterial Fish Pathogens: Disease in Farmed and Wild Fish. 2007, Springer-Praxis, Goldalming, 4
7. Schiewe MH, Trust TJ, Crosa JH: *Vibrio ordalii* sp. nov.: a causative agent of vibriosis in fish. Curr Microbiol. 1981, 6: 343-348. 10.1007/BF01567009.
8. Schmidtke LM, Carson J: Characteristics of *Vagococcus salmoninarum* isolated from diseased salmonid fish. J Appl Bacteriol. 1994, 77: 229-236.
9. Gauger E, Gómez-Chiarri M: 16S ribosomal DNA sequencing confirms the synonymy of *Vibrio harveyi* and *V. carchariae*. Dis Aquat Org. 2002, 52: 39-46. 10.3354/dao052039.
10. Pedersen K, Verdonck L, Austin B, Austin DA, Blanch AR, Grimont PAD, Jofre J, Koblavi S, Larsen JL, Tiainen T, Vigneulle M, Swings J: Taxonomic evidence that *Vibrio carchariae* Grimes et al. 1985 is a junior synonym of *Vibrio harveyi*(Johnson and Shunk 1936) Baumann et al. 1981. Int J Syst Bacteriol. 1998, 48: 749-758. 10.1099/00207713-48-3-749.
11. Thompson FL, Hoste B, Vandemeulebroecke K, Engelbeen K, Denys R, Swings J: *Vibrio trachuri* Iwamoto et al., 1995 is a junior synonym of *Vibrio harveyi* (Johnson and Shunk 1936) Baumann et al. 1981. Int J Syst Evol Microbiol. 2002, 52: 973-976. 10.1099/ijs.0.02046-0.
12. Baumann P, Bang SS, Baumann L: Phenotypic characterization of *Beneckea anguillara* biotypes I and II. Curr Microbiol. 1978, 1: 85-88. 10.1007/BF02605421.
13. MacDonell MT, Colwell RR: Phylogeny of the Vibrionaceae and recommendations for two new genera, *Listonella* and *Shewanella*. Syst Appl Microbiol. 1985, 6: 171-182.
14. Vandamme P, Pot B, Gillis M, de Vos P, Kersters K, Swings J: Polyphasic taxonomy, a consensus approach to bacterial systematics. Microbiol Rev. 1996, 60: 407-438.
15. Toranzo AE, Santos Y, Nieto TP, Barja JL: Evaluation of different assay systems for identification of environmental *Aeromonas* strains. Appl Environ Microbiol. 1986, 51: 652-656.
16. Sneath PHA, Johnson R: The influence of numerical taxonomic similarities of errors in microbiological tests. J Gen Microbiol. 1972, 72: 377-392.
17. Austin B, Embley TM, Goodfellow M: Selective isolation of *Renibacterium salmoninarum*. FEMS Microbiol Lett. 1983, 17: 111-114. 10.1111/j.1574-6968.1983.tb00383.x.
18. Walters GR, Plumb JA: Modified oxidation/fermentation medium for use in the identification of bacterial fish pathogens. J Fish Res Bd Can. 1978, 35: 1629-1630. 10.1139/f78-253.
19. Adams A, Thompson KD: Recent applications of biotechnology to novel diagnostics for aquatic animals. Rev Sci Tech. 2008, 27: 197-209.
20. Goerlich R, Schlüsener HJ, Lehmann J, Greuel E: The application of monoclonal antibodies to diagnosis of *Aeromonas salmonicida* infections in fishes. Bull Europ Assoc Fish Pathol. 1984, 4: 66-
21. Jung TS, Thompson KD, Adams A, Morris DJ, Snedden K: The production of monoclonal antibodies against *Photobacterium damselae* subsp. *piscicida* and their

application in immunohistochemistry. J Fish Dis. 2001, 24: 64-67. 10.1046/j.1365-2761.2001.00260.x.

22. Steiropoulos NA, Yuksel SA, Thompson KD, Adams A, Ferguson HW: Detection of *Rickettsia*-like organisms (RLOs) in European sea bass *(Dicentrarchus labrax*, L) by immunohistochemistry, using rabbit anti-*Rickettsia salmonis* serum. Bull Eur Assoc Fish Pathol. 2002, 22: 428-432.
23. Adams A, Thompson KD, Morris D, Farias C, Chen SC: Development and use of monoclonal antibody probes for immunohistochemistry, ELISA and IFAT to detect bacterial and parasitic fish pathogens. Fish Shellfish Immunol. 1995, 5: 537-547. 10.1016/S1050-4648(95)80040-9.
24. Klesius P, Evans J, Shoemaker C, Yeh H, Goodwin AE, Adams A, Thompson KD: Rapid detection and identification of *Streptococcus iniae* using a monoclonal antibody-based indirect fluorescent antibody technique. Aquaculture. 2005, 258: 180-186. 10.1016/j.aquaculture.2005.06.040.
25. Adams A: Advances in disease diagnosis, vaccine development and other emerging methods to control pathogens. Edited by: Burnell G, Allan G. 2009, New Technologies in Aquaculture, Improving Production, Efficiency, Quality and Environmental Management, Woodhead, England, Chapter 7: 197-211.
26. Saeed MO, Plumb JA: Serological detection of *Edwardsiella ictaluri* Hawke lipopolysaccharide antibody in serum of channel catfish *Ictalurus punctatus* Rafinesque. J Fish Dis. 1987, 10: 205-209. 10.1111/j.1365-2761.1987.tb01062.x.
27. Adkinson MA, Gilad O, Hedricks RP: An enzyme linked immunosorbent assay (ELISA) for detection of antibodies to the Koi herpes virus (KHV) in the serum of Koi *Cyprinus carpio*. Fish Pathol. 2005, 40: 53-62. 10.3147/jsfp.40.53.
28. Mayer H, Masoud H, Urbanik-Sypniewska T, Weckesser J: Lipid A composition and phylogeny of Gram-negative bacteria. Bull Jap Fed Cult Collect. 1989, 5: 19-25.
29. Scheifer KH: Classification of bacteria and Archaea: past, present and future. Syst Appl Microbiol. 2009, 32: 533-542. 10.1016/j.syapm.2009.09.002.
30. Rhodes MW, Kator H, Kotob S, van Berkum P, Kaattari I, Vogelbein W, Quinn F, Floyd MM, Butler WR, Ottinger CA: *Mycobacterium shottsii* sp nov., a slowly growing species isolated from Chesapeake Bay striped bass (*Morone saxatilis)*. Int J Syst Evol Microbiol. 2003, 53: 421-424. 10.1099/ijs.0.02299-0.
31. Tindall BJ, Rosselló-Móra R, Busse HJ, Ludwig W, Kämpfer P: Notes on the characterization of prokaryote strains for taxonomic purposes. Int J Syst Evol Microbiol. 2010, 60: 249-266. 10.1099/ijs.0.016949-0.
32. Williams AM, Fryer JL, Collins MD: *Lactococcus piscium* sp. nov. a new *Lactococcus* species from salmonid fish. FEMS Microbiol Lett. 1990, 68: 109-114. 10. 1111/j.1574-6968.1990.tb04132.x.
33. Backman S, Ferguson HW, Prescott JF, Wilcock BP: Progressive panophthalmitis in chinook salmon, *Oncorhynchus tshawytscha* (Walbaum): a case report. J Fish Dis. 1990, 13: 345-353. 10.1111/j.1365-2761.1990.tb00793.x.
34. Isik K, Chun J, Hah YC, Goodfellow M: *Nocardia salmonicida* nom. rev., a fish pathogen. Int J Syst Bacteriol. 1999, 49: 833-837. 10.1099/00207713-49-2-833.
35. Fiedler F, Draxl R: Biochemical and immunochemical properties of the cell surface of *Renibacterium salmoninarum*. J Bacteriol. 1986, 168: 799-804.
36. Sanders JE, Fryer JL: *Renibacterium salmoninarum* gen. nov., sp. nov., the causative agent of bacterial kidney disease in salmonid fishes. Int J Syst Bacteriol. 1980, 30: 496-502. 10.1099/00207713-30-2-496.
37. Embley TM, Goodfellow M, Minnikin DE, Austin B: Fatty acid, isoprenoid quinone and polar lipid composition in the classification of *Renibacterium salmoninarum*. J Appl Bacteriol. 1983, 55: 31-37.
38. Fox GE, Pechman KR, Woese CR: Comparative cataloging of 16S ribosomal ribonucleic acid: molecular approach to procaryotic systematics. Int J Syst Bacteriol. 1977, 27: 44-57. 10.1099/00207713-27-1-44.

39. Lane DJ, Field KG, Olsen GJ, Pace NR: Reverse transcriptase sequencing of ribosomal RNA for phylogenetic analysis. Meth Enzymol. 1988, 167: 138-144. full_text.
40. Sanger F, Nicklen S, Coulson AR: DNA sequencing with chain-terminating inhibitors. Proc Natl Acad Sci USA. 1977, 74: 5463-5467. 10.1073/pnas.74.12.5463.
41. Saiki RK, Gelfand DH, Stoffel S, Scharf SJ, Higuchi R, Horn GT, Mullis KB, Erlich HA: Primer-directed enzymatic amplification of DNA with a thermostable DNA polymerase. Science. 1988, 239: 487-491. 10.1126/science.2448875.
42. De Ley J, Park IW, Tijtgat R, van Ermengem J: DNA homology and taxonomy of *Pseudomonas* and *Xanthomonas*. J Gen Microbiol. 1966, 42: 43-56.
43. Fox GE, Wisotzkey JD, Jurtshuk P: How close is close: 16S rRNA sequence identity may not be sufficient to guarantee species identity. Int J Syst Bacteriol. 1992, 42: 166-170. 10.1099/00207713-42-1-166.
44. Kvitt H, Colorni A: Strain variation and geographic endemism in *Streptococcus iniae*. Dis Aquat Org. 2004, 61: 67-73. 10.3354/dao061067.
45. Gauthier G, Lafay B, Ruimy R, Breittmayer V, Nicolas JL, Gauthier M, Christen R: Small-subunit rRNA sequences and whole DNA relatedness concur for the re-assignment of *Pasteurella piscicida* (Snieszko et al.) Janssen and Surgalla to the genus *Photobacterium* as *Photobacterium damsela* subsp. *piscicida* comb. nov. Int J Syst Bacteriol. 1995, 45: 139-144. 10.1099/00207713-45-1-139.
46. Gutenberger SK, Giovannoni SJ, Field KG, Fryer JL, Rohovec JS: A phylogenetic comparison of the 16S rRNA sequence of the fish pathogen, *Renibacterium salmoninarum*, to Gram-positive bacteria. FEMS Microbiol Lett. 1991, 77: 151-156. 10.1111/j.1574-6968.1991.tb04339.x.
47. Stackebrandt E, Wehmeyer U, Nader H, Fiedler F: Phylogenetic relationship of the fish pathogenic *Renibacterium salmoninarum* to *Arthrobacter, Micrococcus* and related taxa. FEMS Microbiol Lett. 1988, 50: 117-120. 10.1111/j.1574-6968.1988.tb02922.x.
48. Wiens GD, Rockey DD, Wu ZN, Chang J, Levy R, Crane S, Chen DS, Capri GR, Burnett JR, Sudheesh PS, Schipma MJ, Burd H, Bhattacharyya A, Rhodes LD, Kaul R, Strom MS: Genome sequence of the fish pathogen *Renibacterium salmoninarum* suggests reductive evolution away from an environmental *Arthrobacter* ancestor. J Bacteriol. 2008, 190: 6970-6982. 10.1128/JB.00721-08.
49. Ransangan J, Mustafa S: Identification of *Vibrio harveyi* isolated from diseased Asian sea bass *Lates calcarifer* by use of 16S ribosomal DNA sequencing. J Aquat Anim Health. 2009, 21: 150-155. 10.1577/H09-002.1.
50. Nomoto R, Munasinghe LI, Shimahara Y, Yasuda H, Nakamura A, Misawa N, Itami T, Yoshida T: Lancefield group C *Streptococcus dysgalactiae* infection responsible for fish mortalities in Japan. J Fish Dis. 2004, 27: 679-686. 10.1111/j.1365-2761.2004.00591.x.
51. Doménech A, Fernández-Garayzábal JF, Pascual C, Garcia JA, Cutuli MT, Moreno MA, Collins MD, Dominguez L: Streptococcosis in cultured turbot, *Scophthalmus maximus* (L.), associated with *Streptococcus parauberis*. J Fish Dis. 1996, 19: 33-38. 10.1111/j.1365-2761.1996.tb00117.x.
52. Wallbanks S, Martinez-Murcia AJ, Fryer JL, Phillips BA, Collins MD: 16S rRNA sequence determination for members of the genus *Carnobacterium* and related lactic acid bacteria and description of *Vagococcus salmoninarum* sp. nov. Int J Syst Bacteriol. 1990, 40: 224-230. 10.1099/00207713-40-3-224.
53. Chen SC, Liaw LL, Su HY, Ko SC, Wu CY, Chaung HC, Tsai YH, Yang KL, Chen YC, Chen TH, Lin GR, Cheng SY, Lin YD, Lee JL, Lai CC, Weng YJ, Chu SY: *Lactococcus garvieae*, a cause of disease in grey mullet, *Mugil cephalus* L., in Taiwan. J Fish Dis. 2002, 25: 727-732. 10.1046/j.1365-2761.2002.00415.x.
54. Hiney MP, Smith PR: Validation of polymerase chain reaction-based techniques for proxy detection of bacterial fish pathogens: framework, problems and possible solutions for environmental applications. Aquaculture. 1998, 162: 41-68. 10.1016/S0044-8486(98)00207-5.

55. Knibb W, Colorni A, Ankaoua M, Lindell D, Diamant A, Gordin H: Detection and identification of a pathogenic marine mycobacterium from the European seabass *Dicentrarchus labrax* using polymerase chain reaction and direct sequencing of 16S rRNA sequences. Mol Mar Biol Biotechnol. 1993, 2: 225-232.
56. McCormick JI, Hughes MS, McLoughlin MF: Identification of *Mycobacterium chelonae* in a cichlid oscar *Astronotus ocellatus* Cuvier, by direct cycle sequencing of polymerase chain reaction amplified 16S rRNA gene sequences. J Fish Dis. 1995, 18: 459-461. 10.1111/j.1365-2761.1995.tb00337.x.
57. Zlotkin A, Eldar A, Ghittino C, Bercovier H: Identification of *Lactococcus garvieae* by PCR. J Clin Microbiol. 1998, 36: 983-985.
58. Mata AI, Gibello A, Casamayor A, Blanco MM, Domínquez L, Fernández-Garayzábal JF: Multiplex PCR assay for the detection of bacterial pathogens associated with warm-water streptococcosis in fish. Appl Environ Microbiol. 2004, 70: 3183-3187. 10.1128/AEM.70.5.3183-3187.2004.
59. Aoki T, Park CI, Yamashita H, Hirono I: Species-specific polymerase chain reaction primers for *Lactococcus garvieae*. J Fish Dis. 2000, 23: 1-6. 10.1046/j.1365-2761.2000.00207.x.
60. León G, Maulén N, Figueroa J, Villanueva J, Rodríguez C, Vera MI, Krauskopf M: A PCR based assay for the identification of the fish pathogen *Renibacterium salmoninarum*. FEMS Microbiol Lett. 1994, 115: 131-136.
61. León G, Martinez MA, Etchegaray JP, Vera MI, Figueroa J, Krauskopf M: Specific DNA probes for the identification of the fish pathogen, *Renibacterium salmoninarum*. Wld J Microbiol Biotechnol. 1994, 10: 149-153. 10.1007/BF00360876.
62. Magnüssen HB, Fridjónsson ÓH, Andrésson ÓS, Benediktsdóttir E, Gudmundsdóttir S, Andrésdottír V: *Renibacterium salmoninarum*, the causal agent of bacterial kidney disease in salmonid fish, detected by nested reverse transcription-PCR of 16S rRNA sequences. Appl Environ Microbiol. 1994, 60: 4580-4583.
63. Miyoshi Y, Suzuki S: PCR method to detect *Nocardia seriolae* in fish samples. Fish Pathol. 2003, 38: 93-97. 10.3147/jsfp.38.93.
64. Soria-Carrasco V, Talavera G, Igea J, Castresana J: The K tree score: quantification of differences in the relative branch length and topology of phylogenetic trees. Bioinformatics. 2007, 23: 2954-2956. 10.1093/bioinformatics/btm466.
65. Maher M, Palmer R, Gannon F, Smith T: Relationship of a novel bacterial fish pathogen to *Streptobacillus moniliformis*and the fusobacteria group, based on 16S ribosomal RNA analysis. Syst Appl Microbiol. 1995, 18: 79-84.
66. Sakai M, Konno T, Tassakka ACMAR, Ponpornpisit A, Areechon N, Katagiri T, Yoshida T, Endo M: Characterization of a *Mycobacterium* sp. isolated from guppy *Poecilia reticulata*, using 16S ribosomal RNA and its internal transcribed spacer sequences. Bull Eur Assoc Fish Pathol. 2005, 25: 64-69.
67. Pourahmad F, Cervellione F, Thompson KD, Taggart JB, Adams A, Richards RH: *Mycobacterium stomatepiae* sp. nov., a slowly growing non-chromogenic species isolated from fish. Int J Syst Evol Microbiol. 2008, 58: 2821-2827. 10.1099/ijs.0.2008/001164-0.
68. Martinez-Murcia AJ, Esteve C, Garay E, Collins MD: *Aeromonas allosaccharophila* sp. nov., a new mesophilic member of the genus *Aeromonas*. FEMS Microbiol Lett. 1992, 91: 199-206. 10.1111/j.1574-6968.1992.tb05209.x.
69. Huys G, Kämpfer P, Swings J: New DNA-DNA hybridization and phenotypic data on the species *Aeromonas ichthiosmia* and *Aeromonas allosaccharophila*: A. *ichthiosmia is a later synonym of A. veronii* Hickman-Brenner et al. 1987. Syst Appl Microbiol. 2001, 24: 177-182. 10.1078/0723-2020-00038.
70. Mikalsen J, Olsen AB, Tengs T, Colquohoun DJ: *Francisella philomiragia* subsp *noatunensis* subsp nov., isolated from farmed Atlantic cod (*Gadus morhua* L.). Int J Syst Evol Microbiol. 2007, 57: 1960-1965. 10.1099/ijs.0.64765-0.
71. Nylund A, Ottem KF, Watanabe K, Karlsbakk E, Krossoy B: *Francisella* sp (Family Francisellaceae) causing mortality in Norwegian cod *(Gadus morhua)* farming. Arch Microbiol. 2006, 185: 383-392. 10.1007/s00203-006-0109-5.

72. Birkbeck TH, Laidler LA, Grant AN, Cox DI: *Pasteurella skyensis* sp nov., isolated from Atlantic salmon *(Salmo salar* L.). Int J Syst Evol Microbiol. 2002, 52: 699-704. 10.1099/ijs.0.01884-0.
73. Fryer JL, Lannan CN, Giovannoni SJ, Wood ND: *Piscirickettsia salmonis* gen. nov., the causative agent of an epizootic disease in salmonid fishes. Int J Syst Bacteriol. 1992, 42: 120-126. 10.1099/00207713-42-1-120.
74. Arkush KD, Mcbride AM, Mendonca HL, Okihiro MS, Andree KB, Marshall S, Henriquez V, Hedrick RP: Genetic characterization and experimental pathogenesis of *Piscirickettsia salmonis* isolated from white seabass *Atractascion nobilis*. Dis Aquat Org. 2005, 63: 139-149. 10.3354/dao063139.
75. Corbeil S, Hyatt AD, Crane MStJ: Characterisation of an emerging rickettsia-like organism in Tasmanian farmed Atlantic salmon *Salmo salar*. Dis Aquat Org. 2005, 64: 37-44. 10.3354/dao064037.
76. Nishimori E, Kita-Tsukamoto K, Wakabayashi H: *Pseudomonas plecoglossicida* sp nov., the causative agent of bacterial haemorrhagic ascites of ayu, *Plecoglossus altivelis*. Int J Syst Evol Microbiol. 2000, 50: 83-89. 10.1099/00207713-50-1-83.
77. Romalde JL, Ravelo C, Valdes I, Magariños B, de la Fuente E, Martin CS, Avendano-Herrera R, Toranzo AE: *Streptococcus phocae*, an emerging pathogen for salmonid culture. Vet Microbiol. 2008, 130: 198-207. 10.1016/j.vetmic.2007.12.021.
78. Pineiro-Vidal M, Carballas C, Gomez-Barreiro O, Riaza A, Santos Y: *Tenacibaculum soleae* sp nov., isolated from diseased sole *(Soleae senegalensis* Kaup). Int J Syst Evol Microbiol. 2008, 58: 881-885. 10.1099/ijs.0.65539-0.
79. Lunder T, Sorum H, Holstad G, Steigerwalt AG, Mowinckel P, Brenner DJ: Phenotypic and genotypic characterization of *Vibrio viscosus* sp nov and *Vibrio wodanis* sp nov isolated from Atlantic salmon *(Salmo salar)* with "winter ulcer". Int J Syst Evol Microbiol. 2000, 50: 427-450. 10.1099/00207713-50-2-427.
80. Benediktsdóttir E, Verdonck L, Spröer C, Helgason S, Swings J: Characterization of *Vibrio viscosus* and *Vibrio wodanis*isolated from different geographical locations: a proposal for re-classification of *Vibrio viscosus* as *Moritella viscose*comb. nov. Int J Syst Evol Microbiol. 2000, 50: 479-488. 10.1099/00207713-50-2-479.
81. Benediktsdóttir E, Helgason S, Sigurjónsdóttir H: *Vibrio* spp. isolated from salmonids with shallow skin lesions and reared at low temperature. J Fish Dis. 1988, 21: 19-28.
82. Urakawa H, Kita-Tsukamoto K, Stevens SE, Ohwada K, Colwell RR: A proposal to transfer *Vibrio marinus* (Russell 1891) to a new genus *Moritella* gen. nov. as *Moritella marina* comb. nov. FEMS Microbiol Lett. 1998, 165: 373-378. 10.1111/j.1574-6968.1998.tb13173.x.
83. Del-Pozo J, Crumlish M, Ferguson HW, Turnbull JF: Retrospective cross-sectional study on "*Candidatus arthromitus*" associated rainbow trout gastroenteritis (RTGE) in the UK. Aquaculture. 2009, 290: 22-27. 10.1016/j.aquaculture.2009.02.009.
84. Michel C, Bernardet JF, Daniel P, Chilmonczyk S, Urdaci M, de Kinkelin P: Clinical and aetiological aspects of a summer enteritic syndrome associated with the sporulating segmented filamentous bacterium "*Candidatus Arthromitus*" in farmed rainbow trout, *Oncorhynchus mykiss* (Walbaum). J Fish Dis. 2002, 25: 533-543. 10.1046/j.1365-2761.2002.00400.x.
85. Del-Pozo J, Turnbull J, Ferguson H, Crumlish M: A comparative molecular study of the presence of "*Candidatus arthromitus*" in the digestive system of rainbow trout, *Oncorhynchus mykiss* (Walbaum), healthy and affected with rainbow trout gastroenteritis. J Fish Dis. 2010, 33: 241-250. 10.1111/j.1365-2761.2009.01117.x.
86. Stevenson RMW, Daly JG: Biochemical and serological characteristics of Ontario isolates of *Yersinia ruckeri*. Can J Fish Aquat Sci. 1982, 39: 870-876. 10.1139/f82-118.
87. Karlsen M, Nylund A, Watanabe K, Helvik JV, Nylund S, Plarre H: Characterization of "*Candidatus Clavochlamydia salmonicola*": an intracellular bacterium infecting salmonid fish. Environ Microbiol. 2008, 10: 208-218.
88. Larsen JL, Jensen NJ: An *Aeromonas* species implicated in ulcer-disease of the cod (*Gadus morhua*). Nord Veterinaermed. 1977, 29: 199-211.

89. Orozova P, Barker M, Austin DA, Austin B: Identification and pathogenicity to rainbow trout, *Oncorhynchus mykiss*(Walbaum), of some aeromonads. J Fish Dis. 2009, 32: 865-871. 10.1111/j.1365-2761.2009.01065.x.
90. Carnahan AM, Altwegg M: Taxonomy. The Genus Aeromonas. Edited by: Austin B, Altwegg M, Gosling PJ, Joseph S, Wiley, Chichester. 1996, 1-38.
91. Allen DA, Austin B, Colwell RR: *Aeromonas media*, a new species isolated from river water. Int J Syst Bacteriol. 1983, 33: 599-604. 10.1099/00207713-33-3-599.
92. MacInnes JI, Trust TJ, Crosa JH: Deoxyribonucleic acid relationships among members of the genus *Aeromonas*. Can J Microbiol. 1979, 25: 579-586. 10.1139/m79-083.
93. Esteve C: Numerical taxonomy of *Aeromonadaceae* and *Vibrionaceae* associated with reared fish and surrounding fresh and brackish water. System Appl Microbiol. 1995, 18: 391-402.
94. Esteve C, Biosca EG, Amaro C: Virulence of *Aeromonas hydrophila* and some other bacteria isolated from European eels *Anguilla anguilla* reared in fresh water. Dis Aquat Org. 1993, 16: 15-20. 10.3354/dao016015.
95. Ali A, Carnahan AM, Altwegg M, Lüthy-Hotenstein J, Joseph SW: *Aeromonas bestiarum* sp. nov. (formerly genomospecies DNA group 2 *A. hydrophila)* a new species isolated from non-human sources. Med Microbiol Lett. 1996, 5: 156-165.
96. Toranzo AE, Baya AM, Romalde JJ, Hetrick FM: Association of *Aeromonas sobria* with mortalities of adult gizzard shad, *Dorosoma cepedianum* Lesueur. J Fish Dis. 1989, 12: 439-448. 10.1111/j.1365-2761.1989.tb00555.x.
97. Ma Z, Yang H, Li T, Luo L, Gao JL: Isolation and identification of pathogenic *Aeromonas veronii* isolated from infected Siberian sturgeon (*Acipenser baerii*). Wei Sheng Wu Xue Bao. 2009, 49: 1289-1294. (article in Chinese)
98. Snieszko SF: Genus IV. *Aeromonas* Kluyver and van Niel 1936, Bergey's Manual of Determinative Bacteriology. Edited by: Breed RS, Murray EGD, Smith NR. 1957, Williams and Wilkins, Baltimore, 189-193. 7
99. Colwell RR, MacDonell MT, De Ley J: Proposal to recognize the family Aeromonadaceae fam. nov. Int J Syst Bacteriol. 1986, 36: 473-477. 10.1099/00207713-36-3-473.
100. Griffin PJ, Snieszko SF, Friddle SB: A more comprehensive description of *Bacterium salmonicida*. Trans Am Fish Soc. 1953, 82: 129-138. 10.1577/1548-8659(1952)82[129:AMCDOB]2.0.CO;2.
101. Austin DA, McIntosh D, Austin B: Taxonomy of fish associated *Aeromonas* spp., with the description of *Aeromonas salmonicida* subsp. *smithia* subsp. nov. Syst Appl Microbiol. 1989, 11: 277-290.
102. Duff DCB: Dissociation in *Bacillus salmonicida*, with special reference to the appearance of the G form of culture. J Bacteriol. 1937, 34: 49-67.
103. Pacha RE, Kiehn ED: Characterization and relatedness of marine vibrios pathogenic to fish: physiology, serology and epidemiology. J Bacteriol. 1969, 100: 1242-1247.
104. Koppang EO, Fjølstad M, Melgård B, Vigerust M, Sørum H: non-pigment-producing isolates of *Aeromonas salmonicida*subspecies *salmonicida:* isolation, identification, transmission and pathogenicity in Atlantic salmon, *Salmo salar* L. J Fish Dis. 2000, 23: 39-48. 10.1046/j.1365-2761.2000.00205.x.
105. Nakatsugawa T: Atypical *Aeromonas salmonicida* isolted from cultured shotted halibut. Fish Pathol. 1994, 29: 193-198. 10.3147/jsfp.29.193.
106. Kaku Y, Yamada Y, Wakabayashi H: Characterization of atypical *Aeromonas salmonicida* isolated from an epizootic ulcerative disease in carp (*Cyprinus carpio*). Fish Pathol. 1999, 34: 155-162. 10.3147/jsfp.34.155.
107. Wiklund T, Bylund G: Skin ulcer disease of flounder *Platichthys flesus* in the northern Baltic Sea. Dis Aquat Org. 1993, 17: 165-174. 10.3354/dao017165.
108. Austin B: Recovery of "atypical" isolates of *Aeromonas salmonicida*, which grow at 37°C, from ulcerated non-salmonids in England. J Fish Dis. 1993, 16: 165-168. 10.1111/j.1365-2761.1993.tb00862.x.

109. Goodwin AE, Merry GE: Are all Koi ulcer cases associated with infection by atypical *Aeromonas salmonicida*? Polymerase chain reaction assays of Koi carp skin swabs submitted by hobbyists. J Aquat Anim Health. 2009, 21: 98-103. 10.1577/H08-042.1.
110. Pedersen K, Kofod H, Dalsgaard I, Larsen JL: Isolation of oxidase-negative *Aeromonas salmonicida* from disease turbot *Scophthalmus maximus*. Dis Aquat Org. 1994, 18: 149-154. 10.3354/dao018149.
111. Wiklund T, Dalsgaard I, Eerola E, Olivier G: Characteristics of "atypical" cytochrome-oxidase negative *Aeromonas salmonicida* isolated from ulcerated flounder *(Platichthys flesus* (L.). J Appl Bacteriol. 1994, 76: 511-520.
112. Wiklund T, Dalsgaard I: Atypical *Aeromonas salmonicida* associated with ulcerated flatfish species in the Baltic Sea and the North Sea. J Aquat Anim Health. 1995, 7: 218-224. 10.1577/1548-8667(1995)007<0218:AASAWU>2.3.CO;2.
113. Goldschmidt-Clermont E, Hochwartner O, Demarta A, Caminada AP, Frey J: Outbreaks of an ulcerative and haemorrhagic disease in Arctic char *Salvelinus alpinus* caused by *Aeromonas salmonicida* subsp *smithia*. Dis Aquat Org. 2009, 86: 81-86. 10.3354/dao02110.
114. Austin B, Austin DA, Dalsgaard I, Gudmundsdóttir BK, Høie S, Thornton JM, Larsen JL, O'Hici B, Powell R: Characterization of atypical *Aeromonas salmonicida* by different methods. Syst Appl Microbiol. 1998, 21: 50-64.
115. Høie S, Dalsgaard I, Aase IL, Heum M, Thornton JM, Powell R: Polymerase chain reaction (PCR) based typing analysis of atypical isolates of the fish pathogen *Aeromonas salmonicida*. Syst Appl Microbiol. 1999, 22: 403-411.
116. Lund V, Jenssen LM, Wesmajervi MS: Assessment of genetic variability and relatedness among atypical *Aeromonas salmonicida* from marine fishes, using AFLP fingerprinting. Dis Aquat Org. 2002, 50: 119-126. 10.3354/dao050119.
117. Yamada Y, Kaku Y, Wakabayashi H: Phylogenetic intrarelationships of atypical *Aeromonas salmonicida* isolated in Japan as determined by 16S rDNA sequencing. Fish Pathol. 2000, 35: 35-40. 10.3147/jsfp.35.35.
118. Smith IW: The classification of *Bacterium salmonicida*. J Gen Microbiol. 1963, 33: 263-274.
119. Kimura T: A new subspecies of *Aeromonas salmonicida* as an etiological agent of furunculosis on "Sakuramasu" *(Oncorhynchus masou)* and pink salmon *(O. gorbuscha)* rearing for maturity. Fish Pathol. 1969, 3: 34-44. 10.3147/jsfp.3.2_34. Part 1. On the serological properties
120. Schubert RHW: Genus II. *Aeromonas* Kluyver and van Niel 1936, 398, Bergey's Manual of Determinative Bacteriology. Edited by: Buchanan RE, Gibbons NE. 1974, Williams and Wilkins, Baltimore, 345-348. 8th
121. Popoff M: Bacterioses, generalities - *Aeromonas* et Aeromonose. Mimeograph. 1970, Institut National de la Recherche Agronomique, 1-38.
122. Paterson WD, Douey D, Desautels D: Relationships between selected strains of typical and atypical *Aeromonas salmonicida, Aeromonas hydrophila,* and *Haemophilus piscium*. Can J Microbiol. 1980, 26: 588-598. 10.1139/m80-104.
123. Snieszko SF, Griffin P, Friddle SB: A new bacterium *(Haemophilus piscium* n. sp.) from ulcer disease of trout. J Bacteriol. 1950, 59: 699-710.
124. Thornton JM, Austin DA, Austin B, Powell R: Small subunit rRNA gene sequences of *Aeromonas salmonicida* subsp. *smithia* and *Haemophilus piscium* reveal pronounced similarities with *A. salmonicida* subsp. *salmonicida*. Dis Aquat Org. 1999, 35: 155-158. 10.3354/dao035155.
125. McCarthy DH: Some ecological aspects of the bacterial fish pathogen - *Aeromonas salmonicida*. Aquatic Microbiology Symp Soc Appl Bacteriol. 1980, 6: 299-324.
126. Belland RJ, Trust TJ: DNA:DNA reassociation analysis of *Aeromonas salmonicida*. J Gen Microbiol. 1988, 134: 307-315.
127. Eddy BP: Cephalotrichous, fermentative Gram-negative bacteria: the genus *Aeromonas*. J Appl Bacteriol. 1960, 23: 216-248.

128. Kluyver AJ, Van Niel CB: Prospects for a natural system of classification of bacteria. Z Bakteriol Parasiten, Infekt Hyg, Orig. 1936, 94: 369-403. 2 Abt
129. Cowan ST: Cowan and Steel's Manual for the Identification of Medical Bacteria. 1974, Cambridge University Press, Cambridge, 2
130. Kimura T: A new subspecies of *Aeromonas salmonicida* as an etiological agent of furunculosis on "Sakuramasu" *(Oncorhynchus masou)* and pink salmon *(O. gorbuscha)* rearing for maturity. Part 2. On the morphological and physiological properties. Fish Pathol. 1969, 3: 45-52. 10.3147/jsfp.3.2_45.
131. Liu PV: Observations on the specificities of extra-cellular antigens of the genera *Aeromonas* and *Serratia*. J Gen Microbiol. 1961, 24: 145-153.
132. Popoff M: Étude sur les *Aeromonas salmonicida*. II. Caracterisation des bactériophages actifs sur les "*Aeromonas salmonicida*" et lysotypie. Ann Recherch Vét. 1971, 2: 33-45.
133. Popoff M: Intérêt diagnostique d'un bactériophage spécifique des *Aeromonas salmonicida*. Ann Rech Vét. 1971, 2: 137-139.
134. Kilian M: A taxonomic study of the genus *Haemophilus*, with the proposal of a new species. J Gen Microbiol. 1976, 93: 9-62.
135. Broom AK, Sneath PHA: Numerical taxonomy of *Haemophilus*. J Gen Microbiol. 1981, 126: 123-149.
136. Trust TJ, Ishiguro EE, Atkinson HM: Relationship between *Haemophilus piscium* and *Aeromonas salmonicida* revealed by *Aeromonas hydrophila* bacteriophage. FEMS Microbiol Lett. 1980, 9: 199-201. 10.1111/j.1574-6968.1980.tb05636.x.
137. Dalsgaard I, Gudmundsdóttir BK, Helgason S, Høie S, Thoresen OF, Wichardt UP, Wiklund T: Identification of atypical *Aeromonas salmonicida*: inter-laboratory evaluation and harmonization of methods. J Appl Microbiol. 1998, 84: 999-1006. 10.1046/j.1365-2672.1998.00435.x.
138. Ross AJ, Rucker RR, Ewing WH: Description of a bacterium associated with redmouth disease of rainbow trout (*Salmo gairdneri*). Can J Microbiol. 1966, 12: 763-70. 10.1139/m66-103.
139. Steinum T, Sjastad K, Falk K, Kvellestad A, Colquhuin DJ: An RT PCR-DGGE survey of gill-associated bacteria in Norwegian seawater-reared Atlantic salmon suffering proliferative gill inflammation. Aquaculture. 2009, 293: 172-179. 10.1016/j.aquaculture.2009.05.006.
140. Ewing WH, Ross AJ, Brenner DJ, Fanning GR: *Yersinia ruckeri* sp. nov., the redmouth (RM) bacterium. Int J Syst Bacteriol. 1978, 28: 37-44. 10.1099/00207713-28-1-37.
141. Steigerwalt AG, Fanning GR, Fife-Asbury MA, Brenner DJ: DNA relatedness among species of *Enterobacter* and *Serratia*. Can J Microbiol. 1976, 22: 121-137. 10.1139/m76-018.
142. Bercovier H, Mollaret HH: Genus XIV. *Yersinia* van Loghem 1944, 15AL. Bergey's Manual of Systematic Bacteriology. Edited by: Krieg NR, Holt JG. 1984, Williams and Wilkins, Baltimore, I: 489-506.
143. Austin DA, Robertson PAW, Austin B: Recovery of a new biogroup of *Yersinia ruckeri* from diseased rainbow trout *(Oncorhynchus mykiss*, Walbaum). Syst Appl Microbiol. 2005, 26: 127-131. 10.1078/072320203322337416.
144. Fouz B, Zarza C, Amaro C: First description of non-motile *Yersinia ruckeri* serovar I strains causing disease in rainbow trout, *Oncorhynchus mykiss* (Walbaum), cultured in Spain. J Fish Dis. 2006, 29: 339-346. 10.1111/j.1365-2761.2006.00723.x.
145. Smith IW: A disease of finnock due to *Vibrio anguillarum*. J Gen Microbiol. 1961, 74: 247-252.
146. Kaper JB, Lockman H, Remmers EF, Kristensen K, Colwell RR: Numerical taxonomy of vibrios isolated from estuarine environments. Int J Syst Bacteriol. 1983, 33: 229-255. 10.1099/00207713-33-2-229.
147. Pazos F, Santos Y, Magariños B, Bandín I, Núñez S, Toranzo AE: Phenotypic characteristics and virulence of *Vibrio anguillarum* - related organisms. Appl Environ Microbiol. 1993, 59: 2969-2976.

148. Olsen JE, Larsen JL: Ribotypes and plasmid content of *Vibrio anguillarum* strains in relation to serovar. Appl Environ Microbiol. 1993, 59: 3863-3870.

149. Austin B, Alsina M, Austin DA, Blanch AR, Grimont F, Grimont PAD, Jofre J, Koblavi S, Larsen JL, Pedersen K, Tiainen T, Verdonck L, Swings J: Identification and typing of *Vibrio anguillarum* a comparison of different methods. Syst Appl Microbiol. 1995, 18: 285-302.

150. Austin B, Austin DA, Blanch AR, Cerdà M, Grimont F, Grimont PAD, Jofre J, Koblavi S, Larsen JL, Pedersen K, Tiainen T, Verdonck L, Swings J: A comparison of methods for the typing of fish-pathogenic *Vibrio* spp. Syst Appl Microbiol. 1997, 20: 89-101.

151. Schiewe MH: Taxonomic status of marine vibrios pathogenic for salmonid fish. Dev Biol Stand. 1981, 49: 149-158.

152. Muthiara LW, Raymond BT, Dekievit TR, Stevenson RHW: Antibody specificities of polyclonal rabbit and rainbow trout antisera against *Vibrio ordalii* and serotype O:2 strains of *Vibrio anguillarum*. Can J Microbiol. 1993, 39: 492-499. 10.1139/m93-070.

153. Sørensen UBS, Larsen JL: Serotyping of *Vibrio anguillarum*. Appl Environ Microbiol. 1986, 51: 593-597.

154. Grisez L, Ollevier F: Comparative serology of the marine fish pathogen *Vibrio anguillarum*. Appl Environ Microbiol. 1995, 61: 4367-4373.

155. Pedersen K, Dalsgaard I, Larsen JL: Characterization of atypical *Aeromonas salmonicida* isolates by ribotyping and plasmid profiling. J Appl Bacteriol. 1996, 80: 37-44.

156. Pedersen K, Tiainen T, Larsen JL: Plasmid profiles, restriction length polymorphisms and O-serotypes among *Vibrio anguillarum* isolates. Epidemiol Infect. 1996, 117: 471-478. 10.1017/S0950268800059136.

157. Kitao T, Aoki T, Fukodome M, Kawano K, Wada Y, Mizuno Y: Serotyping of *Vibrio anguillarum* isolated from diseased freshwater fish in Japan. J Fish Dis. 1983, 6: 175-181. 10.1111/j.1365-2761.1983.tb00064.x.

158. Tiainen T, Pedersen K, Larsen JL: Immunological reactivity of *Vibrio anguillarum* sero-subgroups O2a and O2b, and comparison of their lipopolysaccharide profiles. Curr Microbiol. 1997, 34: 38-42. 10.1007/s002849900141.

159. O'Hici B, Olivier G, Powell R: Genetic diversity of the fish pathogen *Aeromonas salmonicida* demonstrated by random amplified polymorphic DNA and pulsed-field gel electrophoresis analyses. Dis Aquat Org. 2000, 39: 109-119.

160. Umelo E, Trust TJ: Identification and molecular characterization of two tandomly located flagellin genes from *Aeromonas salmonicida* A449. J Bacteriol. 1997, 179: 5292-5299.

161. McIntosh D, Austin B: Atypical characteristics of the salmonid pathogen *Aeromonas salmonicida*. J Gen Microbiol. 1991, 137: 1341-1343.

CHAPTER 10

Identification of Bacterial Fish Pathogens in Brazil by Direct Colony PCR and 16S rRNA Gene Sequencing

F. A. Sebastião, L. R. Furlan, D. T. Hashimoto, F. Pilarski*

Aquaculture Center of São Paulo State University—CAUNESP, Universidade Estadual Paulista, Jaboticabal, Brazil

ABSTRACT

Intensive fish farming systems in Brazil have increased the disease incidence, mainly of bacterial origin, due to higher stocking density, high organic matter levels and poor quality of the aquatic environment that causes high mortality rates during outbreaks. The identification of pathogenic species using a fast and reliable method of diagnosis is essential for successful epidemiological studies and disease control. The present study evaluated the use of direct colony PCR in combination with 16S rRNA gene sequencing to diagnose fish bacterial diseases, with the goal of reducing the costs and time necessary for bacterial identification. The method was successful for all 178 isolates tested and produced bands with the same intensity as the standard PCR performed using pure DNA. In conclusion, the genetics methods allowed detecting the most common and important pathogens in Aquaculture, including 12 species of occurrence in Brazilian fish farms. The results of the present study constitute an advance in the available diagnostic methods for bacterial pathogens in fish farms.

Keywords: Direct Colony PCR, 16S rRNA Sequencing, Bacterial Fish Pathogens

1. INTRODUCTION

Due to its high water availability and favorable climate conditions, Brazil displays high potential for the development of fish farming, which is an activity that has been growing substantially over the last few years. According to the Food and Agriculture Organization of the United Nations [1] , Brazil is the second largest aquaculture producer in Latin America and the Caribbean, and freshwater aquaculture (tilapia, carp, and native fish) represented 87% (545,300 ton) of the total aquaculture production in 2011.

The growing interest in this activity and, consequently, the search for higher profitability, have been leading producers to adopt super-intensive production systems. However, the high density of confined fish, inadequate farming management practices, and water contamination by toxic products cause chronic stress and immunosuppression in farmed animals. These effects lead to the occurrence of diseases and epizootic outbreaks caused by pathogens that would not have high expression in natural environments [2] . Knowledge about the etiological agents, pathogenesis, biochemistry, antigenicity, epizootiology, and inter-relationship of stress and environmental factors of bacterial infections affecting fish is essential to avoid and control diseases. However, these factors have not been well studied, especially because fish farming is a recent activity, with its intensification beginning in the 1990s [3] .

Gram-negative bacteria such as Aeromonas, Flavobacterium, Pseudomonas, and Francisella and gram-posi- tive bacteria from the genera Streptococcus and Lactococcus [4] - [11] are some of the pathogens responsible for economic losses in Brazil. They can cause high fish mortality rates up to 72 h after infection [12] .

Although the number of studies focusing on the diagnosis of bacterial etiological agents has increased over the last few years, there are still few available alternatives for the control of fish bacterial infections in Brazil. Therefore, quicker and more effective diagnostic alternatives are necessary, which would help control diseases before they lead to irreversible clinical consequences and high mortality rates. Molecular diagnostic methods use reduced volumes of sample material and exhibit high sensitivity, specificity, and accuracy in pathogen detection [13].

Methods that do not require purified DNA extraction, such as direct colony PCR, are quicker and less expensive and may greatly aid in the early detection of fish pathogens [14] . In addition, because not all microorganism sequences are catalogued in current databases, the use of universal and degenerate primers is a wise strategy. For this reason, methods based on

16S rRNA ribosomal gene amplification and sequencing have been widely explored [13] .

The use of universal PCR primers is based on the hypothesis that the primers used are complementary to conserved regions of genes in the environment, resulting in amplification; in turn, heterogeneity is found inside of the fragments flanked by the primers, in hypervariable regions [15] [16] . This method has been revolutionizing microbial ecology, from studies of non-cultivable bacteria to the correct identification of pathogens for accurate diagnoses.

The aim of the present study was to evaluate the direct colony PCR combined with 16S rRNA gene sequencing as a faster and less expensive method to identify fish bacterial pathogens, compared to the classic PCR protocol. Moreover, we have used these methods to demonstrate the efficiency of genetic approaches for the practical evaluation of the diagnosis of aquaculture diseases in Brazilian fish farms.

2. MATERIALS AND METHODS

2.1. Bacterial Strains and Culture Conditions

178 bacterial isolates were obtained between 2010 and 2014 from the following hosts (n = number of assessed fish): tilapia (Oreochromis niloticus n = 93), tambaqui (Colossoma macropomum n = 10), carp (Cyprinus carpio n = 3), cachara (Pseudoplatystoma reticulatum n = 34), and pacu (Piaractus mesopotamicus n = 8).

The hosts exhibited clinical signs of bacterial diseases, such as skin ulcerative lesions, hemorrhagic septicemia, meningoencephalitis, fin rot, exophthalmia, and were collected at fish farms in different regions of Brazil:Dourados (Mato Grosso do Sul State, 22°13'16"S, 54°48'20"W, n = 38), Rio de Janeiro (Rio de Janeiro State, 22°54'S, 43°10'W, n = 4), Itambaracá (Paraná State, 23°0'49"S, 50°24'7"W, n = 10), Itaju (22°25'37"S, 45°27'11"W), Arealva (22°1'38"S, 48°54'36"W), Porto Ferreira (21°51'18"S, 47°28'45"W), Guaíra (20°19'5"S, 48°18'42"W), Santa Fé do Sul (20°12'43"S, 50°55'38"W), Palmital (22°47'30"S, 50°12'18"W) and Jaboticabal (21°15'19"S, 48°19'21"W—São Paulo State, n = 123).

For the isolation of bacteria, scrapings were performed using sterile swabs on fish kidneys and brain. Gram- negative colonies were plated on TSA (Tryptic Soy Agar-Biolife), and TSB (Tryptic Soy Broth-Biolife) and incubated for 24 h in bacteriological incubator adjusted to 28°C. While gram-positive colonies were seeded in Columbia blood agar (Difco) incubated for 24 - 72 h at 30°C and subcultured in BHI (Brain Heart Infusion Broth, Himedia).

The strains of Palmital (SP) were obtained directly from the Laboratory of Aquatic Animal Disease, APTA, Votuporanga, SP.

2.2. Molecular Identification of Isolates

Two methods of molecular diagnosis were compared in this study aiming to evaluate the efficiency of direct colony PCR (time and cost effectiveness) in relation to the PCR amplification of purified DNA by extraction, both combined with gene sequencing (Table 1).

The standard PCR of purified DNA method followed the steps below.

2.2.1. DNA Extraction

One colony of each isolated was transferred to a tube containing appropriate liquid culture medium (TSB for gram-negative and BHI for gram-positive) and incubated at 28°C until the OD_{600} was between 1 and 1.5. Following incubation, 1.0 mL of the bacteria culture was centrifuged at 12,000×g for 1 min, the supernatant was discarded, and the pellet was frozen at −20°C until DNA extraction. The Axyprep® miniprep kit for bacterial genomic DNA was used according to the manufacturer's instructions (Axygen Biosciences, Union City, CA, USA). DNA was quantified by fluorometry using a Qubit 2.0 fluorometer (Life Technologies, NY, USA).

2.2.2. Standard PCR

PCR was performed in a 25 µL final volume, containing 2.5 µL of 10X buffer (10 mM Tris-HCl, 50 mM KCl), 0.2 µL of 25 mM dNTP, 1.0 µL of 50 mM MgSO4, 0.2 µL of *Taq* High Fidelity (Platinum®Taq DNA Polymerase, Life Technologies, NY, USA), 2.0 µL of each primer (10 pmol), 25 ng of DNA template, and Milli-Q water up to the final volume. The PCR program consisted of 94°C for 2 min; 35 cycles of 94°C for 30 seconds, 55°C for 30 s, and 68°C for 1.5 min; and final extension at 68°C for 10 min. We used the primers 8F/907R (Table 2), specific for the 16S rRNA bacterial gene [15] [17] [18] . The resulting amplicons of approximately 900 bp (base pair) were analyzed by electrophoresis in 1.5% agarose gel stained with ethidium bromide, according to Sambrook et at. [19] .

2.2.3. Purification of PCR Products and Gene Sequencing

PCR products were purified using a MinElute Kit (Qiagen, Crawley, West Sussex, UK) according to the manufacturer's instructions. Purified PCR products were quantified using a Qubit 2.0 fluorometer, and gene sequencing was performed using 50 ng/μL per sample. Sequencing was performed according to Sanger [20] . PCR products were amplified using AmpliTaq polymerase and BigDye Terminator (Applied Biosystems) according to the manufacturer's instructions, using the primer 907R. Sequencing was performed using an ABI PRISM 3730 DNA analyzer (Applied Biosystems).

Table 1. Steps of the two methods compared in this study: Standard PCR of purified DNA and direct colony PCR, both combined with the 16S rRNA gene sequencing.

Steps	Standard PCR	Direct colony PCR
Bacterial isolation	X	X
Replication in broth	X	-
DNA extraction	X	-
PCR	X	X
Electrophoresis	X	X
PCR product purification	X	-
16S rRNA sequencing	X	X
Nucleotide analysis	X	X

Table 2. Sequence of primers used for amplification of the 16S rRNA gene.

Primer	Target sequence 5'- 3'
8F	AGA GTT TGA TYM TGG CTC AG
907R	CCG TCA ATT CMT TTR AGT TT

2.3. Direct Colony PCR

This method allows PCR to be performed on colonies isolated from Petri dishes, without the step of DNA extraction. Colonies (1 - 2 mm diameter) were inoculated by placing a sterile tooth pick at the bottom of a PCR tube (0.2 mL) and incubated at −20°C overnight. The following solution was then added in the PCR tube: 2.0 μL of 10× buffer (10 mM Tris-HCl, 50 mM KCl), 1.2 μL of 50 mM $MgCl_2$, 0.2 μL of 25 mM dNTP, 0.7 μL of each primer 8F/907R (10 pmol/μL), 0.2 μL of *Taq* DNA polymerase (2.5 U), and Milli-Q water up to 20 μL. The PCR program consisted of 95°C for 5 min; 30 cycles of 95°C for 1 min, 54°C for 1.5 min, and 72°C for 1 min; and final extension at 72°C for 5 min. The amplified PCR products, at 50 ng/μL mean concentration, were analyzed by electrophoresis in a 1.5% agarose gel stained with ethidium bromide [19] . The gels were visualized under UV light, using a ChemiDoc MP imaging system (Bio-Rad). Samples were quantified by fluorometry using a Qubit 2.0 fluorometer and sequenced as described above.

After sequencing, samples of both methods had their nucleotides analyzed.

2.4. Analysis of Nucleotide Sequences

The obtained sequences were visualized using the Bio Edit Sequence Alignment Editor software (v. 7.1.11). Phred quality of sequences was determined. The initial and final portions of the sequences were then removed, keeping only the high-quality fragment.

After trim, sequences were exported in FASTA format and compared with the GenBank database (http://www.ncbi.nlm.nih.gov/genbank/) using the Eztaxon algorithm (http://www.ezbiocloud.net/eztaxon/identify). 100% coverage and identity ≥ 98% were considered for specific identification. Sequences were also submitted to Ribosomal Database Project II (http://rdp.cme.msu.edu) for comparison and identification.

The sequences obtained in the present study were deposited at NCBI GenBank under accession numbers KJ560937 to KJ561113. The complete list of species identified, accession numbers, place of origin, fish species, season and year of collection, and size of amplified PCR products were included as Supplementary Material.

The Brazilian isolates tested were *S. agalactiae* (n = 23), *S. iniae* (n = 4), *Lactococcus lactis* (n = 11), *L. raffinolactis* (n = 2), *L. garvieae* (n = 16), *Enterococcus casseliflavus* (n = 16), *E. durans* (n = 2), *E. faecalis* (n = 11), *Edwardsiella tarda* (n = 5), *Aeromonas hydrophila* (n = 16), *A. jandaei* (n = 2), *A. veronii* (n = 15), *Pseudomonas sp.* (n = 15).

A phylogenetic diagram was constructed for validation of the sequencing data, using the 138 isolates listed above from the 178 of the present study. In addition, we used as reference 16 sequences originated from different countries (Table 3), obtained from GenBank database.

The 154 FASTA sequences were aligned using the ClustalW Multiple Alignment tool (BioEdit Sequence Alignment Editor software, v. 7.1.11). The data were then entered in Mega software (v. 5.05) to determine the best substitution model. As a result of the preliminary analysis, a maximum-likelihood phylogenetic diagram was constructed, using the Kimura 2-parameter model, with a gamma-shape parameter with 5 categories, the nearest-neighbor-interchange tree inference option. The stability of internal nodes was assessed by bootstrap analysis with 1000 replicates.

3. RESULTS AND DISCUSSION

3.1. The Comparison of the Two Methods: Direct Colony PCR and Classic PCR Protocol

We found no difference in the band size in agarose electrophoresis, nor in the peaks pattern of electrophero- grams in the two methods evaluated.

Bands resulting from the direct colony PCR exhibited the same intensity as those of the standard PCR of purified DNA, for all 178 isolates tested (Figure 1).

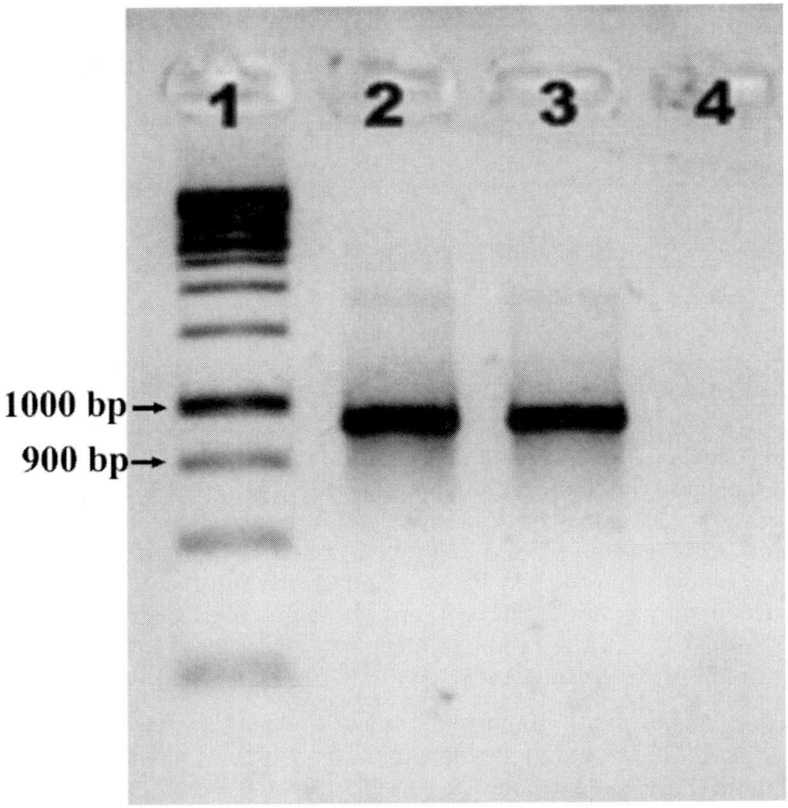

Figure 1. Electropherogram of 1.5% agarose gel stained with ethidium bromide, showing amplification of 16S rRNA gene (primers 8F/907R). Lane 1, Marker 1kb. Lane 2, direct colony PCR. Lane 3, standard PCR. Lane 4, negative control.

Table 3. Reference strains used for the maximum-likelihood phylogenetic analysis and their places of origin.

Genbank access number	Identification	Origin
JX861241	*Aeromonas veronii*	India
ATCC35624	*Aeromonas veronii*	Japan
NR_074841	*Aeromonas hydrophila*	USA
JN644061	*Aeromonas jandaei*	China
ATCC49568	*Aeromonas jandaei*	India
NBRC_105688	*Edwardsiella tarda*	Japan
EU239205	*Pseudomonas fulva*	Korea
KC210866	*Enterococcus casseliflavus*	China
AB530699	*Enterococcus faecalis*	Thailand
NR_036922	*Enterococcus durans*	Germany
KC176716	*Streptococcus agalactiae*	China
NR_027517	*Streptococcus dysgalactiae*	Japan
NR_025148	*Streptococcus iniae*	Israel
NR_044359	*Lactococcus raffinolactis*	South Korea
KC429785	*Lactococcus lactis*	China
ATCC49156	*Lactococcus garvieae*	USA

Electropherograms resulting from the sequencing of both methods exhibited Phred quality scores ≥20. All the isolates had the same results of bacterial identification for both techniques (direct colony PCR and standard PCR of purified DNA). Thus, direct colony PCR was a less expensive and faster diagnostic method, as shown on Table 4. There were 51% savings in cost analysis per sample for direct colony PCR compared to Standard PCR of purified DNA. Moreover, direct colony PCR reduces 2 days in time to issue the final report. After the installation of a bacterial outbreak, fish shoals can be decimated by up to 72 hours. Therefore, rapid diagnosis in aquaculture is a critical point in the production chain, which can be assessed by the genetic tools of the present study.

A faster diagnosis is important, since the one based on classical microbiology techniques (isolation, platting and biochemical tests) can exceed the time for treatment in seven to 15 days and, in many cases, ending up inconclusive. The molecular diagnosis, on the other hand, can provide a faster, low cost, conclusive diagnosis, which is essential to determine the best treatment in fish farming (Table 5).

Besides, in an attempt to control disease outbreaks, in classical scenery in Brazil, producers use multiple antibiotics indiscriminately, selecting resistant strains, contaminating fish, water and raising the risks to consumer health, endorsing the need for rapid and effective diagnosis [21] .

A maximum-likelihood phylogenetic tree was built to validate the sequencing data (Figure 2). The bacterial isolates of the same species or phylogenetic related were correctly grouped into a common branch, as expected. The principle of maximum likelihood for phylogenetic inference evaluates the probability of a given model of evolutionary changes explaining the origin of the data observed. In this method, the initial tree is constructed using the neighbor-joining method, and the length of each branch is adjusted to maximize the likelihood that the information will produce the topology of the tree for the desired evolutionary model [22].

These results confirm and validate the direct colony PCR method to be applied as a reliable tool for the identification of bacterial fish pathogens in aquaculture. Although this method has already been used in previous studies for different purposes [14] [23] , the present study represents the first practical application for the diagnosis of aquaculture diseases, a field lacking in terms of technological advancement.

Table 4. Cost analysis per sample for bacterial identification, performed in university laboratory already equipped.

	Standard PCR		Direct colony PCR	
Steps	US	Time	US	Time
Bacterial isolation on plate	0.31	24 - 72 h	0.31	24 - 72 h
Replication in broth	0.41	24 - 48 h	-	-
DNA extraction	5.6	1 - 3 h	-	-
PCR	2.52	3 h	2.52	3 h
Electrophoresis	0.27	1.5 h	0.27	1.5 h
PCR product purification	2.54	1 h	-	-
16S rRNA sequencing	5.0	24 h	5.0	24 h
Nucleotide analysis	-	1 h	-	1 h
Overall	**16.65**	**4 - 7 days**	**8.1**	**3 - 5 days**

*Isolation times vary depending on the species being cultured.

Table 5. Advantages and disadvantages of each method for aquaculture diagnosis.

	Classical microbiology	Standard PCR	Colony PCR
Advantages	Less technicization	Conclusive diagnosis, faster than the classical Microbiology	Conclusive diagnosis; 51% more economical; 24 - 48 h faster than standard PCR
Disadvantages	Inconclusive and time consuming diagnosis	More costly; It depends on bacterial culture	1% to 3% can fail; Still depends bacterial culture

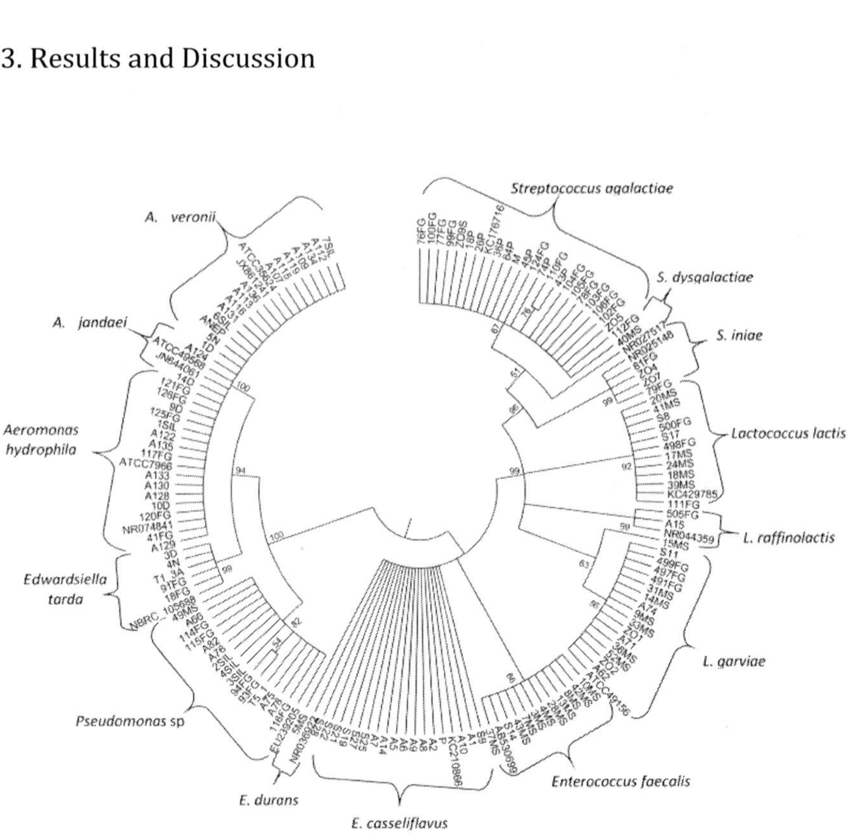

Figure 2. Relationship among different bacteria species using 16S rRNA gene sequences, inferred by maximum-likelihood method. The phylogenetic diagram shows the correct clustering of related fish bacteria isolated in the present study.

3.2. The Analysis of the Common Bacterial Fish Pathogens

Direct colony PCR, combined with gene sequencing, was able to detect the most common and important pathogens in aquaculture, such as Aeromonas hydrophila, Aeromonas veronii, Aeromonas jandaei, Streptococcus agalactiae, Streptococcus iniae, Streptococcus dysgalactiae, Edwardsiella tarda, Pseudomonas sp., Lactococcus garvieae, Citrobacter freundii, Plesiomonas shigelloides, and Enterococcus sp.

As shown in Figure 3, genera related to pathogenic bacteria and with higher frequency among 178 bacterial isolates of this study were *Aeromonas* (31%), *Lactococcus* (23%), *Enterococcus* (22%), *Streptococcus* (20%), *Pseudomonas* (11%), *Citrobacter* (6%), *Edwardsiella* (5%), *Acinetobacter* (3%), *Enterobacter* (2%), *Plesiomonas* (1%) and *Weissela* (1%).

Of the 43 Aeromonas isolates, 53% were identified as A. hydrophila by 16S rRNA gene sequencing. This result is in accordance with previous reports that found this species to be predominant [24] . In turn, A. veronii corresponded to 40% of the isolates. The seasonality was also observed in the present study: at higher temperatures (Spring/Summer) there were higher isolation rates of these pathogens [25] , which causes hemorrhagic septicemia, characterized by small superficial lesions, focal hemorrhages, ulcers, abscesses, and abdominal distension. Internally, there can be ascitic fluid accumulation, anemia, and lesions in the liver and kidneys [26] .

For the genus Lactococcus, the emerging species L. garvieae corresponded to 52% of the total 29 isolates of this genus, followed by L. lactis with 41% incidence in fish originating from the states included in the present study, with higher incidence in P. reticulatum. The species L. garvieae has been isolated from several fish species worldwide, namely in Japan [27] , South Africa [28] , Europe [29] , and Brazil. Its first outbreak was reported in 2009 [7] . Fish with lactococcal infection exhibit lethargy, anorexia, skin darkening and swim closer to the water surface [30] , resulting in considerable economic losses, especially during the summer months when the water temperature increases [4] . Few studies report L. lactis as an opportunistic pathogen. However, L. lactis subsp. lactis has been responsible for a 100% loss of hybrid sturgeons (Huso huso × Acipenser ruthenus) in a fish farm in Taiwan, China [31] .

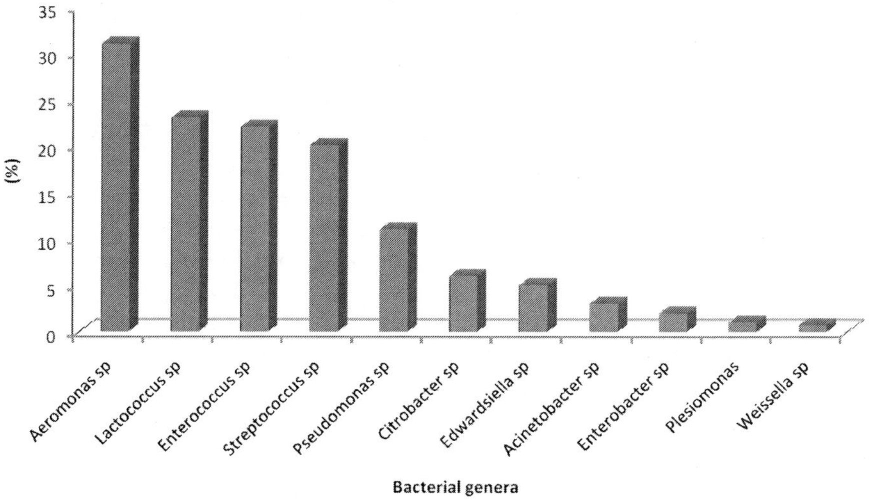

Figure 3. Percentage of bacterial genera identified by 16S rRNA gene of the 178 isolates of this study distributed in the states of Mato Grosso do Sul, São Paulo, Paraná and Rio de Janeiro.

In the present study, 31 Enterococcus strains were isolated from skin and kidney samples. Of these, 55% were E. casseliflavus, 36% E. faecalis, 6% E.durans, and 3% E. sulfureus. The predominance of E. casseliflavus has also been observed among isolates from water and sediment, accounting for 66.7% of a total of 410 Enterococcus sp. isolates in Thailand [32] .

Of the 27 Streptococcus strains originating from the states of Mato Grosso do Sul, Paraná, and São Paulo, 89% corresponded to S. agalactiae; this was previously observed by Netto et al. [33] and Figueiredo et al. [10] . Although infection by S. agalactiae is the main cause of losses in tilapia farming worldwide, this pathogen has also been isolated from "cachara" originating from Mato Grosso do Sul. S. agalactiae has been identified in several other fish species, such as Sparus auratus, Liza klunzingeri [34] , and Pampusargenteus [35] . Infected fish have meningoencephalitis, exophthalmia, erratic swimming, mainly.

The species *P. putida* (27%) and **P. fulva** (20%) were the predominant Pseudomonas species observed (n = 15). Eissa et al. [36] observed an incidence of 30.83% of Pseudomonas species in Nile tilapia in Egypt. Hussain [37] and Zorrilla et al. [38] reported 13.5%, and 9.7% incidence, respectively, of Pseudomonas species in marine fish, values that are similar to the 11% incidence found in the present study. *P. fluorescens, P. angulliseptica, P. aeruginosa* and *P. putida* were identified in various species of fish as causative agents of Pseudomonas septicemia. The disease is characterized by petechial hemorrhage, darkness of the skin, detached scales, abdominal ascitis and exophthalmia [39] .

As the number of isolates from each region was dissimilar and low, it would not be advisable to determine a frequency profile of pathogens by location, neither the prevalence of bacterial genera by fish species, but we emphasize the importance of drawing a regional profile in aquaculture health monitoring programs and preventive management, therefore, in case of disease outbreak, treatment measures are different in each region, since factors such as light, water quality and soil contamination, quantity of parasites, management, etc are also peculiar to each locality.

4. CONCLUSION

Direct colony PCR combined with 16S rRNA gene sequencing constitutes an efficient alternative for diagnosing bacterial fish diseases, with decreased cost and time compared with the classical methods used in Brazil, such as isolation, biochemical tests, and conventional PCR.

ACKNOWLEDGEMENTS

F. A. Sebastião, L. R. Furlan, D. T. Hashimoto, F. Pilarski (FAPESP-Process 2011/07951-5) for the financial support; the Aquaculture Center (CAUNESP/UNESP, Jaboticabal) and the Laboratory of Microbial and Plant Biochemistry, Technology Department (Laboratório de Bioquímica de Microrganismos e Plantas, FCAV/UNESP Jaboticabal), UNESP, for technical support; and Dr. Fabiana Garcia for donating the bacterial strains.

REFERENCES

1. FAO Fishery and Aquaculture Country Profiles. Brazil (2010) Country Profile Fact Sheets. FAO Fisheries and Aquaculture Department [online]. Rome. Updated 1 June 2010.
2. Dash, S.S., Dasi, B.K., Pattnaik, P., Samal, S.K., Sahu, S. and Ghosh, S. (2009) Biochemical and Serological Characte- rization of Flavobacterium columnare from Freshwater Fishes of Eastern India. Journal of World Aquaculture Society, 40, 236-247. http://dx.doi.org/10.1111/j.1749-7345.2009.00246.x
3. Shama, S., Brandão, D.A., Vargas, A.C., Costa, M.M. and Pedrozo, A.F. (2000) Bactérias com potencial patogênico nos rins e lesões externas de jundiás (Rhamdia quelen) cultivados em sistema semi-intensivo. Ciência Rural, 30, 293- 298. http://dx. doi.org/10.1590/S0103-84782000000200016
4. Vendrell, D., Balcazar, J.L., Ruiz-Zarzuela, I., Blas, I.D., Girones, O. and Muzquiz, J.L. (2006) Lactococcus garvieae in Fish: A Review. Comparative Immunology, Microbiology and Infectious Diseases, 29, 177-198. http://dx.doi. org/10. 1016/j.cimid.2006.06.003
5. Olivares-Fuster, O., Klesius, P.H., Evans, J. and Arias, C.R. (2008) Molecular Typing of Streptococcus agalactiae Isolates from Fish. Journal of Fish Diseases, 31, 277-283. http://dx.doi.org/10.1111/j.1365-2761.2007.00900.x
6. Staroscik, A.M., Hunnicutt, D.W., Archibald, K.E. and Nelson, D.R. (2008) Development of Methods for the Genetic Manipulation of Flavobacterium columnare. BMC Microbiology, 8, 115. http://www.biomedcentral.com/1471-2180/8/115
7. Evans, J.J., Klesius, P.H. and Shoemaker, C.A. (2009) First Isolation and Characterization of Lactococcus garvieae from Brazilian Nile Tilapia, Oreochromis niloticus (L.), and Pintado, Pseudoplathystoma corruscans (Spix & Agassiz). Journal of Fish Diseases, 32, 943-951. http://dx.doi.org/10.1111/j.1365-2761.2009.01075.x
8. Birkbeck, T.H., Feist, S.W. and Verner-Jeffreys, D.W. (2011) Francisella Infections in Fish and Shellfish. Journal of Fish Diseases, 34, 173-187. http:// dx.doi.org/10.1111/j.1365-2761.2010.01226.x
9. Burr, S.E., Goldschmidt-Clermont, E., Kuhnert, P. and Frey, J. (2012) Heterogeneity of Aeromonas Populations in Wild and Farmed Perch, Perca fluviatilis L. Journal of Fish Diseases, 35, 607-613. http://dx.doi.org/10.1111/j.1365-2761.2012.01388.x
10. Figueiredo, H.C.P., Nobrega-Netto, L., Leal, C.A.G., Pereira, U.P. and Mian, G.F. (2012) Streptococcus iniae Outbreaks in Brazilian Nile Tilapia (Oreochromis niloticus L.) Farms. Brazilian Journal of Microbiology, 43, 576-580. http://dx. doi.org/ 10.1590/S1517-83822012000200019
11. Beaz-Hidalgo, R. and Figueras, M.J. (2012) Molecular Detection and Characterization of Furunculosis and Other Aeromonas Fish Infections. In: Carvalho, E., Ed., Health

and Environment in Aquaculture, InTech Open Access Publisher, 97-132. http://dx.doi.org/10.5772/29901

12. Silva, B.C., Mouriño, J.L.P., Vieira, F.N., Jatobá, A., Seiffert, W.Q. and Martins, M.L. (2012) Haemorrhagic Septicaemia in the Hybrid Surubim (Pseudoplatystoma corruscans × Pseudoplatystoma fasciatum) Caused by Aeromonas hydrophila. Aquaculture Research, 43, 908-916. http://dx.doi.org/10.1111/j.1365-2109.2011.02905.x
13. Janda, J.M. and Abbott, S.L. (2007) 16S rRNA Gene Sequencing for Bacterial Identification in the Diagnostic Laboratory: Pluses, Perils, and Pitfalls. Journal of Clinical Microbiology, 45, 2761-2764. http://dx.doi.org/10.1128/JCM.01228-07
14. Coton, E. and Coton, M. (2005) Multiplex PCR for Colony Direct Detection of Gram-Positive Histamine- and Tyramine-Producing Bacteria. Journal of Microbiological Methods, 63, 296-304. http://dx.doi.org/10.1016/j.mimet.2005.04.001
15. Ben-Dov, E., Shapiro, O.H., Siboni, N. and Kushmaro, A. (2006) Advantage of Using Inosine at the 3' Termini of 16S rRNA Gene Universal Primers for the Study of Microbial Diversity. Applied and Environmental Microbiology, 72, 6902-6906. http://dx.doi.org/10.1128/AEM.00849-06
16. Claesson, M.J., Wang, Q., O'Sullivan, O., Greene-Diniz, R., Cole, J.R., Ross, R.P. and O'Toole, P.W. (2010) Comparison of Two Next-Generation Sequencing Technologies for Resolving Highly Complex Microbiota Composition Using Tandem Variable 16S rRNA Gene Regions. Nucleic Acids Research, 38, e200. http://dx.doi.org/10.1093/nar/gkq873
17. Lane, D.J., Pace, B., Olsen, G.J., Stahlt, D.A., Sogint, M.L. and Pace, N.R. (1985) Rapid Determination of 16S Ribosomal RNA Sequences for Phylogenetic Analyses. Proceedings of the National Academy of Sciences of the United States of America, 82, 6955-6959. http://dx.doi.org/10.1073/pnas.82.20.6955
18. Felske, A., Rheims, H., Wolterink, A., Stackebrandt, E. and Akkermans, A.D. (1997) Ribosome Analysis Reveals Prominent Activity of an Uncultured Member of the Class Actinobacteria in Grassland Soils. Microbiology, 143, 2983- 2989. http://dx.doi.org/10.1099/00221287-143-9-2983
19. Sambrook, J. and Russel, D.W. (2001) Molecular Cloning. 3rd Edition, Cold Spring Harbor Laboratory Press, New York.
20. Sanger, F., Nicklen, S. and Coulson, A.R. (1977) DNA Sequencing with Chain-Terminating Inhibitors. Proceedings of the National Academy of Sciences of United States of America, 74, 5463-5467. http://dx.doi.org/10.1073/pnas.74.12.5463
21. Meireles, M.A.O.M. (2008) Uso de antimicrobianos e resistência bacteriana: Aspectos socioeconômicos e comporta- mentais e seu impacto clínico e ecológico. Monograph (Microbiology Expert). Universidade Federal de Minas Gerais, Belo Horizonte.
22. Tamura, K., Peterson, D., Peterson, N., Stecher, G., Nei, M. and Kumar, S. (2011) MEGA5: Molecular Evolutionary Genetics Analysis Using Maximum Likelihood, Evolutionary Distance, and Maximum Parsimony Methods. Molecular Biology and Evolution, 10, 2731-2739. http://dx.doi.org/10.1093/molbev/msr121
23. Kong, P., Richardson, P.A. and Hong, C.X. (2005) Direct Colony PCR-SSCP for Detection of Multiple Pythiaceous Oomycetes in Environmental Samples. Journal of Microbiological Methods, 61, 25-32. http://dx.doi.org/10.1016/j.mimet.2004.10.019
24. Belem-Costa, A. and Cyrino, J.E.P. (2006) Antibiotic Resistence of Aeromonas hydrophila Isolated from Piaractus mesopotamicus (Holmberg, 1887) and Oreochromis niloticus (Linnaeus, 1758). Scientia Agricola, 63, 281-284. http://dx.doi.org/10.1590/S0103-90162006000300011
25. Pereira, C.S., Amorim, S.D., Santos, A.F.M., Reis, C.M.F., Theophilo, G.N.D. and Rodrigues, D.P. (2008) Characterization of Aeromonas spp. Isolates from Newborns Hospitalized. Revista da Sociedade Brasileira de Medicina Tropical, 41, 179-182. http://dx.doi.org/10.1590/S0037-86822008000200009
26. Garcia, F., Pilarski, F., Onaka, E.M., Moraes, F.R. and Martins, M.L. (2007) Hematology of Piaractus mesopotamicus Fed Diets Supplemented with Vitamins C and E,

Challenged by Aeromonas hydrophila. Aquaculture, 271, 39-46. http://dx.doi.org/10.1016/j.aquaculture.2007.06.021
27. Nishiki, I., Furukawa, M., Matui, S., Itami, T., Nakai, T. and Yoshida, T. (2011) Epidemiological Study on Lactococcus garvieae Isolates from Fish in Japan. Fisheries Science, 77, 367-373. http://dx.doi.org/10.1007/s12562-011-0332-0
28. Bekker, A., Hugo, C., Albertyn, J., Boucher, C.E. and Bragg, R.R. (2011) Pathogenic Gram-Positive Cocci in South African Rainbow Trout, Oncorhynchus mykiss (Walbaum). Journal of Fish Diseases, 34, 483-487. http://dx.doi.org/10.1111/j.1365-2761.2011.01259.x
29. Eyngor, M., Zlotkin, A., Ghittino, C., Prearo, M., Douet, D.G., Chilmonczyk, S. and Eldar, A. (2004) Clonality and Diversity of the Fish Pathogen Lactococcus garvieae in Mediterranean Countries. Applied and Environmental Microbiology, 70, 5132-5137. http://dx.doi.org/10.1128/AEM.70.9.5132-5137.2004
30. Avci, H., Aydoğan, A., Tanrikul, T.T. and Birincioğlu, S.S. (2010) Pathological and Microbiological Investigations in Rainbow Trout (Oncorhynchus mykiss Walbaum, 1792) Naturally Infected with Lactococcus garvieae. Kafkas Üniversitesi Veteriner Fakültesi Dergisi, 16, S313-S318.
31. Chen, M.H., Hung, S.W., Shyu, C.L., Lin, C.C., Liu, P.C., Chang, C.H., Shia, W.Y., Cheng, C.F., Lin, S.L., Tu, C.Y., Lin, Y.H. and Wang, W.S. (2012) Lactococcus lactis Subsp. Lactis Infection in Bester Sturgeon, a Cultured Hybrid of Huso huso × Acipenser ruthenus, in Taiwan. Research in Veterinary Science, 93, 581-588. http://dx.doi.org/10.1016/j.rvsc.2011.10.007
32. Petersen, A. and Dalsgaard, A. (2003) Species Composition and Antimicrobial Resistance Genes of Enterococcus spp., Isolated from Integrated and Traditional Fish Farms in Thailand. Environmental Microbiology, 5, 395-402. http://dx.doi.org/10.1046/j.1462-2920.2003.00430.x
33. Netto, L.N., Leal, C.A.G. and Figueiredo, H.C.P. (2011) Streptococcus dysgalactiae as an Agent of Septicaemia in Nile Tilapia, Oreochromis niloticus (L.). Journal of Fish Diseases, 34, 251-254. http://dx.doi.org/10.1111/j.1365-2761.2010.01220.x
34. Evans, J.J., Wiedenmayer, A.A. and Klesius, P.H. (2002) A Transport System for Maintenance of Viability of Acinetobacter calcoaceticus, Streptococcus iniae, and Streptococcus agalactiae over Varying Time Periods. Bulletin of the European Association of Fish Pathologists, 22, 238-246.
35. Duremdez, R., Al-Marzouk, A. and Qasem, J.A. (2004) Isolation of Streptococcus agalactiae from Cultured Silver Pomfret, Pampus argenteus (Euphrasen), in Kuwait. Journal of Fish Diseases, 27, 307-310. http://dx.doi.org/10.1111/j.1365-2761.2004.00538.x
36. Eissa, N.M.E., Abou, E.E.N., Shaheen, A.A. and Abbass, A. (2010) Characterization of Pseudomonas Species Isolated from Tilapia "Oreochromis niloticus" in Qaroun and Wadi-El-Rayan Lakes, Egypt. Global Veterinaria, 5, 116-121.
37. Hussain, R.A. (2002) Studies on Some Bacterial Infections Affecting Certain Marine Fishes in the Arabian Gulf of Kingodom of Saudi Arabia. Ph.D. Dissertation., Faculty of Veterinary Medicine and Animal Resources, King Faisal University, Al-Ahsa.
38. Zorrilla, I., Chabrillón, M., Arijo, S., Díaz-Rosales, P., Martínez-Manzanares, E., Balebona, M.C. and Moriñigo, M.A. (2003) Bacteria Recovered from Diseased Cultured Gilthead Sea Bream (Sparus aurata L.) in Southwestern Spain. Aquaculture, 218, 11-20. http://dx.doi.org/10.1016/S0044-8486(02)00309-5
39. Austin, B. and Austin, D.A. (2007) Bacterial Fish Pathogens. Diseases of Farmed and Wild Fish. Springer-Praxis Publishing, Ltd., Chichester.

CHAPTER 11

Strategies and Hurdles using DNA Vaccines to Fish

Linn B Hølvold[1], Anne I Myhr[2] and Roy A Dalmo[1*]

[1]*UiT - The Arctic University of Norway, Faculty of Biosciences, Fisheries & Economics, Norwegian College of Fishery Science, 9037 Tromsø, Norway.*

[2]*Genøk – Centre for Biosafety, The Science Park, Postbox 6418, 9294 Tromsø, Norway.*

ABSTRACT

DNA vaccinations against fish viral diseases as IHNV at commercial level in Canada against VHSV at experimental level are both success stories. DNA vaccination strategies against many other viral diseases have, however, not yet yielded sufficient results in terms of protection. There is an obvious need to combat many other viral diseases within aquaculture where inactivated vaccines fail. There are many explanations to why DNA vaccine strategies against other viral diseases fail to induce protective immune responses in fish. These obstacles include: 1) too low immunogenicity of the transgene, 2) too low expression of the transgene that is supposed to induce protection, 3) suboptimal immune responses, and 4) too high degradation rate of the delivered plasmid DNA. There are also uncertainties with regard distribution and degradation of DNA vaccines that may have implications for safety and regulatory requirements that need to be clarified. By combining plasmid DNA with different kind of adjuvants one can increase the immunogenicity of the transgene antigen – and perhaps increase the vaccine efficacy. By using molecular adjuvants with or without in combination with targeting assemblies one may expect different responses compared with naked DNA. This includes targeting of DNA vaccines to antigen presenting cells as a factor in improving their potencies and efficacies by means of encapsulating the DNA vaccine in certain carriers systems that may increase transgene and MHC expression. This review will focus on DNA vaccine delivery, by the use of biodegradable PLGA particles as vehicles for plasmid DNA mainly in fish.

1. INTRODUCTION

There are quite a limited number of DNA vaccine strategies that have been successful in giving significant protection against fish diseases. The excellent exceptions are DNA vaccination against infectious hematopoietic necrosis virus (IHNV) at commercial level and against viral hemorrhagic septicemia virus (VHSV) at experimental/small scale level [1]. A promising strategy to increase the DNA vaccine efficacy, against other viral diseases, is to target the construct and/transgene to antigen presenting cells. Wang et al. [2] have presented an overview on how DNA vaccines can be targeted to antigen presenting cells (APC) and dendritic cells (DC) using molecular assemblies with the resulting immune responses. Another strategy is to encapsulate the DNA vaccine in certain carriers with the intention to increase transgene and MHC expression. There are many particulate carriers that have been explored to present and target DNA vaccines to desired cells and tissues, one of them being poly (D,L-lactic-co-glycolic)-acid (PLGA) particles.

2. DNA VACCINES AND VACCINATION

The definition of DNA vaccination as provided by The Norwegian Biotechnology Advisory Board [3] is "the intentional transfer of genetic material (DNA or RNA) to somatic cells for the purpose of influencing the immune system". This sets it slightly apart from gene therapy, which in the same report is referred to as an introduction of novel gene(s) for purposes other than influencing the immune system. The mechanism of a DNA vaccine can in many ways be likened to that of a virus, as it requires the same cellular machinery in order to replicate and since it also triggers immune responses normally seen with viral infections [4]. Unlike conventional viral vaccines based on subunits or killed virus, a DNA vaccine may conserve the structure and hence also antigenicity of a transgenic antigen/protein [5].

3. IMMUNE RESPONSES TO DNA VACCINATION – MAINLY IN MICE AND BIRDS

A vital attribute of DNA vaccines is the ability to induce all three arms of adaptive immunity, namely; helper T-cells, cytotoxic T cells (CTLs) and antibodies, although they were initially investigated in the attempt to find ways of delivering antigen to major histocompatibility complex (MHC) class I and induce T helper 1 (Th1) responses [6, 7]. Professional APCs are the cells that primarily contribute to the immune responses to DNA vaccination. Direct transfection of DCs provides the most efficient priming of naïve CTLs, and is perhaps the major mechanism for priming of these cells [8]. APCs are also able to take up exogenous antigens and process them for presentation by either MHC class II, or MHC class I following transfer of the antigen to the cytosol (cross-priming) [9]. These responses are vital in the cellular immune response following DNA vaccination. The expression of reporter genes is not only higher in fish, but also seems to have a longer duration. Transgene expression has been detected at the injection site as long as two years after injection of glass catfish (*Kryptopterus bicirrhus*) [10], and Tonheim et al. detected both supercoiled (sc) DNA and luciferase expression at the injection site 535 days after intramuscular injection of Atlantic salmon (*Salmo salar* L.) with pDNA [11].

The immunogenicity of DNA vaccines stems not only from the expression of the GOI, but also from properties of the plasmid vectors themselves. The ability of foreign nucleic acids to induce interferons (IFNs) in mouse fibroblasts was discovered by two independent research groups as early as in 1963 [12]. The most studied pDNA property in terms of possible adjuvant effects is CpG motifs, regions in the DNA where a cytosine nucleotide occurs next to a guanine nucleotide and the two are linked by a phosphodiester bond. In vertebrates these sequences are highly methylated, whereas they show a much lower methylation frequency in viral and bacterial DNA (such as pDNA). These CpG motifs can act as pathogen-associated molecular pattern molecules (PAMPs) and are recognized as danger signals by the vertebrate immune system, resulting in a release of cytokines, macrophage activation, a differentiation of Th1 effector cells as well as B-cell proliferation and secretion of antibodies [13]. Plasmid DNA containing optimized CpG content may be likely used in DNA vaccine strategies.

4. DNA VACCINATION AGAINST NOVIRHABDOVIRUSES

The first DNA vaccination of fish took place in 1996, when Anderson et al. immunized rainbow trout (*Onchorhyncus mykiss*) against IHNV [14]. Since then several trials have been performed for a wide variety of fish species and pathogens and in 2005 a vaccine against IHNV infection in salmonids (Apex-IHN®, Novartis Animal Health) was also one of the first DNA vaccine ever to be cleared for marketing (by the Canadian Food Inspection Agency). In 1999 the injection of Atlantic salmon with pCMV4-G (plasmid-encoded glycoprotein) from a rainbow trout IHNV isolate induced significant protection against challenges with IHNV, even though the salmon were much larger than the rainbow trout in previous challenge studies [15]. DNA vaccination of fish has been shown to induce both innate and adaptive immune responses similar to what is seen in mammalian species, and seems especially efficient against novirhabdoviruses (like VHSV and IHNV). These are simple RNA viruses with either five or six genes and a single viral surface protein (glycoprotein, or G protein) that acts as the protective antigen [16].

In rainbow trout an immunization against VHSV enables the induction of cell-mediated immune responses encompassing both CTLs and natural killer (NK) cells and has also been shown to significantly reduce the replication of virus during challenge [17]. Interestingly, when Cuesta et al. looked at the innate and adaptive responses in vaccinated and control fish after challenge, they found the highest increase in vaccinated fish to be that of innate immune responses [18]. Lorenzen et al. [19] have also demonstrated the importance of innate responses in early antiviral defense, wherein rainbow trout were subjected to VHSV challenge following vaccination with pIHN-G (plasmid-encoded IHNV glycoprotein). Whereas protection at late stages of the challenge could only be conferred by previous immunization with pVHS-G, the two vaccines induced similar levels of immune responses and protection during the first week following challenge.

In conclusion, anti novirhabdovirus response governed by DNA vaccines relies much on the immunogenicity of the G-protein resulting in long-lasting protection mediated by cellular and humoral responses. In addition, there is an innate and specific response following DNA vaccination that may be protective in a short-term perspective.

Other RNA viruses, or the larger DNA viruses, often offer more difficulty in identifying a protective antigen, although viral surface protein genes are almost always chosen for DNA vaccines [2]. Some pathogens, such as infectious pancreatic necrosis virus (IPNV), primarily cause disease in fry.

Whereas vaccination by injection is highly impractical at this stage [20], the vaccination of post-smolts has been shown to induce protection upon challenge [21]. Good protection did, however, require the use of plasmids encoding all the large poly-proteins of the IPNV.

5. RECENT DNA VACCINATION LABORATORY TRIALS

DNA vaccines and their effects against several viral and bacterial diseases in fish have been reviewed by Tonheim et al.[22], Kurath [23], Redding and Weiner [24], and Gomez-Casado et al. [25]. Since these reviews were published, several new results using DNA vaccine strategies have been reported (Table 1).

Table 1 Experimental DNA vaccines and their protection in fish following experimental infection

Pathogen	Gene inserted	Host	Administration route/adjuvant	Protection	References
IHNV	IHNV-G plus suicidal gene	Rainbow trout	I.m/none	Yes	[26]
IHNV	IHNV-G; different genogroups	Rainbow trout	I.m/none	Yes; cross-protection	[27]
IHNV	IHNV-G	Rainbow trout	Oral/PLGA	No	[28]
VHSV	*E. tarda* as delivery vehicle of the vaccine	Olive/Japanese flounder (*Paralichthys olivaceus*)	I.m	Yes	[29]
IPNV	VP2		Oral/alginate	Yes	[30]
IPNV	VP2; Segment A of TA strain	Atlantic salmon	I.m	No	[31]
SAV	E1 and E2	Atlantic salmon	I.m	No	[32]
Megalocytivirus	86-residue VP	Turbot (*Scophthalmus maximus*)	I.m	Yes	[33]
Edwardsiella tarda	Eta6-FliC chimeric protein	Japanese flounder (*Paralichthys olivaceus*)	I.m	Yes	[34]
E. tarda	D15-like surface antigen	Japanese flounder	I.m	Yes	[35]
E. tarda	Eta2	Japanese flounder	I.m	Yes	[36]
Streptococcus iniae	sagF, sagG and sagI	Japanese flounder	I.m	Yes	[37]
S. iniae	Sia10 delivered by *E. tarda*	Japanese flounder	Oral/alginate, immersion boosted	Yes	[38]
Vibrio harveyi	FlaA	Yellow grouper (*Epinephelus awoara*)	I.m	Yes	[39]
V. harveyi	DegQ or/and Vhp1	Japanese flounder	I.m	Yes	[40]
V. alginolyticus	FlaA	Red snapper (*Lutjanus campechanus*)	I.m	Yes	[41]
S. iniae and *V. anguillarum*	Sia10 and/or OmpU	Turbot	I.m	Yes, cross-protection	[37]
Flavobacterium psychrophilum	Hsp60, hsp70	Rainbow trout	I.m	No	[42]
Cryptocaryon irritans	iAg	Orange spotted grouper (*E. coioides*)	I.m	Yes	[43]
Cryptobia salmositica	Metalloprotease	Atlantic salmon and rainbow trout	I.m	Partly	[44]
Ichthyophthirius multifiliis	Immunobilization antigens and cystein protease	Rainbow trout	I.m, gene gun and air pressure	No	[45]

Apparently, DNA vaccination may also confer protection against bacteria and parasites – but not against all. Bacteria and parasites may express and harbor numerous different antigens dependent on their life cycle, this make a DNA vaccination strategy more complicated. There are developed "traditional" oil adjuvanted vaccines against a number of bacterial diseases, where the strategies may be quite efficient preventing infection and disease. This is in contrast to some parasitic infections/attachment such as sea lice - where no immune prophylaxis exists at present at an industrial scale.

6. ADMINISTRATION AND DISTRIBUTION OF DNA VACCINES

6.1. Intramuscular injection

Intramuscular injection is widely applied for pDNA delivery in fish and generally results in strong expressions of transgene at the injection site [22]. Studies in mice have found the dispersion of pDNA immediately following intramuscular injection to take place primarily between the muscle body and epimysium (connective tissue that enclosing the entire muscle). Myocytes and mononuclear cells take up pDNA after administration [22], but despite a rapid initiation of uptake the subsequent uptake is slow and cells along the muscle fibers have been shown to be transfected over a period of hours following injection. With very small fish this initial dispersion of a vaccine might be enough to ensure the perfusion of intact pDNA to more distant tissues, while in large fish the injected volume will mainly rest along the needle trajectory [22]. The transportation of pDNA to and from blood to other tissues has been reported for various fish species [22]. Plasmids have been recovered from sites such as liver, spleen, head-kidney, heart and intestine for some time after injection, but mainly persist at the site of injection. Degradation of the pDNA starts within five minutes following injection of mice, with as much as 95-99% of the initial pDNA amount degraded within 90 min [22]. The rate of degradation in the tissue of cold-water fish remains to be determined. The extent of histopathological changes at the injection site following intramuscular DNA delivery in fish appear to increase with an increase in vaccination dose [46], but vaccination will generally induce only moderate local tissue damage in form of degeneration of myocytes, hemorrhages and a transient influx of inflammatory cells [20].

6.2. Other routes of delivery

Other routes of pDNA administration that have been investigated in fish are intravenous, intraperitoneal, oral delivery, and particle bombardment [1, 21]. Accumulation of naked pDNA took place primarily in the heart, kidney and liver following intravenous administration [22], whereas oral delivery resulted in a recovery of DNA fragments from the pyloric region, kidney, spleen and liver – assessed by revers-transcription polymerase chain reaction (RT-PCR) [28].

7. UPTAKE OF PLASMID DNA IN FISH AND MAMMALIAN SPECIES

A wide variety of mammalian cell types has been shown to take up pDNA – reviewed by Budker et al. [47], but so far uptake in fish has only been reported for myocytes, head kidney macrophages and endocardial endothelial cells (EECs) [22]. The exact mechanisms by which myocytes take up pDNA remains to be determined, but several suggestions have been made [48]. It was previously theorized that direct injections caused temporary membrane disruptions and/or pores that allowed for the entry of pDNA [5], but studies have shown that such disruptions in fact work to abolish transfection [12].

Several PPRs have been shown to bind nucleic acids including plasmid DNA, reviewed by Desmet et al. [49]. Of the number of cell-surface receptors investigated in terms of DNA binding and uptake, scavenger receptors (SRs) in particular have been a subject of interest. These receptors comprise a broad family of membrane proteins capable of binding a wide range of anionic ligands and are present on several different cell types [50]. The uptake of pDNA by SRs has been demonstrated in Atlantic cod (*Gadus morhua*) atrial EECs [51], but although the SRs may bind DNA they appear not to be essential for the immunostimulatory activity of CpG DNA [52]. This indicates that SRs activity may be essential for the transgene expression – as they may bind and facilitate degradation of high amount of plasmid DNA before expression occurs.

8. FACTORS INFLUENCING TRANSFECTION AND TRANSGENE EXPRESSION

Nucleic acids have poor intrinsic transfection efficiency due to their large size and negative charge [53]. The processes of gene transcription and translation are carried out by the cell's own machinery, and may be influenced by a variety of factors such as pDNA vector design, pDNA concentrations, administration volumes, age and size of the fish, water temperatures and method as well as route of administration.

It has been well documented that the DNA topoform has a strong influence on the efficiency of transfection [54]. Supercoiled DNA is frequently reported as the most potent topoform followed by open circular (oc) forms, whereas linearization of the DNA has been shown to nearly abolish the expression of transgene and linearized pDNA. The choice of promoter also has a profound influence on the level of transgene expression [55] and the cytomegalovirus (CMV) promoter has often been reported as the most effective. Because of this, the CMV promoter is the most commonly applied in gene transfection studies and its potency has been demonstrated also in fish [5]. The use of an intron and poly-adenylation (termination) signals further improves expression [55].

Despite the inherent adjuvant effects of DNA vaccines, it has been shown that immune responses triggered by DNA vaccination may in fact limit the duration of transgene expression [55]. The lack of anti-Luc antibodies means that both levels and duration of expression are commonly higher in studies applying luciferase compared to more immunogenic antigens, and immune responses have only been observed with the application of large pDNA doses and potent adjuvants such as Freund´s complete adjuvant (FCA) [56]. A co-administration of vectors encoding Luc and immunogenic proteins (β-gal, G protein) decreases both the strength and duration of luciferase expression, as the initiated antigen specific cytotoxic responses work to eliminate transfected cells expressing the transgenic protein [56]. The stimulation of PRRs may also induce responses that can be detrimental to transgene expression. The hallmark cytokines of the inflammatory response, TNF-α and IL-1β, have both been shown to inhibit transgene expression in vitro and in vivo [57], as have the Th1 associated cytokines IFN-γ and IFN-α [58]. Levels of inhibition have in fact been shown to correlate with the levels of CpG-induced cytokines, and synergistic effects were also observed. The inhibitory effect takes place at the mRNA (post-transcriptional) level, hence not causing vector degradation, inhibition of total cellular protein synthesis or elimination of infected/transfected cells [58].

Whereas intramuscular injection is widely acknowledged as the superior method of administration to achieve high levels of transgene expression [1], the result still depends on factors such as dose, volume and fish size and age [59]. Transgene expression has been found to be higher in young and growing fish [5], and small fish sizes appear to favor not only the distribution of pDNA throughout tissue but the distribution of transgene expression as well [5]. For example expression of luciferase in thymus, gills, spleen and kidney has been reported for small fish (< 5 g), although the highest expression is consistently detected at the site of injection in myocytes, infiltrating cells and epithelial cells lining small capillaries [20, 60].

The first reporter gene study in fish indicated the existence of a maximum above where there would be no further increase in expression [60]. This observation is supported by later findings in zebra fish (*Danio rerio*) and rainbow trout [5], and there are indications that excessive DNA concentrations may actually reduce transgene expression [21]. The injection of large volumes might contribute to a spatial distribution by creating temporary gaps between fibers [61], which appears to induce higher expression as well as reduce the variations commonly observed in in vivo transfection studies [5]. A pre-injection of an isotonic solution created the same effect, and lead to both higher and more equal transgene expression among individuals after DNA injection [61].

9. ADVANTAGES, DISADVANTAGES AND CHALLENGES OF DNA VACCINES

The potency of DNA vaccines for inducing the different branches of both innate and adaptive immunity has already been described. DNA vaccines also show high efficiency when given at early life stages [62], and provide the benefit of inducing protective immunity over a wide span of temperatures [63].

The advantages of DNA vaccination still stretch beyond merely the immunological capacities. Looking at the concept from a manufacturer or/and investor's standpoint, DNA vaccines are relatively inexpensive and easy to produce. The processes required for production are identical for all DNA vaccines, and the ease of cloning also enables rapid modifications in a way that is generally not obtainable with conventional vaccine preparations [60].

Potential side effects include e.g. risk of autoimmunity, immune tolerance against the expressed antigen, too high CTL response resulting in myositis, chromosomal integration, injection site inflammation and tissue destruction [1, 64].

10. APPLICATION OF MOLECULAR ADJUVANTS TO INCREASE TRANSGENE IMMUNOGENICITY

To increase immunogenicity of a given DNA vaccine one may consider optimization by including co-injected adjuvant, either being traditional aluminum salts, polysaccharides (e.g. zymosan, glucans, chitosan), different liposomes, synthetic polymers and TLR agonists. Moreover, plasmid encoded cytokine adjuvants may also be one approach to increase the immunogenicity. IL-2, IFN-γ, IL-12, GM-CSF and IL-15 have been shown to modulate immune responses when co-encoded by the DNA vaccine [65]. One may also apply the concept of immune modulating effects chemokines, transcription factors and/or co-stimulatory factors assembled into the plasmid vectors. Using fish models, only one report on the use of cytokine adjuvants encoded by the same plasmid as the expression plasmid has been published. Caipang et al. used Japanese flounder interferon regulatory factor-1 (IRF-1) cloned into a plasmid DNA vector containing the major capsid protein gene of sea bream iridovirus. The antibody levels of fish injected with this vaccine were not significant higher with the control plasmid without IRF-1 gene [66]. Unfortunately, no experimental virus challenge was carried out to monitor vaccine efficacy. To increase vaccine potency and efficacy of poor performing DNA vaccines one should explore strategies such as inclusion of molecular adjuvants perhaps in combination with targeting carrier systems such as nano- and microparticles. Although transfection levels are often low, studies have demonstrated a persistence of transgene expression at the injection site that might coincide with a time when the fish would normally be ready for slaughter [67]. There have been experiments with suicidal DNA vaccines for fish, where the plasmid vector includes a protein to induce apoptosis after an immune response has been triggered [26].

11. PLGA PARTICLES AS CARRIER SYSTEMS FOR DNA VACCINES – FOCUS ON FISH

Besides the proven efficacy of G-protein expressing plasmids against novirhabdoviruses, there is an urgent need to develop high performing vaccines against other viral diseases. In this aspect, DNA vaccines delivered by several kinds of particles may show promise. There is considerable research being done on the application of carrier systems and/or DNA-complexes to facilitate higher levels of entry and reduce degradation, reviewed by Saade et al. [65]. The potential of PLGA particles as adjuvants and carriers for DNA vaccine delivery has received considerable attention in mammalian studies [68]. In spite of this, reports on the use of PLGA particles for DNA delivery to fish are nearly non-existent. In 2008, Tian et al. were the first to report on the use of PLGA microcapsules containing a plasmid vaccine for the oral immunization of Japanese flounder against lymphocystis disease virus (LCDV) [69]. Following immunization they detected transgene expression in gills, intestine, spleen and kidney from fish vaccinated with encapsulated pDNA. The encapsulated pDNA also induced higher levels of antibodies compared to control fish injected with naked pDNA. Tian and Yu later demonstrated a significant increase in resistance to LCDV infection after oral administration of a pDNA vaccine encapsulated into PLGA nanoparticles [70]. Finally, the last study so far reported was by Adomako et al., who used PLGA nanoparticles incorporated into feed for oral delivery of a DNA vaccine against IHNV to rainbow trout [28]. They found that particles were mostly taken up in the posterior gut, but that a significant number of fish that showed uptake in gastrointestinal epithelial cells did not demonstrate detectable levels of transgene expression. Although antibody responses could be detected in fish given high doses of pDNA, the RPS at a six weeks post-vaccination challenge was still only 22%. No VHSV/IHNV DNA vaccine trials using nano- or micro-particle vehicles have been carried out so far. It would have been very interesting to compare the vaccine efficacy of naked plasmid encoding for VHSV and IHNV G-protein to corresponding particle delivered vaccines.

The exploitation of PLGA encapsulated DNA vaccines in fish is new, currently there are no information whether this strategy is a way to go to increase efficacy of DNA vaccines for fish. There are, however, indications that pDNA encapsulated in PLGA nanoparticles induce a antiviral immune response in salmon - at a higher level than what is achieved with only pDNA [71].

12. ISSUES OF MAKING DNA ENCAPSULATED PLGA PARTICLES

The w1/o/w2 method described previously is frequently used for the encapsulation of pDNA into PLGA particles, but results are highly variable with regard to encapsulation efficiency and loading as well as DNA degradation and release. The hydrophilic nature of DNA complicates the process of encapsulation as it increases the risk of plasmid diffusion into the w2 phase during solvent evaporation [72], and the encapsulation process also offers multiple challenges in terms of preservation of bioactivity – i.e. keeping the supercoiled DNA topoform intact.

Among the factors known to affect the integrity of the pDNA to be encapsulated are the polymer composition and molecular weight (Mw), shear force, preparation temperatures, solvents and the concentration and Mw of the applied stabilizer [73]. In general, polymers with a high Mw also result in the highest encapsulation efficiencies and lower the burst release of plasmid DNA [73].

13. TRANSGENE EXPRESSION AND IMMUNE RESPONSES BY PLGA-ENCAPSULATED PDNA

Encapsulated DNA has been shown to be more potent than naked DNA at mediating transgene expression in vitro in a variety of mammalian cells types [74]. However, in vivo studies report on a superiority of naked DNA in eliciting transgene expression not only compared to PLGA particles, but also to other formulated DNA vaccines as well [74].

Small particles (< 100 nm) are often shown to be internalized more rapidly, and also show the highest transfection efficiencies [75].

The adsorption of plasmid DNA onto the surface of PLGA particles, rather than encapsulation, has been reported to result in a higher transgene expression, but the expression declined more rapidly [48].

In addition to the pro-inflammatory cytokines induced by empty PLGA nano- and micro-particles (IL-1β and TNF-α) [76], the use of different particle sizes might influence the resulting cytokine profile after administration of encapsulated or particle-bound CpG DNA [77]. In addition to pro-inflammatory cytokines, nanoparticles have been shown to also enable an induction of antiviral cytokines such as type I IFNs in addition [77]. The encapsulation of pDNA encoding antigens has also been shown to elicit CTL-responses, even with pDNA-amounts that elicited no such responses after naked administration [77]. Moreover, encapsulated pDNA also enhanced the total antibody response at high doses, while inducing a more rapid and complete sero-conversion when lower doses were applied [78].

The injection of PLG microspheres into muscle has been shown to result in a foreign body response, with a large influx of different inflammatory cells that appear largely related to microspheres especially at later time-points

[79]. These infiltrating cells were also the ones that were primarily transfected, an observation similar to that found in another study [80].

14. OTHER PARTICLES IN VACCINE DELIVERY TO FISH

The extensive use of PLGA as vaccine delivery vehicles can largely be attributed to their high biocompatibility as well as the ease with which the particles can be prepared. There are other means to increase the level of transgene expression that may increase vaccine efficacy, given that the transgenes produced display fair immunogenic properties, by using tailored particle systems – as outlined by Nguyen et al. [73]. Other particles have also been investigated as carrier systems for vaccination of fish (Table 2), either on their own or in combination.

Table 2 Compounds tested as adjuvants and gene vaccine delivery systems to fish species, published from 2007 to date

Particles	Carrier molecule(s)	Assembly	Species
Alginate	β(1,4)-D-mannuronic and α(1,4)-L-guluronic acid residues	Encapsulation	Japanese flounder [81]
Chitosan	β(1,4)-D-glucosamine and N-acetyl-D-glucosamine copolymer	Encapsulation	Nile tilapia (*Oreochromis niloticus*) [82]
			Japanese flounder [83]
			Asian sea bass (*Lates calcifer*) [84]
			Turbot [85]
Liposomes	Artificial lipid bilayer vesicles	Encapsulation	Kelp grouper (*E. bruneus*) [86]
Polycaprolactone	Biodegradable, synthetic polymer	Coating	Indian major carp (*Labeo rohito*) [87]
Calcium phosphate	Inorganic, biodegradable and biocompatible material	Coating	Indian major carp [88]

15. CURRENT CHALLENGES IN THE USE OF PLGA PARTICLES AS CARRIERS/ADJUVANTS

Despite a growing number of optimistic reports on the adjuvant/carrier properties of PLGA particles, there are still many hindrances to be overcome. One of these is the preparation method, where different drugs/vaccines require different conditions. While some compounds are easily encapsulated, others – like DNA vaccines – are more difficult to encapsulate in an efficient manner. Low encapsulation efficiencies not only result in low antigen loading, but also mean that a large amount of drug/vaccine goes to waste during particle preparation. The detrimental effects that encapsulation may have on certain antigens such as pDNA are also major limiting factors at the moment, and need to be resolved. Particle preparation is also low-scale work as of yet, with considerable effort to be made before the process is optimized on a large scale suitable for mass-production.

16. CONCERNS REGARDING PLGA NANO- AND MICRO-PARTICLES

Any construct/compound, when brought down to sub-micron sizes, will exhibit new and potentially harmful characteristics [89]. The small size means they can interact with biological membranes in an entirely new way, thus inducing responses not seen with larger constructs. As with all new applications it takes time to fully survey the potential side effects. In vitro studies using PLGA nano- and micro-particles have so far not revealed any toxic effects, even at large doses [90]. The various concerns regarding the use of PLGA particles in aquaculture vaccines have recently been reviewed by Nielsen et al. [91].

17. SAFETY AND REGULATORY ASPECTS BY DNA VACCINES

Safety aspects include potential effects on the vaccinated animals, the environment and the consumer [1, 64, 92]. Other safety issues include potential shed of the vaccine to the environment from the vaccinated animal and by predatory animals [93]. Human safety does also, although with low probability, include potential effects by self-injection by vaccinators. These safety aspects need to be taken into consideration by relevant authorities when safety aspects are to be documented.

The Canadian Food Inspection Agency (CFIA) that gives authority to the Veterinary Biologics Section (VBS) of the Animal Health and Production Division (AHPD) approved the IHNV DNA vaccine for commercial use in Canada in 2005 [94]. Five aspects were taken into consideration before the vaccine was approved: (i) public perception and acceptance, (ii) regulatory and environmental concerns, (iii) risk-benefit, (iv) feasibility of producing the vaccine at a scale and cost appropriate for the fish industry, and (v) intellectual property issues [91]. At present there has not been approved any DNA vaccine to be used in aquaculture in Europe. To obtain marketing authorization within Europe, a new veterinary medicine has to meet the criteria and requirements of the EU pharmaceutical legislation for both medical and veterinary applications. To achieve a marketing application a

MAA (marketing authorization application) has to be submitted to the European Medicines Agency (EMA). The EMA has drafted guidelines for the veterinary use of DNA vaccines [95]. The European guidelines include several aspects to be considered in order to conduct a risk assessment of DNA vaccines. This includes i) the possibility of pDNA integrating into the chromosome, ii) concerns about possible adverse effects on the immune system, iii) risks posed by the additional use of genes encoding cytokines or co-stimulatory molecules or iv) undesirable biological activity by the expressed antigen itself. At present there are work undertaken within EMA with the intention to revise and update the guidance for DNA vaccines and the goal is to provide a specific guidance document [96]. This work is especially concerned to the use of DNA vaccines to humans and if the outcome will have influence on the veterinary use of DNA vaccines is therefore uncertain. The guidelines prepared by the FDA (US) points to some of the same areas as in EU, and recommends that safety testing should include tests on vaccine immunogenicity, effects from cytokines and other immunomodulatory genes, autoimmunity, local reactogenicity and systemic toxicity and studies of bio-distribution, persistence and integration [97].

The issue of plasmid persistence and chromosomal integration of DNA vaccine is of relevance for both safety and policy [92, 93]. Norwegian authorities may for example, due to uncertainties with regard to whether DNA vaccines persist degradation in tissues and organs label DNA vaccinated fish as a GMO [98]. If a DNA vaccinated animal is considered to be a GMO the producers has also to meet the requirements of the EU environmental legislation on the deliberate release of GMOs (Directive 2001/18/EC). The objective of an environmental risk assessment (ERA) in accordance with Directive 2001/18/EC is to identify and assess on a case-by-case basis the potential harmful effects of a GMO for humans, animals (domestic and wildlife), plants, microorganisms and the environment. The EMA has developed two specific guidelines for the preparation of ERA to facilitate adoptions of the requirements and the methodology of the Directive to GMO-containing medical products. Although these guidelines have mainly been used for MAA of GM vaccines, they may provide necessary information and procedures that can also be used to perform ERA for DNA vaccines and vaccinated animals.

The process of obtaining a market authorization for a new fish vaccine is a both time consuming and expensive. To overcome some of the legal uncertainties there is a need to assess potential consequences on a case-by-case basis and to be aware of that a number of factors such as the inserted gene, other gene inserts of the pDNA, injection site and amount of vaccine injected are of importance. Safety and regulatory uncertainties are related to distribution and degradation of the DNA after injection and includes that there is a need of research on the (i) the stability of the DNA vaccine, (ii) plasmid persistence, (iii) unintended immunological impacts, and (iv) potential for integration of the pDNA into the chromosome of the recipient organism [93]. These uncertainties are also of relevance for consumer acceptance and the markets confidence in safety.

The GMO issues may only be of concern for countries that have specific GMO legislation and that demands that GMOs and products thereof need to be labeled. In USA and Canada there are no requirements for labeling of food containing GMOs, and they do not have specific GMO legislation. In Europe due to the uncertainties with regard to the persistence of a DNA vaccine, the vaccinated fish may need to be labeled as a GMO. For example European countries that employ restrictions on GMO may not import the DNA vaccinated fish, and if the fish need to be labeled as a GMO this will have an influence on consumer willingness to buy it. Ultimately may these uncertainties prevent the use of DNA vaccines [92].

18. CONCLUSIONS

There is a crucial need to increase efficacy of DNA vaccines against persistent and hard-to-combat viral infections, this can be met by: (i) Application of vaccine carriers to increase the uptake in antigen presenting cells – followed by enhanced presentation of transgene peptides/antigens, (ii) Use of nano-scale particles to increase the level of cross presentation by such cells – this may also be beneficial to produce antibody response as well as cell mediated immunity, and (iii) Employing additional adjuvants such as TLR ligands, other than RNA and/or DNA, ligands to boost the response considerably. Safety and regulatory uncertainties are related to distribution and degradation of the DNA after injection and more effort needs to be put into gaining understanding of the mechanisms of pDNA uptake, from the moment of administration until the stage of transcription and translation in the nucleus.

19. ACKNOWLEDGEMENTS

This project was funded by the Research Council of Norway (project nos. 182035 and 183204/S40) and Tromsø Research Foundation ("Induction and assessment of T cell immunity to virus antigens in salmonids).

20. REFERENCES

1. Evensen Ø, Leong JA: DNA vaccines against viral diseases of farmed fish. Fish Shellfish Immun. 2013, 35: 1751-1758. 10.1016/j.fsi.2013.10.021.
2. Wang G, Pan L, Zhang Y: Approaches to improved targeting of DNA vaccines. Hum Vaccin. 2011, 7: 1271-1281. 10.4161/hv.7.12.17983.
3. Board TNBA: Regulation of DNA vaccines and gene therapy on animals. 2003, [http://www.bion.no/filarkiv/2010/07/2006_03_regulation_of_DNA_vaccines.pdf]
4. Ballesteros NA, Saint-Jean SS, Perez-Prieto SI, Coll JM: Trout oral VP2 DNA vaccination mimics transcriptional responses occurring after infection with infectious pancreatic necrosis virus (IPNV). Fish Shellfish Immunol. 2012, 33: 1249-1257. 10.1016/j.fsi.2012.09.004.
5. Heppell J, Lorenzen N, Armstrong NK, Wu T, Lorenzen E, Einer-Jensen K, Schorr J, Davis HL: Development of DNA vaccines for fish: vector design, intramuscular injection and antigen expression using viral haemorrhagic septicaemia virus genes as model. Fish Shellfish Immunol. 1998, 8: 271-286. 10.1006/fsim.1997.0133.
6. Liu MA: DNA vaccines: an historical perspective and view to the future. Immunol Rev. 2011, 239: 62-84. 10.1111/j.1600-065X.2010.00980.x.
7. Raz ETH, Sato Y, Corr M, Dudler JA, Swain SL, Spiegelberg HL, Varson DA: Preferential induction of a Th1 immune response and inhibition of specific IgE antibody formation by plasmid DNA immunization. Proc Natl Acad Sci U S A. 1996, 93: 5141-5145. 10.1073/pnas.93.10.5141.
8. Sbai H, Schneider J, Hill AV, Whalen RG: Role of transfection in the priming of cytotoxic T-cells by DNA-mediated immunization. Vaccine. 2002, 20: 3137-3147. 10.1016/S0264-410X(02)00251-7.
9. Dixon B, Stet RJM: The relationship between major histocompatibility receptors and innate immunity in teleost fish. Dev Comp Immunol. 2001, 25: 683-699. 10.1016/S0145-305X(01)00030-1.
10. Dijkstra JM, Okamoto H, Ototake M, Nakanishi T: Luciferase expression 2 years after DNA injection in glass catfish (*Kryptopterus bicirrhus*). Fish Shellfish Immunol. 2001, 11: 199-202. 10.1006/fsim.2000.0303.
11. Tonheim TC, Leirvik J, Løvoll M, Myhr AI, Bøgwald J, Dalmo RA: Detection of supercoiled plasmid DNA and luciferase expression in Atlantic salmon (*Salmo salar* L.) 535 days after injection. Fish Shellfish Immunol. 2007, 23: 867-876. 10.1016/j.fsi.2007.03.015.
12. Rotem Z, Cox RA, Isaacs A: Inhibition of virus multiplication by foreign nucleic acid. Nature. 1963, 197: 564-566. 10.1038/197564a0.
13. Klinman DM: Immunotherapeutic uses of CpG oligodeoxynucleotides. Nat Rev Immunol. 2004, 4: 248-257.
14. Anderson ED, Mourich DV, Fahrenkrug SC, LaPatra S, Shepherd J, Leong JAC: Genetic immunization of rainbow trout (*Onchorhynchus mykiss*) against infectious hematopoietic necrosis virus. Mol Mar Biol Biotechnol. 1996, 5: 114-122.
15. Traxler GS, Anderson E, LaPatra SE, Richard J, Shewmaker B, Kurath G: Naked DNA vaccination of Atlantic salmon *Salmo salar* against IHNV. Dis Aquat Organ. 1999, 38: 183-190.
16. Purcell MK, Laing KJ, Winton JR: Immunity to fish rhabdoviruses. Viruses. 2012, 4: 140-166. 10.3390/v4010140.
17. Purcell MK, Nichols KM, Winton JR, Kurath G, Thorgaard GH, Wheeler P, Hansen JD, Herwig RP, Park LK: Comprehensive gene expression profiling following DNA vaccination of rainbow trout against infectious hematopoietic necrosis virus. Mol Immunol. 2006, 43: 2089-2106. 10.1016/j.molimm.2005.12.005.
18. Cuesta A, Tafalla C: Transcription of immune genes upon challenge with viral hemorrhagic septicemia virus (VHSV) in DNA vaccinated rainbow trout (*Oncorhynchus mykiss*). Vaccine. 2009, 27: 280-289. 10.1016/j.vaccine.2008.10.029.

19. Lorenzen N, Lorenzen E, Einer-Jensen K, LaPatra SE: Immunity induced shortly after DNA vaccination of rainbow trout against rhabdoviruses protects against heterologous virus but not against bacterial pathogens. Dev Comp Immunol. 2002, 26: 173-179. 10.1016/S0145-305X(01)00059-3.
20. Lorenzen E, Lorenzen N, Einer-Jensen K, Brudeseth B, Evensen Ø: Time course study of in situ expression of antigens following DNA-vaccination against VHS in rainbow trout (*Oncorhynchus mykiss* Walbaum) fry. Fish Shellfish Immunol. 2005, 19: 27-41. 10.1016/j.fsi.2004.10.009.
21. Gomez-Chiarri M, Livingston SK, Muro-Cacho C, Sanders S, Levine RP: Introduction of foreign genes into the tissue of live fish by direct injection and particle bombardment. Dis Aquat Organ. 1996, 27: 5-12.
22. Tonheim TC, Bøgwald J, Dalmo RA: What happens to the DNA vaccine in fish? A review of current knowledge. Fish Shellfish Immunol. 2008, 25: 1-18. 10.1016/j.fsi.2008.03.007.
23. Kurath G: Biotechnology and DNA vaccines for aquatic animals. Rev Sci Tech. 2008, 27: 175-196.
24. Redding L, Weiner DB: DNA vaccines in veterinary use. Expert Rev Vaccines. 2009, 8: 1251-1276. 10.1586/erv.09.77.
25. Gomez-Casado E, Estepa A, Coll JM: A comparative review on European-farmed finfish RNA viruses and their vaccines. Vaccine. 2011, 29: 2657-2671. 10.1016/j.vaccine.2011.01.097.
26. Alonso M, Chiou PP, Leong JA: Development of a suicidal DNA vaccine for infectious hematopoietic necrosis virus (IHNV). Fish Shellfish Immunol. 2011, 30: 815-823. 10.1016/j.fsi.2011.01.001.
27. Penaranda MMD, LaPatra SE, Kurath G: Specificity of DNA vaccines against the U and M genogroups of infectious hematopoietic necrosis virus (IHNV) in rainbow trout (*Oncorhynchus mykiss*). Fish Shellfish Immunol. 2011, 31: 43-51. 10.1016/j.fsi.2011.03.003.
28. Adomako M, St-Hilaire S, Zheng Y, Eley J, Marcum RD, Sealey W, Donahower BC, LaPatra S, Sheridan PP: Oral DNA vaccination of rainbow trout, *Oncorhynchus mykiss* (Walbaum), against infectious haematopoietic necrosis virus using PLGA [Poly(D, L-Lactic-Co-Glycolic Acid)] nanoparticles. J Fish Dis. 2012, 35: 203-214. 10.1111/j.1365-2761.2011.01338.x.
29. Choi SH, Kim MS, Kim KH: Immunization of olive flounder (*Paralichthys olivaceus)* with an auxotrophic *Edwardsiella tarda* mutant harboring the VHSV DNA vaccine. Fish Shellfish Immunol. 2012, 33: 569-574. 10.1016/j.fsi.2012.06.013.
30. Heras AI D l, Rodríguez Saint-Jean S, Perez-Prieto SI: Immunogenic and protective effects of an oral DNA vaccine against infectious pancreatic necrosis virus in fish. Fish Shellfish Immunol. 2010, 28: 562-570. 10.1016/j.fsi.2009.12.006.
31. Munang'andu HM, Fredriksen BN, Mutoloki S, Brudeseth B, Kuo T-Y, Marjara IS, Dalmo RA, Evensen Ø: Comparison of vaccine efficacy for different antigen delivery systems for infectious pancreatic necrosis virus vaccines in Atlantic salmon (*Salmo salar* L.) in a cohabitation challenge model. Vaccine. 2012, 30: 4007-4016. 10.1016/j.vaccine.2012.04.039.
32. Xu C, Mutoloki S, Evensen Ø: Superior protection conferred by inactivated whole virus vaccine over subunit and DNA vaccines against salmonid alphavirus infection in Atlantic salmon (*Salmo salar* L.). Vaccine. 2012, 30: 3918-3928. 10.1016/j.vaccine.2012.03.081.
33. Zhang M, Hu YH, Xiao ZZ, Sun Y, Sun L: Construction and analysis of experimental DNA vaccines against megalocytivirus. Fish Shellfish Immunol. 2012, 33: 1192-1198. 10.1016/j.fsi.2012.09.010.
34. Jiao XD, Zhang M, Hu YH, Sun L: Construction and evaluation of DNA vaccines encoding *Edwardsiella tarda* antigens. Vaccine. 2009, 27: 5195-5202. 10.1016/j.vaccine.2009.06.071.

35. Sun Y, Liu CS, Sun L: Construction and analysis of the immune effect of an *Edwardsiella tarda* DNA vaccine encoding a D15-like surface antigen. Fish Shellfish Immunol. 2011, 30: 273-279. 10.1016/j.fsi.2010.10.020.
36. Sun Y, Liu CS, Sun L: Comparative study of the immune effect of an *Edwardsiella tarda* antigen in two forms: Subunit vaccine vs DNA vaccine. Vaccine. 2011, 29: 2051-2057. 10.1016/j.vaccine.2011.01.013.
37. Sun Y, Hu YH, Liu CS, Sun L: Construction and comparative study of monovalent and multivalent DNA vaccines against *Streptococcus iniae*. Fish Shellfish Immunol. 2012, 33: 1303-1310. 10.1016/j.fsi.2012.10.004.
38. Sun Y, Hu YH, Liu CS, Sun L: A *Streptococcus iniae* DNA vaccine delivered by a live attenuated *Edwardsiella tarda* via natural infection induces cross-genus protection. Lett Appl Microbiol. 2012, 55: 420-426.
39. Qin YX, Su YQ, Wang SF, Yan QP: Immunogenicity and protective efficacy of *Vibrio harveyi* pcFlaA DNA vaccine in *Epinephelus awoara*. Chin J Oceanol Limn. 2009, 27: 769-774.
40. Hu YH, Sun L: A bivalent *Vibrio harveyi* DNA vaccine induces strong protection in Japanese flounder (*Paralichthys olivaceus*). Vaccine. 2011, 29: 4328-4333. 10.1016/j.vaccine.2011.04.021.
41. Liang HY, Wu ZH, Jian JC, Huang YC: Protection of red snapper (*Lutjanus sanguineus*) against *Vibrio alginolyticus* with a DNA vaccine containing flagellin flaA gene. Lett Appl Microbiol. 2011, 52: 156-161. 10.1111/j.1472-765X.2010.02981.x.
42. Plant KP, LaPatra SE, Cain KD: Vaccination of rainbow trout, *Oncorhynchus mykiss* (Walbaum), with recombinant and DNA vaccines produced to *Flavobacterium psychrophilum* heat shock proteins 60 and 70. J Fish Dis. 2009, 32: 521-534. 10.1111/j.1365-2761.2009.01008.x.
43. Priya TAJ, Lin YH, Wang YC, Yang CS, Chang PS, Song YL: Codon changed immobilization antigen (iAg), a potent DNA vaccine in fish against *Cryptocaryon irritans* infection. Vaccine. 2012, 30: 893-903. 10.1016/j.vaccine.2011.11.102.
44. Tan CW, Jesudhasan P, Woo PTK: Towards a metalloprotease-DNA vaccine against piscine cryptobiosis caused by *Cryptobia salmositica*. Parasitol Res. 2008, 102: 265-275.
45. Jorgensen LV, Sigh J, Kania PW, Holten-Andersen L, Buchmann K, Clark T, Rasmussen JS, Einer-Jensen K, Lorenzen N: Approaches towards DNA vaccination against a skin ciliate parasite in fish. PloS One. 2012, 7: e48129-10.1371/journal.pone.0048129.
46. Garver KA, Conway CM, Elliott DG, Kurath G: Analysis of DNA-vaccinated fish reveals viral antigen in muscle, kidney and thymus, and transient histopathologic changes. Mar Biotechnol. 2005, 7: 540-553. 10.1007/s10126-004-5129-z.
47. Budker V, Budker T, Zhang GF, Subbotin V, Loomis A, Wolff JA: Hypothesis: naked plasmid DNA is taken up by cells *in vivo* by a receptor-mediated process. J Gene Med. 2000, 2: 76-88.
48. Zhong Q, Chinta DM, Pamujula S, Wang H, Yao X, Mandal TK, Luftig RB: Optimization of DNA delivery by three classes of hybrid nanoparticle/DNA complexes. J Nanobiotechnol. 2010, 8: 6-10.1186/1477-3155-8-6.
49. Desmet CJ, Ishii KJ: Nucleic acid sensing at the interface between innate and adaptive immunity in vaccination. Nat Rev Immunol. 2012, 12: 479-491. 10.1038/nri3247.
50. Mukhopadhyay S, Gordon S: The role of scavenger receptors in pathogen recognition and innate immunity. Immunobiology. 2004, 209: 39-49. 10.1016/j.imbio.2004.02.004.
51. Seternes T, Tonheim TC, Løvoll M, Bøgwald J, Dalmo RA: Specific endocytosis and degradation of naked DNA in the endocardial cells of cod (*Gadus morhua* L.). J Exp Biol. 2007, 210: 2091-2103. 10.1242/jeb.003186.
52. Zhu FG, Reich CF, Pisetsky DS: The role of the macrophage scavenger receptor in immune stimulation by bacterial DNA and synthetic oligonucleotides. Immunology. 2001, 103: 226-234. 10.1046/j.1365-2567.2001.01222.x.

53. Zhang XX, McIntosh TJ, Grinstaff MW: Functional lipids and lipoplexes for improved gene delivery. Biochimie. 2012, 94: 42-58. 10.1016/j.biochi.2011.05.005.
54. Weintraub H, Cheng PF, Conrad K: Expression of transfected DNA depends on DNA topology. Cell. 1986, 46: 115-122. 10.1016/0092-8674(86)90865-2.
55. Williams JA, Carnes AE, Hodgson CP: Plasmid DNA vaccine vector design: impact on efficacy, safety and upstream production. Biotechnol Adv. 2009, 27: 353-370. 10.1016/j.biotechadv.2009.02.003.
56. Davis HL, Millan CLB, Watkins SC: Immune-mediated destruction of transfected muscle fibers after direct gene transfer with antigen-expressing plasmid DNA. Gene Ther. 1997, 4: 181-188. 10.1038/sj.gt.3300380.
57. Tan Y, Li S, Pitt BR, Huang L: The inhibitory role of CpG immunostimulatory motifs in cationic lipid vector-mediated transgene expression in vivo. Hum Gene Ther. 1999, 10: 2153-2161. 10.1089/10430349950017149.
58. Qin LH, Ding YZ, Pahud DR, Chang E, Imperiale MJ, Bromberg JS: Promoter attenuation in gene therapy: interferon-gamma and tumor necrosis factor-alpha inhibit transgene expression. Hum Gene Ther. 1997, 8: 2019-2029. 10.1089/hum.1997.8.17-2019.
59. McLauchlan PE, Collet B, Ingerslev E, Secombes CJ, Lorenzen N, Ellis AE: DNA vaccination against viral haemorrhagic septicaemia (VHS) in rainbow trout: size, dose, route of injection and duration of protection-early protection correlates with Mx expression. Fish Shellfish Immunol. 2003, 15: 39-50. 10.1016/S1050-4648(02)00137-7.
60. Hansen E, Fernandes K, Goldspink G, Butterworth P, Umeda PK, Chang KC: Strong expression of foreign genes following direct injection into fish muscle. FEBS Lett. 1991, 290: 73-76. 10.1016/0014-5793(91)81229-2.
61. Doh SG, Vahlsing HL, Hartikka J, Liang X, Manthorpe M:Spatial-temporal patterns of gene expression in mouse skeletal muscle after injection of lacZ plasmid DNA. Gene Ther. 1997, 4: 648-663. 10.1038/sj.gt.3300460.
62. Corbeil S, LaPatra SE, Anderson ED, Kurath G: Nanogram quantities of a DNA vaccine protect rainbow trout fry against heterologous strains of infectious hematopoietic necrosis virus. Vaccine. 2000, 18: 2817-2824. 10.1016/S0264-410X(00)00078-5.
63. Lorenzen E, Einer-Jensen K, Rasmussen JS, Kjaer TE, Collet B, Secombes CJ, Lorenzen N: The protective mechanisms induced by a fish rhabdovirus DNA vaccine depend on temperature. Vaccine. 2009, 27: 3870-3880. 10.1016/j.vaccine.2009.04.012.
64. Myhr A, Dalmo RA: Introduction of genetic engineering in aquaculture: Ecological and ethical implications for science and governance. Aquaculture. 2005, 250: 542-554. 10.1016/j.aquaculture.2004.12.032.
65. Saade F, Petrovsky N: Technologies for enhanced efficacy of DNA vaccines. Expert Rev Vaccines. 2012, 11: 189-209. 10.1586/erv.11.188.
66. Caipang CMA, Hirono I, Aoki T: Modulation of the early immune response against viruses by a teleostean interferon regulatory factor-1 (IRF-1). Comp Biochem Physiol A Mol Integr Physiol. 2009, 152: 440-446. 10.1016/j.cbpa.2008.11.022.
67. Tonheim T, Dalmo RA, Bøgwald J, Seternes T: Specific uptake of plasmid DNA without reporter gene expression in Atlantic salmon (*Salmo salar* L.) kidney after intramuscular administration. Fish Shellfish Immunol. 2008, 24: 90-101. 10.1016/j.fsi.2007.09.006.
68. Tinsley-Bown AM, Fretwell R, Dowsett AB, Davis SL, Farrar GH: Formulation of poly(D,L-lactic-co-glycolic acid) microparticles for rapid plasmid DNA delivery. J Control Release. 2000, 66: 229-241. 10.1016/S0168-3659(99)00275-8.
69. Tian J, Sun X, Chen X, Yuc J, Qu L, Wang L: The formulation and immunisation of oral poly (DL-lactide-co-glycolide) microcapsules containing a plasmid vaccine against lymphocystis disease virus in Japanese flounder (*Paralichthys olivaceus*). Int Immunopharmacol. 2008, 8: 900-908. 10.1016/j.intimp.2008.02.006.
70. Tian J, Yu J: Poly(lactic-co-glycolic acid) nanoparticles as candidate DNA vaccine carrier for oral immunization of Japanese flounder (*Paralichthys olivaceus*) against

lymphocystis disease virus. Fish Shellfish Immunol. 2011, 30: 109-117. 10.1016/j.fsi.2010.09.016.

71. Hølvold LB, Fredriksen BN, Bøgwald J, Dalmo RA: Transgene and immune gene expression following intramuscular injection of Atlantic salmon (*Salmo salar* L.) with DNA-releasing PLGA nano- and microparticles. Fish Shellfish Immunol. 2013, 35: 890-899. 10.1016/j.fsi.2013.06.030.
72. Malyala P, O'Hagan DT, Singh M: Enhancing the therapeutic efficacy of CpG oligonucleotides using biodegradable microparticles. Adv Drug Deliv Rev. 2009, 61: 218-225. 10.1016/j.addr.2008.12.009.
73. Nguyen DN, Green JJ, Chan JM, Longer R, Anderson DG: Polymeric materials for gene delivery and DNA vaccination. Adv Mater. 2009, 21: 847-867. 10.1002/adma.200801478.
74. Cohen H, Levy RJ, Gao J, Fishbein I, Kousaev V, Sosnowski S, Slomkowski S, Golomb G: Sustained delivery and expression of DNA encapsulated in polymeric nanoparticles. Gene Ther. 2000, 7: 1896-1905. 10.1038/sj.gt.3301318.
75. Prabha S, Zhou WZ, Panyam J, Labhasetwar V: Size-dependency of nanoparticle-mediated gene transfection: studies with fractionated nanoparticles. Int J Pharm. 2002, 244: 105-115. 10.1016/S0378-5173(02)00315-0.
76. Sharp FA, Ruane D, Claass B, Creagh E, Harris J, Malyala P, Singh M, O'Hagan DT, Petrilli V, Tschopp J, O'Neill LA, Lavelle EC: Uptake of particulate vaccine adjuvants by dendritic cells activates the NALP3 inflammasome. Proc Natl Acad Sci U S A. 2009, 106: 870-875. 10.1073/pnas.0804897106.
77. Hedley ML, Curley J, Urban R: Microspheres containing plasmid-encoded antigens elicit cytotoxic T-cell responses. Nat Med. 1998, 4: 365-368. 10.1038/nm0398-365.
78. Singh M, Briones M, Ott G, O'Hagan D: Cationic microparticles: a potent delivery system for DNA vaccines. Proc Natl Acad Sci U S A. 2000, 97: 811-816. 10.1073/pnas.97.2.811.
79. Jang JH, Shea LD: Intramuscular delivery of DNA releasing microspheres: microsphere properties and transgene expression. J Control Release. 2006, 112: 120-128. 10.1016/j.jconrel.2006.01.013.
80. Labhasetwar V, Bonadio J, Goldstein SA, Levy JR: Gene transfection using biodegradable nanospheres: results in tissue culture and a rat osteotomy model. Colloid Surface B. 1999, 16: 281-290. 10.1016/S0927-7765(99)00079-X.
81. Tian JY, Sun XQ, Chen XG: Formation and oral administration of alginate microspheres loaded with pDNAcoding for lymphocystis disease virus (LCDV) to Japanese flounder. Fish Shellfish Immunol. 2008, 24: 592-599. 10.1016/j.fsi.2008.01.009.
82. Leal CAG, Carvalho-Castro GA, Sacchetin PSC, Lopes CO, Moraes AM, Figueiredo HCP: Oral and parenteral vaccines against *Flavobacterium columnare*: evaluation of humoral immune response by ELISA and in vivo efficiency in Nile tilapia (*Oreochromis niloticus*). Aquacult Int. 2010, 18: 657-666. 10.1007/s10499-009-9287-x.
83. Tian J, Yu J, Sun X: Chitosan microspheres as candidate plasmid vaccine carrier for oral immunisation of Japanese flounder (*Paralichthys olivaceus*). Vet Immunol Immunopathol. 2008, 126: 220-229. 10.1016/j.vetimm.2008.07.002.
84. Kumar SR, Ahmed VPI, Parameswaran V, Sudhakaran R, Babu VS, Hameed ASS: Potential use of chitosan nanoparticles for oral delivery of DNA vaccine in Asian sea bass (*Lates calcarifer*) to protect from *Vibrio* (*Listonella*) *anguillarum*. Fish Shellfish Immunol. 2008, 25: 47-56. 10.1016/j.fsi.2007.12.004.
85. Leon-Rodriguez L, Luzardo-Alvarez A, Blanco-Mendez J, Lamas J, Leiro J: Biodegradable microparticles covalently linked to surface antigens of the scuticociliate parasite *P. dicentrarchi* promote innate immune responses *in vitro*. Fish Shellfish Immunol. 2013, 34: 236-243. 10.1016/j.fsi.2012.10.029.
86. Harikrishnan R, Kim JS, Balasundaram C, Heo MS: Vaccination effect of liposomes entrapped whole cell bacterial vaccine on immune response and disease protection in *Epinephelus bruneus* against *Vibrio harveyi*. Aquaculture. 2012, 342: 69-74.

87. Behera T, Swain P: Antigen adsorbed surface modified poly-epsilon-caprolactone microspheres stimulates both adaptive and innate immune response in fish. Vaccine. 2012, 30: 5278-5284. 10.1016/j.vaccine.2012.05.028.
88. Behera T, Swain P: Antigen adsorbed calcium phosphate nanoparticles stimulate both innate and adaptive immune response in fish, *Labeo rohita* H. Cell Immunol. 2011, 271: 350-359. 10.1016/j.cellimm.2011.07.015.
89. Myhr AI, Dalmo RA: Nanotechnology and risk: what are the issues?. Nanoethics: The Ethical and Social Implications of Nanotechnology. Edited by: Allhoff FLP, Moor J, Weckert J. 2007, New Jersey: John Wiley & Sons Inc, 149-161.
90. Gvili K, Benny O, Danino D, Machluf M: Poly(D,L-lactide-co-glycolide acid) nanoparticles for DNA delivery: waiving preparation complexity and increasing efficiency. Biopolymers. 2007, 85: 379-391. 10.1002/bip.20697.
91. Nielsen KN, Fredriksen BN, Myhr AI: Mapping uncertainties in the upstream: the case of PLGA nanoparticles in salmon vaccines. Nanoethics. 2011, 5: 57-71. 10.1007/s11569-011-0111-5.
92. Lorenzen N, LaPatra SE: DNA vaccines for aquacultured fish. Rev Sci Tech. 2005, 24: 201-213.
93. Gillund F, Tonheim T, Seternes T, Dalmo RA, Myhr AI: DNA vaccination in aquaculture –Expert judgements of impact on environment and fish health. Aquaculture. 2008, 284: 25-34. 10.1016/j.aquaculture.2008.07.044.
94. Salonius K, Simard N, Harland R, Ulmer JB: The road to licensure of a DNA vaccine. Curr Opin Investig Drugs. 2007, 8: 635-641.
95. EMEA, Committee for Veterinary Medicinal Products: Note for guidance: DNA vaccines non-amplifiable in Eukaryotic cells for veterinary use. 2001, [http://www.ema.europa.eu/docs/en_GB/document_library/Scientific_guideline/2009/10/WC500004614.pdf]
96. EMEA, Committee for the Medicinal Products for Human Use: Concept paper on guidance for DNA vaccines. 2012, [http://www.ema.europa.eu/docs/en_GB/document_library/Scientific_guideline/2012/03/WC500124898.pdf]
97. Food and Drug Administration (FDA), Center for Biologics Evaluation and Research: Guidance for Industry. Considerations for Plasmid DNA Vaccines for Infectious Disease indications. 2007, [http://www.fda.gov/downloads /biologi csblood vaccines/guidancecomplianceregulatoryinformation/guidances/vaccines/ucm091968.pdf]
98. Foss GS, Rogne S: Gene medication or genetic modification? The devil is in the details. Nat Biotechnol. 2003, 21: 1280-1281. 10.1038/nbt1103-1280.

CHAPTER 12

The Genome of *Spironucleus salmonicida* Highlights a Fish Pathogen Adapted to Fluctuating Environments

Feifei Xu, Jon Jerlstrom-Hultqvist, Elin Einarsson, A sgeir A stvaldsson, Staffan G. Svard,

Department of Cell and Molecular Biology, Science for Life Laboratory, Uppsala University, BMC, Uppsala, Sweden

ABSTRACT

Spironucleus salmonicida causes systemic infections in salmonid fish. It belongs to the group diplomonads, binucleated heterotrophic flagellates adapted to micro-aerobic environments. Recently we identified energy-producing hydrogenosomes in *S. salmonicida*. Here we present a genome analysis of the fish parasite with a focus on the comparison to the more studied diplomonad *Giardia intestinalis*. We annotated 8067 protein coding genes in the ~12.9 Mbp *S. salmonicida* genome. Unlike *G. intestinalis*, promoter-like motifs were found upstream of genes which are correlated with gene expression, suggesting a more elaborate transcriptional regulation. *S. salmonicida* can utilise more carbohydrates as energy sources, has an extended amino acid and sulfur metabolism, and more enzymes involved in scavenging of reactive oxygen species compared to *G. intestinalis*. Both genomes have large families of cysteine-rich membrane proteins. A cluster analysis indicated large divergence of these families in the two diplomonads. Nevertheless, one of *S. salmonicida* cysteine-rich proteins was localised to the plasma membrane similar to *G.*

intestinalis variant-surface proteins. We identified *S. salmonicida* homologs to cyst wall proteins and showed that one of these is functional when expressed in *Giardia*. This suggests that the fish parasite is transmitted as a cyst between hosts. The extended metabolic repertoire and more extensive gene regulation compared to *G. intestinalis* suggest that the fish parasite is more adapted to cope with environmental fluctuations. Our genome analyses indicate that *S. salmonicida* is a well-adapted pathogen that can colonize different sites in the host.

INTRODUCTION

Most of the eukaryotic diversity is represented by microbial organisms [1]. Yet, the eukaryotic genomic efforts are very biased because only a minority of the microbial groups have been sampled on the genomic level, whereas there is a multitude of animal, plant and fungi genome projects. To understand the true diversity of eukaryotes we need to study the whole eukaryotic diversity. In this study we explore one previously understudied eukaryotic group, the diplomonads, in order to understand the diversity within the group and broaden the knowledge of eukaryotes in general.

Diplomonads are a group of anaerobic, flagellated protists, classified within Fornicata in the supergroup Excavata [1]. They lack aerobic mitochondria [2], although reduced mitochondria (mitosomes) have been identified in the most studied diplomonad *Giardia intestinalis* [3], and recently we identified hydrogenosomes in *Spironucleus salmonicida* [4], the focus of this study. Diplomonads have two diploid nuclei and most likely a sexual or parasexual life cycle [5]–[7], and there is an on-going metabolic adaptation by acquisition of mainly prokaryotic genes [8]–[10]. Together these findings refute earlier suggestions that diplomonads represent a primitive bacterial-like eukaryotic group [11]. There are free-living members of diplomonads, such as *Trepomonas*, as well as commensals or parasites of various animals [2]. For example, *G. intestinalis* causes diarrhea in humans and other animals [12], [13] and members of the genus *Spironucleus* can cause severe infections in ornamental and farmed fish [14]. Diplomonads with different life-styles are intermixed in the diplomonad phylogeny, even within *Spironucleus*, suggesting that transitions between lifestyles have happened multiple times in the group [15].

Aquaculture is a fast growing food sector in the world. The diplomonad *S. salmonicida* ("the salmonid killer") is a threat to sustainable aquaculture because it is able to cause systemic infections in farmed Atlantic salmon, Chinook salmon and Arctic char [16], [17]. Gross pathologies of *S. salmonicida* include internal haemorrhaging, splenomegaly and granulomatous lesions in the liver and spleen. In Northern Norway, outbreaks of spironucleosis in farmed Atlantic salmon, *Salmo salar*, is a recurring problem and causes mass mortality and economical loss. Drug treatment is not possible, making studies of the parasite important to develop alternative strategies [14]. The pathogenic *S. salmonicida* is genetically different from the morphologically indistinguishable diplomonad *Spironucleus barkhanus* which is a commensal in wild freshwater populations of Arctic char and grayling *Thymallus thymallus* [17], [18]. The parasite has recently been identified in both wild Arctic char and brown trout but no indications of disease were observed [19]. This suggests that wild salmonids might be asymptomatic reservoir hosts and that *S. salmonicida* is an opportunistic pathogen. We have indeed a very limited knowledge about how this important fish parasite is transmitted between hosts or the life cycle in general, how it is able to form the deadly lesions, the virulence genes responsible for the invasive infections and how it avoids the fish immune system during infection.

We have developed a stable transfection system for *S. salmonicida* to study the parasite [20]. Here we present a thoroughly annotated genome sequence and comparative analyses to *G. intestinalis*. We identify large differences in transcriptional regulation, the metabolic capacity and candidate variable surface proteins, which is in agreement with phenotypic differences between the species. In contrast, conservation of genes involved in encystation suggests similar machineries for that stage in the life cycle. The development of *S. salmonicida* into a model system contributes to the understanding of the pathogenicity and evolution of this enigmatic eukaryotic group, as well as eukaryotes in general.

RESULTS AND DISCUSSION

Sequencing and assembly

The genome of *S. salmonicida* (ATCC 50377) was characterized using the optical mapping method provided by OpGen. The optical maps indicate a genome size of 12.6 Mbp distributed in nine chromosomes. The size is in good agreement with earlier estimates using flow cytometry [18], and similar to *G. intestinalis* (Table 1). We sequenced the genome *de novo*using a complimentary approach of 454 FLX to get large scaffolds and Illumina to increase sequence quality. The Illumina technology was also applied on RNA to get RNA-Seq data. This yielded a draft assembly of the genome containing 452 contigs in 233 scaffolds with a total length of 12.9 Mbp. The number of contigs and scaffolds are slightly more than the first *G. intestinalis* genome, but less than the two subsequently published genomes [10], [21], [22]. The largest scaffold is 0.56 Mbp in size, and the scaffold N50 is 0.15 Mbp. The average coverage of 454 and Illumina reads were 40× and 280× in the selected draft assembly, respectively.

Table 1. Comparison of the *S. salmonicida* and *G. intestinalis* genomes.

	S. salmonicida	*G. intestinalis*
Size (Mbp)	12.9	11.7
Chromosomes	9	5
G+C content (%)	33.4	49.0
Proteins annotated	8067	5901
Mean gene length (aa)	373	530
Gene density per kbp	0.63	0.50
Coding percentage (%)	72.1	78.2
Mean intergenic distance (bp)	421	481
Introns	3	6
tRNAs	145	63

The *S. salmonicida* genome is not very repetitive. Only 5.2% of the genome was masked by RepeatMasker (http://www.repeatmasker.org/), with 4.8% of genome as low complexity. The allelic sequence heterozygosity is ~0.15%, which is much lower than the *G. intestinalis* GS genome, but higher than *G. intestinalis* WB [10], [21]. 64.8% of the draft genome can be mapped onto the optical maps. The scaffolds which could not be mapped have mostly sizes below 40 kbp, which is the limit for a sequence to be able to uniquely map to an optical map due to the expected frequencies of restriction sites. The relatively low number of scaffolds and the good agreement with the optical maps suggest that our selected draft assembly is of high quality and suitable to perform whole genome analyses of this diplomonad and compare the results to previously published genomes.

Annotation and analysis of coding and intergenic regions

We developed an annotation pipeline in-house during this project, which combines results from various sources (details described in methods and Protocol S1). Using the pipeline we annotated and manually inspected 8067 genes with an addition of 267 partial genes and 21 pseudogenes.

Since the *S. salmonicida* genome is divergent from previously sequenced genomes, only 3164 of the genes have functional annotations including 879 genes annotated with domain information. The remaining 4903 genes code for hypothetical proteins, with 847 of those displaying similarity to genes in other species. The RNA-Seq data mapped well to open reading frames with clear boundaries (Figure 1A). This correlation was used as an indicator of functional genes during the manual review of the annotation.

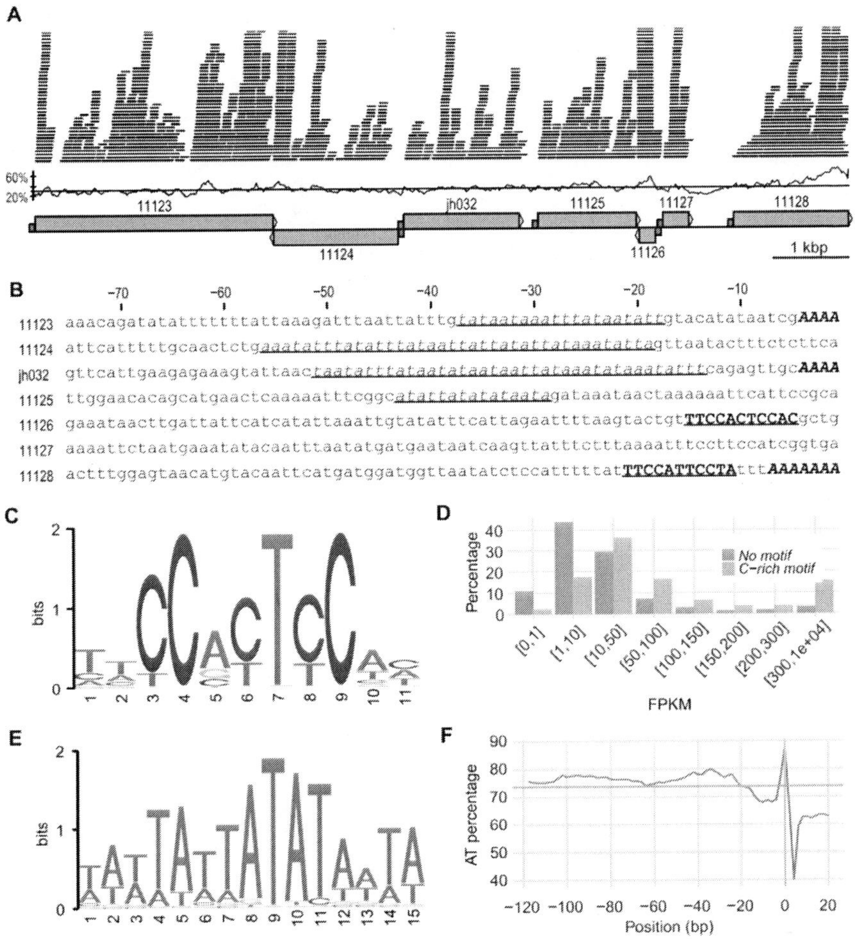

Figure 1. A 10 kbp genomic region with promoter info. A 10 kbp long genomic region located on scaffold scf7180000020498. The first part shows the Illumina RNA-Seq reads mapped onto the region. A coverage cutoff 40 was used for a better display. The GC content in 100 bp windows with a step size of 20 bp is shown. The average GC content in the region (30.5%) is indicated by a line. Green boxes with arrows indicate position and direction

of annotated genes. Half sized grey boxes indicate 75 bp promoter regions with sequences shown in B. Numbers refer to protein IDs: 11123, Phospholipid-transporting ATPase; 11124, Hypothetical protein; jh032, Hypothetical protein; 11125, Long-flagella protein, kinase, CMGC RCK; 11126, Ribosomal protein S30; 11127, Prefoldin subunit 6; 11128, ATP-dependent RNA helicase. **B**. 75 bp promoter sequences of the genes shown in A. TATA-box motifs are in underlined italic font, C-rich motifs are in bold underlined upper cases, and the multiple As before start codon are in bold italic upper cases. **C**. Sequence logo of C-rich motif. **D**.Comparison of expression between genes with and without C-rich motif. Genes are divided into eight categories based on their FPKM-values. The y-axis represents the percentage of genes within each category. **E**. Sequence logo of TATA-box motif. **F**. AT contents in percentage of the 20 bp C-terminus of the genes with their 120 bp promoter regions drawn with window size of 3 and step size of 2. X-axis shows the positions, while y-axis shows the AT percentage. Green line indicates the average AT percentage of the regions. Plot was drawn in R.

The low frequency of introns (see below) makes gene identification easier in this genome than in many other eukaryotes. On the other hand, *S. salmonicida* uses an alternative genetic code in which only a single termination codon is used [23]. This leads to few termination codons and a high frequency of random open reading frames. In combination with the large genetic distance to the closest previously sequenced genome, this makes gene identification challenging. We believe that our annotation is of fairly high standard also for the hypothetical genes since 80% (3909 out of 4903) of them have support from RNA-Seq data and the average length of hypothetical genes is 345 bp close to 417 bp for the genes with functional annotation. Together this suggests that most genes annotated as hypothetical genes are functional genes.

Interestingly, *S. salmonicida* has around 3000 more annotated protein-coding genes than *G. intestinalis*, but with genes overall shorter at an average length of 373 aa, the percentage of the coding regions in the genome is still less, 72% versus 78% (Table 1). There are 47 cases of overlapping genes with an average overlap length of 38 bp. The genome has an average GC content of 33.4% and the coding regions have on average higher GC content (36.2%) compared to the intergenic regions (25.4%). 3268 *S. salmonicida* genes are found to be shared with 3089 *G. intestinalis* genes, whereas 4799 lack detectable homologs in the human parasite (Table S1). The average level of amino acid identity is 39.4% for orthologous genes (Figure S1). Similarity searches against available sequences in the public databases identified homologs for 182 of the *S. salmonicida* genes missing in *G. intestinalis*. The remaining 4617 genes lack detectable homologs, making 57.2% of the genes unique to *S. salmonicida* among the available sequenced genomes. We found little conserved synteny between *S. salmonicida*and *G. intestinalis* using the shared genes identified.

The largest clusters of *S. salmonicida* proteins consist of cysteine-rich proteins (discussed below) and proteins with protein kinase domains. Using a combination of three approaches we identified 138 putative protein kinases. These were classified according to Kinase.com database (http://kinase.com/) and the result was compared to the *G. intestinalis* kinome [24](Table S2). One distinct difference between the two diplomonads was observed. While *Giardia*devotes 71% (198 out of 278) of its kinome to the NIMA (Never in Mitosis Gene A)-Related Kinase (NEK) family [24], *S. salmonicida* has only 18. NEK kinase family is known to regulate entry to mitosis [25] and flagella length [24]. It is universally present in eukaryotes, but typically found in fewer than 10 copies [24], thus is only slightly expanded in *S. salmonicida*. Thus, NEK kinases have expanded in the *G. intestinalis* lineage and the biological function of this massive expansion is likely not shared with *S. salmonicida*.

In contrast to NEK kinases, tRNAs are more abundant in *S. salmonicida*; there are 145 tRNA genes in the fish parasite genome, compared to 63 in the *G. intestinalis* genome. The tRNAs code for all 20 standard amino acids as well as one selenocysteine tRNA, and all expected tRNA synthetases were found during the annotation. Five 5S ribosomal RNA (rRNA) genes were identified in different locations of the genome, and one copy of 5.8S, 18S and 28S rRNAs were found in a single small contig which most likely is a collapse of repetitive reads since the contig has a ten times higher read coverage than the genomic average. Homologs to all ribosomal proteins found in *G. intestinalis* are present in the *S. salmonicida* genome. Three additional short ribosomal proteins, S30, L29e and L39, were identified which lack recognizable homologs in *G. intestinalis*. Another difference is that *S. salmonicida* encodes multiple copies of a dozen ribosomal proteins, whereas *G. intestinalis* only has a single gene for each subunit. Thus, *S. salmonicida* has more genes devoted to the core protein synthesis machinery than the previously studied diplomonad.

Diplomonad genomes retain few and different introns

Although the splicing machinery is ancestrally present in *G. intestinalis* [26], only six introns have been identified in the three sequenced *G. intestinalis* genomes [27], [28]. Using our RNA-Seq data we could identify four putative introns in the *S. salmonicida* genome. Three of these were confirmed using RT-PCR. One is in a gene coding for ribosomal protein S24, one in the gene for ribosomal protein L30, and one in a gene for an unknown protein. The fourth intron in a hypothetical protein makes a short extension on the N-terminus, and the extension is only weakly expressed according to the RNA-Seq data. This could be an intron on its way to be lost, leading to a shorter protein.

S. salmonicida introns show the canonical GT/AG splice sites and are similar to the ones in *Giardia* and *Trichomonas vaginalis*. *S. salmonicida* introns contain a conserved AC-repeat motif, ACTAACAAACTAG, similar to ACTAAC ACACAG in *T. vaginalis* [29] and [AC]CT[GA]AC[AC]CACAG in *Giardia* [10] (Figure S2). This indicates that excavates likely have a shared intron splicing mechanism, strongly supporting an ancient presence probably followed by extensive intron loss in the two diplomonad lineages. Introns are found in different genes in *G. intestinalis* and *S. salmonicida*. Thus, the intron loss may eventually go to completion because there might not be a single diplomonad gene that requires the presence of an intron. Three split introns have been found in two genes in the *G. intestinalis* genome [27], [28], in addition to the canonical introns. Genes containing split introns are encoded from different loci in the genome and the transcripts from these are trans-spliced into a single mRNA used in translation. The homologs to the genes containing split introns in *G. intestinalis* [28] were found intact without introns in *S. salmonicida*. Further attempts failed to reveal split introns in *S. salmonicida* from the currently available data. The splicing machinery in *G. intestinalis* is highly reduced [10]. When we analyzed the machinery in *S. salmonicida* very similar results were obtained; nine Sm-like proteins were identified, as was putative Prp8, 22, 28 and 43 proteins.

Divergent signal peptides in *S. salmonicida*

Signal peptides are present in the N-terminal of newly synthesized proteins destined to the secretory pathway. We used SignalP, version 4.1 [30], to identify putative signal peptides in the diplomonad genomes. The method predicted 381 and 109 proteins in the *G. intestinalis* and *S. salmonicida* genomes, respectively. Looking into five orthologous groups which had members with predicted signal peptides in both organisms, we realized that certain *S. salmonicida*orthologs carry weaker signals. For example, a group of sugar transporters contains multiple *S. salmonicida* proteins with predicted signal peptides as well as members that share the characteristic pattern in the SignalP analysis, but score below the threshold (Figure S3). A similar case was found in the cyst wall proteins (see below). This indicates that *S. salmonicida*has more signal peptides than predicted, with some signal peptides being divergent and therefore are not recognized by the currently available methods. An updated profile could be used in future improved searches for signal peptides in this genome, given that experimental studies confirm the function for these signal peptides scoring below the threshold using the available profile.

The signal recognition particle (SRP) binds the signal peptide when it emerges from the exit site of the translating ribosome [31]. The complex of the translating ribosome and the SRP particle docks to the signal recognition receptor in the ER membrane and the nascent protein is translocated through the Sec61 channel. Most eukaryotic SRPs contain six proteins and the 7S RNA, divided into the Alu and S domains [31]. The two proteins in the Alu domain are absent from the *S. salmonicida* genome, and the Alu-domain is missing in the 7S RNA (S. Svärd, unpublished results). The role of the Alu domain is to arrest translation elongation just after the signal sequence emerges from the ribosome [31] to provide a time window for translocation of the nascent chain into the ER. The lack of an Alu domain in the *S. salmonicida* and *G. intestinalis* [22] SRP suggests that this process is regulated differently in diplomonads. The SRP 72 protein of the S domain could neither be identified; nor could the Sec61-β and Sec61-γ subunits. Thus, SRP and its interacting proteins are highly diverged in *S. salmonicida*, in line with the diverged signal peptides.

Identification of an abundant C-rich motif that is a putative promoter in *S. salmonicida*

Transcription has been found to be loosely regulated in *G. intestinalis* with fuzzy boundaries of gene transcripts and a relative high fraction of anti-sense transcription [27], [32]. This is coupled to an absence of conserved promoter motifs in *G. intestinalis* except AT-rich sequences at the transcription start sites [12]. The picture is different in *S. salmonicida* which show more specific boundaries of transcripts for most genes (Figure 1A). Putative regulatory elements included a TATA-box motif, a C-rich motif and an enrichment of As which probably served as transcription initiator element (Figure 1B).

We identified a conserved 11-nt C-rich motif in the upstream regions of annotated genes (Figure 1C) shared by 16.7% of the *S. salmonicida* genes. This motif is preferentially found in a position around 10 bp upstream of the initiation codon and is a strong candidate for being part of a promoter. Indeed, genes with a C-rich motif in the promoter region are more often observed to be highly expressed as measured by RNA-Seq reads (Figure 1D). Conserved house-keeping genes are over-represented among the genes that have this C-rich motif, for example 71 out of 82 ribosomal proteins are connected with the motif. These observations suggest that the C-rich motif is a strong candidate for a promoter sequence in *S. salmonicida* which should be tested further experimentally.

Similar C-rich motifs are found in the two other diplomonads with sequence data available, *S. vortens* and *G. intestinalis*, as well as the parabasalid *T. vaginalis*, but with much lower frequencies. For example, there are only around 1% of *G. intestinalis* genes that have similar C-rich motifs, and majority of those are hypothetical genes. For *S. vortens* and *T. vaginalis*, we found C-rich motifs in 5% and 1% of the analysed genes, respectively (Figure S4A). Thus, the motif is most frequent in the *Spironucleus* genomes, suggesting similarities in gene regulation. Alternative motifs were found upstream of smaller subsets of the annotated genes. For example, cyst wall proteins shared a putative promoter motif (see below).

A divergent TATA-binding protein (TBP) is found in *G. intestinalis*, although no conserved TATA-motifs are present [33]. The putative TBP is even more divergent in the *S. salmonicida* genome. It was identified as hypothetical protein that contained a divergent TBP domain, the amino acid sequence did not show significant similarities to the *Giardia* or any other TBPs in standard searches. Nevertheless, this putative protein has many potential binding sites within the genome because a clear TATA-box motif was detected in the promoter regions of 80.7% of the genes in *S. salmonicida* (Figure 1E). This adds to the picture of the presence of a more elaborate transcriptional regulation in this organism.

A third distinct sequence pattern was found using the sequence logo method on the 5′ end of the genes. There is an enrichment of As immediately upstream of the start codon in the *S. salmonicida* genes (Figure S4B). In fact, in 33.0% of the *S. salmonicida* genes, we observed at least three As right before ATG start codon. This AAA signal is probably part of the transcription initiator element (Inr). We see no clear difference on transcription expression levels between genes with and without the AAA using RNA-Seq data.

We analysed the AT content upstream of all protein-coding genes. The location of TATA-box motif corresponds well to the AT percentage peak between −60 bp to −20 bp, the C-rich motif corresponds to a dip around −10 bp, and the putative Inr-element are shown as an increase of the AT-content close to the start codon (Figure 1F). *S. vortens* has similar AT percentage dip as well as similar C-rich motif, whereas *G. intestinalis* without the general C-rich motif does not have the AT percentage dip upstream of the start codon (Figure S3CD).

The distinct motifs upstream of genes can only function as regulatory elements if they are recognized by DNA-binding proteins. Therefore we searched the proteome for homologs of such protein families previously analysed [34]. We found that there was a large expansion of proteins containing Myb-like DNA binding domains; there are 107 such proteins annotated in the *S. salmonicida* genome, compared to 8 in the *G.*

intestinalis genome. Myb domain-containing proteins are also expanded in *T. vaginalis* [34]. Furthermore, *S. salmonicida* has several other putative transcription factors similar to C2-H2 Zn-finger, E2F and DP1 transcription factors. Taken together with the observation of several putative promoter motifs (Figure 1), it is very likely that there are differences in the regulation on the transcriptional level in *S. salmonicida* compared to in *G. intestinalis.*

Tight regulation of genes is important for organisms living in fluctuating environments, and probably a feature of free-living ancestors of these parasites. The extent of regulation on the post-transcriptional level is not well-known from any diplomonad, but here we report observations that suggest differences of the potential for transcriptional level regulation which may be coupled to different life styles. The retention of extensive transcriptional regulation in *S. salmonicida*, but not in *G. intestinalis*, could be the basis for its ability for systematic infections in the fish during which it is likely to permit radically different microenvironments. *G. intestinalis*, on the other hand, can only grow within a defined part of the intestine of its host, a comparatively stable environment with less need of transcriptional regulation.

Overlap of the polyadenylation signal and termination codon may lead to codon reassignment

In sequence surveys of *S. salmonicida* and *S. barkhanus*, a conserved motif was found around the termination codon connected with the presence of a polyA tail around 14 bp downstream, suggesting that the termination codon and the polyadenylation signal overlap [8], [18]. Here we show that this signal is not restricted to highly expressed genes. The sequence logo analysis of 3′ end of the genes reveals a dominance of A and T two positions upstream and G right before the only stop codon TGA used in *S. salmonicida* (Figure S5A). This putative signal, AGTGA, shows a similar pattern to the most used polyadenylation signal, AGTAAA, in *Giardia* [12], [27]. A 4-bp polyadenylation signal, TAAA, was recently functionally identified in *T. vaginalis* [35]. Strikingly, TAA is used in ~90% of the genes in that genome and in more than half of the analysed genes the position of the polyadenylation and termination overlapped.

The fact that the termination codon serves as the core motif in the polyadenylation signal could indeed explain how a change of genetic code could happen. If only a single termination codon can serve as a polyadenylation signal, there could be a preferential use of that codon at the 3′ end of all genes in a genome, provided there is selection for an overlap of termination codon and polyadenylation signal. As a consequence, the other two codons will be free to adapt new functions. This could indeed be an explanation for the code re-assignment in a subset of the diplomonads [23].

The polyadenylation machinery in *Giardia* is highly reduced compared to the corresponding machinery in yeasts [10]. An analysis of the polyadenylation machinery in *S. salmonicida* shows the same picture (Figure S5B) with only a few of the proteins identified. Interestingly, *Giardia*and *Spironucleus* seem to be missing the same proteins, which have been lost or degenerated to such a degree that they cannot be identified using sequence-based searches.

Putative encystation pathway

Transmission routes of piscine *Spironucleus* species have not been mapped and there are only anecdotal reports of cysts. *Spironucleus vortens* has recently been shown to survive for more than 30 days outside the host in feces [36] but this has not been observed in *S. salmonicida*. Cysts have been detected in the terrestrial species *Spironucleus muris* and *Spironucleus meleagridis* and their cysts show immunological cross-reactivity to the cyst wall of *Giardia* cysts [37], [38]. Here we use comparative genomics tools in combination with functional characterizations to study the putative presence of a cyst stage in *S. salmonicida*.

The production of the environmentally resistant cyst wall that protects the cyst from the harsh environment outside the host has been extensively studied in *Giardia*. The cyst wall is composed of the aminosugar N-acetylgalactosamine (GalNAc) as well as three cyst wall proteins (CWP 1–3) [13]. The five enzymes needed to produce GalNAc from fructose 6-phosphate are all present in *S. salmonicida*. The analysis also uncovered a family of eight highly similar candidate cyst wall proteins with homology to *Giardia* CWP-1 (Figure 2A). We also identified another three potential cyst wall proteins that displayed a higher degree of divergence. The former family of genes included a potential homolog of *G. intestinalis* CWP-2 which also carries a short basic extension at the C-terminal similar to that protein, although the extension is substantially shorter in the *Spironucleus* protein (Figure 2A). The presence of regulatory motif upstream of encystation-inducible genes has been noted in *G. intestinalis* [39]. We have also found a conserved motif in the promoter regions of the conserved encystation related proteins in *S. salmonicida* (Figure 2B). The comparative genomic analyses strongly suggested a cyst stage in the *S. salmonicida* life cycle. This was further tested experimentally.

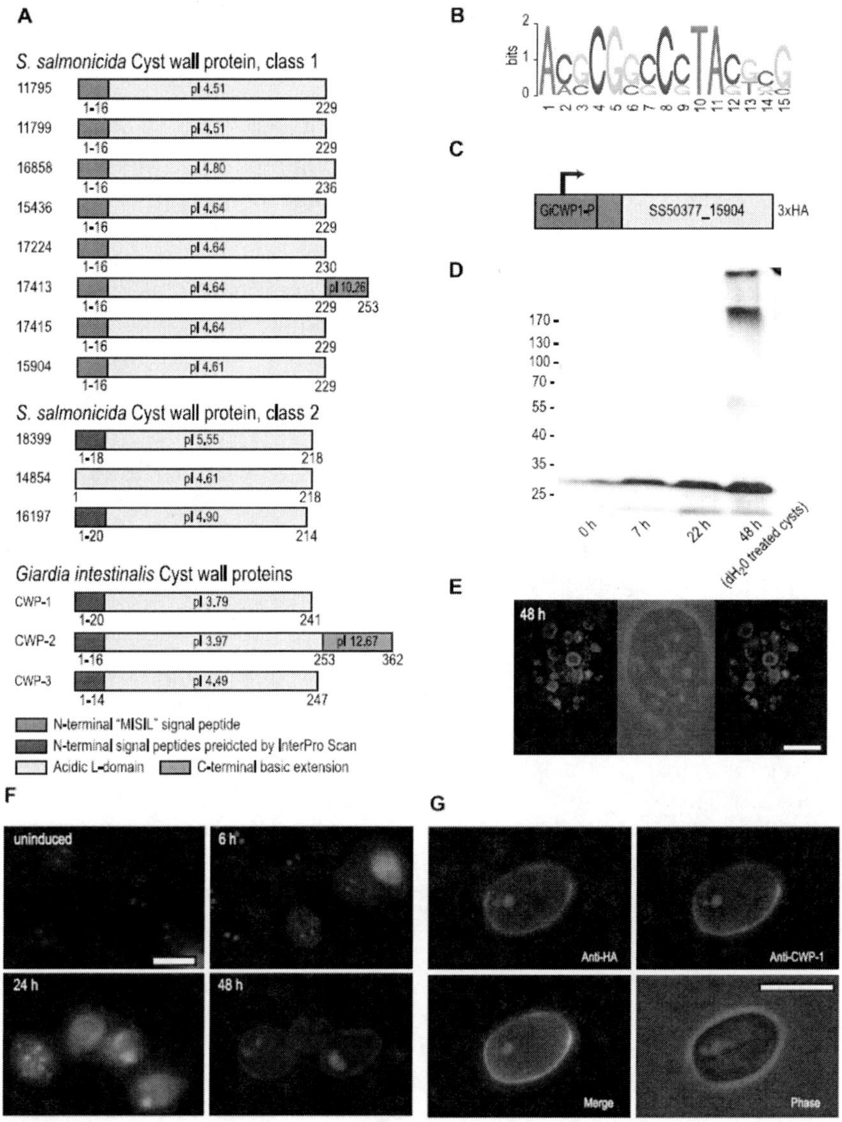

Figure 2. *S. salmonicida* cyst wall proteins traffic in *G. intestinalis* ESVs and incorporate into the cyst wall. **A**. Schematic representation of candidate *S. salmonicida* cyst wall proteins (CWPs). Numbers refer to protein IDs and conserved features are shown by coloured boxes with the amino acid positions indicated. The isoelectric point (pI) is indicated. **B**. Sequence logo of motifs upstream of the eight class 1 CWPs, glucosamine-6 phosphate deaminase, two glucose 6-phosphate N-acetyltransferase and two UDP-glucose 4-epimerases. **C**. The construct used to express the

candidate *S. salmonicida* class 1 CWP in *G. intestinalis* during encystation. The red box indicates the promoter region of *G. intestinalis* CWP-1. **D.** Western blot of samples taken from *G. intestinalis*transfectants carrying the *S. salmonicida* CWP construct. Expected size of the epitope-tagged SS50377_15904 is 28.4 kDa. **E** and **F.** Immunofluorescence analysis of *G. intestinalis* transfectants at different time points into encystation. The protein was detected using anti-HA conjugated to AlexaFluor488 (green) and the nuclei were labelled using DAPI (blue). Scale bars, 5 μm. **G.** Immunofluorescence micrograph of a *G. intestinalis* transfectant at 48 h post inducation of encystation following water-treatment. The cysts were stained by rabbit anti-HA and detected by anti-rabbit Alexa Fluor 594 (red), and probed using anti-CWP-1 conjugated to FITC (green). Scale bar, 10 μm.

S. salmonicida cyst wall proteins traffic in *Giardia* ESVs and incorporate into the cyst wall

Attempts of inducing encystation in *S. salmonicida* using cues traditionally employed in *G. intestinalis* (increased bile concentration and pH or cholesterol-deprivation) have not been successful at triggering cyst formation in *S. salmonicida*. Neither are cyst-like objects observed during routine *in vitro* passaging of the parasite. However, *S. salmonicida* cyst wall proteins display similar characteristics as *G. intestinalis* cyst wall proteins in analyses for signal peptides, even though they score below the threshold (Figure S6). This suggests that they may be functional in *G. intestinalis*. To test this we fused a *S. salmonicida* cyst wall protein to the *G. intestinalis* CWP-1 promoter, inserted a C-terminal 3×HA epitope tag and established *G. intestinalis* transfectants carrying an episomal plasmid with the construct (Figure 2C). This particular *S. salmonicida* cyst wall protein does not contain any TAG or TAA codons that in *S. salmonicida* encode glutamine. We proceeded to induce encystation in *Giardia* and studied the expression of the *S. salmonicida* cyst wall protein by Western blot and immunofluorescence (Figure 2DEFG). *G. intestinalis* transfectants carrying epitope-tagged SS50377_15904 under transcriptional control of the CWP-1 promoter were encysted and samples were taken at 0, 7, 22 and 48 h for analysis by Western blot. The sample at 48 h was water-treated to yield water-resistant cysts. The SS50377_15904 construct is induced upon encystation and is found to be present in high molecular weight protein species resistant to non-reducing conditions at 48 h post induction (Figure 2D). We further studied the construct by immunofluorescence at different time points into encystation (Figure 2EF). At 6 h post induction the cells show expression of the protein in the cytosol. At 24 h some cells show partial sorting of the protein into maturing encystation-specific vesicles

(ESVs). In some cells the protein was present in the doughnut-shaped fluid-phase of the ESV, in a similar way as *Giardia* CWP-1 [40] (Figure 2EF). This pattern is more pronounced in some cells at 48 h (Figure 2E). At 48 h post induction the protein can also be detected as incorporated in the cyst wall in a subpopulation of the cells (Figure 2F), as judged by the co-localization with *G. intestinalis* CWP-1 (Figure 2G). These results show that *S. salmonicida* has a cyst wall protein that is functional in *G. intestinalis*, and supports the hypothesis that the parasite has a cyst stage in the life cycle.

Proteases

S. salmonicida has to be able to degrade the host tissue to invade different organs in the fish. We identified that the *S. salmonicida* degradome consists of 111 protease homologs. The *S. salmonicida* proteases are divided into four catalytic classes and 26 families according to Merops protease classification [41] (Table S3). It includes 50 cysteine proteases belonging to 7 families; 34 metallo proteases belonging to 13 families, 8 serine proteases belonging to 4 families, and 14 threonine proteases belonging to the T1 family. Cysteine proteases from parasites are important virulence factors and known to degrade the host's extracellular matrix during invasion [42]. *S. salmonicida* has slightly more proteases in this category than *Giardia*, which is in agreement with its invasive phenotype [16].

S. salmonicida contains novel classes of cysteine-rich membrane proteins

Pathogens need to constantly avoid the host immune system to be able to strive within the host. A common way to do this is by antigenic variation by frequent changes of the cell surface which is the part exposed to the host immune system. In *G. intestinalis* this is done by a protein family of variant-specific surface proteins (VSPs) [43]. Hundreds of genes encoding VSPs are found in the *G. intestinalis* genomes and these are among the most divergent protein families within the *Giardia* genomes, both within and between isolates [22], [43]. Only one VSP is expressed at a time, and expression is switched from one VSP to another every 6 to 13 generations [44]. In this way a new VSP will be exposed on the surface of the parasite before it has been recognized by the adaptive immune system. *Giardia* VSPs are cysteine-rich (~12% cysteine) with frequent CXXC motifs and a conserved C-terminal membrane domain which is followed by a hydrophilic cytoplasmic tail with a conserved five amino acid CRGKA signature sequence [43].

S. salmonicida also harbours many cysteine-rich proteins. In a subset of these we identified a [KR][KR]X[KR][KR] motif (Figure 3A, Figure S7) towards the C-terminal which is reminiscent of the *Giardia* CRGKA signature sequence in VSPs. We classified the *S. salmonicida* cysteine-rich proteins into three groups based on the presence and absence of CXXC, CXC, KKXKK motifs and a C-terminal transmembrane (TM) domain (Figure 3B, Table S4). We named the groups cysteine-rich membrane protein 1 (CRMP-1), cysteine-rich membrane protein 2 (CRMP-2) and cysteine-rich protein (CRP).

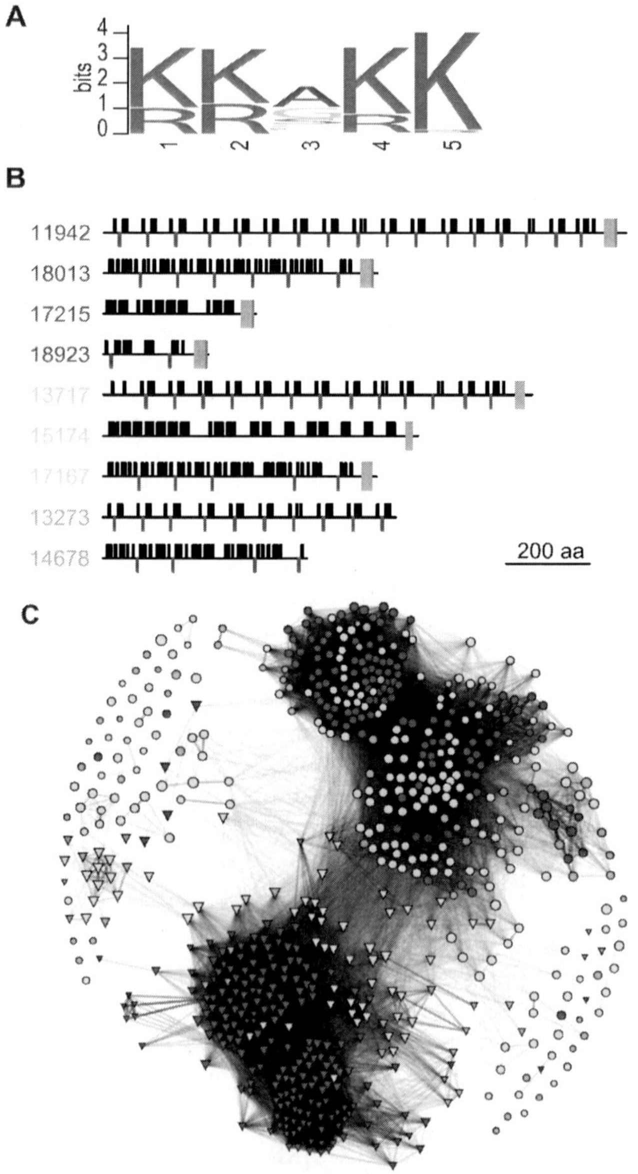

D

E

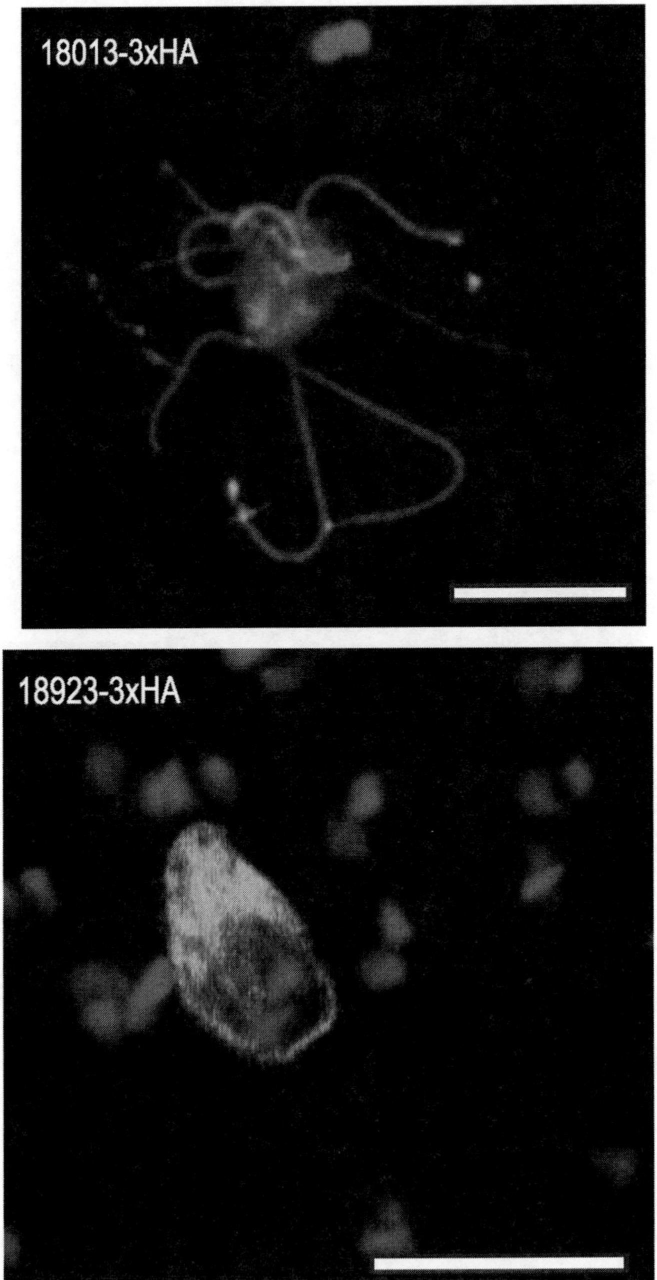

F

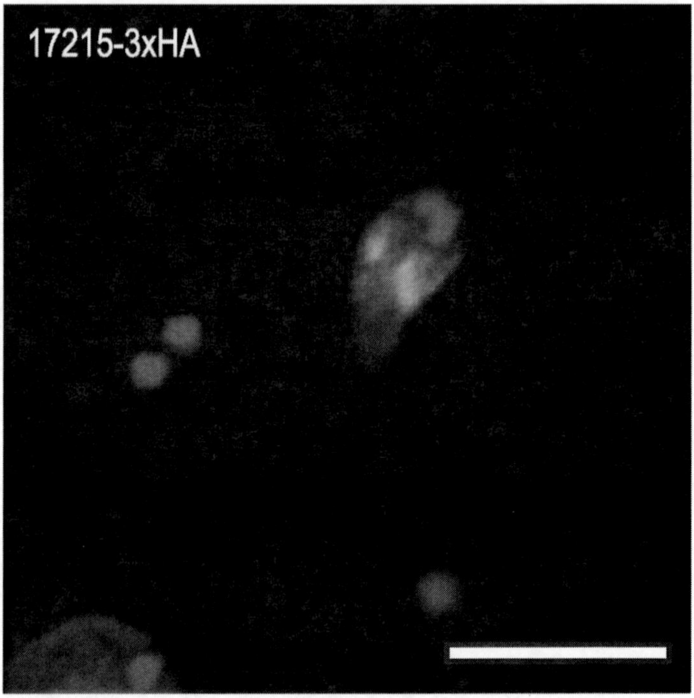

Figure 3. Diverse families of *S. salmonicida* cysteine-rich proteins. A. Sequence logo of a five amino acid motif shared by CRMP-1. **B**. Schematic view of selected *S. salmonicida* cysteine-rich proteins. Numbers refer to protein IDs and blue font indicates CRMP-1, pink CRMP-2 and green CRP. Black bars above and blue bars below the line indicate CXXC and CXC domains, respectively. Green and red boxes indicate TM domain and the conserved five amino acid motif shown in A, respectively, the basis for the classification. **C**. Network analysis of cysteine-rich proteins. *S. salmonicida* genes are represented in circle whereas *G. intestinalis* genes are shown in triangle. Blue indicates VSP or CRMP-1, pink indicates HCMP or CRMP-2, and green indicates HCP or CRP. Edges are weighted and scaled by reciprocal BLAST scores. Node sizes are scaled by protein sizes. **DEF**.Members of the *S. salmonicida* CRMP-1 protein family localise to different membrane domains. Stably transfected *S. salmonicida*carrying the pSpiro-PAC-18013-3×HA (D), pSpiro-PAC-18923-3×HA (E) and pSpiro-PAC-17215-3×HA (F), episomal plasmids.

The CRMP-1 group contains 125 membrane proteins with a conserved C-terminal five amino acid KKXKK motif, and three or more CXXC and CXC motifs. This group of proteins are similar to *Giardia* VSPs, but with a slightly different C-terminal sequence (Figure 3B). Some of the proteins contain an extra tail after this motif (Figure S7). The similarity to *Giardia* VSPs suggests that they may function as variable surface proteins in *S. salmonicida*. However, di-lysine motifs in the C-terminal of type I integral membrane proteins have been found to be both necessary and sufficient for endoplasmic reticulum (ER)-retention in other eukaryotes [45]. Protein disulfide isomerase-2 is a type-I integral membrane protein that carries a consensus C-terminal di-lysine motif for (KKAKKSE) and localises to the ER in *S. salmonicida* [20]. It is possible that the C-terminal tail of CRMP-1 proteins function as an ER retention signal.

The CRMP-2 group contains 195 cysteine-rich proteins which lack the conserved five amino acids signature sequence, but contain three of more CXXC or CXC motifs as well as a C-terminal TM domain (Figure 3B). The combination of CXXC and CXC motifs and a transmembrane domain, but lack of the five amino acid signature sequence found in CRMP-1, make the CRMP-2 more similar to the class of high cysteine membrane proteins (HCMP) in *G. intestinalis*. A study in *Giardia* showed that HCMPs were structurally similar to that of VSPs, but the only characterized HCMP was regulated and expressed similar to a cyst wall protein (CWP) [46].

We have also annotated 52 CRPs which contain more than 10% cysteines but do not belong to the other two categories (Figure 3B).

We performed a network analysis based on bi-directional BLAST hits to determine the relationship between and within the groups of *S. salmonicida* and *G. intestinalis* cysteine-rich proteins (Figure 3C). Strikingly, the cysteine-rich proteins from the two parasites are found in two distinct clusters with very little overlap. Some of the proteins show very little sequence similarity within the protein family and are found outside these clusters. Two different evolutionary scenarios could create such a pattern. It could be that the families of cysteine-rich proteins have expanded independently in the two parasites, or there is a high gene-turnover rate in which new cysteine-rich proteins are created via duplication events with a similar rate as genes are lost.

The patterns within the species-specific clusters are different (Figure 3C). The majority of the *G. intestinalis* VSPs are found in two clusters, with the HCMPs and high cysteine proteins (HCPs) loosely associated with these. The clustering within the *S. salmonicida* cluster is less distinct. CRMP-1 and CRMP-2 tend to cluster within the same class, but there are also intermixing between groups. Thus, the network analysis suggests that the classification into VSPs and HCMPs according to the presence of the C-terminal motif has a stronger correlation with overall primary sequence similarity in the proteins in *G. intestinalis* than in *S. salmonicida*.

The presence of transmembrane domains of the CRMP-1 and CRMP-2 proteins suggests that they are localised to either internal or external membranes in the cell. To test this, we epitope tagged three selected CRMP-1 proteins and studied their cellular localization (Figure 3DEF). This analysis shows one that localises like VSPs to the surface of the cellular body and flagella (Figure 3D), one that localises to the cellular body (Figure 3E) and one that localises to ER-like structures (Figure 3F). It remains to be determined whether different CRMPs are exposed at different time points of infection, as expected if they are responsible for antigenic variation in the fish parasite. Likewise, additional data are needed to determine if these proteins are under positive selection for variation because the available sequences are too divergent for such analyses.

It has been determined that the expression of VSPs in *Giardia* is regulated by components of the RNA interference pathway [47]. However, we failed to find Dicer, Argonaute and RNA-dependent RNA polymerase, key component of the RNA interference pathway, in the *S. salmonicida* genome. This indicates that CRMP expression in *S. salmonicida* is regulated by some other unknown means. We neither found any secretion signal peptides in the N-terminal of the protein as in *Giardia*. Either the signal peptides used by *S. salmonicida* are very divergent from what is known, or the secretion of CRMP proteins to the surface of the cell is regulated by some unknown means. 74% of the *Giardia* VSPs share Inr of PyAatgTT [43], while in *S. salmonicida*, 85 out of 125 CRMP-1s have the common AAAatg Inr without any clear conserved bases after start codon. Obviously, the *S. salmonicida* CRMPs are regulated differently from *G. intestinalis* VSPs, although the CRMPs show structural similarities and localise to the surface of the cell as VSPs. It may be that the two parasites independently have developed different mechanisms for expression of large families of cysteine-rich membrane proteins as an adaptation to a parasitic life style.

S. salmonicida has a more extensive metabolism and more transporters than *G. intestinalis*

Diplomonads were previously viewed as primitive eukaryotes harbouring bacterial-like metabolism [11]. The prokaryotic features of the enzymes later turned out to be explained by recent gene acquisitions rather than ancient retention of the genes by genomic studies of *G. intestinalis* and *S. salmonicida* [8]–[10], [48]. This metabolic adaptation via lateral gene transfer has, for example, contributed to the anaerobic metabolism of these organisms. With the complete *S. salmonicida* genome we can have more insights into the similarities and differences in the metabolism of these two pathogenic diplomonads.

S. salmonicida and *G. intestinalis* genes were classified into functional categories (Table 2). *G. intestinalis* has more enzyme functions represented in only two categories: slightly more enzymes involved in glycan metabolism were detected, and *S. salmonicida* lacks the mevalonate pathway (in Metabolism of terpenoids and polyketides) which is present in *Giardia*. In the other nine categories *S. salmonicida* has more enzymes. The largest differences in the metabolism are observed for the categories carbohydrate metabolism, energy metabolism and amino acid metabolism in which *S. salmonicida* has 32, 21 and 26 more enzymes than *G. intestinalis*, respectively.

Table 2. Metabolic enzymes identified in the KAAS analysis.

	S. salmonicida	*G. intestinalis*
Carbohydrate metabolism	66	34
Energy metabolism	49	28
Lipid metabolism	26	16
Nucleotide metabolism	55	48
Amino acid metabolism	41	15
Metabolism of other amino acids	15	6
Glycan biosynthesis and metabolism	10	11
Metabolism of cofactors and vitamins	29	17
Metabolism of terpenoids and polyketides	5	13
Biosynthesis of other secondary metabolites	7	2
Xenobiotics biodegradation and metabolism	12	7

The higher number of metabolic genes in *S. salmonicida* suggests that this parasite can utilise more metabolites than *G. intestinalis*. We classified the putative transporter proteins into families to test if it also has a higher capacity for transport of metabolites. In total 219 putative transporters were identified in *S. salmonicida* compared to 138 in *G. intestinalis* (Table S5). The three most common families are the Major facilitator superfamily (MFS), the ATP-binding cassette (ABC) superfamily, and the Amino

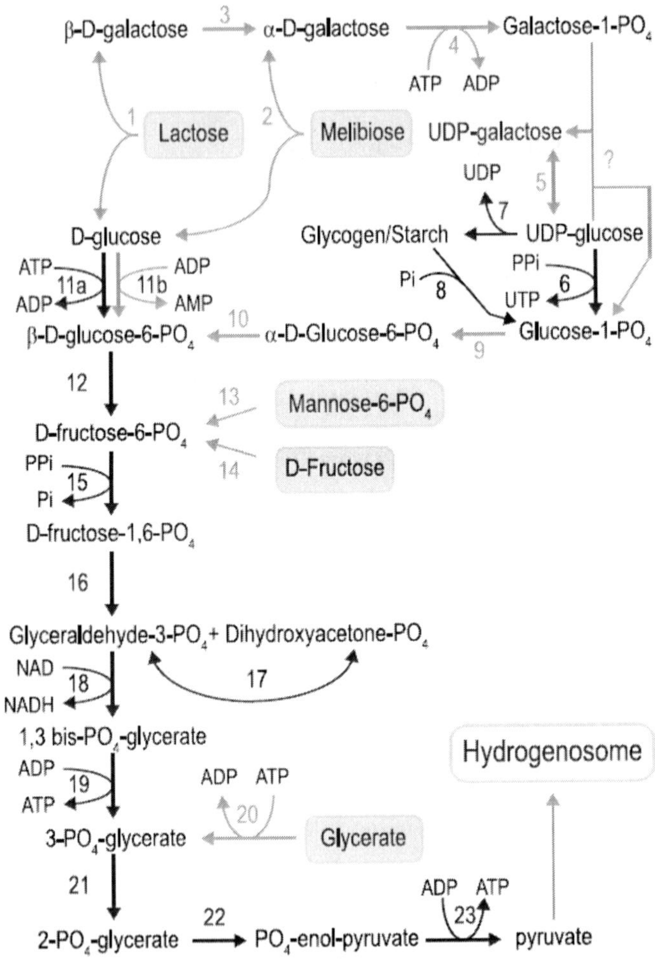

Figure 4. Carbohydrate metabolism in *S. salmonicida*. Black arrows indicate enzymatic functions present in both *G. intestinalis* and *S. salmonicida*, and green arrows indicate functions not detected in *G. intestinalis*. Key to enzymes: 1. β-galactosidase, 2. α-galactosidase, 3. galactose mutarotase, 4. galactokinase, 5. UDP-glucose 4′ epimerase, 6. UTP-glucose-1 phosphate uridylyltransferase, 7. glycogen synthase, 8. glycogen phosphorylase, 9. phosphoglucomutase, 10. glucose-6-phosphate 1-epimerase, 11a. glucokinase, 11b. ADP-specific glucokinase, 12. glucose phosphate isomerase, 13. phosphomannose isomerase, 14. Fructokinase, 15. pyrophosphate-dependent phosphofructokinase, 16. fructose bisphosphate aldolase, 17. triosephosphate isomerase, 18. glyceraldehyde 3-phosphate dehydrogenase, 19. phosphoglycerate kinase, 20. glycerate kinase, 21. phosphoglyceromutase, 22. enolase, 23. pyruvate kinase.

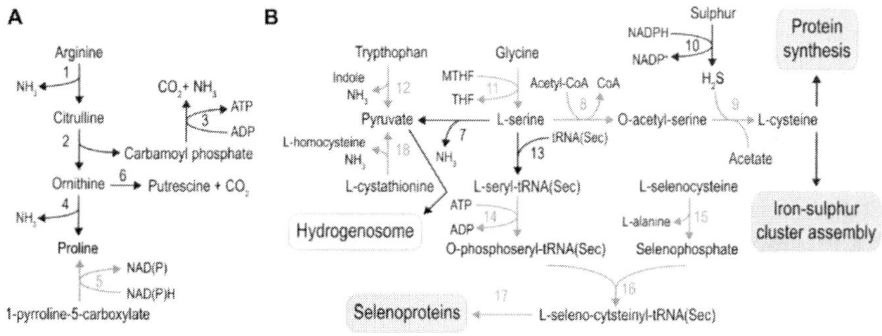

Figure 5. Amino acid metabolism in *S. salmonicida*. A. Arginine dihydrolase pathway and proline metabolism, and **B**. serine, cysteine, sulfur and selenium metabolism. Black arrows indicate enzymatic functions present in both *G. intestinalis* and *S. salmonicida*, and green arrows indicate functions absent in *G. intestinalis*. Key to enzymes: 1. arginine deiminase, 2. ornithine carbamoyl transferase, 3. carbamate kinase, 4. ornithine cyclodeaminase, 5. pyrroline-5-carboxylate reductase, 6. ornithine decarboxylase, 7. L-serine dehydratase, 8. serine O-acetyltransferase, 9. cysteine synthase, 10. sulfide dehydrogenase, 11. serine hydroxymethyltransferase. 12. tryptophanase, 13. seryl-tRNA synthetase, 14. *O*-phosphoseryl tRNA(Sec) kinase, 15. selenophosphate synthetase - NifS fusion protein, 16. *O*-phosphoseryl-tRNA(Sec) selenium transferase, 17. selenocysteine (Sec)-specific elongation factor, 18. cystathionine β-lyase.

acid/auxin permease (AAAP) family. The MFS and SBC superfamilies have broad specificities including metabolites such as sugars and amino acids [49], [50]. Together these two superfamilies have 67 and 43 members in the *S. salmonicida* and *G. intestinalis* genomes, respectively (Table S5). The AAAP family, in contrast, transport only single or multiple amino acids [51]. Twenty AAAPs are found in the *S. salmonicida* genome, compared to nine in the *G. intestinalis* genome. These observations suggest that the fish parasite is able to transport a larger variety of metabolites than *G. intestinalis*, especially amino acids and sugars.

We examined the metabolic capacity in more detail for some of the functional categories to understand the metabolic differences between the two diplomonads on a finer scale (Figures 4and 5, Figure S8).

Carbohydrates as energy sources

A major metabolic difference between the two diplomonads is that *S. salmonicida* harbours hydrogenosomes [4]. 20 proteins have been experimentally confirmed to be localised to the organelle, serving functions such as iron-sulfur cluster biogenesis, protein translocation, hydrogenase maturation and metabolic enzymes. We suggested a potential pathway for ATP production from pyruvate via the concerted action of hydrogenosomal pyruvate:ferrodoxin oxidoreductase (PFOR), [FeFe] hydrogenases, ferredoxins and a potential acetyl-CoA synthetase (ADP-forming) [4]. The hydrogenosomal presence of serine hydroxymethyltransferase and a putative H-protein of the glycine cleavage system argue for that at least parts of the amino acid metabolism is localised to the hydrogenosome. Thus, the hydrogenosomes in *S. salmonicida* likely produce ATP from pyruvate, the end product of glycolysis. This provides *S. salmonicida* with an additional way of converting pyruvate into energy, compared to the hydrogenosome-lacking *G. intestinalis* (Figure S8).

S. salmonicida encodes an almost identical set of enzymes as *G. intestinalis* to perform glycolysis (Figure 4). However, the fish parasite has a more extensive capability to use different metabolites to feed into the pathway. Mannose-6-phosphate and fructose can be converted into fructose 6-phosphate by phosphomannose isomerase and fructokinase, and *S. salmonicida*appears to be able to utilise glycerate by the action of glycerate kinase to generate 2-PO_4-glycerate, the substrate for enolase in the penultimate step of glycolysis (Figure 4). We also identified homologs to α-galactosidase and β-galactosidase, two enzymes which catalyse the hydrolysis of galactosides into monosaccharides. The enzymes have glycolipids and glycoproteins as substrates, as well as the disaccharides lactose and melibiose (Figure 4). The glucose generated when these enzymes act on disaccharides shuttles directly into glycolysis whereas the galactose needs to be metabolised by a specialised set of enzymes (Figure 4). The Leloir pathway converts galactose to glucose-1-phosphate. We detected three out of the four enzymes (galactose mutarotase, galactokinase and UDP-glucose 4-epimerase) of the pathway in *S. salmonicida*. The third enzyme of the pathway, galactose 1-phosphate uridylyltransferase, could not be identified. This function is probably performed by an unidentified protein. Glucose-1-phosphate is then shuttled into the glycolysis via two additional enzymes detected in *S. salmonicida*, but absent in *G. intestinalis*: a phosphoglucomutase and glucose-6-phosphate 1-epimerase. Our bioinformatic analysis indicates that *S. salmonicida*likely can use five additional carbohydrates compared to *G. intestinalis* to feed into the glycolysis (Figure 4). However, experimental studies of the fish parasite are needed to test this hypothesis.

There are a large amount of *S. salmonicida* proteins putatively involved in end product synthesis in *S. salmonicida* (Figure S8). The pyruvate generated in the glycolysis can be converted to acetyl-CoA by the action of the five PFORs identified in the genome, shuttling electrons to [FeFe] hydrogenase via ferredoxin (Figure S8). The two [2Fe-2S] type ferredoxins in *S. salmonicida* have been localised to the hydrogenosome. *S. salmonicida* encodes seven iron-only hydrogenases, at least two of which are located in the hydrogenosome. In addition to these enzymes we identified eight flavodoxins in the genome. These are bacterial flavoproteins containing one molecule FMN that typically can replace the role of ferredoxin in electron-transfer functions. The presence of multiple paralogs of the proteins in these pathways and their different localizations [4] suggest that hydrogen and energy generation from pyruvate probably is occurring both in the hydrogenosome and the cytosol.

Amino acids as energy sources

The capacity to metabolise amino acids appears to be greater in *S. salmonicida* than in *G. intestinalis*, and several of the differences suggest that the fish parasite utilise a variety of amino acids as energy sources (Figure 5). We identify all the enzymes of the arginine dihydrolase pathway, a rare pathway in eukaryotes that is present in both *G. intestinalis* and *T. vaginalis* [52], [53] and allows the utilization of arginine as an energy source (Figure 5A). Proline is potentially synthesized not only from ornithine cyclodeaminase but also from pyrroline 5-carboxylate using pyrroline 5-carboxylate reductase (ProC). The activity of ProC is dependent on cofactor F_{420} and we detect the presence of a putative F_{420}:gamma-glutamyl ligase in the *S. salmonicida* genome that might be involved in the synthesis of the final cofactor. Tryptophan could also serve as an energy source in *S. salmonicida* due to the presence of three copies of a bacterial-like tryptophanase that generate pyruvate from tryptophan with the concomitant production of indole and NH_3 (Figure 5). We previously identified a potential homolog of the H-protein of the glycine cleavage system in the hydrogenosomes of *S. salmonicida* as well as serine hydroxymethyltransferase. Consequently, *S. salmonicida* might employ a glycine cleavage system in the metabolism of glycine and serine. Serine can be used to generate pyruvate by L-serine dehydratase. Serine is also used in the synthesis of selenocysteine and in the *de novo* synthesis of cysteine via serine *O*-acetyl transferase and cysteine synthase A (Figure 5). Cystathionine might be used to generate pyruvate by the formation of homocysteine employing a cystathionine β-lyase. Energy generation is however not the only purpose of the *S. salmonicida* amino acid metabolism. In addition, the fish parasite has an extended set of enzymes for incorporation of cysteine and selenocysteine into proteins.

Selenium metabolism

Selenium metabolism is a trait absent in *G. intestinalis* and *T. vaginalis*, but present in *Spironucleus* [18]. The tRNA(Sec) and the four enzymes needed to incorporate selenocysteine into protein with selenophosphate and L-serine as precursors were found in the *S. salmonicida*genome (Figure 5). The identified polypeptide coding for selenophosphate synthetase is fused with a NifS-like protein in its C-terminal end. It has been shown that NifS-like proteins can function as a selenocysteine lyase in *Escherichia coli* and *Arabidopsis* which delivers selenium to selenophosphate synthetase [54], [55]. The fusion suggests that the NifS-like part has a selenocysteine lyase activity, which would enable *S. salmonicida* to utilise selenocysteine for selenophosphate biosynthesis (Figure 5). This fusion has only been found in the bacterium *Caldithrix abyssi* (GI: 493985699); it probably replaced the cognate selenophosphate synthetase and provided *S. salmonicida* the ability to use ingested selenoproteins as a selenium source. Only three putative selenoproteins could be found, all selenoprotein W paralogs. However, selenium may be incorporated into proteins independent of selenocysteine in *S. salmonicida*. Selenophosphate synthetase has sometimes been found in prokaryotic genomes that lack other selenium utilization genes [56], [57]. A gene uniquely shared between such genomes was detected and speculated to be involved in incorporation of selenium into protein independent of selenocysteine [56], [57]. *S. salmonicida* has three identical homologs of this protein which previously has not been detected in any eukaryotic genome. Thus, it may be that selenium is used in other proteins than the identified canonical selenoproteins in the *S. salmonicida* genome.

Sulfur metabolism in *S. salmonicida*

The large amount of cysteine-rich and iron-sulfur cluster containing proteins in *S. salmonicida*provides high demand of available cysteine. Sulfur is used in many metabolites and is essential for growth of all organisms, and inorganic sulfur is assimilated by photosynthetic organisms and fungi [58]–[60]. Other organisms often rely on uptake of reduced sulfur compounds from the environment [61]. *G. intestinalis* indeed seems to lack genes for biosynthesis of cysteine and methionine, sulfur-containing amino acids [10], [12]. On the other hand, protein-coding genes with sequence similarity to prokaryotic sulfide dehydrogenase [62], [63] have been found in *G. intestinalis* and *S. salmonicida* [64] (Figure 5). Sulfide dehydrogenase has been proposed to be part of fermentation of organic compounds in archaea with sulfur as the electron acceptor [65], [66]. It may be that this bacterial acquisition has a similar role in diplomonads by oxidation of NADPH to $NADP^+$ (Figure 5). Here we identified two additional enzymes, cysteine synthase and serine *O*-acetyltransferase which enable the *S. salmonicida* to biosynthesize cysteine from sulfur and serine (Figure 5). Consequently, in the presence of sulfur or sulfide the parasite is not dependent on a cysteine-rich diet for the synthesis of key enzymes containing iron-sulfur clusters. However, these key enzymes are sensitive to oxygen damage.

An expanded repertoire of genes involved in oxidative stress response

The conventional enzymes for oxidative stress response, superoxide dismutase, catalase and glutathione peroxidase have not been found in *G. intestinalis* neither using experimental [67]nor bioinformatic approaches [10]. Instead an O_2-scavaging NADH oxidase [48], [67], [68], superoxide reductase [69] and flavodiiron proteins [70] have been found to be involved in the antioxidative response. *S. salmonicida* causes systemic infections and thus needs a more efficient oxygen scavenging system than *Giardia*. Experimental data suggest that *S. vortens*has an elaborate system that consumes O_2 for several hours in the absence of added substrates [71]. We indeed identified many more genes involved in protection against reactive oxygen species (ROS) in *S. salmonicida* than in *G. intestinalis* (Table 3). We cannot find any enzymes for glutathione synthesis and recycling in *S. salmonicida*, suggesting that cysteine is the major intracellular thiol, as in *Giardia* and *Entamoeba histolytica*. Future experiments will show if expression of the cysteine-rich CRMP proteins in external and internal membranes (Figure S9) is important in ROS protection, similar to the cysteine-rich metallothioneins in mammalian cells [72]. The importance of the antioxidative response for *S. salmonicida* is underscored by the redundancy of oxygen detoxification mechanisms and the presence of multiple orthologs for many of the proteins (Table 3, Figure S9). Many of the enzymes have bacterial origin and can also be found in *E. histolytica* and *T. vaginalis* [8], [9], [73]. However, the nitric oxide protecting giardial enzyme flavohemoglobin [74] could not be found in the *S. salmonicida* genome. The amino acid methionine (Met) can be oxidized to methionine sulfoxide (MetO) but methionine sulfoxide reductases A (MsrA) and B (MsrB) reduce MetO back to Met, reactivating the oxidized proteins (Figure S9). We detected two MsrA and two MsrB genes in the *S. salmonicida* genome (Table 3), showing that these processes are active in the parasite. Interestingly, certain forms of MRS contain redox-active selenocysteine residues and it is possible that the *S. salmonicida* MRS proteins are selenocysteine proteins. This analysis shows that *S. salmonicida* has an extensive oxygen detoxification system, well-adapted for coping with changing O_2-levels during infection and transmission.

Table 3. Enzymes involved in the oxidative stress response.

Protein	#Spiro	#*Giardia*	Function
FAD/FMN dependent oxidoreductase	3	1	O_2 to $O_2^{\cdot-}$
NADH oxidase	4	1	O_2 to $O_2^{\cdot-}$
NADPH oxidoreductase	3	1	O_2 to $O_2^{\cdot-}$
Nitroreductase	3	1	O_2 to $O_2^{\cdot-}$
A-type flavoprotein	7	1	O_2 to H_2O
Superoxide reductase	1	1	$O_2^{\cdot-}$ to H_2O_2
Hybrid cluster protein	2	1	H_2O_2 to H_2O
Peroxiredoxin	4	3	H_2O_2 to H_2O
Protein containing alkyl hydroperoxide reductase	1	3	H_2O_2 to H_2O
Rubrerythrin 1	5	0	H_2O_2 to H_2O
Peptide methionine sulfoxide reductase MsrA	2	1	repair of oxidative damaged proteins
Peptide methionine sulfoxide reductase MsrB	2	1	repair of oxidative damaged proteins

CONCLUSIONS

The analyses of the *S. salmonicida* genome have provided insights into the biology of this fish parasite. *S. salmonicida* is capable of infecting a large number of different tissues, which have different micro environments, and thereby causing systemic infections. Our analyses have revealed an organism adapted to such fluctuating environments. More regulatory elements, for example putative promoters in many of the genes in the genomes, were found in *S. salmonicida* compared to *G. intestinalis*, suggesting that the fish parasite has a larger potential for regulation on the transcriptional level. The fish parasite encodes enzymes for several more carbohydrates and amino acids for energy production than *G. intestinalis*. Efficient transcriptional regulation of these enzymes may enable *S. salmonicida* to utilise different metabolites during infection of various tissues. The oxygen levels are fluctuating during systematic infections. *S. salmonicida* has a larger number of genes involved in oxidative stress response compared to *G. intestinalis*. The presence of these proteins probably enables *S. salmonicida* to use oxygen-sensitive iron-sulfur cluster containing enzymes for energy production throughout the infection.

The bioinformatic and functional studies indicate that *S. salmonicida* spread between hosts via a cyst stage, although we were unable to encyst the parasite *in vitro*. The conservation of the encystation genes between *G. intestinalis* and *S. salmonicida* suggests that this process was present in the diplomonad ancestor. Both characterized diplomonads have large repertoires of cysteine-rich proteins. The cluster analysis indicated that these protein families are very divergent between the two diplomonads. A subset of the cysteine-rich proteins is used for antigenic variation within *G. intestinalis* and a similar role in *S. salmonicida* appears likely. If so, the mechanisms for antigenic variation are probably rather different in the two diplomonads. Accordingly, convergent evolution, rather than shared ancestry, may have resulted in similar functions for cysteine-rich proteins in *G. intestinalis* and *S. salmonicida*. Thus, the question if the ancestral diplomonad was a free-living organism or a parasite remains open.

Our analyses have uncovered large functional differences within the group diplomonads which provide insights into the flexibility of eukaryotic genomes. We believe that the combination of a draft genome with high-quality annotation and the ability to perform functional studies could turn *S. salmonicida* into a powerful model organism. Not only for comparative studies to the important human parasite *G. intestinalis*, but also for eukaryotes in general.

MATERIALS AND METHODS

Material and sequencing

S. salmonicida (ATCC 50377), previously known as *S. barkhanus* [17], was isolated from a muscle abscess in Atlantic salmon grown in Vesterålen Sea in northern Norway. Cells were obtained from American Type Culture Collection (ATCC) and grown in axenic culture following the ATCC protocol. *S. salmonicida* was cultivated in LYI media in tightly capped slanted culture tubes (Nunc) at 16°C according to reference [20]. *G. intestinalis* WB/C6 (ATCC 50803) was cultivated according to reference [75] in TYDK media in tightly capped slanted culture tubes at 37°C. Total genomic DNA was isolated from trophozoites using standard methods. Total RNA was harvested from two batches of trophozoites exponential and stationary stages of growth using standard methods. Amplification with PCR primers specific to bacterial ribosomal RNA indicated no signs of bacterial contamination in the cultures. Equal amounts of total RNA from the two growth stages were pooled and mRNA was isolated using polyA-selection.

Total genomic DNA was sequenced using a Genome Sequencer FLX with GS FLX Titanium series reagents, one run with single shotgun reads and one run from a sequencing library with 3 kbp inserts, yielded 454 reads of 55× genome coverage. The genomic DNA was also sequenced with the Illumina Genome Analyzer IIx instrument, one run with paired-end reads with 100 bp in read length and pairs are 350 bp apart, which yielded Illumina reads of over 200× coverage. The same Illumina instrument was used to sequence *S. salmonicida* mRNA, with paired end reads with insert sizes of

175 bp. Raw DNA and RNA sequence reads are archived at NCBI Sequence Read Archive (SRA) under accession number SRA091283.

Optical mapping

S. salmonicida cells (10^9 cells) were harvested by chilling on ice followed by pelleting at 500× *g*, 5 min, and 4°C. The cells were washed two times with 10 pellet volumes of PBS. The resulting cell pellet was resuspended in 500 μl 200 mM NaCl, 100 mM EDTA, 10 mM Tris pH 7.2. The suspension was combined with 1% InCert agarose (Cat. No. 50121, Lonza) prepared in ddH_2O and 100 μl plugs were moulded by incubation at 4°C for 30 min. Cells were lysed by incubating the plugs in 5 mL NDSK solution (1% *N*-lauroylsarcosine, 2 mg/ml Proteinase K in 0.5 M EDTA pH 9.5) at 50°C for 8 h in an upright 50 mL Falcon tube. After 8 h the NDSK solution was replaced by of 5 mL fresh NDSK solution and the plugs were incubated another 18 h. The final lysed nearly transparent plugs were stored in 0.5 M EDTA, pH 9.5 and shipped to OpGen for optical map determination employing the *Nhe*I restriction enzyme. MapSolver v3.2.0 provided by OpGen was used to map assembly sequences to the optical maps.

Genome assembly

Celera Assembler (CA) v6.0 [76] was used to generate the selected genome assembly using 454 single and mate pair reads. The Illumina DNA reads were mapped to 454 assembly using BWA v0.5.9 [77] and Nesoni v0.40 (http://bioinformatics.net.au/software.nesoni.shtml) was used to correct 454 homopolymer errors based on the mapped bam file. The ribosomal RNAs were not present in the initial assembly, likely due to high coverage. They were then found in one 5.8 kb degenerate contig assembled with Celera Assembler by searching against Rfam 10.0 [78] using infernal v1.0.2 [79]. This extra contig were then included in the final assembly. The degenerate contigs were also used to search against the UniProt database using BLAST [80]. No biologically meaningful data were found in these contigs. Further details of the genome assembly are found in Protocol S1.

This Whole Genome Shotgun project has been deposited at DDBJ/EMBL/GenBank under the accession AUWU00000000. The version described in this paper is version AUWU01000000. The data will also be included in a 2014 release of GiardiaDB (http://giardiadb.org).

Repeat detection

RepeatMasker version open-3.3.0 (http://www.repeatmasker.org/) was used to screen genome repeats. It was run with default settings and sequence comparison in RepeatMasker was performed by the program cross_match version 1.080812. RepeatMasker library used was RM database version 20110920 with RepBase Update 20110920.

Heterozygosity estimation

Samtools [81] mpileup with B flag was used to generate pileup file from Illumina DNA reads mapped bam file. SNP sites were called in positions of base coverage more than 20 reads with an alternative base in more than 10% of the reads.

Genome annotation

An in-house annotation pipeline was implemented to annotate the genome. The annotation pipeline consists of structural and functional annotation. For the structural annotation, EvidenceModeler (EVM) r03062010 [82] was used to combine *ab initio* gene predictions, domain information and transcript alignments in weighted manner to arrive at a consensus gene structure. GlimmerHMM v3.0.1 [83], Prodigal v2.50 [84] and Glimmer3 v.3.02 [85] were the gene prediction programs used. Among them, GlimmerHMM performed best and was weighted highest. Domain information was from Pfam 25.0 [86] and TIGRFAM 10.0 [87] hits using HMMER3 3.0 (http://hmmer.janelia.org/). Transcript alignments include RNA-Seq reads as well as the *Spironucleus* ESTs from dbEST were mapped to the draft assembly. RNA-Seq mapping was done by BWA [77] and EST mapping was done by BLAST. RNA-Seq was weighted most among all the information used. The consensus genes from EVM were then functionally annotated using BLAST results against UniprotKB 20111005 [88] as well as Pfam and TIGRFAM domain hits. Both the functional and structural annotations were then inspected manually with Artemis [89], and adjustments were done to improve the quality of the annotation. More details of the annotation pipeline are found in the Protocol S1.

Shared and unique genes

The *G. intestinalis* genome refers to the genome from isolate WB [10] in the context of this paper unless otherwise stated. The *G. intestinalis* genome sequence and gene annotation were downloaded from *Giardia*DB 3.1 [90].

OrthoMCL v2.0.2 [91] was run with match cutoff of 50% and e-value cutoff of 1e-10, which resulted in 1349 shared core groups between *S. salmonicida* and *G. intestinalis* including 1718 genes from *S. salmonicida* and 1431 genes from *G. intestinalis.* On top of the OrthoMCL grouping, a pair of genes is considered to be shared between the two species if reciprocal BLAST hits have e-value <1e-03. This approach adds 1550 more *S. salmonicida* genes and 1658 more *G. intestinalis* genes into the shared pool. The rest of the genes were then considered to be unique to each other, which gives 4799 unique *S. salmonicida* genes compared to *G. intestinalis* and 2812 unique *G. intestinalis* genes. The 4799 *S. salmonicida*genes were then used in searches against UniProt KB 20130403 database. The same e-value cutoff was used to estimate *S. salmonicida* unique genes in comparison to all available sequenced genomes. Protein identities were extracted from the reciprocal BLAST results between *S. salmonicida* and *G. intestinalis.* 1147 OrthoMCL orthologous groups with only one member from each species were included in the analysis.

Synteny similarity between *G. intestinalis* and *S. salmonicida* was studied using the 1349 shared core groups based on a sliding window approach. Only homologous groups with less than five members in each species were included to reduce the noise from large protein families. We used a sliding window size of 20 kbp with step size of 1 kbp. Any window containing genes from at least three different homologous groups in both species were indicated and manually examined in ACT [92]. Among the 230 regions indicated, we failed to see any promising synteny blocks between the two species.

Protein kinases

Protein kinases were identified in combination of three different approaches. 129 genes were found to contain significant Pfam Pkinase domain (PF00069) (score >25). OrthoMCL [91] gene clusters shared between *S. salmonicida* and *G. intestinalis* was used to annotate 67 protein kinases, adding 7 extra kinases.

Protein kinases are categorized into group, family and subfamily. To assign all the protein kinases into group, family and subfamily, we constructed HMM profiles for different group, family and subfamily domains using the alignment files downloaded from the Kinase.com database (http:// kinase.com/). Hmmbuild from HMMER 3.0 (http:// hmmer.janelia.org/) was used to build HMM profiles from the alignments; hmmsearch to search the protein sequences against HMM profiles. Cutoff score 40 was used for the proteins with Pkinase domain, and score 50 was used for the rest of proteins. This annotated 137 protein kinases, adding 3 extra. The final description of the 138 protein kinases were decided in combination of orthology and HMMER results with manual efforts, and followed name convention in *Giardia*. *G. intestinalis* protein kinase annotations were taken from reference [24].

tRNAs and rRNAs

tRNAs were predicted by tRNAScan-SE v1.23 [93]. The most sensitive co-variance model was used. 5S, 18S and 28S rRNAs were predicted with RNAmmer v1.2 [94]. 5.8S were predicted by similarity search against Rfam 10.0 [78] using infernal v1.0.2 [79]. 5.8S and 28S overlapped in the initial prediction, and the 28S start was then adjusted according to alignment to other annotated 28S in NCBI.

Introns

Potential introns were validated by PCR using genomic DNA or cDNA as template, and the PCR products were sequenced with Sanger sequencing. The conserved AC-repeat motif in the intron was also used to search in the whole genome in attempt to look for potential introns or split introns, and all the potential cases were inspected and tested experimentally. To search for split introns, we collected the RNA-Seq reads which aligned to two different positions in the genome for inspection. However, due to the abundance of chimeric reads from sequencing, it was difficult to identify the true signal of split introns, and we did not see any obvious case which could indicate a split intron.

Promoters

MEME suite v4.8.1 [95] was used for promoter motif analysis. A maximum of 400 bp upstream of annotated genes (including 3′ partial genes) were used in search of potential promoter motifs using MEME. If the intergenic region was shorter than 400 bp, the longest possible sequence was used; if the intergenic regions were shorter than 8 bp, the sequence was ignored. 8208 promoter sequences were used to search for motifs. MEME was set to search for 10 most likely motifs with sizes from 6 bp to 30 bp. FIMO was then used to search in *S. vortens*, *G. intestinalis*and *T. vaginalis* promoter regions for similar motifs. *T. vaginalis* sequences and annotations were from TrichDB 1.3 and *S. vortens* data were from JGI (http://www.jgi.doe.gov/). BLAST searches were used to identify 1726 *S. vortens* homologs to 1253 *S. salmonicida* genes with BLASTX e-value <1e-10, >60% alignment match, <20% length difference, and with proper start and stop codon.

MEME was used to search for promoter motifs upstream of 19 putative *S. salmonicida* cyst-related genes (3 identical CWPs were excluded).

Gene expression levels

Expression level was measured by mapped RNA-Seq reads in term of FPKM (Fragments Per Kilobase of transcript per Million mapped reads). Cufflinks v2.0.2 [96] was used to calculate the FPKM. The reference annotation file was provided to estimate the expression level of the annotated genes.

Cloning and heterologous expression in *G. intestinalis* of a *S. salmonicida* cyst wall protein

The fusion of the CWP-1 promoter of *G. intestinalis* and the cyst wall protein was constructed by the creation of *Eco*RV site in the N-terminal of the SS50377_15904 gene. The mutated base is shown in bold font in the primer sequences below. The introduction of this unique restriction site did not alter the resulting amino acid sequence of the SS50377_15904 gene and created a seamless fusion to the CWP-1 promoter. The *S. salmonicida* cyst wall protein (SS50377_15904) was amplified by PCR from *S. salmonicida* genomic DNA as described in [20] using primers CWPE-F-EcoRV CCCGATATCTATCCTGGCAGTCCTCACACAGC and CWP-E-R CCCGCGGCCG CTGTCTAACGTAGACGCCGCAGTC. The *G. intestinalis* CWP-1 promoter was amplified by PCR from genomic DNA of *G. intestinalis* isolate WB/C6 using primers CWP1-P-HindIII CCCAAGCTTCAGAGGCATTGGACTTTGTCATG and CWP1-P-EcoRV CCCGATATCATCCCTGATATTTTATTTCTGTGTTTCTTG.
Underlined sequences denote introduced restriction enzyme sites. The SS50377_15904 and CWP1-P PCR products were gel purified by QIAquick Gel Extraction kit (Qiagen) and digested using *Hind*III and *Eco*RV and *Eco*RV and *Not*I. All restriction enzymes were of the FastDigest type from Fermentas. The fragments were purified by QIAquick PCR purification kit (Qiagen) and ligated using T4 DNA ligase into digested (*Hind*III and *Not*I) and FastAP (Fermentas) dephosphorylated PHA-5 vector (Jerlström-Hultqvist, unpublished results). The vector was transformed into DH5α *E. coli* cells and correct clones were identified by restriction digestion of purified plasmids. The sequence of identified clones was verified by Sanger sequencing.

Cloning and immunofluorescence of *S. salmonicida* cysteine-rich proteins

S. salmonicida cysteine-rich proteins were amplified by PCR from *S. salmonicida* genomic DNA using primers-pairs 17215-F ATATGCTAGCG CTTCATTGTAACATTTAATAAATTATCTCGCACATC 17215-R TATGCGG CCGCCATT GGATTTTGAACCATTCTACGACATT 18013-F ATATGCTAGC TAAACTTCGTATGATATGCAATAAACGGC 18013-R TATGCGGCC GCCACCAA AGTACGTTACTAAGTGGCTCA 18923-F ATATGCTA GCTAATTA TGGTTC TGCAGTGAGGAGTG 18923-R TATGCGGCCGCCACTGTTACTCCACTCTCTT CTTAGCC respectively as described previously [20]. The PCR products were gel-purified, digested by *Nhe*I and *Not*I and inserted into pSpiro-PAC-3×HA-C vector linearized with the above mentioned enzymes. Correct plasmids were recovered and sequences verified as described above.

Transfection of G. intestinalis or S. salmonicida

Plasmid DNA for transfection of *S. salmonicida* or *G. intestinalis* was prepared as described in reference [20]. Culture and transfection of *S. salmonicida* and *G. intestinalis* WB/C6 were according to reference [20] and reference [75], respectively. Transfectants of either organism were selected and maintained using 50 μg/ml puromycin (A.G Scientific). The *G. intestinalis*transfectants were induced to encyst by replacing the normal growth media of 70–80% confluent cultures with encystation media with pH 7.8 and 1.25 mg/ml of bovine bile. *In vitro*generated cysts were harvested 48 h post induction by centrifugation at 500× *g* and kept in water at 4°C for 24 h.

Immunofluorescence labelling and western blot of cells

S. salmonicida or encysting *G. intestinalis* transfectants cells (0, 6, 22 and 48 h post induction) as well as water-treated cysts were collected for immunofluorescence and Western blot as described [20], [75]. Cells were fixed, permeabilised and blocked [20]. The cells were stained either alone or with appropriate combinations anti-HA-Alexa Fluor 488 antibody (A488-101L, Covance; 1:250 dilution), anti-CWP1-FITC antibody (Waterborne Inc.; 1:20 dilution) or rabbit monoclonal HA-tag (C29F4) antibody (Cat. No: 3724, Cell signalling technologies; 1:1600 dilution). The rabbit monoclonal was detected using goat anti-rabbit conjugated to AlexaFluor 594 (A-11037, Invitrogen; 1:250 dilution). The stained cells were mounted in VectaShield with DAPI (Cat. No: H-1200, Vector Laboratories) and viewed using either a Zeiss Axioplan 2 epifluoresence microscope or a Zeiss 510 laser scanning confocal microscope. AxioVision LE 4.8.2.0 or Zen 2011 v7.0.0.285 (Carl Zeiss GmBH) was used to process images.

Cells carrying CRMP-1 constructs were fixed using 2% PFA, blocked with 2% BSA and stained using mouse anti-HA-Alexa Fluor 488 (1:250), mounted in VectaShield medium containing DAPI and viewed using a Zeiss 510 laser scanning confocal microscope.

Proteases

Peptidase and peptidase inhibitor sequences were obtained from the MEROPS database [41](release 9.6). Predicted *S. salmonicida* genes were used to search against all the peptidase protein sequences using BLASTP with e-value cutoff of 1e-03, and to search against Pfam domains (v25.0) [86] using HMMER 3.0 (http://hmmer.janelia.org/). A gene was assigned as a peptidase if any of the following three conditions satisfied: if a gene was supported by both a BLAST hit with e-value <1e-03 and Pfam peptidase domain hit with score >20; if a gene was supported by a good domain hit with score >50; if a gene was supported by a good BLAST hit with e-value <1e-10 and was already manually annotated as protease. The catalytic type and protease family were predicted in accordance with the classification in MEROPS. The same strategy was used to assign the proteases of *G. intestinalis*. The assignment results were similar to the information from the MEROPS database.

Cysteine-rich proteins

We aligned proteins with more than 5% cysteine using MUSCLE (v3.8.31) [97], and detected transmembrane domains, a conserved [KR][KR]X[KR][KR] signature motif, and repeated motifs of CXXC and CXC. Three types of cysteine-rich proteins were defined depending on the presence and absence of the transmembrane domain and sequence motifs within the cysteine-rich proteins. We used gene networks [98] to investigate the diversity of those three types together with the VSPs, HCMPs and HCPs from *G. intestinalis*. Python igraph was used to plot the gene networks. Each node represents a gene sequence, and sequences are connected if they share a sequence homology which was determined by the reciprocal BLASTP e-value <1e-05.

Analysis of pathways

Pathways were predicted using KEGG Automatic Annotation Server (KAAS) [99] with default cutoff 60. The predictions for *S. salmonicida* and *G.*

intestinalis were compared. The differences were manually examined and expanded using the literature on *G. intestinalis* metabolism [10], [12].

Transporter proteins

To identify potential transporter proteins in *S. salmonicida*, we started from proteins with at least one transmembrane domain predicted by TMHMM v2.0 [100]. Transporter family information was downloaded from the Transporter classification database (TCDB; http://www.tcdb.org/). Transporters were assigned using a combination of BLASTP hits with <1e-03 against TCDB collections of transporters, and Pfam domain hits with score >25. The same strategy was applied to assign *G. intestinalis* transporters. The transporter assignments were then verified with orthologous groups between the two organisms and proved to be consistent.

REFERENCES

1. Adl SM, Simpson AG, Lane CE, Lukes J, Bass D, et al. (2012) The revised classification of eukaryotes. J Eukaryot Microbiol 59: 429–514.
2. Brugerolle G, Lee JJ (2002) Order Diplomonadida. In: Lee JJ, Leedale GF, Bradbury P, editors. An Illustrated Guide to the Protozoa, 2nd edn. Lawrence, Kansas: Society of Protozoologists. pp. 1125–1135.
3. Tovar J, León-Avila G, Sánchez LB, Sutak R, Tachezy J, et al. (2003) Mitochondrial remnant organelles of *Giardia* function in iron-sulphur protein maturation. Nature 426: 172–176.
4. Jerlström-Hultqvist J, Einarsson E, Xu F, Hjort K, Ek B, et al. (2013) Hydrogenosomes in the diplomonad *Spironucleus salmonicida*. Nat Commun 4: 2493.
5. Ramesh MA, Malik SB, Logsdon JM Jr (2005) A phylogenomic inventory of meiotic genes; evidence for sex in *Giardia* and an early eukaryotic origin of meiosis. Curr Biol 15: 185–191.
6. Cooper MA, Adam RD, Worobey M, Sterling CR (2007) Population genetics provides evidence for recombination in *Giardia*. Curr Biol 17: 1984–1988.
7. Andersson JO (2012) Double peaks reveal rare diplomonad sex. Trends Parasitol 28: 46–52.
8. Andersson JO, Sjögren ÅM, Horner DS, Murphy CA, Dyal PL, et al. (2007) A genomic survey of the fish parasite *Spironucleus salmonicida* indicates genomic plasticity among diplomonads and significant lateral gene transfer in eukaryote genome evolution. BMC Genomics 8: 51.
9. Andersson JO, Sjögren ÅM, Davis LAM, Embley TM, Roger AJ (2003) Phylogenetic analyses of diplomonad genes reveal frequent lateral gene transfers affecting eukaryotes. Curr Biol 13: 94–104.
10. Morrison HG, McArthur AG, Gillin FD, Aley SB, Adam RD, et al. (2007) Genomic minimalism in the early diverging intestinal parasite *Giardia lamblia*. Science 317: 1921–1926.
11. Upcroft J, Upcroft P (1998) My favorite cell: *Giardia*. Bioessays 20: 256–263.
12. Adam RD (2001) Biology of *Giardia lamblia*. Clin Microbiol Rev 14: 447–475.

13. Ankarklev J, Jerlström-Hultqvist J, Ringqvist E, Troell K, Svärd SG (2010) Behind the smile: cell biology and disease mechanisms of *Giardia* species. Nat Rev Microbiol 8: 413–422.
14. Williams CF, Lloyd D, Poynton SL, Jorgensen A, Millet COM, et al. (2011) *Spironucleus*species: economically-important fish pathogens and enigmatic single-celled eukaryotes. J Aquac Res Development S2.
15. Kolisko M, Cepicka I, Hampl V, Leigh J, Roger AJ, et al. (2008) Molecular phylogeny of diplomonads and enteromonads based on SSU rRNA, alpha-tubulin and HSP90 genes: implications for the evolutionary history of the double karyomastigont of diplomonads. BMC Evol Biol 8: 205.
16. Kent ML, Ellis J, Fournie JW, Dawe SC, Bagshaw JW, et al. (1992) Systemic hexamitid (Protozoa, Diplomonadida) infection in seawater pen-reared Chinook salmon *Oncorhynchus tshawytscha*. Dis Aquat Organ 14: 81–89.
17. Jørgensen A, Sterud E (2006) The marine pathogenic genotype of *Spironucleus barkhanus* from farmed salmonids redescribed as *Spironucleus salmonicida* n. sp. J Eukaryot Microbiol 53: 531–541.
18. Roxström-Lindquist K, Jerlström-Hultqvist J, Jørgensen A, Troell K, Svärd SG, et al. (2010) Large genomic differences between the morphologically indistinguishable diplomonads *Spironucleus barkhanus* and *Spironucleus salmonicida*. BMC Genomics 11: 258.
19. Jørgensen A, Torp K, Bjorland MA, Poppe TT (2011) Wild arctic char *Salvelinus alpinus*and trout *Salmo trutta*: hosts and reservoir of the salmonid pathogen *Spironucleus salmonicida* (Diplomonadida; Hexamitidae). Dis Aquat Organ 97: 57–63.
20. Jerlström-Hultqvist J, Einarsson E, Svärd SG (2012) Stable transfection of the diplomonad parasite *Spironucleus salmonicida*. Eukaryot Cell 11: 1353–1361.
21. Franzén O, Jerlström-Hultqvist J, Castro E, Sherwood E, Ankarklev J, et al. (2009) Draft genome sequencing of *Giardia intestinalis* assemblage B isolate GS: are human giardiasis caused by two different species? PLoS Pathog 5 (8) e1000560.
22. Jerlström-Hultqvist J, Franzén O, Ankarklev J, Xu F, Nohynkova E, et al. (2010) Genome analysis and comparative genomics of a *Giardia intestinalis* assemblage E isolate. BMC Genomics 11: 543.
23. Keeling PJ, Doolittle WF (1997) Widespread and ancient distribution of a noncanonical genetic code in diplomonads. Mol Biol Evol 14: 895–901.
24. Manning G, Reiner DS, Lauwaet T, Dacre M, Smith A, et al. (2011) The minimal kinome of *Giardia lamblia* illuminates early kinase evolution and unique parasite biology. Genome Biol 12: R66.
25. O'Connell MJ, Krien MJ, Hunter T (2003) Never say never. The NIMA-related protein kinases in mitotic control. Trends Cell Biol 13: 221–228.
26. Collins L, Penny D (2005) Complex spliceosomal organization ancestral to extant eukaryotes. Mol Biol Evol 22: 1053–1066.
27. Franzen O, Jerlström-Hultqvist J, Einarsson E, Ankarklev J, Ferella M, et al. (2013) Transcriptome profiling of *Giardia intestinalis* using strand-specific RNA-seq. PLoS Comput Biol 9: e1003000.
28. Kamikawa R, Inagaki Y, Tokoro M, Roger AJ, Hashimoto T (2011) Split introns in the genome of *Giardia intestinalis* are excised by spliceosome-mediated trans-splicing. Curr Biol 21: 311–315.
29. Carlton JM, Hirt RP, Silva JC, Delcher AL, Schatz M, et al. (2007) Draft genome sequence of the sexually transmitted pathogen *Trichomonas vaginalis*. Science 315: 207–212.
30. Petersen TN, Brunak S, von Heijne G, Nielsen H (2011) SignalP 4.0: discriminating signal peptides from transmembrane regions. Nat Methods 8: 785–786.
31. Akopian D, Shen K, Zhang X, Shan SO (2013) Signal recognition particle: an essential protein-targeting machine. Annu Rev Biochem 82: 693–721.
32. Teodorovic S, Walls CD, Elmendorf HG (2007) Bidirectional transcription is an inherent feature of *Giardia lamblia* promoters and contributes to an abundance of

sterile antisense transcripts throughout the genome. Nucleic Acids Res 35: 2544–2553.

33. Best AA, Morrison HG, McArthur AG, Sogin ML, Olsen GJ (2004) Evolution of eukaryotic transcription: insights from the genome of *Giardia lamblia*. Genome Res 14: 1537–1547.
34. Iyer LM, Anantharaman V, Wolf MY, Aravind L (2008) Comparative genomics of transcription factors and chromatin proteins in parasitic protists and other eukaryotes. Int J Parasitol 38: 1–31.
35. Fuentes V, Barrera G, Sanchez J, Hernandez R, Lopez-Villasenor I (2012) Functional analysis of sequence motifs involved in the polyadenylation of *Trichomonas vaginalis*mRNAs. Eukaryot Cell 11: 725–734.
36. Williams CF, Vacca AR, Lloyd D, Schelkle B, Cable J (2013) Non-invasive investigation of *Spironucleus vortens* transmission in freshwater angelfish *Pterophyllum scalare*. Dis Aquat Organ 105: 211–223.
37. Januschka MM, Erlandsen SL, Bemrick WJ, Schupp DG, Feely DE (1988) A comparison of *Giardia microti* and *Spironucleus muris* cysts in the vole: an immunocytochemical, light, and electron microscopic study. J Parasitol 74: 452–458.
38. Wood AM, Smith HV (2005) Spironucleosis (Hexamitiasis, Hexamitosis) in the ring-necked pheasant (*Phasianus colchicus*): detection of cysts and description of *Spironucleus meleagridis* in stained smears. Avian Dis 49: 138–143.
39. Morf L, Spycher C, Rehrauer H, Fournier CA, Morrison HG, et al. (2010) The transcriptional response to encystation stimuli in *Giardia lamblia* is restricted to a small set of genes. Eukaryot Cell 9: 1566–1576.
40. Konrad C, Spycher C, Hehl AB (2010) Selective condensation drives partitioning and sequential secretion of cyst wall proteins in differentiating *Giardia lamblia*. PLoS Pathog 6: e1000835.
41. Rawlings ND, Morton FR, Kok CY, Kong J, Barrett AJ (2008) MEROPS: the peptidase database. Nucleic Acids Res 36: D320–325.
42. Sajid M, McKerrow JH (2002) Cysteine proteases of parasitic organ. Mol Biochem Parasitol 120: 1–21.
43. Adam RD, Nigam A, Seshadri V, Martens CA, Farneth GA, et al. (2010) The *Giardia lamblia vsp* gene repertoire: characteristics, genomic organization, and evolution. BMC Genomics 11: 424.
44. Nash TE, Banks SM, Alling DW, Merritt JW Jr, Conrad JT (1990) Frequency of variant antigens in *Giardia lamblia*. Exp Parasitol 71: 415–421.
45. Pagny S, Lerouge P, Faye L, Gomord V (1999) Signals and mechanisms for protein retention in the endoplasmic reticulum. J Exp Bot 50: 157–164.
46. Davids BJ, Reiner DS, Birkeland SR, Preheim SP, Cipriano MJ, et al. (2006) A new family of giardial cysteine-rich non-VSP protein genes and a novel cyst protein. PLoS ONE 1: e44.
47. Prucca CG, Slavin I, Quiroga R, Elias EV, Rivero FD, et al. (2008) Antigenic variation in *Giardia lamblia* is regulated by RNA interference. Nature 456: 750–754.
48. Nixon JEJ, Wang A, Field J, Morrison HG, McArthur AG, et al. (2002) Evidence for lateral transfer of genes encoding ferredoxins, nitroreductases, NADH oxidase, and alcohol dehydrogenase 3 from anaerobic prokaryotes to *Giardia lamblia* and *Entamoeba histolytica*. Eukaryot Cell 1: 181–190.
49. Law CJ, Maloney PC, Wang DN (2008) Ins and outs of major facilitator superfamily antiporters. Annu Rev Microbiol 62: 289–305.
50. Davidson AL, Dassa E, Orelle C, Chen J (2008) Structure, function, and evolution of bacterial ATP-binding cassette systems. Microbiol Mol Biol Rev 72: 317–364.
51. Young GB, Jack DL, Smith DW, Saier MH Jr (1999) The amino acid/auxin: proton symport permease family. Biochim Biophys Acta 1415: 306–322.
52. Schofield PJ, Costello M, Edwards MR, O'Sullivan WJ (1990) The arginine dihydrolase pathway is present in *Giardia intestinalis*. Int J Parasitol 20: 697–699.

53. Yarlett N, Martinez MP, Moharrami MA, Tachezy J (1996) The contribution of the arginine dihydrolase pathway to energy metabolism by *Trichomonas vaginalis*. Mol Biochem Parasitol 78: 117–125.
54. Lacourciere GM, Mihara H, Kurihara T, Esaki N, Stadtman TC (2000) *Escherichia coli*NifS-like proteins provide selenium in the pathway for the biosynthesis of selenophosphate. J Biol Chem 275: 23769–23773.
55. Lacourciere GM (2002) Selenium is mobilized *in vivo* from free selenocysteine and is incorporated specifically into formate dehydrogenase H and tRNA nucleosides. J Bacteriol 184: 1940–1946.
56. Haft DH, Self WT (2008) Orphan SelD proteins and selenium-dependent molybdenum hydroxylases. Biol Direct 3: 4.
57. Zhang Y, Turanov AA, Hatfield DL, Gladyshev VN (2008) *In silico* identification of genes involved in selenium metabolism: evidence for a third selenium utilization trait. BMC Genomics 9: 251.
58. Kopriva S, Patron NJ, Keeling P, Leustek T (2008) Phylogenetic analysis of sulfate assimilation and cysteine biosynthesis in phototrophic organisms. In: Hell R, Dahl C, Knaff DB, Leustek T, editors. Sulfur Metabolism in Phototrophic Organisms: Springer Netherlands. pp. 31–58.
59. Takahashi H, Kopriva S, Giordano M, Saito K, Hell R (2011) Sulfur assimilation in photosynthetic organisms: molecular functions and regulations of transporters and assimilatory enzymes. Annu Rev Plant Biol 62: 157–184.
60. Hérbert A, Casaregola S, Beckerich J-M (2011) Biodiversity in sulfurmetabolism in hemiascomycetous yeasts. FEMS Yeast Res 11: 366–378.
61. Payne SH, Loomis WF (2006) Retention and loss of amino acid biosynthetic pathways based on analysis of whole-genome sequences. Eukaryot Cell 5: 272–276.
62. Ma K, Adams MWW (1994) Sulfide dehydrogenase from the hyperthermophilic archaeon *Pyrococcus furiosus*: a new multifunctional enzyme involved in the reduction of elemental sulfur. J Bacteriol 176: 6509–6517.
63. Hagen WR, Silva PJ, Amorim MA, Hagedoorn P-L, Wassink H, et al. (2000) Novel structure and redox chemistry of the prosthetic groups of the iron-sulfur flavoprotein sulfide dehydrogenase from *Pyrococcus furiosus*; evidence for a [2Fe-2S] cluster with Asp(Cys)3 ligands. J Biol Inorg Chem 5: 527–534.
64. Andersson JO, Roger AJ (2002) Evolutionary analyses of the small subunit of glutamate synthase: gene order conservation, gene fusions and prokaryote-to-eukaryote lateral gene transfers. Eukaryot Cell 1: 304–310.
65. Bridger SL, Clarkson SM, Stirrett K, DeBarry MB, Lipscomb GL, et al. (2011) Deletion strains reveal metabolic roles for key elemental sulfur-responsive proteins in *Pyrococcus furiosus*. J Bacteriol 193: 6498–6504.
66. Liu Y, Beer LL, Whitman WB (2012) Sulfur metabolism in archaea reveals novel processes. Environ Microbiol 14: 2632–2644.
67. Brown DM, Upcroft JA, Upcroft P (1995) Free radical detoxification in *Giardia duodenalis*. Mol Biochem Parasitol 72: 47–56.
68. Brown DM, Upcroft JA, Upcroft P (1996) A H_2O-producing NADH oxidase from the protozoan parasite *Giardia duodenalis*. Eur J Biochem 241: 155–161.
69. Testa F, Mastronicola D, Cabelli DE, Bordi E, Pucillo LP, et al. (2011) The superoxide reductase from the early diverging eukaryote *Giardia intestinalis*. Free Radic Biol Med 51: 1567–1574.
70. Vicente JB, Testa F, Mastronicola D, Forte E, Sarti P, et al. (2009) Redox properties of the oxygen-detoxifying flavodiiron protein from the human parasite *Giardia intestinalis*. Arch Biochem Biophys 488: 9–13.
71. Millet CO, Cable J, Lloyd D (2010) The diplomonad fish parasite *Spironucleus vortens*produces hydrogen. J Eukaryot Microbiol 57: 400–404.
72. Babula P, Masarik M, Adam V, Eckschlager T, Stiborova M, et al. (2012) Mammalian metallothioneins: properties and functions. Metallomics 4: 739–750.

73. Andersson JO, Hirt RP, Foster PG, Roger AJ (2006) Evolution of four gene families with patchy phylogenetic distribution: influx of genes into protist genomes. BMC Evol Biol 6: 27.
74. Mastronicola D, Testa F, Forte E, Bordi E, Pucillo LP, et al. (2010) Flavohemoglobin and nitric oxide detoxification in the human protozoan parasite *Giardia intestinalis*. Biochem Biophys Res Commun 399: 654–658.
75. Jerlström-Hultqvist J, Stadelmann B, Birkestedt S, Hellman U, Svärd SG (2012) Plasmid vectors for proteomic analyses in *Giardia*: purification of virulence factors and analysis of the proteasome. Eukaryot Cell 11: 864–873.
76. Miller JR, Delcher AL, Koren S, Venter E, Walenz BP, et al. (2008) Aggressive assembly of pyrosequencing reads with mates. Bioinformatics 24: 2818–2824.
77. Li H, Durbin R (2009) Fast and accurate short read alignment with Burrows-Wheeler transform. Bioinformatics 25: 1754–1760.
78. Gardner PP, Daub J, Tate J, Moore BL, Osuch IH, et al. (2011) Rfam: Wikipedia, clans and the "decimal" release. Nucleic Acids Res 39: D141–145.
79. Nawrocki EP, Kolbe DL, Eddy SR (2009) Infernal 1.0: inference of RNA alignments. Bioinformatics 25: 1335–1337.
80. Consortium U (2012) Reorganizing the protein space at the Universal Protein Resource (UniProt). Nucleic Acids Res 40: D71–75.
81. Li H, Handsaker B, Wysoker A, Fennell T, Ruan J, et al. (2009) The Sequence Alignment/Map format and SAMtools. Bioinformatics 25: 2078–2079.
82. Haas BJ, Salzberg SL, Zhu W, Pertea M, Allen JE, et al. (2008) Automated eukaryotic gene structure annotation using EVidenceModeler and the Program to Assemble Spliced Alignments. Genome Biol 9: R7.
83. Majoros WH, Pertea M, Salzberg SL (2004) TigrScan and GlimmerHMM: two open source ab initio eukaryotic gene-finders. Bioinformatics 20: 2878–2879.
84. Hyatt D, Chen GL, Locascio PF, Land ML, Larimer FW, et al. (2010) Prodigal: prokaryotic gene recognition and translation initiation site identification. BMC Bioinformatics 11: 119.
85. Delcher AL, Bratke KA, Powers EC, Salzberg SL (2007) Identifying bacterial genes and endosymbiont DNA with Glimmer. Bioinformatics 23: 673–679.
86. Finn RD, Mistry J, Tate J, Coggill P, Heger A, et al. (2010) The Pfam protein families database. Nucleic Acids Res 38: D211–222.
87. Selengut JD, Haft DH, Davidsen T, Ganapathy A, Gwinn-Giglio M, et al. (2007) TIGRFAMs and Genome Properties: tools for the assignment of molecular function and biological process in prokaryotic genomes. Nucleic Acids Res 35: D260–264.
88. UniProt C (2012) Reorganizing the protein space at the Universal Protein Resource (UniProt). Nucleic Acids Res 40: D71–75.
89. Rutherford K, Parkhill J, Crook J, Horsnell T, Rice P, et al. (2000) Artemis: sequence visualization and annotation. Bioinformatics 16: 944–945.
90. Aurrecoechea C, Brestelli J, Brunk BP, Carlton JM, Dommer J, et al. (2009) GiardiaDB and TrichDB: integrated genomic resources for the eukaryotic protist pathogens *Giardia lamblia* and *Trichomonas vaginalis*. Nucleic Acids Res 37: D526–530.
91. Li L, Stoeckert CJ Jr, Roos DS (2003) OrthoMCL: identification of ortholog groups for eukaryotic genomes. Genome Res 13: 2178–2189.
92. Carver TJ, Rutherford KM, Berriman M, Rajandream MA, Barrell BG, et al. (2005) ACT: the Artemis Comparison Tool. Bioinformatics 21: 3422–3423.
93. Schattner P, Brooks AN, Lowe TM (2005) The tRNAscan-SE, snoscan and snoGPS web servers for the detection of tRNAs and snoRNAs. Nucleic Acids Res 33: W686–689.
94. Lagesen K, Hallin P, Rodland EA, Staerfeldt HH, Rognes T, et al. (2007) RNAmmer: consistent and rapid annotation of ribosomal RNA genes. Nucleic Acids Res 35: 3100–3108.
95. Bailey TL, Elkan C (1994) Fitting a mixture model by expectation maximization to discover motifs in biopolymers. Proc Int Conf Intell Syst Mol Biol 2: 28–36.

96. Trapnell C, Williams BA, Pertea G, Mortazavi A, Kwan G, et al. (2010) Transcript assembly and quantification by RNA-Seq reveals unannotated transcripts and isoform switching during cell differentiation. Nat Biotechnol 28: 511–515.
97. Edgar RC (2004) MUSCLE: a multiple sequence alignment method with reduced time and space complexity. BMC Bioinformatics 5: 113.
98. Beauregard-Racine J, Bicep C, Schliep K, Lopez P, Lapointe FJ, et al. (2011) Of woods and webs: possible alternatives to the tree of life for studying genomic fluidity in *E. coli*. Biol Direct 6: 39.
99. Moriya Y, Itoh M, Okuda S, Yoshizawa AC, Kanehisa M (2007) KAAS: an automatic genome annotation and pathway reconstruction server. Nucleic Acids Res 35: W182–185.
100. Krogh A, Larsson B, von Heijne G, Sonnhammer EL (2001) Predicting transmembrane protein topology with a hidden Markov model: application to complete genomes. J Mol Biol 305: 567–580.
101. Waterhouse AM, Procter JB, Martin DM, Clamp M, Barton GJ (2009) Jalview Version 2 - a multiple sequence alignment editor and analysis workbench. Bioinformatics 25: 1189–1191.

CHAPTER 13

Effect of B-Glucan on The Immune Response of Early Stage of Anabas Testudineus (Bloch) Challenged with Fungus Saprolegnia Parasitica

Basanta Kumar Das*, Pranati Pattnaik, Chandan Debnath, Dipak Kumar Swain and Jyotirmayee Pradhan

Fish Health Management Division, Central Institute of Freshwater Aquaculture (CIFA), P. O. Kausalyagnaga, Bhubaneswar, Odisha 751 002, India

ABSTRACT

The present study was carried out to study the effect of different dosages of β-glucan suspension on immune response and disease resistance in *Anabas testudineus* spawns against a fungal pathogen *Saprolegnia parasitica*. Eight day old spawns were exposed for 3 h in four different dosages of β-glucan suspension in phosphate buffered saline at the rate of 0, 5, 10, 15 mg l^{-1}. The cell suspension of spawn was assayed for total protein, acid phosphate, lysozyme, bactericidal and NBT activity. The spawns were then challenged with $3x10^5$ CFU ml^{-1} of *S. parasitica*. The survivability percentage and immunological parameters were assayed upto day 7. Exposure of fish for 3h to all the concentrations of β-glucan recorded a significant enhancement in the immunological parameters such as lysozyme activity, bactericidal activity and NBT activity by the end of day 7. The challenge study indicated least mortality in the groups exposed to 10 mgl^{-1} and 15 mgl^{-1} but 15 mgl^{-1} gives little higher survivability. Thus 3 h exposure of β-glucan suspension could reduce the mortality and increase the immunity of *A. testudineus* spawns against *S. parasitica*.

Keywords: *Anabas testudineus Saprolegnia parasitica* β-glucan Immunity Bactericidal

INTRODUCTION

Saprolegnia is ubiquitous in freshwater ecosystems and is considered as the main genus of water molds responsible for significant fungal infections of freshwater fish and eggs (Noga 1996), especially from the egg stage through smoltification (Bruno & Wood 1999; Pickering 1994). *Saprolegnia* is an opportunist facultative parasite (Neish 1977), which is saprotrophic and necrotrophic (Bruno & Wood 1999). On fish, *Saprolegnia* invades epidermal tissues visible as white or grey patches of filamentous mycelium (Bruno & Wood 1999; Beakes et al.1994), generally beginning on the head or fins (Neish 1977; Willoughby 1994) and can spread over the entire surface of the body. *Saprolegnia* is characterized by an external, cotton-like appearance that radiates out in a circular, crescent-shaped or whorled pattern. The fungal spores may be transmitted by hatchery fish, wild fish, eggs, water supplies, and equipment (Bruno & Wood 1999).

A number of workers have reported *Saprolegnia* infection in fishes; salmonids (Beakes et al. 1994; Hatai & Hoshiai , Hatai & Hoshiai Hatai & Hoshiai 1994), teleosts (Bruno & Wood 1999), Channel catfish (Howe et al. 1999), pike (Willoughby 1985), bass (Noga 1996), elver and suckers (Roberts 1989), roach, orfe, carp, tench, lamprey, sturgeon, barramundi, tilapia, and mullet (Bruno & Wood 1999). It has also been associated with tropical fish, including the kissing gourami, guppy, swordfish and platyfish (Willoughby 1994; Roberts 1989). The air breathing teleost, *Anabas testudineus*, from an Indian river was also found to carry *S. parasitica* infections (Mohanta & Patra 1992).

Saprolegnia generally invades fish that have been stressed or otherwise have weakened immune systems (Bruno & Wood 1999). A number of chemicals such as malachite green (Willoughby & Roberts 1992; Bruno & Wood , Bruno & Wood Bruno & Wood 1999), 37% formaldehyde (Van Waters 1988), hydrogen peroxide (Fitzpatrick et al. 1995; Marking et al. 1994), sodium chloride at high concentrations (Pickering 1994;Marking et al. 1994), etc. are available for treatment of fungal infection but a few are approved for use in aquaculture. The reduction of stress and boosting up immune response by application of immunostimulants appears to be the single greatest factor to help fish resist saprolegniasis.

β-glucan are a group of glucose polymers which are the main structural components of cell-wall in fungi, plants and some bacteria (Brown & Gordon 2003). They can be derived from the cell walls of yeast, bacteria, fungi, and cereals such as oats, barley, and rye. They have a backbone of β (1–3)-linked β -D-glucopyranosyl units with side chains of different lengths. A variety of cell surface receptors bind β -glucan, including lectins, scavenger receptors, and integrins on monocyte/macrophages, neutrophils, and natural killer (NK) cells and various lymphocyte subpopulations (Brown & Gordon 2003). Engagement of these receptors by β -glucan may induce activation of leukocytes, phagocytic activity, production of inflammatory cytokines and chemokines, microbial killing, and initiate the development of adaptive immunity, all of which contribute to the anti-infective and antitumorigenic properties of β -glucan (Brown et al. 2003; Misra *et al.* 2006). β -Glucan has also been shown to enhance resistance of fish and crustacea against bacterial and viral infections (Chang et al. 2003; Misra et al. 2004; Misra et al. 2006).

The present experiment was designed to study the effect of β –glucan immersion in three different doses on immune response and survival of eight day old spawn of *Anabas testudineus* challenged with *Saprolegnia parasitica*.

MATERIALS AND METHODS

Fish

Anabas testudineus spawns of 8 days old were obtained from the farm of Central Institute of Freshwater Aquaculture (CIFA), Bhubaneswar, Orissa, India. The spawns were acclimatized in a circular plastic container of 20 l capacity in the laboratory of Fish Health Management Division, CIFA for two days. The spawns were fed with a formulated diet obtained from CIFA at 5% of the total body weight for two times a day.

β–glucan

β–glucan obtained from Barley (Sigma) in powder form was added to phosphate buffer saline (pH 7.4) and subsequently sonicated (Artek Sonic Dismembrator Model 150) using a microtip at a relative output of 0.6 and duty cycle of 40%. The concentration was adjusted as 5 mgl^{-1}, 10 mgl^{-1} and

15 mgl^{-1} respectively by using sterile milliQ water (Millipore Corporation, India). Immersion was selected as the route of administration of β–glucan.

Experimental design

The spawns after acclimatization were divided into four major groups for exposure to different concentrations of β–glucan, such as Group A (0 mg l^{-1}, control), Group B (5 mg l^{-1}), Group C (10 mg l^{-1}) and Group D (15 mg l^{-1}). Fifty spawns were kept in each of the duplicate beaker (1L) for each treatment group. After bath exposure with β –glucan suspension for 3 h, four spawns were collected from each tank after a time interval of 3 h for evaluating changes in lysozyme activity, acid phosphate (AcP) activity, total protein content, bactericidal activity and superoxide anion assay. As it was difficult to draw blood from spawn with indistinctly developed internal organs, the entire body of four spawns taken at a time was aseptically placed in a macerator containing Hank's Balanced Salt Solution (HBSS) and anticoagulant, heparin. The macerated sample was layered on to preform continuous gradients of 51% percoll (Pharmacia, Uppsala, Sweden) in 8.5 g l^{-1} NaCl and centrifuged at 800 g for 20 min at 4°C. The white cell band formed at the interface of the cell suspension and percoll layer was harvested with a Pasteur pipette, diluted ten folds with HBSS and recentrifuged at 800 g for 10 min at 4°C to remove residual percoll. The resulting pellet was again washed twice and centrifuged at 800 g for 10 min at 4°C. The resultant white band cell suspension was used for assaying lysozyme, acid phosphatase, bactericidal, superoxide anion activities and total protein content. After the third wash, the concentration and viability of leucocytes cell suspensions were determined in 0.2% trypan blue. The total protein content of the cell lysate was analysed using protein kit (Bangalore Genei Protein Kit, Bradford macro method) and bovine serum albumin as standard as per the manufacturers instructions.

Challenge

After exposing in β–glucan suspension for 3 h, the spawns were removed from the glucan treated/untreated group and placed in normal freshwater. The spawns were divided into two sub-groups (32 nos. each) and each treatment group was challenged by immersing in viable fungal suspension of *Saprolegnia parasitica* at a concentration of 3×10^5 CFU ml^{-1} for seven days. Four spawns from the first sub-group of each treatment challenged with *S. parasitica* were taken on day 3 and day 7 post challenges for evaluating changes in lysozyme, acid phosphate and total protein content. Each test was repeated four times and mean of the sample was taken.

Lysozyme activity of the cell suspension

Lysozyme activity of cell suspension in each subgroup was measured by turbidimetric method described by Parry *et al.*(1965) using 0.2 mg ml^{-1} lyophilized *Micrococcus luteus* ATCC 49732 (DIFCO, BBL- Qualis lab) as the substrate in 0.05 M phosphate buffer (pH 6.2). Various amount of cell suspension (25-100 µL) were added to 2 ml of the bacterial cell and the absorbance was measured at 0.5 and 4.5 min intervals at 530 nm (25°C). One unit of lysozyme activity is defined as the amount of sample causing a decrease in absorbance of 0.001 min^{-1}.

Bactericidal activity of the cell suspension

Bactericidal activity of the cell suspension was estimated by following the procedure of Kajita et al. (1990) with little modification. An equal volume (100 µl) of the spawn cell suspension and bacterial cell was mixed and incubated for 1h at 25°C. A blank control was prepared by replacing the cell suspension with sterile PBS. The mixture was then diluted with sterile PBS at a ratio of 1: 10. The diluted mixture (100 µl) was pour plated in nutrient agar and plates were incubated for 24 h at 37°C. The number of viable bacteria was determined by counting the colonies grown in nutrient agar plates.

Bactericidal activity in % = (No. of colonies in control- No. of colonies in sample × 100) / No. of colonies in control

NBT activity of the cell suspension

The NBT activity was measured according to Chung & Secombes (1988) with some modifications. At first, 96-well flat bottom microtitre plate was taken and coated with 100 µl poly-L-lysine solution (0.2% Sigma). To this, 100 µl cell suspension sample was added in 5 of the 96 wells of the microtitre plate and then incubated at 25°C for 2 h and then washed with HBSS. Then 100 µl of NBT solution (0.02 gm NBT + 0.050 µl methanol + 0.94 µl PBS + 0.010 µl bacterial suspension) was added, which contained 10^6 numbers of *A. hydrophila* cells/ml. Subsequently it was incubated at 25°C for 1h. Then the medium was removed and the reaction was terminated. Again washing was done (3 times) by addition of methanol, which was followed by air drying. Subsequently, the formazan in each well was dissolved with 120 µl of 2 M KOH and 140 µl of DMSO. The absorbance was measured at 655 nm in the ELISA reader (Biorad), with 405 nm as reference.

Acid Phosphatase activity

The acid phosphatase activity was measured using Acid phosphatase kit (Batch No. 14407/14507 manufactured by Accurex Biomedical Pvt. Ltd.) following the method of Hillman (1971). The working solution was prepared by dissolving the contents of the substrate bottle with diluent. One ml of the AcP working solution was added to 100 µl of sample. Absorbance was measured after 5 min at 405 nm and subsequently three more readings were recorded after 1 min interval. The concentration of AcP was calculated as u/l by using the formulae i.e., Conc. in 1u/l = 743 × Δ Abs./ min

Statistical analysis

The mean values of these parameters were recorded over five days and ANOVA followed by Duncan's multiple range tests were performed on the data (SAS version 9.1) to find the difference at 5% ($P \leq 0.05$) level. Simultaneously, the percentage survival of spawn in the different groups of each challenge treatment was recorded daily over 7 days.

RESULTS

There were changes in ACP activity, total protein content, lysozyme activity, bactericidal activity and superoxide anion assay in different glucan treated/ untreated groups of *A testudineus* spawns, though the number of leucocytes were adjusted to a fixed level of 10^5 cells ml^{-1}.

The total protein content of cell suspension lysate was significantly increased in Group C fish after 3 h β-glucan post-treatment (Figure 1). There was significant increase in protein content in Groups B and C after day 3 post-challenge with *S. paracitica.* However, the protein content was reduced significantly in Groups B and C after day 5 post-challenge. In case of group D fish the protein content was higher than the control group of fish.

The lysozyme enzyme activity was significantly increased in groups B and D after 3h exposure to β-glucan over control. There was a significant increase in lysozyme activity in Group B and D over the control on day 5 post-challenge treatment with *S. parasitica* (Figure 2). However, the lysozyme enzyme activity was significantly decreased in Group C invariable of treatment with β-glucan and post- challenge with *S. parasitica.*

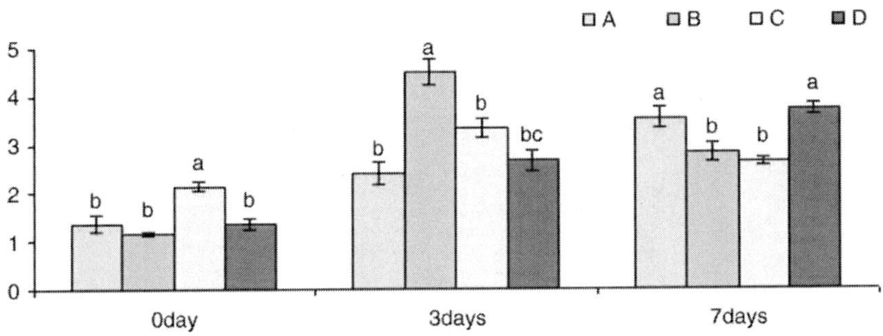

Figure 1 Pre-challenge and post-challenge total protein (mgml $^{-1}$) content of the cell suspension collected after different time of exposure with betaglucan. Bars represent pooled mean values. **A**, 0 mgl^{-1} glucan: **B**, 5 mgl^{-1} glucan: **C**, 10 mgl^{-1}glucan: **D**, 15 mgl^{-1} glucan. Bars bearing common superscript are not significant at 5% level in comparison to each other (n = 6).

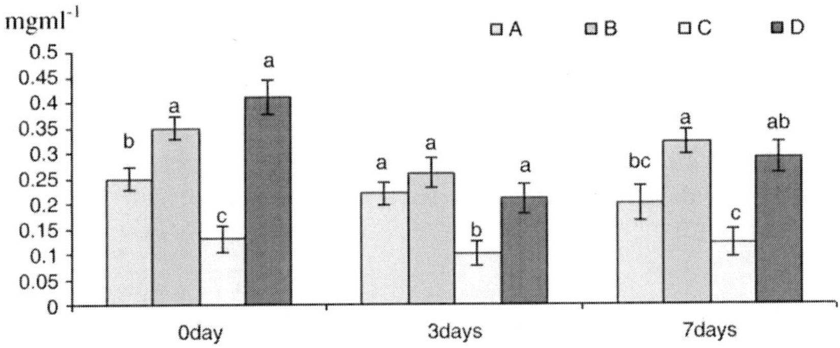

Figure 2 Pre-challenge and post-challenge lysozyme (Ul $^{-1}$) content of the cell suspension collected after different time of exposure with betaglucan. Bars represent pooled mean values. **A**, 0 mgl^{-1} glucan: **B**, 5 mgl^{-1} glucan: **C**, 10 mgl^{-1} glucan: **D**, 15 mgl^{-1} glucan. Bars bearing common superscript are not significant at 5% level in comparison to each other (n = 6).

A significant increase in bactericidal activity was observed in Groups B and C exposed for 3 h in β-glucan suspension as compared to control (Figure 3). However, the bactericidal activity was significantly reduced in all the groups on day 3 and day 5 of post-challenge treatment with *S. parasitica* over the control group expect for Group C on day 5 post-challenge treatment which showed a significant increase in bactericidal activity.

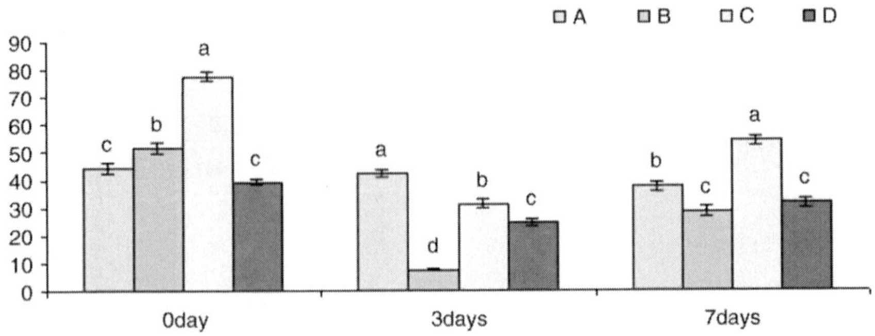

Figure 3 Pre-challenge and post-challenge bactricidal (%) activity of the cell suspension collected after different time of exposure with betaglucan. Bars represent pooled mean values. **A**, 0 mgl^{-1} glucan: **B**, 5 mgl^{-1} glucan: **C**, 10 mgl^{-1} glucan: **D**, 15 mgl^{-1} glucan. Bars bearing common superscript are not significant at 5% level in comparison to each other (n = 6).

The leucocyte cell suspension of spawns of *Anabas* showed a marginal increase in superoxide anion activity in groups exposed to 3 h β-glucan over the control group. There was a significant increase in superoxide anion activity in Groups C and D on day 3 post-challenge treatment with *S. parasitica.* Similarly, significant increase was observed in all the Groups (B,C, D) on day 5 post-challenge treatment with *S. parasitica* (Figure 4).

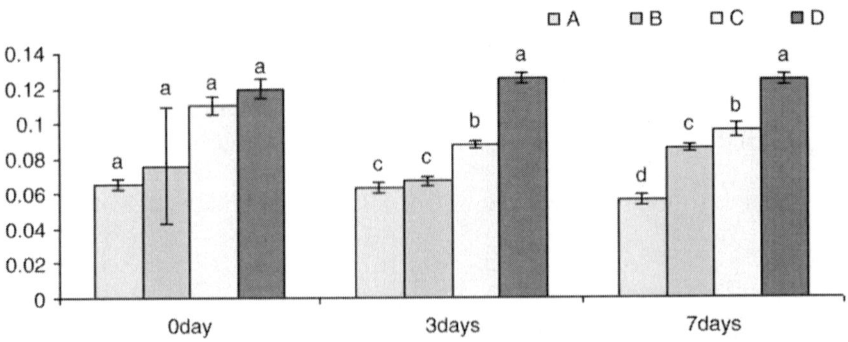

Figure 4 Pre-challenge and post-challenge NBT (O.D.) activity of the cell suspension collected after different time of exposure with beta-glucan. Bars represent pooled mean values. **A**, 0 mgl^{-1} glucan: **B**, 5 mgl^{-1} glucan: **C**, 10 mgl^{-1}glucan: **D**, 15 mgl^{-1} glucan. Bars bearing common superscript are not significant at 5% level in comparison to each other (n = 6).

The AcP activity recorded a marginal increase in Group C (9.83%) over the control group after 3 h β-glucan exposure in pre-challenge treatment. There was significant increase in the AcP activity in Group B on day 3 post-challenge treatment with *S. parasitica.* Groups C and D recorded an increase of 23.75% and 19.28% respectively over the control group A on day 3 post-challenge treatment (Figure 5). The AcP activity was significantly increased in all groups (B,C,D) on day 5 post-challenge treatment over the control group A.

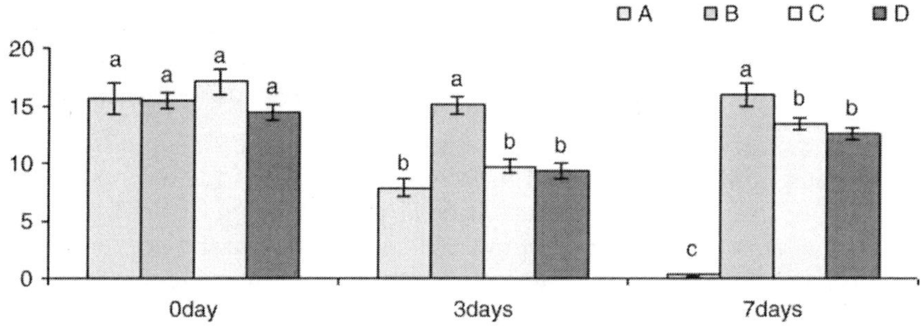

Figure 5 Pre-challenge and post-challenge acid phosphatase (Ul $^{-1}$) activity of the cell suspension collected after different time of exposure with beta-glucan. Bars represent pooled mean values. **A**, 0 mgl^{-1} glucan: **B**, 5 mgl^{-1} glucan: **C**, 10 mgl^{-1} glucan: **D**, 15 mgl^{-1} glucan. Bars bearing common superscript are not significant at 5% level in comparison to each other (n = 6).

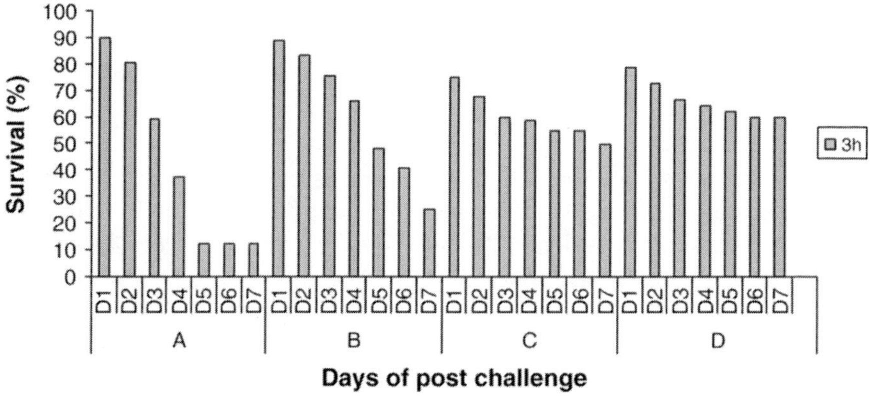

Figure 6 Percent survival of spawn after immersion challenge with *Saprolegnia parasitica* (n = 32). **A**, 0 mgl^{-1} glucan: **B**, 5 mgl^{-1} glucan: **C**, 10 mgl^{-1} glucan: **D**, 15 mgl^{-1} glucan.

The survivability percentage of post challenge treatment group is indicated in Figure 6. It was noticed that survival percentage was maximum in Group D fish exposed to 3 h post immersion with β–glucan suspension after day 7 (Figure 6). Survivability was only 12.5 percent in control groups.

DISCUSSION

The present work demonstrates the effect of multiple immersion dosages of β-glucan suspension on immune response and disease resistance in *A. testudineus* spawns against a fungal pathogen *Saprolegnia parasitica*.

As a first line of defense, mucous, epithelium from skin, gills and intestine and various peptides, lysozomal enzymes, complement factors and other lytic factors present in the serum prevent adherence and colonization of microorganisms (Alexander & Ingram 1992), resulting in the prevention of infection and diseases. In addition, macrophages which are innate immune cells present at hatching stage (Baulny et al. 1996) and innate immune factors constitute a second barrier against invading pathogens. The enhancements of lysozyme activity complement activity and bactericidal activity due to administration of β-glucan for a longer duration have been reported (Trinder 1969 Doumas et al. 1971 Bradford 1976 Dalmo et al. 1996). The effect of β-glucan on immunity and survival of *Anabas testudineus* against *A. hydrophila* was studied by Das et al. (2009). Based on previous observations, for exposure durations of 2 h and 3 h, it was found that the exposure of *A. testudineus* spawn to β-glucan at $10 mgl^{-1}$ for 3 h followed by challenge with *A. hydrophila* provided better immunity and protection (Das et al. 2009). The present study was undertaken to document, the effect of β-glucan suspension on immune response and disease resistance of early stages of fish against fungal pathogen *S. parasitica*.

It has been shown that injection of laminarina β-1, 3 glucan typical of those present in fungal cell walls, induces immune responses in both adult and larval locusts (Goldsworthy et al. 2002; Mullen and Goldsworthy 2003). According to them, the activation of immune response varies in emerging adult as that of mature adults.

The total protein content of the cell suspension was found to change upon exposure to β glucan suspension of various concentrations over 3 h exposures as reported by Das et al. (2009). Following challenge with *S. parasitica*, the protein concentration was increased in all the groups as compared to control on day 3 and decreased significantly in groups B and C on day 7.

There was increase in lysozyme activity in all the treatment groups as compared to the control as reported in our previous work (Das et al. 2009). The post- challenge groups of fishes showed increased lysozomal enzymes but the increase was marginal as compared to challenge with *A. hydrophila*(Das et al. 2009). This indicates that lysozomal enzymes produced by the leucocytic cell suspension may be spared to counter act the fungal challenge.

Bactericidal activity of cell suspension was also increased upon exposure to 3 h β-glucan suspension at all doses but decreased followed by post challenge fish on day 3 and day 7 except experimental group C on day 7. Increased bactericidal activity was reported by Das et al. (2009) following β-glucan immunostimulation. Interestingly, *A. hydrophila* post-challenge increased the bactericidal activity (Das et al. 2009) and *S. parasitica* post-challenge decreased the bactericidal activity. Such decreased bactericidal properties of fish cell suspension involved in some innate and/or adaptive immune response are depressed in the cell suspension at the early stages which are utilized to fight against fungal infection.

Interestingly there is a rise in total NBT positive cells after exposure of β-glucan suspension for 3 h followed by fungal infection as reported with bacterial challenge (Das et al. 2009). Misra *et al.*(2005) reported a peak O_2^- production during the NBT assay after feeding β–glucan at a dose of 500 mg kg^{-1} diet for 42 days. The role of β-glucan in our study was visible from the mortality pattern in the spawn for 7 days post challenge with *S. parasitica*which could be attributed to the increased activities of different peptides (Couso et al. 2003; Misra et al. 2005). Post challenged group showed increased phosphatase activities indicates higher breakdown of energy reserves, which was utilized by the growth and survival of fish (Das et al. 2009). Though, the acidphosphatase activity was decreased in prechallenged glucan stimulation at 5 and 15 mg ml^{-1} which might be corroborated with the findings of Sahu et al. (2008) and Das et al. (2009).

Exposure of β-glucan at different dosages in our study showed reduced mortality on the fish on the 7th day post challenge through bath immersion with the fungal *pathogen S. parasitica*. The protective effect of β-glucan against the fungus noticed in the challenge study irrespective of dosages. Further, it can be inferred from the bath challenge study that the increased protection against the fungus could be due to the enhanced immunity as evidenced by the increase in different immune parameters of the leucocytes cell suspension.

It is thus evident from the present experiment that β-glucan also protect the fish from fungal infection of *S. parasitica*at the 10 and 15 mgl^{-1} concentrations exposure for 3 h as noticed from the survival rate. But in earlier work, we have reported that β-glucan at a dose of 10 mgl^{-1} for 3 h is sufficient enough for giving protection against *Aeromonas hydrophila* infection. Though 15 mgl^{-1} gives little higher survival but 10 mgl^{-1} is equally good enough for protection (60% and 50% respectively). Especially spawn of 8 days old receives protection to both the bacteria and fungi and increased nonspecific immunity as evidenced from the leucocytes suspension parameters. This will have a great impact in the nursery phase rearing of this species where spawns could be exposed to glucan suspension for preventing loss due to fungal and bacterial infections and will provide healthy spawn for grow out culture practices to the farmers.

ACKNOWLEDGEMENTS

The authors are gratefully acknowledged the Director, Central Institute of Freshwater Aquaculture for providing all possible assistance and cooperation for conducting this investigation.

REFERENCES

1. Alexander JB, Ingram GA: Noncellular non-specific defence mechanism of fish. *Ann Rev Fish Dis* 1992, 2: 249-279.
2. Baulny MOD, Quentel C, Fournier V, Lamour F, Gourvello RL: Effect of long term oral administration of β- glucan as an immunostimulant or an adjuvant on some non-specific parameters of the immune response of turbot, *Scophthalmus maximus* . *Dis Aquat Org* 1996, 26: 139-147.
3. Beakes GW, Wood SE, Burr AW: Features which characterize *Saprolegnia* isolates from salmonid fish lesions – A review. In *Salmon Saprolegniasis*. Edited by: Mueller GJ. U.S. Department of Energy, Bonneville Power Administration, Portland, Oregon; 1994:33-66.
4. Bradford MM: A rapid and sensitive method for the quantification of microgram quantities of protein. *Anal Biochem* 1976, 72: 248. 10.1016/0003-2697(76)90527-3
5. Brown GD, Gordon S: Fungal beta-glucans and mammalian immunity. *Immunity* 2003, 19: 311-315. 10.1016/S1074-7613(03)00233-4
6. Brown GD, Herre J, Williams DL, Willment JA, Marshall AS, Gordon S: Dectin-1 mediates the biological effects of beta-glucans. *J Exp Med* 2003, 197: 1119-1124. 10.1084/jem.20021890
7. Bruno DW, Wood BP: *Saprolegnia* and other *Oomycetes* . In *Fish Diseases and Disorders, vol. III, Viral, Bacterial and Fungal Infections*. Edited by: Woo PTK, Bruno DW. CABI Publishing, Wallingford, Owon, United Kingdom; 1999:599-659.
8. Chang CF, Su MS, Chen HY, Liao IC: Dietary beta- 1,3-glucan effectively improves immunity and survival of *Penaeus monodon* challenged with white spot syndrome virus. *Fish Shellfish Immunol* 2003, 115: 297-310.
9. Chung S, Secombes CJ: Analysis of events occurring within teleost macrophages during the respiratory burst. *Comp. Bio-chem Physiol* 1988, 89(B):539-544.
10. Couso N, Castro R, Magarinos B, Obach A, Lamas J: Effect of oral administration of glucans on the resistance of gilthead seabream pasteurellosis. *Aquaculture* 2003, 219: 99-109. 10.1016/S0044-8486(03)00019-X

11. Dalmo RA, Boywald J, Ingebrigstsen K, Seljelid R: The immunomodulatory effect of laminaran [β(1,3)-D-glucan] on Atlantic salmon, *Salmon salar* L, anterior kidney leucocytes after intraperitoneal, peroral and peranal administration. *J Fish Dis* 1996, 19: 449-457. 10.1111/j.1365-2761.1996.tb00385.x
12. Das BK, Debnath C, Patnaik P, Swain DK, Kumar K, Mishra BK: Effect β- glucan on immunity and survival of early stage of *Anabas testudineus* (Bloch). *Fish Shellfish Immunol* 2009, 27: 678-683. 10.1016/j.fsi.2009.08.002
13. Doumas BT, Watson WA, Biggs HG: Albumin standards and the measurement of serum albumin with bromocresol green. *Clinical Chimica Acta* 1971, 31: 87-96. 10.1016/0009-8981(71)90365-2
14. Fitzpatrick MS, Schreck CB, Chitwood RL: Evaluation of three candidate fungicides for treatment of adult spring chinook salmon. *Prog Fish-Cul* 1995, 57: 153-155. 10.1577/1548-8640(1995)057<0153:TNEOTC>2.3.CO;2
15. Goldsworthy G, Opoku-Ware K, Mullen L: Adipokinetic hormone enhances laminarin and bacterial lipopolysaccharide-induced activation of the prophenoloxidase cascade in the African migratory locust, *Locusta migratoria* . *J Insect Physiol* 2002, 48: 601-608. 10.1016/S0022-1910(02)00085-9
16. Hatai K, Hoshiai GI: Pathogenicity of Saprolegnia parasitica coker. In *Salmon Saprolegniasis*. Edited by: Mueller GJ. U.S. Department of Energy, Bonneville Power Administration, Portland, Oregon; 1994:87-98.
17. Hillman G: Continuous photmetric measurement of prostrate acid phosphatase activity. *Z Klin Chem Klin Biochem* 1971, 9: 273.
18. Howe GE, Gingerich WH, Dawson VK, Olson JJ: Efficacy of hydrogen peroxide for treating Saprolegniasis in channel catfish. *J Aquat Anim Health* 1999, 11: 222-230. 10.1577/1548-8667(1999)011<0222:EOHPFT>2.0.CO;2
19. Kajita Y, Sakai M, Atsuta S, Kobayash M: The immunomodulatory effects of levamisole on rainbow trout, *Oncorhynchus mykiss* . *Fish Pathol* 1990, 25: 93-98. 10.3147/jsfp.25.93
20. Marking LL, Rach JJ, Schreier TM: Evaluation of antifungal agents for fish culture. *Prog. Fish-Cult* 1994, 56: 225-231. 10.1577/1548-8640(1994)056 <0225:AFSEOA>2.3.CO;2
21. Misra CK, Das BK, Mukherjee SC, Pattnaik P: Effect of multiple injections of β- glucan on non-specific immune response and disease resistance in *Labeo rohita* fingerlings. *Fish Shellfish Immunol* 2005, 20: 305-319.
22. Misra CK, Das BK, Mukherjee SC, Pattnaik P: Effcet of long term administration of dietary β- glucan on immunity, growth and survival of *Labeo rohita* fingerlings. *Aquaculture* 2006, 255: 82-94. 10.1016/j.aquaculture.2005.12.009
23. Misra CK, Das BK, Pradhan J, Pattnaik P, Sethi S, Mukherjee SC: Changes in lysosomal enzyme activity and protection against *Vibrio* infection in *Macrobrachium rosenbergii* (De Man) post larvae after bath immunostimulation with β- glucan. *Fish Shellfish Immunol* 2004, 17: 389-395. 10.1016/j.fsi.2004.04.008
24. Mohanta BK, Patra AK: Ulcerative fish disease in tropical air breathing teleost, *Anadas testudineus* (Bloch). *J Freshwater Biol* 1992, 4: 295-297.
25. Mullen L, Goldsworthy GJ: Changes in lipophorins are related to the activation of phenoloxidase in the haemolymph of *Locusta* migratoriain response to injection of immunogens. *Insect Biochem Mol Biol* 2003, 33: 661-670. 10.1016/S0965-1748(03)00045-6
26. Neish GA: Observations on saprolegniasis of adult sockeye salmon, *Oncorhynchus nerka* (Walbaum). *J Fish Biol* 1977, 10: 513-522. 10.1111/j.1095-8649.1977 .tb04084.x
27. Noga EJ: *Fish Disease Diagnosis and Treatment*. Mosby-Year Book, Inc, St. Louis, MO; 1996:367.
28. Parry RM, Chandan RC, Shahani KM: A rapid and sensitive assay of muramidase. *Proc Soc Exp Biol Ned* 1965, 119: 384-386.
29. Pickering AD: *Factors which predispose salmonid fish to Saprolegniasis*. Portland, Oregon, U.S. Department of Energy, Bonneville Power Administration; 1994:67-84.

30. Roberts RJ: *Fish Pathology*. 2nd edition. Bailliere Tindall Publishers. p, London, England; 1989. 467

31. Sahu S, Das BK, Mishra BK, Pradhan J, Samal SK, Sarangi N: Effect of dietary *Curcuma longa* on enzymatic and immunological profiles of rohu, *Labeo rohita* (Ham.), infected with *Aeromonas hydrophila* . *Aquacult Res* 2008, 39: 1720-1730. 10.1111/j.1365-2109.2008.02048.x

32. Trinder P: Determination of glucose in blood using glucose oxidase with an alternative oxidase acceptor. *Ann Clin Biochem* 1969, 6: 2-4.

33. Van Waters : *Rogers Inc. Material safety data sheet.* Van Waters and Rogers, Inc editors, Seattle, WA; 1988.

34. Willoughby LG: Rapid preliminary screening of *Saprolegnia* on fish. *J Fish Diseases* 1985, 8: 473-476. 10.1111/j.1365-2761.1985.tb01282.x

35. Willoughby LG, Roberts RJ: Towards strategic use of fungicides against *Saprolegnia parasitica* in salmonid fish hatcheries. *J Fish Dis* 1992, 15: 1-13. 10.1111/j.1365-2761.1992.tb00631.x

36. Willoughby LG: *Fungi and Fish Diseases*. Pisces Press, Stirling, Scotland; 1994. 57 p

CHAPTER 14

Farm-Level Risk Factors for Fish-Borne Zoonotic Trematode Infection in Integrated Small-Scale Fish Farms in Northern Vietnam

Van Thi Phan[1,2*], Annette Kjær Ersbøll[1], Khue Viet Nguyen[2], Henry Madsen[3], Anders Dalsgaard[4]

[1] *Department of Large Animal Sciences, Section of Veterinary Epidemiology, Faculty of Life Sciences, University of Copenhagen, Frederiksberg, Denmark,*

[2] *Centre for Environment and Disease Monitoring in Aquaculture, Research Institute for Aquaculture No.1, Tu Son, Vietnam,*

[3] *Centre for Health Research and Development, Department of Veterinary Disease Biology, Faculty of Life Sciences, University of Copenhagen, Frederiksberg, Denmark,*

[4] *Department of Veterinary Disease Biology, Faculty of Life Sciences, University of Copenhagen, Frederiksberg, Denmark*

ABSTRACT

Background: Northern Vietnam is an endemic region for fish-borne zoonotic trematodes (FZT), including liver and intestinal flukes. Humans acquire the FZT infection by eating raw or inadequately cooked fish. The production of FZT-free fish in aquaculture is a key component in establishing a sustainable program to prevent and control the FZT transmission to humans. Interventions in aquaculture should be based on knowledge of the main risk factors associated with FZT transmission.

Methodology/Principal Findings: A longitudinal study was carried out from June 2006 to May 2007 in Nam Dinh province, Red River Delta to investigate the development and risk factors of FZT infections in freshwater cultured fish. A total of 3820 fish were sampled six times at two-month intervals from 96 fish farms. Logistic analysis with repeated measurements was used to evaluate potential risk factors based on information collected through questionnaire interviews with 61 fish farm owners. The results showed that the FZT infections significantly increased from first sampling in June to July 2006 (65%) to sixth sampling in April to May, 2007 (76%). The liver fluke, *Clonorchis sinensis* and different zoonotic intestinal flukes including *Haplochis pumilio, H. taichui, H. yokogawai, Centrocestus formosanus* and *Procerovum varium* were found in sampled fish. Duration of fish cultured (sampling times), mebendazole drug self-medication of household members, presence of snails in the pond, and feeding fish with green vegetation collected outside fish farms all had a significant effect on the development of FZT prevalence in the fish.

Conclusions/Significance: The FZT prevalence in fish increased by 11 percentage points during a one-year culture period and the risk factors for the development of infection were identified. Results also highlight that the young fish are already highly infected when stocked into the grow-out systems. This knowledge should be incorporated into control programs of FZT transmission in integrated small-scale aquaculture nursery and grow-out systems in Vietnam.

Author Summary

Fish are the second intermediate host for fish-borne zoonotic trematodes (FZT). Humans acquire the FZT by eating raw or inadequately cooked fish. Therefore any sustainable program to prevent and control human FZT infection should consider how to minimize the FZT prevalence in fish. Understanding the development in prevalence of FZT in fish and the risk factors involved are of key importance in order to plan and implement an FZT prevention program successfully. During a one-year production cycle in integrated small-scale aquaculture grow-out systems, the FZT prevalence in fish increased by 11 percentage points. Three risk factors associated with the development of the prevalence of FZT infection in fish were identified including presence of snails in the pond, feeding fish with green vegetation collected outside fish farms, and mebendazole medication by the household members. Aquaculture management solutions addressing these three risk factors as well as the high level of FZT infection in juvenile fish stocked in grow-out systems should be found in order to produce fish free of FZT, which are safe for human consumption.

INTRODUCTION

Fish-borne zoonotic trematodes (FZT) belonging to the Opisthorchiidae and Heterophyidae families are important emerging and re-emerging pathogens causing liver and intestinal fluke diseases in human [1], [2]. The life cycle of FZT involves humans and animals such as dog, cat, pig and fish-eating birds as final hosts. The FZT live and develop to adult flukes in the liver or intestines of the final host and produce eggs that are excreted into the environment through faeces. The eggs are ingested by fresh water snails, where they develop to a stage called cercariae. Free swimming cercariae are shed to the water before they penetrate into fish. Inside the fish tissues, cercariae transform into encysted metacercariae [2]–[4]. Humans and animals acquire the FZT infection through consumption of raw, inadequately cooked, dried, salted or pickled fish that harbor infective metacercariae stages [4]–[6]. Keiser and Utzinger [4], [5]estimated that about 681 million people worldwide are at risk of infection and more than 46 million people are infected with liver flukes (*Clonorchis sinensis* and *Opisthorchis* sp.). There are no such estimates available of the number of people infected with intestinal flukes worldwide [4], but the prevalence of infection is believed to be high, e.g. the Red River Delta, Northern Vietnam is endemic for FZT, including both liver and intestinal trematode infections in humans and animals [7]–[9].

The fisheries sector in Vietnam plays an important role in providing food for domestic consumption and generating foreign currencies through export of seafood with aquaculture products contributing significantly to such export. From 1998 to 2008, the aquaculture sector in Vietnam expanded rapidly with the production volume increasing about 5.8 times from 425,000 tons to about 2.5 million tons in 2008. In the same period, the overall export value of seafood increased from 858 million USD to 4,510 million USD [10]. Vietnamese aquaculture systems include marine, brackish water and freshwater aquaculture. The freshwater aquaculture production environments include ponds, ditches, cages, net enclosures and pens in reservoirs, lakes, rivers and channels, and paddy fields. In the Red River Delta of Northern Vietnam, small-scale polyculture farming with several carps species and tilapia is the most important farming system [11]. At the study site in Nam Dinh province, fish are stocked continuously in small-scale grow-out ponds and partial harvest is done to generate extra income or for household consumption. The main stocking season for grow-out ponds is late Spring and early Summer where juveniles are available from the nurseries.

In Nam Dinh province in particular, is mainly practiced as small-scale family-based systems. The cultured fish are a main protein source for the households and consumed by the families at any time of the year. In Nam

Dinh province, as in many other areas in Northern Vietnam, freshwater fish ponds are stocked with multiple fish species and raised in a continuous production cycle. The aquaculture activities are often integrated with livestock, and vegetable production. The grow-out ponds are earthen ponds and often located in backyard of households (Figure 1.). These systems are conducive for the life cycle of FZT as contamination with FZT eggs from animal and human hosts is high and the snail intermediate host are often present in high numbers. There have been a number of educational campaigns to stop people in Nam Dinh province from eating raw or inadequate cooked fish, however, the FZT prevalence in humans in Nam Dinh remains high [8]. It is an arduous task to convince people to stop eating raw and inadequate cooked fish because there is a long tradition for eating such dishes. Therefore, producing fish free of FZT for human consumption is important to prevent such infection in human.

Figure 1. A grow-out pond in Nam Dinh province. Shown a typical grow-out pond in Nam Dinh province. The pond is earthen pond and located near the households.

The present study was carried out during a period of 1-year and aimed to investigate the prevalence and the development of FZT infection in fish during a 1-year culture cycle and to assess risk factors for FZT infection in fish.

METHODS

Ethics statement

The study protocol was approved by the Aquatic animal scientific committee of Research Institute for Aquaculture No.1, Dinh Bang, Tu Son, Bac Ninh, Vietnam. Informed consent was obtained in writing from fish farmers as they submitted a signed request form to participate in the study.

Figure 2. Sampling fish by using cast net. Shown a throw of cast net when sampling fish from a grow-out pond in Nam Dinh.

Study area, design and unit

Nam Dinh province is well- known as an FZT endemic province with high FZT prevalences in humans (64.9%), cats (70.2%) and dogs (56.9%) [8], [9]. The study was conducted in Nghia Lac and Nghia Phu communes in Nam Dinh province, which is located in the Red River Delta in Northern Vietnam (Figure 2). The main stocking season is late Spring and beginning of summer. During other times of the year, juveniles are stocked at convenience by farmers. The stocked juveniles mainly originate from local nurseries. The main species stocked are grass carp (*Ctenopharyngodon idellus*), silver carp (*Hypophthalmichthys molitrix*), common carp (*Cyprinus carpio*), Rohu (*Labeo rohita*), and Mrigal (*Cirrhinus mrigala*). So-called self-recruiting species of fish (SRS fish) are occasionally accidently introduced to the ponds during intake of water from nearby canals. Depending on species, the stocked fingerlings are grown for 8 to 12 months to reach harvest size.

This study was carried out during a 1-year fish production cycle from May 2006 to April 2007 and designed as a longitudinal study with 6 repeated samplings from the same ponds at 2-month intervals. The study unit was the pond at each sampling time.

Farm selection and fish sampling

Fish farms were selected from a list of households obtained from the local health clinic in each of the two communes. Farms with no fish ponds and farms that participated in a previous human FZT prevalence study reported by Dung *et al.* [8] were excluded because infected individuals were given drug treatment. Farms were randomly selected from the adjusted lists of households. If the farm had more than one pond, then one pond was randomly selected and used for fish sampling throughout the study period. A total of 96 fish farms were investigated.

Each farm was given one cast net for fish sampling in order to avoid risk of contamination between ponds. In each pond, a cast net was thrown in each of the four corners and in the middle of the pond (Figure 3.). All fish caught from the five throws were put in a bucket and 10 fish were randomly selected (irrespective of fish species). However, if less than 15 fish were caught in a pond then the all fish were selected for analyses.

Analysis for FZT metacercariae in fish

Fish samples were preserved on ice and transported within 12 hours to the parasitological laboratory of the Centre for Environment and Disease Monitoring in Aquaculture (CEDMA), Research Institute for Aquaculture No.1 (RIA1), Dinh Bang, Tu Son, Vietnam. In the laboratory, fish samples were kept in a refrigerator at 4°C for no more than five days before being processed. The length and weight of each fish were recorded before being processed and digested in 1% pepsin for the release of metacercariae following procedures previously described [12] and modified as described by Chi *et al.* [13]. Small fish (less than 200 g in weight) were ground and digested whole. For larger fish (>200 g), the whole fish was ground, mixed well and a 50 g sub-sample was obtained and digested for recovery of metacercariae.

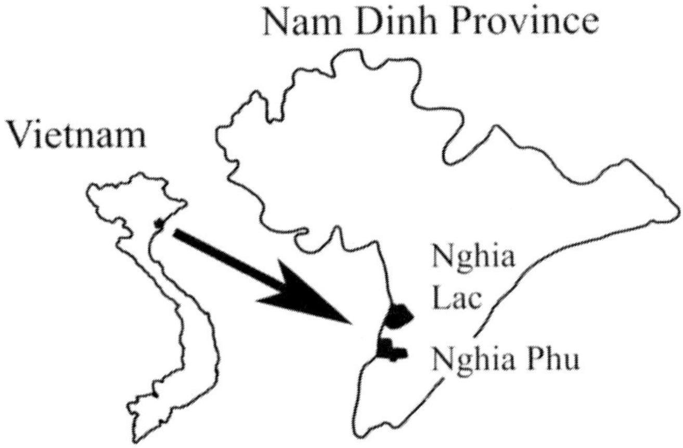

Figure 3. Map of study area. Nam Dinh province is located in the Southern Red River Delta, bordered by Ha Nam province to the North, the Gulf of Tonkin in the South-East, Thai Binh province to the East and Ninh Binh province on the West. Nghia Lac and Nghia Phu communes belong to Nghia Hung district, Nam Dinh province.

Metacercariae identification skills among staff were evaluated at the beginning of the study to ensure that FZT species were correctly identified by providing test samples for each staff to identify. Three times during the study period, the identification of the metacercariae was verified by experimental infection of mice and subsequent identification of adult flukes recovered. The experimental infections and identification of adult flukes followed previously described procedures and criteria [14]. The identity of metacercariae verified in animal experiments was subsequently confirmed to be correctly identified.

Snail sampling and cercariae analysis

Data on snails were extracted from Dung *et al.* [15]. Snail sampling was conducted by the same person in all ponds during the morning hours for 30 minutes per pond using a scooping method and/or hand-picking from the pond. Snails were transferred to plastic containers and transported alive to the laboratory where they were identified according to keys by Brandt [16], Thanh [17] and Frandsen and Christensen [18]. Snails were then examined for trematode infections using one or more of three methods depending on snail size, i.e. shedding, crushing and cutting, and cercariae were identified only to major group [15].

Questionnaire development and interviewing

A questionnaire including 33 questions was formulated based on information obtained during group discussions with international experts, scientists at CEDMA and fishery staff at provincial level about potential risk factors for FZP infection at farm level. Information was obtained about farm practices in relation to the pond management, animal reservoir host and human waste such as how often and how farmers clean the pond, water source, practices for keeping animals such as dogs, cats, pigs, chicken and ducks, toilet facilities and others. The draft questionnaire was pretested by interviewing aquaculture research staff at CEDMA and five farmers in each commune with the farmers subsequently being excluded from the study. Needed adjustments were then made to the questionnaire which also included a section for recording observations made during the farm visit.

Four persons were involved in interviewing the managers of the farms. Before the interviews were initiated, the four persons discussed and agreed upon the method of interviewing and filling out the questionnaires. Each farm was interviewed one time in the connection with the first sampling by two persons. One person was always the interviewer and the second person was observing the interview and took notes about observed farm management practices, presence of animal and toilet facilities etc. in order to validate the answers given by the manager of the farm.

Most of the questions in the questionnaire were closed with 2 to 4 possible answers. Open questions were used to obtain information about pond area, number of fish stocked and yield. The questionnaire can be obtained from corresponding author.

Derived questionnaire variables

All questionnaire variables were initially evaluated using frequency distributions in order to obtain variables for further analysis. For some questions, most farmers gave the same answers. Hence these variables (questions) were excluded from the analysis because they did not contribute to explaining differences in transmission due to skewed distribution. Furthermore, many of the questions were not independent. Therefore, new variables were derived based on biological similarity and normal daily practices. As an example, a variable "animal fed with live fish" was derived based on specific information about how pigs, dogs and cats were fed or seen eating live fish. Area of pond was categorized into less than 300 m^2 and ponds equal to or more than 300 m^2. The source of green vegetation for feeding grass carp was originally one question in the questionnaire. However, two new variables were derived as feeding grass carp with additional vegetation originating from inside the garden (yes/no) and feeding grass carp with additional green vegetation originating from outside the garden (yes/no). The term of "additional green vegetation" here includes some types of grasses (sometimes from flooded areas) or aquatic plants collected from other habitats. The remaining variables were derived by aggregating some of the answers possibilities.

Statistical analysis

Data were analysed using the Statistical Analysis System (SAS®, version 9.1). The dependent variable was number of FZT infected fish out of number of examined fish in the pond at each sampling time. Descriptive analysis was performed by means of number of farms, number of infected fish, number of sampled fish, overall FZT prevalence by sampling times and potential risk factors for FZT transmission at the farms. Initially, two-by-two tables were used to evaluate confounding between potential farm-level risk factors.

The effect of potential risk factors on the prevalence of FZT infection in fish was evaluated using a logistic analysis with repeated measurements. The potential risk factors were included in the analysis. Farm nested within commune was included as a random effect. The autocorrelation between repeated measurements was taken into account by introducing a 1storder autoregressive correlation structure in the model. Backward elimination was applied to obtain the main effect resulting model. Interactions (2-way) between variables in the main effect resulting model were evaluated. In order to evaluate confounding between excluded potential risk factors with risk factors in the resulting model, each excluded variable was then added to the resulting model one by one for testing significance. Odds ratios were calculated only for significant effects. A P-value <0.05 was considered significant.

RESULTS

Descriptive analysis

Initially, there were 96 farms included in the study. However, during the 1-year study period some farms terminated fish culture with only 61 farms being sampled five times while 52 farms were sampled six times. A total of 3820 fish originating from 61 farms were examined for FZT.

FZT in fish and snail species. Among the fish species sampled for FZT examination, Rohu, Silver carp, Mrigal and Grass carp were the dominating species comprising 85% of the total number of fish analyzed. Table 1 lists the different fish species, the percentage infected with FZT, and the parasite species found. The weight of fish sampled ranged from 8 to 1400 g depending on fish species. The liver fluke, *Clonorchis sinensis,* was only found in 1 of 1185 Silver carps analyzed. Different intestinal flukes including *Haplochis pumilio, H. taichui, H. yokogawai, Centrocestus formosanus* and *Procerovum varium* were commonly found in most fish species. *Haplochis pumilio* was recovered from all 18 fish species and the main cultured fish species (Grass carp, Rohu, Mrigal, Silver carp and Pacu) were infected by five different zoonotic intestinal trematodes. Non-zoonotic Exorchis spp. were found as metacercariae in fish in this study area (unpublished data).

Table 1. FZT diversity and prevalence in fish from small-scale farms in Nam Dinh, Vietnam.

				FZT species found[c]					
Fish species (*Latin name*)	N_f (%)[a]	N_f (%)[a]	N_{FZT}/N (%)[b]	*Cs*	*Ht*	*Hp*	*Hy*	*Pr*	*Ce*
Rohu (*Labeo rohita*)	1290(33.75)	1290(33.8)	745 (58)	–	+	+	+	+	+
Silver carp (*Hypophthalmichthys molitrix*)	1185(31.0)	1185(31)	1024 (86)	+	+	+	+	+	+
Mrigal (*Cirrhinus mrigala*)	424(11.09)	424(11.1)	313 (74)	–	+	+	+	+	+
Grass carp (*Ctenopharyngodon idellus*)	351(9.18)	351(9.2)	305 (87)	–	+	+	+	+	+
Crucian carp (*Carrasius auratus*)	196(5.13)	196(5.1)	127 (65)	–	–	+	+	–	–
Barbel chub (*Squaliobarbus curriculus*)	121(3.17)	121(3.2)	86 (71)	–	+	+	–	+	+
Pacu (*Piaractus brachypomum*)	75(1.96)	75(2.0)	55 (73)	–	+	+	+	+	+
Tilapia (*Oreochromis niloticus*)	54(1.41)	54(1.4)	13 (24)	–	–	+	–	–	–
Common carp (*Cyprinus carpio*)	44(1.15)	44(1.2)	36 (82)	–	–	+	+	–	+
Mud carp (*Cirrhinus molitorella*)	22(0.57)	22(0.6)	8 (38)	–	–	+	–	–	–
Catfish (*Clarias batrachus*)	9(0.24)	9(0.2)	7 (78)	–	–	+	+	+	+
Snake-head fish (*Channa orientalis*)	8(0.21)	8(0.2)	7 (88)	–	–	+	–	–	+
Climbing perch (*Anabas testudineus*)	31(0.81)	31(0.8)	29 (94)	–	–	+	–	+	+
Silver barb (*Barbonymus gonionotus*)	6(0.16)	6(0.2)	3 (50)	–	–	+	–	+	–
Sharpbelly (*Hemiculter leucisculus*)	3(0.08)	3(0.1)	1 (33)	–	–	+	–	–	–
Big head carp (*Hypophthalmichthys nobilis*)	1(0.03)	1(0.03)	1 (100)	–	–	+	–	–	–
Bronze featherback (*Notopterus notopterus*)	1(0.03)	1(0.03)	1 (100)	–	–	+	–	–	–
Unidentified	1(0.03)	1(0.03)	1 (100)	–	–	+	–	+	–
TOTAL	3822(100)	3822(100)	2762 (72)						

[a]Nf: Number of specific fish species, %: percentage of each fish species.
[b]N_{FZT} is number of FZT infected fish; N is number of fish,%: prevalence of FZT in each fish species.
[c]*Cs* is *Clonorchis sinensis*; *Ht* is *Haplochis taichui*; *Hp* is *H. pumilio*; *Hy* is *H. yokogawai*; *Pr* is *Procerovum varium*; *Ce* is *Centrocestus formosanus*.

The snail fauna in ponds was dominated by species of the Viviparidae family (53% of all snails collected) and Thiaridae family (41%), but species of the Ampullaridae family (5.5%) and Bithynidae (0.4%) were also present. *Stenothyra messageri* and different pulmonate snail species were occasionally found. Among Viviparidae, only *Angulyagra polyzonata* was found while within the Thiaridae, *Melanoides tuberculata* was dominating (75%); the other species in this family were *Thiara scabra* (22%), *Tarebia granifera* (1.5%) and *Sermyla requetii* (1.7%). The cercariae types recorded from the ponds were echinostomes, gymnocephalous, monostome, parapleurolophocercous (no pleurolophocercous cercariae were found in ponds), and xiphidio cercariae. The trematode species found in the fishes all belonged to the Heterophyidae and these produce cercariae of the parapleurolophocercous (or pleurolophocercous) group, which was the most common type found in the ponds (37% of all infections found). In the ponds, these cercariae were found exclusively in species of the Thiaridae.

FZT development in the 1-year study period. The FZT prevalence in fish sampled increased 11 percentage points over the period, from 65% in the initial sample to 76% in the final sample (Table 2). The increase in FZT prevalence in Nghia Lac and Nghia Phu communes were 9 and 12 percentage points, respectively.

Table 2 shows the descriptive analysis of potential risk factors for FZT infection in fish during the 1-year study period. The FZT prevalence increased by 12 percentage point increase in fish fed additional vegetation collected outside the farm garden, but only 3 percentage points when fed with green vegetation collected within the farm garden. Fish cultured in ponds with snails had an 18 percentage points increase in FZT prevalence, while fish from ponds without snails only showed a 6 percentage points increase in FZT prevalence. In farm households that kept cat(s), the FZT prevalence in fish increased by 10 percentage points, while in farms that did not keep cats the increase was 6 percentage points. If household members had taken a standard round worm treatment with mebendazole during the previous year, the FZT prevalence in fish increased by 13 percentage points, while in those that had not taken treatment showed a 6 percentage points increase in FZT prevalence.

Risk factors for FZP infection in freshwater fish cultured in small-scale integrated ponds

Table 3 presents the final model of risk factors for FZT infection in fish at farm level. Duration of culture period (sampling time) had a significant effect on the development of FZT infected fish (P=0.012). The odds of FZT prevalence in fish at sampling time 3, 5 and 6 was more than 1.6 times higher than at sampling time 1. Farms with household members that took self-medication with antihelminthic worm tablets had 1.52 time higher risk of FZT infection than farms with household members that did not take such treatment (P=0.015).

Table 2. Descriptive analysis of potential risk factors for FZT infection in fish from small-scale farms.

			*Sampling 1 (Jun–Jul)	Sampling 2 (Aug–Sep)	Sampling 3 (Oct–Nov)	Sampling 4 (Dec–Jan)	Sampling 5 (Feb–Mar)	Sampling 6 (Apr–May)	
Potential risk factor	**Level**	N_{hh}	N_{FZT}/N_f (%)	N_{FZT}/N_f (%)	N_{FZT}/N_f (%)	N_{FZT}/N_f (%)	N_{FZT}/N_f (%)	N_{hh}	N_{FZT}/N_f (%)
Overall		61	408/624 (65)	395/577 (68)	428/568 (75)	429/580 (74)	423/563 (75)	52	357/473 (76)
Commune	Nghia Lac	34	244/368 (66)	224/334 (67)	231/317 (73)	212/311 (68)	229/317 (72)	31	215/286 (75)
	Nghia Phu	27	164/256 (64)	171/243 (70)	197/251 (79)	217/269 (81)	194/246 (79)	21	142/187 (76)
Fish pond management									
Area of pond	≤300 m²	30	218/314 (69)	192/271 (71)	196/269 (73)	219/292 (75)	217/284 (76)	24	160/215 (74)
	>300 m²	31	190/310 (61)	203/306 (66)	232/299 (76)	210/288 (73)	206/279 (74)	28	197/258 (76)
Source of water used in pond	Canal	47	311/488 (64)	302/457 (66)	338/450 (75)	356/475 (75)	338/446 (76)	39	273/356 (77)
	Rain	14	97/136 (71)	93/120 (76)	90/118 (76)	73/105 (70)	85/117 (73)	13	84/117 (72)
Frequency of pond preparation	Yearly	22	156/243 (64)	143/208 (69)	142/197 (72)	166/217 (77)	152/190 (80)	21	133/179 (74)
	>1 year	36	224/343 (65)	223/334 (67)	247/331 (75)	237/328 (72)	246/331 (74)	29	199/267 (75)
Use lime in pond	Yes	32	227/333 (68)	213/301 (71)	223/302 (74)	211/293 (72)	227/288 (79)	27	199/264 (75)
	No	26	153/253(61)	153/241 (64)	166/226 (75)	192/252 (76)	170/233 (73)	23	133/182 (73)
Source of fish fry	Local	29	204/316 (65)	204/289 (71)	240/309 (78)	224/309 (73)	214/281 (76)	25	195/248 (79)
	Don't know	23	129/209 (62)	142/212 (67)	135/182 (74)	158/203 (78)	152/209 (73)	18	99/132 (75)
Origin of additional green vegetation**	Outside farm	51	344/519 (66)	336/483 (70)	365/469 (78)	361/480 (75)	355/468 (76)	43	308/396 (78)
	Inside farm	10	64/105 (61)	59/94 (63)	63/99 (64)	68/100 (68)	68/95 (72)	9	49/77 (64)
Presence of snail**	No	42	275/422 (65)	254/385 (66)	253/363 (70)	277/384 (72)	282/381 (74)	36	219/309 (71)
	Yes	19	133/202 (66)	141/192 (73)	175/205 (85)	152/196 (78)	141/182 (78)	16	138/164 (84)
Animal reservoir host									
Rear pig	No	13	84/121 (69)	91/123 (74)	92/113 (81)	103/131 (79)	80/105 (76)	11	72/96 (75)
	Yes	48	324/503 (64)	304/454 (67)	336/455 (74)	326/449 (73)	343/458 (75)	41	285/377 (76)
Have cat	No	15	103/169 (61)	98/139 (71)	98/140 (70)	92/131 (70)	94/126 (75)	13	75/113 (66)
	Yes	46	305/455 (67)	297/438 (68)	330/428 (77)	337/449 (75)	329/437 (75)	39	282/360 (78)
Have dog	No	9	53/83 (64)	68/94 (72)	65/82 (79)	75/90 (83)	58/71 (82)	9	62/76 (82)
	Yes	52	355/541 (66)	327/483 (68)	363/486 (75)	354/490 (72)	365/492 (74)	43	295/397 (74)
Rear duck	No	47	315/471 (67)	316/469 (67)	330/431 (77)	332/440 (76)	320/435 (74)	38	250/332 (75)
	Yes	14	93/153 (61)	79/108 (73)	98/137 (72)	97/140 (69)	103/128 (81)	14	107/141 (76)
Rear chicken	No	21	154/235 (66)	155/226 (69)	184/232 (79)	174/232 (75)	161/221 (73)	20	123/170 (72)
	Yes	40	254/389 (65)	240/351 (68)	244/336 (73)	255/348 (73)	262/342 (77)	32	234/303 (77)
Use animal manure	Pond, garden	16	116/170 (68)	114/160 (71)	127/157 (81)	148/186 (80)	108/147 (74)	13	87/112 (78)
	Others	43	281/433 (65)	267/396 (67)	283/389 (73)	267/375 (71)	302/398 (76)	37	256/339 (76)
Seen animal eat live fish	No	41	274/414 (66)	269/377 (71)	285/377 (76)	267/365 (73)	280/377 (74)	34	230/302 (76)
	Yes	20	134/210 (64)	126/200 (63)	143/191 (75)	162/215 (75)	143/186 (77)	18	127/171 (74)
Feeding animal fish waste	No	30	209/313 (67)	206/291 (71)	229/300 (76)	224/299 (75)	207/284 (73)	27	189/244 (78)
	Yes	31	199/311 (64)	189/286 (66)	199/268 (74)	205/281 (73)	216/279 (77)	25	168/229 (73)
Human waste and other									
Type of toilet	Closed	38	240/383 (63)	251/259 (70)	269/356 (76)	282/377 (75)	276/368 (75)	31	211/264 (80)
	Open	23	168/241 (70)	144/218 (66)	159/212 (75)	147/203 (72)	147/195 (75)	21	146/209 (70)
Discharging waste from toilet	Canal	4	13/27 (48)	26/32 (81)	23/30 (77)	32/42 (76)	23/31 (74)	3	19/23 (83)
	Pond	13	99/146 (68)	98/137 (72)	130/152 (86)	119/146 (82)	108/138 (78)	12	100/122 (82)
	Garden	20	126/205 (62)	129/185 (70)	114/164 (70)	133/183 (73)	132/185 (71)	15	89/117 (76)
Heard about intestinal fluke	No	43	293/435 (67)	288/394 (73)	321/406 (79)	317/404 (75)	310/411 (75)	34	245/319 (77)
	Yes	18	115/189 (61)	107/183 (59)	107/162 (66)	112/156 (72)	113/152 (74)	18	112/154 (73)
Mebendazole drug self-medication**	No	37	237/385 (62)	243/372 (65)	270/358 (75)	260/372 (70)	243/345 (70)	30	215/290 (74)
	Yes	24	171/239 (72)	152/205 (74)	158/210 (75)	169/208 (81)	180/218 (83)	22	142/183 (78)

*N_{hh}: Number of households; N_f: number of fish sample; N_{FZT}: number of fish infected with FZT; %: prevalence of infected fish.
**Identified risk factors.

Table 3. Parameter estimate of farm-level risk factors for FZT infection in fish from small-scale farms.

Variable	Level	Estimate*	SE	OR	95% CI	*P* value
Intercept		−0.013	0.237			
Sampling times						0.012
	1 (Jun–Jul)	0^{a}	-	1.00	-	
	2 (Aug–Sep)	0.141^{ab}	0.150	1.15	0.87–1.53	
	3 (Oct–Nov)	0.503^{c}	0.166	1.65	1.19–2.29	
	4 (Dec–Jan)	0.401^{bc}	0.165	1.49	1.08–2.06	
	5 (Feb–Mar)	0.496^{c}	0.169	1.64	1.18–2.28	
	6 (Apr–May)	0.480^{bc}	0.178	1.62	1.14–2.89	
Snails present in pond						0.022
	Yes	0.419	0.177	1.52	1.07–2.15	
	No	0	-	1.00	-	
Origin of additional green vegetation fed to grass carp						0.034
	Inside farm	0	-	1.00	1.05–2.41	0.034
	Outside farm	0.462	0.213	1.59	-	
Mebendazole drug self-medication						0.015
	Yes	0.424	0.169	1.53	1.1–2.13	0.015
	No	0	-	1.00	-	
Autocorrelation		0.195	0.076			

*Parameter estimate with the same superscript letter are not significant at a 5% significant level.
SE is standard error; OR is odds ratio; 95% CI is 95% confidence interval for OR; *P* is overall P-value.

Odds of FZT infection in fish from ponds with snails was 1.52 times higher (*P*=0.022) than if no snail was seen present. Farms feeding green vegetation supplied from outside their garden had significantly higher FZT infection (OR=1.59) compared to fish from farms that fed green vegetation originating from their own garden.

DISCUSSION

The data we present here form the first study of risk factors for FZT development in fish based on a longitudinal study design. Prevalence of FZT in fish was high in these grow-out ponds and we identified a number of factors that promote transmission of these parasites to fish.

The FZT prevalence was found to be high at the beginning of main stocking season of fingerlings with a subsequent increase in FZT prevalence in grow-out ponds increased from 65% to 76% during the 1-year culture period. The juveniles fish obtained from the local nurseries were highly infected when stocked in the grow-out ponds studied (Van Thi Phan, personal communication). Therefore, it is important that nurseries are a focal point for interventions to prevent juvenile fish from becoming infected with FZT. The larger fish in grow-out ponds seem to be more resistant to FZT infection as there was only 11 percentage points increase in infection during the 1-year culture period. Minimizing the FZT transmission in both juveniles and ready-to-harvest fish is important in preventing FZT transmission to humans since it is normally recognized very difficult to change people's eating habits and stop them from eating raw fish, in particular when such habits are rooted in old cultural traditions.It was also found that the increase in FZT prevalence was faster (3 to 7%) during the initial culture period from June–November and subsequently remained relatively lower and stable (1.2 to1.4% through the remaining period at sampling intervals 4, 5 and 6). Sampling time 4 was at the end of lunar year where the farm household members begin harvesting fish to be consumed during the New Year celebration, but also the time where new fingerlings are stocked into the ponds. At this time of the year, the water temperatures in Northern Vietnam usually range between 14–20°C compared, to the summer period (May to October), when temperatures normally range from 28–35°C. It has previously been shown that snails stop shedding cercariae into water at low water temperatures [19]. Further studies are needed to assess the impact of changing water temperatures on presence of snails and their shedding of cercariae.

We only found *C. sinensis* (liver fluke) in a single fish which was surprising since Dung *et al.* [8]based on faecal (stool) examinations as well as identification of collected adult flukes reported high prevalence of *C. sinensis* in people from the same study area in Nam Dinh province. This could mean that people acquire liver flukes infections from eating raw or inadequately cooked fish from other sources, e.g. from natural habitats or other places than Nam Dinh province. Intestinal fluke species recovered from this study were similar with those reported in Nghe An province [13] and elsewhere in Southern provinces [20], [21], except that *Stellantchasmus falcatus* were not found in our study.

Snails are the first intermediate host for FZT and its presence is essential for further transmission and infection of fish. Snail species belonging to Thiaridae family are a common first intermediate host for intestinal FZT [19], [22] while species of the bithynidae are hosts for *C. sinensis* [15]. In the present study, species of snails of the Thiaridae family, important vectors of FZT, were one of the dominating snail types found in ponds. According to Schell (1985) only pleurolophocercous cercariae are produced within the families Opistorchidae and Cryptogonimidae, while both parapleurolophocercous and pleurolophocercous cercariae are found within the Heterophyidae (intestinal trematodes). In other habitats such as canals and rice fields associated with the fish ponds in Nam Dinh, species of the Bithynidae and Stenothyridae are abundant, but also in these habitats parapleurolophocercous cercariae is the most common type found in snails [15]. Dung *et al.* [15] found that parapleurolophocercous constituted 40.3% of all infections found in snails, while pleurolophocercous only constituted 0.3%, but these could thus potentially belong to three different families. One species belonging to the Cryptogonimidae, Exorchis spp. (not zoonotic), has been found as metacercariae in fish in this area (unpublished). Identification by PCR of cercariae similar to the most common type of parapleurolophocercous cercariae found in our study showed it to be *H. pumilio* [23]. Clearly, there is a need for more detailed identification of cercariae using molecular techniques to clarify host parasite relationships in the study area and elsewhere.

It is a common practise of farmers to feed additional green vegetation to grass carps that are cultured together with other fish species in the same pond. Collection of green vegetation from external sources can introduce infected snails attached to such vegetation. FZT eggs originating from nearby wild animals and humans may also be introduced as such eggs could be present in the mud attached to roots of floating plants. In any case, the practice of adding vegetation to ponds was associated with a significant increased risk of FZT infection in fish. Water intake to the ponds is also a potential risk factor for introducing snails and cercariae into the ponds, but it was not found significantly associated with FZT infection in this study. This may be because the farmers studied use primarily rain water for their ponds, although water from nearby canals may be introduced into the ponds during flooding in rainy season. Thus, further studies are needed to assess the relative importance of introducing snails and cercariae into the ponds through contaminated green fodder and surface run-off into the ponds and through different water sources used to fill the ponds. Similarly, the relative impact of interventions to control such potential risk factors in FZT transmission, e.g. feeding only pelleted feed, preventing surface run-off into ponds, lining of pond embankments to control vegetation and snail populations needs to be assessed. The management of fish feed has previously been proposed as one intervention to control the introduction of snails [24].

The common standard treatment among people in the study area against infection with soil-transmitted round worms is the anthelmintic mebendazole. However, this drug does not have an effect on FZT. Fish cultured in household ponds where the members took this drug had a significantly higher risk of FZT infection compared to fish at farms, where the household members did not take such treatment.. The higher risk of FZT infection in fish associated with mebendazole treatment of household members is difficult to explain. This risk may occur simply by chance. However, one of the most common side effects of mebendazole is diarrhea [25]. Diarrhea would lead to more frequent defecation and this combined with more loose stools and a common open defaecation practice in the study area could increase the risk of faecal (egg) contamination of ponds and other habitats. It may also be speculated that although adult intestinal flukes may be inactivated at some level by mebendazole, eggs may still be viable and excreted at higher rates [26]. Further investigations on this are needed.

Even though the presence of cats on the farm did not have a significant effect on FZT prevelence in fish in this study, it should still be considered to include control measures for cats as the farms that kept cats showed twice as high increase in FZT prevalence in fish compared to farms without cats. Cats have been shown as an important final host for sustaining the life cycle of FZT [9], but their importance in FZT transmission may be questioned as cats tend to deposit faeces away from the pond [26].

Cast net were used in the study to ensure that the fish were randomly sampled and this technique is considered as a good technique for random sampling of fish in aquaculture [27]. However, aquatic plants were often present in varying quantities in the small-scale farm ponds studied. Therefore in some cases, selection bias may have occurred as the fish could escape when the net could not reach the bottom of the ponds due to the presence of aquatic plants.

From the final model of risk factors, it is suggested that snail control either by better pond preparation practices, biological control measures (e.g. stocking of snail eating fish), and ensuring that green vegetation fodder is collected from FZT free areas are important interventions to reduce FZT transmission. Further, better general human hygenic behaviour should be promoted, in particular when individuals take mebendazole drugs, to avoid shedding FZT eggs into the pond environment. However, any practical control strategy for FZT infection in fish farms should be pre-tested and evaluated before being introduced to farmers to apply at their farms.

The results from the present study provide together with findings from other studies in Vietnam on risk factors for FZT infection in humans and animals [7]–[9] important knowledge about FZT transmission that can be used to establish sustainable measures to reduce and control FZT infection in fish raised in grow-out ponds. As juveniles fish already showed high FZT prevalence, any interventions should focus on controlling FZT transmission in nurseries as they appear to be hot spots for FZT infection.

ACKNOWLEDGMENTS

The authors would like to thank staff of Centre for Environment and Disease Monitoring in Aquaculture (CEDMA), Research Institute for Aquaculture No.1 (RIA1), for their support in field and lab works. The farmers in Nghia Lac and Nghia Phu communes, Nghia hung district, Nam Dinh province are thanked for their participation in the study. Special thanks are given to Ms Bui Thi Dung for generation of snail data.

REFERENCES

1. WHO (2004) pp. 1–58.WHO2004158Report of Joint WHO/FAO workshop on food-borne trematode infections in Asia, Ha Noi, Vietnam. Report of Joint WHO/FAO workshop on food-borne trematode infections in Asia, Ha Noi, Vietnam.
2. Chai JY, Murrell KD, Lymbery AJ (2005) Fish-borne parasitic zoonoses: Status and issues. International Journal for Parasitology 35: 1233–1254.JY ChaiKD MurrellAJ Lymbery2005Fish-borne parasitic zoonoses: Status and issues.International Journal for Parasitology3512331254
3. Rim H, Hillyer G, V, Hopla C (1982) Clonorchiasis. Parasitic zoonoses Volume 3 17–32.H. RimG, V. HillyerC. Hopla1982Clonorchiasis.Parasitic zoonoses Volume 31732
4. Keiser J, Utzinger J (2009) Food-Borne Trematodiases. Clinical Microbiology Reviews 22: 466–483.J. KeiserJ. Utzinger2009Food-Borne Trematodiases.Clinical Microbiology Reviews22466483
5. Keiser J, Utzinger J (2005) Emerging foodborne trematodiasis. Emerging Infectious Diseases 11: 1507–1514.J. KeiserJ. Utzinger2005Emerging foodborne trematodiasis.Emerging Infectious Diseases1115071514
6. Lun ZR, Gasser RB, Lai DH, Li AX, Zhu XQ, et al. (2005) Clonorchiasis: a key foodborne zoonosis in China. Lancet Infectious Diseases 5: 31–41.ZR LunRB GasserDH LaiAX LiXQ Zhu2005Clonorchiasis: a key foodborne zoonosis in China.Lancet Infectious Diseases53141
7. De NV, Murrell KD, Cong LD, Cam PD, Chau LV, et al. (2003) The food-borne trematode zoonoses of Vietnam. Journal of Tropical Medicine and Public Health 34: 12–34.NV DeKD MurrellLD CongPD CamLV Chau2003The food-borne trematode zoonoses of Vietnam.Journal of Tropical Medicine and Public Health341234
8. Dung DT, Van De N, Waikagul J, Dalsgaard A, Chai JY, et al. (2007) Fishborne zoonotic intestinal trematodes, Vietnam. Emerging Infectious Diseases 13: 1828–1833.DT DungN. Van DeJ. WaikagulA. DalsgaardJY Chai2007Fishborne zoonotic intestinal trematodes, Vietnam.Emerging Infectious Diseases1318281833
9. Anh NTL, Phuong NT, Johansen MV, Murrell KD, Van PT, et al. (2009) Prevalence and risks for fishborne zoonotic trematode infections in domestic animals in a highly endemic area of North Vietnam. Acta Tropica 112: 198–203.NTL AnhNT PhuongMV JohansenKD MurrellPT Van2009Prevalence and risks for fishborne zoonotic trematode infections in domestic animals in a highly endemic area of North Vietnam.Acta Tropica112198203
10. General Statistics Office of Vietnam (2010) General Statistics Office of Vietnam2010Fisheries production (1998–2008). Fisheries production (1998–2008).
11. FAO (2005) National Aquaculture Sector Overview, Viet Nam.FAO2005National Aquaculture Sector Overview, Viet Nam.National Aquaculture Sector Overview Fact Sheets. National Aquaculture Sector Overview Fact Sheets.
12. WHO (1995) Control of foodborne trematode infections. pp. 1–157.WHO1995Control of foodborne trematode infections.1157Report of a WHO Study Group. Report of a WHO Study Group.
13. Chi TTK, Dalsgaard A, Turnbull JF, Tuan PA, Murrell KD (2008) Prevalence of zoonotic trematodes in fish from a Vietnamese fish-farming community. Journal of Parasitology 94: 423–428.TTK ChiA. DalsgaardJF TurnbullPA TuanKD Murrell2008Prevalence of zoonotic trematodes in fish from a Vietnamese fish-farming community.Journal of Parasitology94423428
14. Kay H, Murrell KD, Hansen AK, Madsen H, Nguyen TTT, et al. (2009) Optimization of An Experimental Model for the Recovery of Adult Haplorchis Pumilio (Heterophyidae: Digenea). Journal of Parasitology 95: 629–633.H. KayKD MurrellAK HansenH. MadsenTTT Nguyen2009Optimization of An Experimental Model for the Recovery of Adult Haplorchis Pumilio (Heterophyidae: Digenea).Journal of Parasitology95629633

15. Dung BT, Madsen H, The DT (2010) Distribution of freshwater snails in family-based VAC ponds and associated waterbodies with special reference to intermediate hosts of fish-borne zoonotic trematodes in Nam Dinh Province, Vietnam. Acta Tropica. BT DungH. MadsenDT The2010Distribution of freshwater snails in family-based VAC ponds and associated waterbodies with special reference to intermediate hosts of fish-borne zoonotic trematodes in Nam Dinh Province, Vietnam.Acta TropicaIn press. In press.
16. Brandt RAM (1974) The non-marine aquatic Mollusca of Thailand. Achiv für Molluskenkunde 105: 1–423.RAM Brandt1974The non-marine aquatic Mollusca of Thailand.Achiv für Molluskenkunde1051423
17. Thanh DN (1980) Identification of freshwater invertebrates in Northern Vietnam. Science and Technology Press (in Vietnamese).. 490 p.DN Thanh1980Identification of freshwater invertebrates in Northern VietnamScience and Technology Press (in Vietnamese).490
18. Frandsen F, Christensen NO (1984) An Introductory Guide to the Identification of Cercariae from African Fresh-Water Snails with Special Reference to Cercariae of Trematode Species of Medical and Veterinary Importance. Acta Tropica 41: 181–202.F. FrandsenNO Christensen1984An Introductory Guide to the Identification of Cercariae from African Fresh-Water Snails with Special Reference to Cercariae of Trematode Species of Medical and Veterinary Importance.Acta Tropica41181202
19. Sen-Hai Y, Long-Qi X (2005) Intestinal trematode infections in China. In: Arizono N, Chai JY, Nawa N, Takahashi Y, editors. Asian Parasitology (Vol.1), Food-borne Helminthiasis in Asia. Chiba, Japan: Federation of Asian Parasitologists, Chiba, Japan. pp. 61–78.Y. Sen-HaiX. Long-Qi2005Intestinal trematode infections in China.N. ArizonoJY ChaiN. NawaY. TakahashiAsian Parasitology (Vol.1), Food-borne Helminthiasis in AsiaChiba, JapanFederation of Asian Parasitologists, Chiba, Japan6178
20. Thu ND, Dalsgaard A, Loan LTT, Murrell K (2007) Survey for zoonotic liver and intestinal trematode metacercariae in cultured and wild fish in An Giang Province, Vietnam. Korean Journal of Parasitology 45: 45–54.ND ThuA. DalsgaardLTT LoanK. Murrell2007Survey for zoonotic liver and intestinal trematode metacercariae in cultured and wild fish in An Giang Province, Vietnam.Korean Journal of Parasitology454554
21. Thien PC, Dalsgaard A, Thanh BN, Olsen A, Murrell KD (2007) Prevalence of fishborne zoonotic parasites in important cultured fish species in the Mekong Delta, Vietnam. Parasitology Research 101: 1277–1284.PC ThienA. DalsgaardBN ThanhA. OlsenKD Murrell2007Prevalence of fishborne zoonotic parasites in important cultured fish species in the Mekong Delta, Vietnam.Parasitology Research10112771284
22. Long-Qi X, Sen-Hai Y, Ying-Dan C (2005) Clonorchiasis sinensis in Japan. In: Arizono NCJNYT, editor. Asian Parasitology 1, Food-borne helminthiasis in Asia. Chiba, Japan: The Federation of Asian Parasitologists. pp. 27–33.X. Long-QiY. Sen-HaiC. Ying-Dan2005Clonorchiasis sinensis in Japan.NCJNYT ArizonoAsian Parasitology 1, Food-borne helminthiasis in AsiaChiba, JapanThe Federation of Asian Parasitologists2733
23. Skov J, Kania PW, Dalsgaard A, Jørgensen TR, Buchmann K (2008) Life cycle stages of heterophyid trematodes in Vietnamese freshwater fishes traced by molecular and morphometric methods. Veterinay Parasitology 160: 66–75.J. SkovPW KaniaA. DalsgaardTR JørgensenK. Buchmann2008Life cycle stages of heterophyid trematodes in Vietnamese freshwater fishes traced by molecular and morphometric methods.Veterinay Parasitology1606675
24. Khamboonraung C, Keawvichit R, Wongworapat K, Suwanrangsi S, Hongpromyart M, et al. (1997) Application of hazard analysis critical control point (HACCP) as a possible control measure for Opisthorchis viverrini infection in cultured carp (Puntius gonionotus). Southeast Asia Journal of Tropical Medicine and Public Health 28(1): 65–72.C. KhamboonraungR. KeawvichitK. WongworapatS. SuwanrangsiM.

Hongpromyart1997Application of hazard analysis critical control point (HACCP) as a possible control measure for Opisthorchis viverrini infection in cultured carp (Puntius gonionotus).Southeast Asia Journal of Tropical Medicine and Public Health28(1)6572

25. Albonico M, Smith PG, Hall A, Chwaya HM, Alawi KS, et al. (1994) A randomised controlled trial comparing Mebendazole and Albendazole against Ascaris, Trichuris and the hookworms. Transactions of the Royal Society of Tropical Medicine and Hygiene 88: 585–589.M. AlbonicoPG SmithA. HallHM ChwayaKS Alawi1994A randomised controlled trial comparing Mebendazole and Albendazole against Ascaris, Trichuris and the hookworms.Transactions of the Royal Society of Tropical Medicine and Hygiene88585589

26. Waikagul J, Watthanakulpanich D, Muennoo C, Maipanich W, Sa-nguankiat S, et al. (2005) Efficacy of benzimidazole carbamate on an intestinal fluke co-infected with nematodes. Journal of the Medical Association of Thailand 88: 233–237.J. WaikagulD. WatthanakulpanichC. MuennooW. MaipanichS. Sa-nguankiat 2005Efficacy of benzimidazole carbamate on an intestinal fluke co-infected with nematodes.Journal of the Medical Association of Thailand88233237

27. Liberg O (1980) Spacing Patterns in A Population of Rural Free Roaming Domestic Cats. Oikos 35: 336–349.O. Liberg1980Spacing Patterns in A Population of Rural Free Roaming Domestic Cats.Oikos35336349

28. Cameron A (2002) Survey toolbox for aquatic animal diseases: a practical manual and software package. ACIAR Monograph Series 94: 1–375.A. Cameron2002Survey toolbox for aquatic animal diseases: a practical manual and software package.ACIAR Monograph Series941375

CHAPTER 15

Skin Lesions in Fish: Causes and Solutions

Brit Tørud[1] and Tore Håstein*[2]

[1]Aakvik settefisk AS, Halsanaustan, Norway and [2]Fiskeri-og kystdepartementet/Veterinærinstituttet, Oslo, Norway

INTRODUCTION

In recent years, the welfare of fish has received increasing attention and attempts have been made to control welfare in farmed fish. This may be achieved through regulation of management practices, including stocking density, etc. Skin lesions in farmed fish are an important challenge to the fish farmer. Fish with ulcers, fin erosions or impaired eyes are very conspicuous among the healthy fish swimming around in the tank or net pen. If only few fish are affected it is not an economical problem, but the welfare of these fish are not taken care of in an acceptable manner. Despite giving the fish the best environmental conditions, gentle handling or an adequate treatment if necessary, some members of such large populations will suffer because of skin lesions. To get these individuals out of the production may be a great challenge.

Internationally, the World Animal Health Organization (OIE) is currently working to establish international accepted guidelines on welfare of fish at slaughter, killing for disposal and land/sea transport [1].

For the purpose of this meeting, only some key diseases causing skin lesions will be covered, as it will be impossible to cover all diseases that may lead to welfare problems in fish. Of the following conditions causing skin lesions will be discussed in more detail.

Management

Impaired function and thus impaired welfare in fish may be caused both by external abiotic factors such as water quality criteria as well as by biotic factors such as starvation, stocking density, invasive species, etc.

A variety of aetiological factors may cause lesions in skin, fins and eyes of fish under farming conditions. Poor husbandry and rapid changes in environmental conditions may lead to health problems and diseases that will have an impact on fish welfare. Handling and grading may impair the mucus and skin if not carried out in a gentle way.

Lesions of mechanical origin may be caused by aggression, particularly in the hatchery stage, because of suboptimal conditions such as insufficient feeding. Sharp protrusions in tanks and cages as well as predators (birds, mink, otter and seal), bacteria and parasites also are important factors impairing fish health and welfare.

Using nets to protect the fish against birds may be a great risk for the fish themselves if the net is not thoroughly tied up.

Winter ulcer disease

Winter ulcer disease is a bacterial condition affecting in particular ongrowing Atlantic salmon (*Salmo salar*) in seawater and pre-smolts in hatcheries that use seawater [2]. The disease is caused by *Moritella viscosa* [3] and it occurs at low water temperatures during the winter months [2]. In the initial phase of the disease, the fish dies without development of lesions in the skin, but in more protracted cases, it is characterised by skin ulcers that may extend to cover large parts of the body. Mortality may extend 40% over a prolonged period, which in addition to downgrading of fish for slaughter may cause considerable economic losses in affected farms.

The main welfare aspect of the winter ulcers is due to osmo-regulatory problems associated with the ulcers at low water temperature as well as the septicaemia involved during a disease outbreak.

In order to combat the disease, medicinal therapy may be used, but antimicrobial treatments are not always effective. This is partly due to the low temperature and difficulties in fish taking enough feed (anorrhexia). From a prevention point of view, commercial vaccines are available, but are not reported to be fully effective [4]. When the temperature rises above 10°C, the disease disappears by itself.

Crustacean parasites (Lepeophteirus salmonis, Caligus elongatus)

Under farming conditions where fish are kept at high densities, this eases the ability of a parasite to find a host organism and clearly influences the level of parasitic infestation [5]. Parasites may thus have an impact on the welfare of an affected fish as pathological changes may occur as a result of the attachment and movement of parasites on the fish.

Sea lice (*Lepeophteirus salmonis, Caligus elongatus*) are crustacean ectoparasites that may cause severe infestations in salmonids. Not only farmed fish in seawater but also in wild salmonid populations in seawater period such as Atlantic salmon and sea trout [6, 7]. The lesions caused by sea lice may be localised or extensive depending on the size of the fish and on the mobility and number of parasites. Infestations can result in itching, skin irritation and ulcerations, starvation, reduced growth, secondary infections and mortality.

Sea lice may also be implicated in the transmission of a number of viral and bacterial diseases, i.e. infectious salmon anaemia (ISA). Sea lice can transfer between farmed and wild populations and have an impact on both.

Treatments are available for farmed salmon, but there are serious practical difficulties in their application as well as concerns regarding environmental impacts. The risk of development of resistance to the limited range of effective therapeutants is a serious challenge.

Several species of fish such as ballan wrasse (*Labrus berggylta*), goldsinny wrasse (*Ctenolabrus rupestris*), sea patridge (*Symphodus melops*), rock cock (*Centrolabrus exoletus*) are also used for delousing of Atlantic salmon. One problem, when using these species as cleaning fish, is that they may attack the Atlantic salmon when the nets are clean and only small numbers of lice are available and thus also causing ulcerations in the skin of the salmon.

No commercial vaccines against sea lice are yet available, but work is ongoing to develop such a vaccine.

In addition to osmoregulatory problems associated with the ulcerations caused by the lice, secondary bacterial infections may hamper the welfare of affected fish unless treated properly.

Jellyfish

Several jellyfish species may cause problems under farming conditions of Atlantic salmon. In Scotland, the four species listed below have been associated with mortalities:

- *Aurelia aurita*
- *Cyaneae capillata*
- *Phialella quadrata*
- *Phialidium sp.*

In Norway mortality in Atlantic salmon has mainly been associated with:

- *Apolemia uvaria*
- *Bolinopsis infundibulum*
- *Cyanea capillata/C. lamarcki*
- *Lepomedusae* spp
- *Muggiaea atlantica* (Siphonophora)

Mass occurrence of jellyfish swarms (blooms) may occur throughout the year, but is most common in the period spring through summer to autumn.

In seawater, mass mortalities may occur if large swarms of jellyfish are carried into areas where Atlantic salmon and rainbow trout are being farmed. While small jellyfish may enter the cages and reach the fish inside the pens, larger jellyfish tend to cling to the nets and tentacles or parts of tentacles enter the cage and sting the fish. In some cases the quantity of jellyfish may cause anoxia in the cages while under other circumstances, the jellyfish may affect the fish by stinging on the body surface, in the eyes or gills or through ingestion. Serious stings can cause rapid death but less severe stinging may result in ulcer formation and secondary infections. There are currently no methods available to prevent jellyfish "attacks". The welfare issues of jellyfish attacks are both the effects of the toxins associated with the attacks and secondary bacterial infections.

From a preventive point of view, measures in order to avoid jellyfish to enter the cages should be carried out.

REFERENCES

1. Håstein T: **OIE involvement in aquatic animal welfare: The need for development of guidelines based on welfare for farming, transport and slaughter purposes in aquatic animals.** in press.
2. Lunder T, Evensen Ø, Holstad G, Håstein T: **«Winter ulcer» in the Atlantic salmon *Salmo salar*. Pathological and bacteriological investigations and transmission experiments.** *Dis Aquat Org* 1995, **23**:39-49.
3. Benediktsdottir E, Verdonck L, Sproer C, Helgason S, Swings J: **Characterisation of *Vibrio viscosus* and *Vibrio woodanis* isolated at different geographical

locations: a proposal for reclassification of *Vibrio viscosus* as *Moritella viscosa*. *comb nov Int J System Evol Microbiol* 2000, **50:**479-488.

4. Håstein T, Gudding R, Evensen Ø: **Bacterial vaccines for fish – an update of the current situation world wide.** In *Progress in Fish Vaccinology, Dev Biol Volume 121*. Edited by: *Midtlyng PJ. Basel, Karger*; 2005:55-74.
5. Barber I: **Parasites, behaviour and welfare in fish.** *Appl Anim Behaviour Sci* 2007, **104:**251-264.
6. Anon: **Fiskehelse og fiskesykdommer.** Edited by: *Poppe TT. Universitetsforlaget*; 1999:411. (in Norwegian).
7. Roberts RJ: **Fish Pathology.** Third edition. *W.B. Saunders*; 2001:472.

CHAPTER 16

Comparative Pathogenomics of Bacteria Causing Infectious Diseases in Fish

Ponnerassery S. Sudheesh, Aliya Al-Ghabshi, Nashwa Al-Mazrooei, and Saoud Al-Habsi

Microbiology Laboratory, Fishery Quality Control Center, Ministry of Agriculture and Fisheries Wealth, P.O. Box 427, 100 Muscat, Oman

ABSTRACT

Fish living in the wild as well as reared in the aquaculture facilities are susceptible to infectious diseases caused by a phylogenetically diverse collection of bacterial pathogens. Control and treatment options using vaccines and drugs are either inadequate, inefficient, or impracticable. The classical approach in studying fish bacterial pathogens has been looking at individual or few virulence factors. Recently, genome sequencing of a number of bacterial fish pathogens has tremendously increased our understanding of the biology, host adaptation, and virulence factors of these important pathogens. This paper attempts to compile the scattered literature on genome sequence information of fish pathogenic bacteria published and available to date. The genome sequencing has uncovered several complex adaptive evolutionary strategies mediated by horizontal gene transfer, insertion sequence elements, mutations and prophage sequences operating in fish pathogens, and how their genomes evolved from generalist environmental strains to highly virulent obligatory pathogens. In addition, the comparative genomics has allowed the identification of unique pathogen-specific gene clusters. The paper focuses on the comparative analysis of the virulogenomes of important fish bacterial pathogens, and the genes involved in their evolutionary adaptation to different ecological niches. The paper also proposes some new directions on finding novel vaccine and chemotherapeutic targets in the genomes of bacterial pathogens of fish.

1. INTRODUCTION

Genome sequencing has provided us with powerful insights into the genetic makeup of the microbial world. The microbial genomics today has progressed from the long drawnout individual genome sequencing projects in the past to a level of technological advancement, where sequencing and comparing the genomes of several strains of a single pathogen is accomplished in a very short period of time [1, 2]. We are currently passing through a period of explosive developments in the field and an overwhelming glut in the genome sequence data of microorganisms. To date, over 1800 microbial genomes have been published and the sequencing of more than 5200 microbial genome are in different stages of completion (http://www.ncbi.nlm.nih.gov/genomes/lproks.cgi).

The genomics information has categorically disproved the earlier thinking that microbial genomes are static and has demonstrated that genomic evolutionary processes are much more flexible and dynamic than previously thought. This has led to the emergence of new ideas such as "uprooting the tree of life" and the concept of "horizontal genomics" [3–8]. This new thinking about microbial genome evolution has emerged from the observations of lineage-specific genome reduction and horizontal gene transfer (HGT), frequently occurring in bacterial genomes. Increasingly, genome sequencing projects have identified an unexpected level of diversity among bacteria, which can often be linked to recombination and gene transfer between a variety of prokaryotic organisms.

There is large variation in size and content of bacterial genomes between different genera and species, and also among strains of the same species. Known genome sizes of bacteria range from under 0.6 to 10 megabases (Mb). The smallest bacterial genomes reported are for the mycoplasmas and related bacteria, with sizes as low as 530 kilobases [9]. It has been emphasized that the adaptive capability ("versatility") of bacteria directly correlates with genome size [10].

Genome sequencing of bacterial pathogens has produced exciting information on evolutionary relationships between pathogenic and nonpathogenic species and has demonstrated how each has developed special adaptations advantageous for each of their unique infectious lifestyles. In the longer term, an understanding of their genome and biology will enable scientists to design means of disrupting their infectious lifestyles.

The genomes of bacteria are made up of circular or linear chromosomes, extrachromosomal linear or circular plasmids as well as different combinations of these molecules. The functionally related genes are clustered together in very close proximity to each other, and those genes located on the "core" part of the chromosome present a relatively uniform G+C content and a specific codon usage. Closely related bacteria generally have very similar genomes [11].

The stability and integrity of the "core" sequences of the genome, however, is often interrupted by the presence of DNA fragments with a G+C content and a codon usage markedly different from those of the "core" genome. The "flexible" gene pool or the so-called "mobilome" [12], is created by the acquisition of strain-specific "assortments" of genetic information mainly represented by mobile genetic elements (MGE), such as plasmids, bacteriophages, genomic/pathogenicity islands (GEIs/PAIs), integrons, IS elements (ISEs), and transposons (see Figure 1). The flexible genes scattered in the genome provide the microbes with an additional repertoire of arsenal, for example, resistance to antibiotics, production of toxic compounds as well as other virulence factors [13].

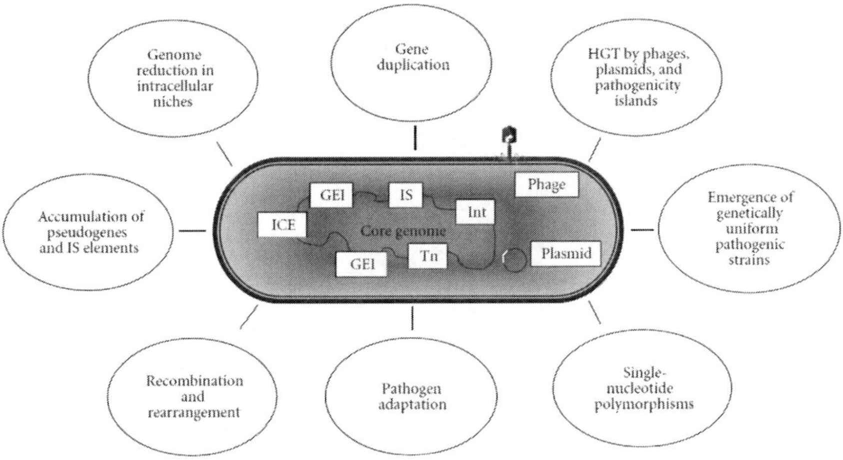

Figure 1: Major factors responsible for the pathogenomic evolution of bacteria (modified from [14, 15]; HGT: horizontal gene transfer, GEIs: genomic islands, ICEs: integrative conjugative elements, Int: integrons, Tn: conjugative transposons, IS: IS elements.

A fundamental question in biology is to define the minimum number of genes or functions to support cellular life. The size of bacterial genomes is primarily the result of two counteracting processes: the acquisition of new genes by gene duplication or by horizontal gene transfer; the deletion of nonessential genes. Genomic flux created by these gains and losses of genetic information can substantially alter gene content. This process drives divergence of bacterial species and eventually adaptation to new ecological niches [16].

Bacterial pathogens are a major cause of infectious diseases and mortality in wild fish stocks and fish reared in confined conditions. Disease problems constitute the largest single cause of economic losses in aquaculture [17]. Concurrent with the rapid growth and intensification of aquaculture, increased use of water bodies, pollution, globalization, and transboundary movement of aquatic fauna, the list of new pathogenic bacterial species isolated from fish has been steadily increasing [18]. In addition, the virulence and host range of existing pathogens has also been increasing, posing considerable challenge to fish health researchers, who are actively looking for more efficient vaccines and therapeutic drugs to combat bacterial fish diseases. The current treatment methods are ineffective and have many practical difficulties.

At the level of host-pathogen interaction, there is considerable pressure on pathogens to adapt to the harsh host environment as well as to adapt and evolve along with the ever changing external environment. The interplay between the host and the pathogen is a complex one, each driven by the need to secure the success of the species. Adaptations by one partner to exploit new environments will often stimulate the other to modify its characteristics to take advantage of the change. As a consequence of this cycle of interaction created by changing environments, new strains of pathogen will evolve. Over time, these strains may emerge as new species with characteristic disease symptoms.

The use of antibiotics to control fish diseases has met with limited success and has the potential danger of antibiotic resistance development in aquatic bacteria (World Health Organization antimicrobial resistance fact sheet 194, http://www.who.int/inf-fs/en/fact194.html) [19]. As aquaculture is one of the fastest growing food production industries in the world, demand for sustainable ways of combating fish diseases is gaining significance. There is tremendous scope for developing novel vaccines and therapeutic drugs against bacterial fish pathogens.

Genomic evolution and adaptive strategies of bacterial fish pathogens are poorly understood and lags far behind that of human and terrestrial animal pathogens. A detailed knowledge of the genome sequences of bacterial fish pathogens and how the genomes of the pathogenic species or strains evolved from nonpathogenic ancestors or counterparts will help us better understand their pathogenicity mechanisms and strategies of host adaptations. This information will help identifying novel vaccine and drug targets in the genomes of pathogens.

Recently, genome sequencing of a number of bacteria pathogenic to fish and other aquatic organisms have been completed. The genome sequence and genome characteristics of important bacterial fish pathogens completed and published to date are summarized in Table 1.

Table 1: Currently sequenced genomes of bacterial pathogens of fish.

Organisms	Size (Mb)	CDS**	Unknown/ Hypothetical genes (%)	Pseudogenes prophages, ISE/GEI	% GC	Chromosomes	Plasmids
Vibrio anguillarum 775 serotype O1	4.117	3880	26	92	44.3	2	1
Vibrio anguillarum 96F serotype O1	4.065	3766	26	38	42	2	0
Vibrio anguillarum RV22 serotype O2*β*	4.022	3949	26	68	43.1	2	0
Vibrio ordalii ATCC 33509	3.415	3281	—	31	43.3	2	0
Vibrio harveyi ATCC BAA-1116*	6.054	—	—	—	45.4	2	1
Vibrio vulnificus YJ016 biotype 1	5.26	5028	34	—	46.1	2	1
Vibrio splendidus strain LGP32	4.974	4498	24.8	—	43.8	2	0
Aliivibrio salmonicida strain LFI1238	4.655	4286	—	1179	38.3	2	4
Flavobacterium psychrophilum JIP02/86	2.862	2432	45.3	94	32.5	1	1
Flavobacterium branchiophilum FL-15	3.56	2867	—	54	32.9	1	—
Flavobacterium columnare ATCC 49512*	3.2	2896	—	—	32.0	1	—
Edwardsiella tarda EIB202	3.76	3486	28	97	59.7	1	1
Edwardsiella ictaluri 93–146*	3.812	3783	—	100	57.4	1	—
Aeromonas hydrophila ATCC 7966	4.744	5195	27.7	7	61.5	1	0
Aeromonas salmonicida A449	4.702	4437	—	258	58.5	1	5
Aeromonas veronii Strain B565	4.551	4057	—	—	58.7	1	—
A. caviae Ae398	4.43	—	—	6	61.4	1	1
Renibacterium salmoninarum ATCC 33209	3.155	3507	25.3	151	56.3	1	0
Streptococcus parauberis	2.143	2641	21.3	—	35.6	1	0
Lactococcus garvieae UNIUD074	2.172	2101	21.8	224	38.7	1	0
Mycobacterium marinum M	6.636	5424	26	65	62.5	1	1

* Unpublished.
** Coding sequences.

The main aim of this paper is to put together and summarize the scattered genome sequencing information on important bacterial fish pathogens available in the literature to date. We sincerely believe that this paper will provide a genomic perspective on the adaptive evolutionary strategies of bacterial fish pathogens in different ecological niches and will help better understand the virulence mechanisms and pathogenesis of infections. It is hoped that this will lead to finding the most appropriate vaccine and therapeutic drug targets in the genomes and developing efficient control and treatment methods for fish diseases.

2. BACTERIAL PATHOGENS OF FISH

Although pathogenic species representing majority of existing bacterial taxa have been implicated in fish diseases, only a relatively small number of pathogens are responsible for important economic losses in cultured fish worldwide. Major bacterial pathogens responsible for infectious disease outbreaks in different species of fish are listed in Table 2. Major groups of bacteria causing infectious diseases in fish and the important genome characteristics of these bacteria are described in the following sections.

Table 2: Major bacterial pathogens of economically important fish.

Causative agent/species	Disease	Main host fish
Gram-negatives		
Vibrio anguillarum	Vibriosis	Salmonids, turbot, sea bass, striped bass, eel, ayu, cod, and red sea bream
Aliivibrio salmonicida (formerly *Vibrio salmonicida*)	Vibriosis	Atlantic salmon, cod
Vibrio vulnificus	Vibriosis	Eels, tilapia
Vibrio ordalii	Vibriosis	Salmonids
Vibrio carchariae (syn.: *Vibrio harveyi*)	Vibriosis, infectious gastroenteritis	Shark, abalone, red drum, sea bream, sea bass, cobia, and flounder
Moritella viscosa (formerly *Vibrio viscosus*)	Winter ulcer	Atlantic salmon
Photobacterium damselae subsp. piscicida (formerly *Pasteurella piscicida*)	Photobacteriosis (pasteurellosis)	Sea bream, sea bass, sole, striped bass, and yellowtail
Pasteurella skyensis	Pasteurellosis	Salmonids and turbot
Tenacibaculum maritimum (formerly *Flexibacter maritimus*)	Flexibacteriosis	Turbot, salmonids, sole, sea bass, gilthead sea bream, red sea bream, and flounder
Flavobacterium psychrophilum	Coldwater disease	Salmonids, carp, eel, tench, perch, ayu
Flavobacterium branchiophila	Bacterial gill disease	A broad range of cultured cold water and warm water salmonid and nonsalmonid fishes
Flavobacterium columnare	Columnaris disease	cyprinids, salmonids, silurids, eel, and sturgeon
Pseudomonas anguilliseptica	Pseudomonadiasis, winter disease	Sea bream, eel, turbot, and ayu
Aeromonas salmonicida	Furunculosis	salmon, trout, goldfish, koi and a variety of other fish species
Aeromonas hydrophila *Aeromonas veronii* Biovar Sobria *Aeromonas sobria* Biovar Sobria (Motile aeromonads)	Motile aeromonas septicemia (MAS), hemorrhagic septicemia, ulcer disease or red-sore disease, and epizootic ulcerative syndrome (EUS)	A wide variety of salmonid and nonsalmonid fish, sturgeon, tilapia, catfish, striped bass, and eel
Edwardsiella ictaluri	Enteric septicemia	Catfish and tilapia
Edwardsiella tarda	Edwardsiellosis	Salmon, carps, tilapia, catfish, striped bass, flounder, and yellowtail
Yersinia ruckeri	Enteric redmouth	Salmonids, eel, minnows, sturgeon, and crustaceans
Piscirickettsia salmonis	Piscirickettsiosis	Salmonids
Gram-positives		
Lactococcus garvieae (formerly *Enterococcus seriolicida*)	Streptococcosis or lactococcosis	Yellowtail and eel
Streptococcus iniae	Streptococcosis	Yellowtail, flounder, sea bass, and barramundi
Streptococcus parauberis	Streptococcosis	Turbot
Streptococcus phocae	Streptococcosis	Atlantic salmon
Renibacterium salmoninarum	Bacterial kidney disease	Salmonids
Mycobacterium marinum	Mycobacteriosis	Sea bass, turbot, and Atlantic salmon

3. VIBRIOS

Bacteria in the genus Vibrio are mainly pathogenic to marine and brackish water fish. However, they are occasionally reported in freshwater species as well [20, 21]. The distribution of vibriosis is worldwide and causes great economic loss to the aquaculture industry [22]. Vibriosis, one of the major bacterial diseases affecting fish, bivalves, and crustaceans, is mainly caused by pathogenic species such as Vibrio anguillarum, V. harveyii (Syn. V. carchariae), V. ordalii, and Aliivibrio salmonicida (formerly Vibrio salmonicida) [23, 24].

Other species such as V. vulnificus [25, 26] and Moritella viscosa (formerly Vibrio viscosus) [27] have been implicated in fish diseases such as septicemia and winter ulcer, respectively; more pathogenic species have been isolated frequently and reported in the literature [28].

Genome sequences of four major fish pathogenic vibrios, V. anguillarum, V. ordalii, Aliivibrio salmonicida, and V.vulnificus have been completed and published [29–31]. Generally, they have two chromosomes, one larger and one smaller. The majority of genes that encode cell functions and pathogenic factors are located in the large one. The small chromosome usually contains genes for environmental adaptation.

Vibrio anguillarum is the most studied aetiological agent of vibriosis [32]. V. anguillarum typically causes a hemorrhagic septicemia. The O1 and O2 serotypes are the virulent strains frequently isolated from diseased fish [33, 34]. Many O1 serotype strains harbor 65 kb pJM1-type plasmids, which carry the siderophore anguibactin biosynthesis and transport genes, a main virulence factor of V. anguillarum, while one of the O1 serotype strains and other serotypes, such as all of the O2 strains, are plasmidless [28, 35, 36]. The O1 serotype strains cause disease in salmonid fish, whereas O2 β strains are usually isolated from cod and other nonsalmonids [28, 32].

Vibrio ordalii is a very close relative of V. anguillarum [37] and was previously recognized as V. anguillarumbiotype 2. Vibriosis caused by these two species are strikingly different based on histological evidences [38]. V. anguillarum has a special affinity for blood and loose connective tissue, whereas V. ordali is mostly present as aggregates in skeletal and cardiac muscles. V. ordalii has a lesser affinity for blood and develops bacteremia only at late stages of disease.

Genomic sequences of three different strains of V. anguillarum (the strain 775 containing plasmid pJM1, serotype O1 strain 96F, and plasmidless serotype O2 β strain RV22) and V. ordali have recently been published [31]. The pJM1 plasmid in the strain 775 contains 65 genes including the anguibactin biosynthesis and transport genes that are unique for the strain.

V. anguillarum 775 contains more transposase genes (about 53) than 96F (about 23), RV22 (about 42), and V. ordalii (about 18).

The genome comparison of V. anguillarum serotypes has revealed some interesting differences in the genomic composition, indicating horizontal acquisition of virulence genes and the evolution of different potential virulence mechanisms among the closely related serotypes [31]. The V. anguillarum 96F strain has a type III secretion system 2 (T3SS2) cluster, which is absent in the 775 strain. The T3SS2 genes are highly conserved with other T3SS2 genes reported in V. parahaemolyticus, V. cholera, and V. mimicus [39–41]. In the 775 strain, three transposase genes are present at the T3SS2 chromosomal location, one of which probably originated from the pJM1, indicating that the gene cluster is inactivated by a transposition, deletion, or inversion event [31]. The 775 strain also contains 10 genomic islands including integrase, transposase, and some novel sequences conferring genomic plasticity to adapt to specific ecological niches.

The strain RV22 genome contains the toxin-antitoxin systems, and genes encoding the accessory V. choleraeenterotoxin (Ace) and the Zonula occludens toxin (Zot), which is not present in the 775 strain. The yersiniabactin-like siderophore cluster, which is highly conserved in many Vibrio species and Photobacterium damselae subspecies piscicida [42], is present in strain RV22 and V. ordalii.

A striking feature of V. ordali genome is its significant reduction in size (3.4 Mb) compared to the V. anguillarum strains 775 (4.1 Mb), 96F (4.0 Mb), and RV22 (4.0 Mb). V. ordali lacks the ABC transporter genes, the type VI secretion systems, and the gene for microbial collagenase. The Syp biofilm formation cluster, which is conserved in many Vibrio species such as V. fischeri, V. vulnificus, and V. parahaemolyticus [43, 44], is present only in V. ordalii. Thus, it is probable that the transition of V. anguillarum to V. ordalii is mediated by genome reductive evolution to become an endosymbiotic organism; V. ordali has the smallest genome of all vibrios.

Vibrio vulnificus includes three distinct biotypes. Biotype 1 strains cause human disease, while biotype 2 infects primarily eels, and biotype 3 infections has been associated with persons handling Tilapia, although the source and reservoir of biotype 3 have yet to be identified [45]. In another classification the terms clade 1 and clade 2 are used based on the multilocus sequence typing (MLST) [46]. Biotype 1 strains are present in both clades, whereas biotype 2 strains are present only in clade 1, and biotype 3 strains appear to be a hybrid between clades 1 and 2. Clade 1 strains are most often isolated from environmental samples, while clade 2 strains are mostly associated with human disease and are considered more virulent. Recent comparative genomic analysis of these biotypes or clades has clearly differentiated them based on the possession of an array of clade-specific unique genes including the presence of a virulence-associated genomic island XII in the highly virulent strains [30].

Aliivibrio salmonicida (formerly Vibrio salmonicida) causes coldwater vibriosis in marine fish such as farmed Atlantic salmon (Salmo salar), sea-farmed rainbow trout (Oncorhynchus mykiss), and captive Atlantic cod

(Gadus morhua) [47]. The Gram-negative bacterium causes tissue degradation, hemolysis, and sepsis in vivo. Genome sequencing of Aliivibrio salmonicida has revealed a mosaic structure of the genome caused by large intrachromosomal rearrangements, gene acquisition, deletion, and duplication of DNA within the chromosomes and between the chromosomes and the plasmids [29].

The genome has many genes that appear to be recently acquired by HGT, and large sections of over 300 coding sequences (CDS) are disrupted by IS elements or contain point mutations causing frame shifts or premature stop codons [29]. The genomic islands (GIs) identified in the bacteria include major virulence-related genes encoding T6SS and Flp-type pilus and genes that appear to provide new functions to the bacteria. The Tad system has been proposed to represent a new subtype of T2SS and is essential for biofilm formation, colonization, and pathogenesis [48].

The genome analysis has unequivocally confirmed that Aliivibrio salmonicida has undergone extensive rearrangement of its genome by losing massive functional genes and acquiring new genes and become host-restricted, allowing the pathogen to adapt to new niches. IS expansion has been related to genome reduction in the evolution and emergence of pathogenicity [49], and accumulation of pseudogenes has been described for several other host restricted pathogens [50, 51].

4. AEROMONADS

Aeromonas hydrophila and other motile aeromonads are among the most common bacteria in a variety of aquatic environments worldwide, including bottled water, chlorinated water, well water, sewage, and heavily polluted waters, and are frequently associated with severe disease among cultured and feral fishes, amphibians, reptiles, and birds [52]. Aeromonads are also considered serious emerging pathogens of human beings [53]. Determination of the etiology of diseases involving aeromonad infections has been complicated by the genetic, biochemical, and antigenic heterogeneity of members of this group.

The genus Aeromonas has been conveniently divided into a group of nonmotile, psychrophilic species, prominently represented by Aeromonas salmonicida, which is an obligate fish pathogen and a second group of mostly human pathogenic, motile, and mesophilic species including A. hydrophila.

Genome sequencing of A. hydrophila ATCC 7966^T, A. salmonicida subsp. salmonicida A449, A. veronii strain B565, and A. caviae [54–57] has helped in resolving their taxonomic confusion and has brought new insights into the way these bacteria adapt to a myriad of ecological niches, their host adaptive evolution and virulence mechanisms.

Aeromonas salmonicida, the causative agent of furunculosis in salmonid and nonsalmonid fish, is a non-motile, Gram-negative bacterium; furunculosis is an important disease in wild and cultured stocks of fish inflicting heavy losses to aquaculture industry worldwide [58, 59]. A. hydrophila causes a septicemic disease in fish known variously as "motile aeromonas septicemia" (MAS), "hemorrhagic septicemia," "ulcer disease," or "red-sore disease" [60]. The disease caused by this bacterium primarily affects freshwater fish such as catfish, several species of bass, and many species of tropical or ornamental fish. A. veronii is the causative agent of bacterial hemorrhagic septicemia in fish and is becoming a major economic problem in the fish-farming industry [23].

Genome sequencing of the fish pathogen A. salmonicida A449 has confirmed the presence of fully functional genes for a type III secretion system (T3SS) that has been shown to be required for virulence in A. salmonicida[61], and genes for a type VI secretion system (T6SS), which is disrupted by an IS element [55]. The ancestral state of the T3SS in A. salmonicida A449 is ambiguous because of the absence of the genes in A. hydrophilaATCC 7966^T, while other A. hydrophila strains carry T3SS operons on the chromosome [62]. The genome contains a multitude of virulence-related genes including several types of adhesins (e.g., surface layer, flagella, and pili), toxin genes (aerolysin, hemolysin, repeats in toxin (RTX) protein, and cytolytic delta-endotoxin), secreted enzymes (protease, phospholipase, nuclease, amylase, pullulanase, and chitinase), antibiotic resistance genes (tetA, β-lactamase gene, and efflux pumps), and genes involved in iron acquisition and quorum sensing.

Most of the above genes are present in A. hydrophila ATCC 7966^T genome and an expansion of gene families (paralogs) of ABC transporters, two-component signal transduction systems (TCSs), transcriptional regulators, FeS cluster-binding proteins involved in energy transduction at the membrane, and methyl-accepting chemotaxis proteins (MCPs). Interestingly, transposase, resolvase, or insertion sequence element sequences were not discovered in the A. hydrophila ATCC 7966^T genome, whereas these have been identified in A. salmonicida and A. caviae genomes. A. salmonicida possesses 88 copies of 10 different IS elements whereas A. caviae Ae398 has only five different IS elements, and A. hydrophila completely lacks IS elements.

Although A. hydrophila ATCC 7966^T has been demonstrated to be the second most virulent species among Aeromonas [63], a very important virulence determinant, T3SS, which is present in A. salmonicida A449 is strikingly absent in A. hydrophila ATCC 7966^T genome. A. caviae contains many putative virulence genes, including those encoding a type 2 secretion system, an RTX toxin, and polar flagella.

The genome of A. veronii strain B565 contains some putative virulence factors, such as chitinase, RTX protein, adhesion factor, flagella, and mannose-sensitive hemagglutinin (MSHA), all of which are shared with A. hydrophila ATCC 7966^T and A. salmonicida A449. On the other hand, 346 genes including some important putative virulence factors such as hemolysins and the type III secretion protein, which are shared by the latter two species are absent in A. veronii strain B565.

Many unique genes in A. hydrophila ATCC 7966^T and A. salmonicida A449 are virulence genes and often form large clusters, such as the rtx cluster in ATCC 7966^T and the flagellar gene cluster in A449, or are involved in mobile elements such as phages and transposons, highlighting their lateral transfer history [56].

The A. hydrophila ATCC 7966^T and A. salmonicida A449 genomes appear to be very closely related, encoding similar number of proteins with only 9% difference in gene content. However, there are many transposons, phage-related genes, and unique CDS in A. salmonicida A449 genome that are different from A. hydrophilaATCC 7966^T sequences, showing their distinct lineages and adaptive evolution that occurred while segregating into different species of the genus.

In sharp contrast to A. hydrophila ATCC 7966^T genome, the A. salmonicida A449 genome is characterized by the presence of large numbers of several different types of IS elements in multiple copies, with more than 20 genes being interrupted by IS elements. A. hydrophila ATCC 7966^T genome has no IS elements.

There is a higher tendency for genomic reduction in A. salmonicida A449 with the formation of many pseudogenes, and A. hydrophila ATCC 7966^T has only seven pseudogenes. The formation of pseudogenes has resulted in the loss of function of many genes including flagella and type IV pili, transcriptional regulators, genes encoding carbohydrate synthesis, and modification enzymes and genes for basic metabolic pathways, which are some characteristic features of pathogenomic evolution.

Thus, A. salmonicida A449 appears to have evolved much faster than A. hydrophila ATCC 7966^T through genetic rearrangements, genomic reduction, and HGT from common ancestral lineages by acquiring and forming multiple plasmids, prophages, a battery of IS elements, pseudogenes, and several individual genes and operons.

5. FLAVOBACTERIUM

The genus Flavobacterium includes over 30 species of which Flavobacterium psychrophilum, F. branchiophilum, and F. columnare are important disease agents for salmonids, catfish, and other cultured species [64, 65]. Flavobacteria are significant as they are ubiquitous in the soil, freshwater, and marine environments and are noted for their novel gliding motility and ability to degrade polymeric organic matter such as hydrocarbons [66].

F. psychrophilum is the etiological agent of bacterial coldwater disease (BCWD). It is a serious fish pathogen causing substantial economic losses and rearing difficulties to both commercial and conservation aquaculture. F. psychrophilum infections are found throughout the world. Juvenile rainbow trout and coho salmon are particularly susceptible to BCWD. However, F. psychrophilum infections have been reported in a wide range of hosts, Anguilla japonica, A. anguilla, Cyprinus carpio, Carassius carassius, Tinca tinca, Plecoglossus altivelis, Perca fluviatilis, and Rutilus rutilus [64, 67]. Fry and fingerlings with BCWD often have skin ulcerations on the peduncle, anterior to the dorsal fin, at the anus, or on the lower jaw and mortalities can go up to 70% [68].

F. branchiophilum is the causative organism of bacterial gill disease (BGD) in several parts of the world [69]. This disease is characterized by explosive morbidity and mortality rates attributable to massive bacterial colonization of gill lamellar surfaces and progressive branchial pathology stemming from high rates of lamellar epithelial necrosis [70].

F. columnare (formerly Cytophaga columnaris; Flexibacter columnaris) is the causative agent of columnaris disease of salmonids and other fishes in commercial aquaculture, the ornamental fish industry, and wild fish populations worldwide [71]. Classically, during outbreaks, its morbidity and mortality rates escalate more gradually than for BGD. Additionally, unlike the pattern of necrosis in BGD, fish with columnaris will have severe necrosis of all parts of the gill as the bacterium invades inwardly [72].

The taxonomy of the three species was initially based on phenotypic characteristics and has been revised several times during the years. The latest classification based on G+C content, DNA-ribosomal ribonucleic acid (rRNA) hybridisation, and fatty acid and protein profiles, has confirmed that all the three species now belong to the phylum/division Cytophaga-Flavobacterium-Bacteroides, family Flavobacteriaceae, and genus Flavobacterium [73].

The whole genome sequences of F. psychrophilum and F. branchiophilum have been published [74, 75]. The F. columnare genome sequence is yet to be completed and published [76].

Prominent features of F. psychrophilum infection include the strong adhesion to fish epithelial tissues followed by gliding motility, rapid and mass tissue destruction, and severe muscle tissue ulcerations. Hence, the identification of multiple genes encoding secreted proteases, adhesins, and gliding motility (gld) genes in F. psychrophilum genome indicates their possible involvement in the virulence of the pathogen. However, the gene sequence of a secreted collagenase was disrupted by an insertion sequence of the IS256 family in several strains isolated from rainbow trout [74] indicating the clonal dissemination of strains containing the disrupted gene. The F. psychrophilum seems to have horizontally acquired virulence associated genes from other unrelated bacteria. It has a hemolysin similar to the toxin VAH5, which is a virulence factor in Vibrio anguillarum [77]. It also has a gene encoding a protein that is similar to domains 1–3 of thiol-activated cytolysin family of pore-forming toxins (TACYs), which has been implicated in the pathogenicity of several Gram-positive bacteria [78]. Interestingly, F. psychrophilum lacks the type III and IV secretion systems usually present in Gram-negative pathogens; but, it has genes encoding PorT and PorR proteins, which are involved in transport and anchoring of virulence factors of the bacteria [79, 80]. In addition, the F. psychrophilum genome contains a large repertoire of genes involved in aerobic respiration, psychrotolerance, and stress response.

The sequencing of F. branchiophilum genome has revealed the existence of virulence mechanisms distinctly different from the closest species, F. psychrophilum. The F. branchiophilum genome has the first cholera-like toxin in a nonproteobacteria and an array of adhesins. A comparative analysis of its genome with genomes of other Flavobacterium species revealed a smaller genome size, large differences in chromosome organization, and fewer rRNA and tRNA genes, fitting with its more fastidious growth. In addition, identification of certain virulence factors, genomic islands, and CRISPR (clustered regularly interspaced short palindromic repeats) systems points to the adaptive evolution of F. branchiophilum by horizontal acquisition of genes.

6. EDWARDSIELLA

The genus Edwardsiella belongs to subgroup 3 of γ-proteobacteria, encompassing a group of Gram-negative enteric bacteria pathogenic to a variety of animals [81]. Two very closely related species, Edwardsiella tarda and E. ictaluri are important fish pathogens. Both are Gram-negative motile rods that are cytochrome oxidase negative and ferment glucose with production of acid and gas. The two species can be differentiated biochemically in that E. tarda produces both indol and hydrogen sulfide, whereas E. ictaluri produces neither. Moreover, the two species do not cross-react serologically. E. tarda has been isolated from many warm water fishes and some coldwater fishes, whereas E. ictaluri has been isolated only from a few species of warm water fishes (Table 2). Additionally, E. tarda causes disease in such other animals as marine mammals, pigs, turtles, alligators, ostriches, skunks, and snakes [81]. It has also occasionally infected humans [82, 83]. In contrast, E. ictaluri is limited to fish, and survivors of epizootics probably become carriers. The geographic range of E. tardais worldwide, whereas that of E. ictaluri is still confined to the catfish growing areas in the United States [84].

E. tarda causes a disease condition in fish called systemic hemorrhagic septicemia with swelling skin lesions as well as ulcer and necrosis in internal organs such as liver, kidney, spleen, and musculature [85]. It has the ability of invading and multiplying in epithelial cells and macrophages in order to subvert the host immune system and to survive in the fish [86].

E. ictaluri is the causative agent of enteric septicemia of catfish (ESC), a major disease affecting the catfish industry. The disease can manifest as an acute form that is characterized by hemorrhagic enteritis and septicemia and a chronic disease that is characterized by meningoencephalitis [87]. Gross external symptoms include hemorrhages on the body, especially around the mouth and fins. Other signs include pale gills, exophthalmia, and small ulcerations on the body [84].

The whole genome sequencing of the two species has recently been completed and published allowing comparative genomic analysis of these very important fish pathogens [88, 89]. The genome sequencing of the two closely related species E. tarda and E. ictaluri has revealed a high level of genomic plasticity with a high content of mobile genetic elements, IS elements, genomic islands, phage-like products, integrases, or recombinases. E. ictaluri displays high biochemical homogeneity with only one serotype, but possess many IS elements in the genome. In addition, highly variable G+C content and a large quantity of variable number of tandem repeats (VNTRs) or direct repeat sequences were identified in the E. tarda genome indicating the rapid genomic evolution undergoing in the species [88]. An interesting feature is the identification of insertion sequence IS Saen1 of Salmonella enterica serovar Enteritidis [90] in both E. tarda EIB202 and E. ictaluri 93–146 genomes. Conversely, the difference in genomic islands among the three species may partially explain their rapid evolutionary changes and diverging lineage from a common ancestor.

The E. tarda genome has a gene cluster sharing high similarities to the pvsABCDE-psuA-pvuA operon, which encodes the proteins for the synthesis and utilization of vibrioferrin, an unusual type of siderophore requiring nonribosomal peptide synthetase (NRPS) independent synthetases (NIS) and usually mediating the iron uptake systems in V. parahaemolyticus and V. alginolyticus [91, 92]. But E. ictaluri genome lacks siderophore biosynthesis genes, even though it possesses heme binding/transport genes.

E. tarda genome is smaller than that of E. ictaluri and other sequenced genomes of Enterobacteriaceae, justifying the hypothesis that E. tarda may not be present as a free living microorganism in natural waters but multiply intracellularly in protozoans and transmitted to fish, reptile, and other animals or humans [81].

The E. tarda and E. ictaluri genomes have a multitude of virulence factors including P pilus, type 1 fimbriae, nonfimbrial adhesins, invasins and hemagglutinins and various secretion pathways including sec-dependent transport system, the components of the main terminal branch of the general secretory pathway (GSP), the signal recognition particle (SRP), and the sec-independent twin arginine transport (Tat), T1SS, TTSS, and T6SS indicating their evolutionary fitness and ability to adapt to a variety of demanding ecological niches and harsh host intracellular environments.

7. YERSINIA RUCKERI

Yersinia ruckeri, the causal agent of enteric redmouth (ERM) disease, which is a systemic bacterial infection of fishes, but is principally known for its occurrence in rainbow trout, Salmo gairdneri [93]. Y. ruckeri was initially isolated from rainbow trout in the Hagerman Valley, Idaho, USA, in the 1950s [94] and is now widely found in fish populations throughout North America, Australia, South Africa, and Europe [95]. Outbreaks of ERM usually begin with low mortalities which slowly escalate and may result in high losses. The problem may become large-scaled if chronically infected fish are exposed to stressful conditions such as high stocking densities and poor water quality [96]. Y. ruckeri is a nonspore-forming bacterium which does not possess a capsule, but often has a flagellum [97].

Historically, Y. ruckeri is fairly homogenous in biochemical reactions. However, Y. ruckeri strains have recently been grouped into clonal types on the basis of biotype, serotype, and outer membrane protein (OMP) profiles [98]. Strains of serovars I and II [99], equivalent to serotypes O1a and O2b, respectively [100], cause most epizootic outbreaks in cultured salmonids, serovar I being predominant in rainbow trout [101]. Within serovar I, six clonal OMP types have been recognized, but only two are associated with major disease outbreaks: clonal group 5, which includes the so-called Hagerman strain and clonal group 2 [98, 102]. Clonal group 5 comprises the majority of isolates, all of them motile and with a widespread distribution (Europe, North America, and South Africa). Clonal group 2 includes only nonmotile strains isolated in the UK.

More recently, multilocus sequence typing has revealed distinct phylogenetic divergence of Y. ruckeri from the rest of the Yersinia genus raising doubts about its taxonomic position [103]. This view has gained credibility after the genome sequencing of Y. ruckeri, which has a substantially reduced total genome size (3.58 to 3.89 Mb), compared with the 4.6 to 4.8 Mb seen in the genus generally [104]. In addition, Y. ruckeri was found to be the most evolutionarily distant member of the genus with a number of features distinct from other members of the genus.

Several common Yersinia genes were missing in Y. ruckeri. These included genes involved in xylose utilization, urease activity, B12-related metabolism, and the mtnKADCBEU gene cluster that comprises the majority of the methionine salvage pathway [104]. The genomic reduction achieved by losing these and other genes is suggestive of its means of adaptation to an obligatory life style in fish hosts.

8. RENIBACTERIUM SALMONINARUM

Renibacterium salmoninarum is a small Gram-positive diplobacillus, and the causative agent of bacterial kidney disease (BKD), which is a slowly progressive, systemic infection in salmonid fishes with a protracted course and an insidious nature [105]. The pathogen can be transmitted from fish to fish [106] or from adults to their progeny via eggs [107]. Infected fish may take months to show signs of disease. bacterial kidney disease is one of the most difficult bacterial diseases of fish to treat [108], mainly due to its ability to evade phagocytosis and invade and survive in host cells [109, 110]. R. salmoninarum is very slow growing, and it is extremely difficult to apply genetic manipulation techniques to study its gene functions.

R. salmoninarum, despite being an obligate intracellular pathogen of fish, is phylogenetically closest to the non-pathogenic environmental Arthrobacter species [51]. Based on 16S rRNA phylogenetic analysis, R. salmoninarum has been included in the actinomycetes subdivision and was found related to a subgroup harboring morphologically and chemotaxonomically rather heterogeneous taxa, including Arthrobacter, Micrococcus, Cellulomonas, Jonesia, Promicromonospora, Stomatococcus, and Brevibacterium [111]. In fact, Arthrobacter davidanieli is commercially used as a vaccine (commercially known as Renogen) and can provide significant cross-protection in Atlantic salmon, though not in Pacific salmon [112]. The genome sequencing of R. salmoninarum ATCC 33209 strain and two Arthrobacter strains, the TC1 and FB24, has revealed many interesting aspects of how this obligates fish pathogen evolved, via genomic reduction and horizontal gene acquisition, from members of the nonpathogenic genus Arthrobacter [51, 113]. A total of 1562 ORF clusters were similar in R. salmoninarum and Arthrobacter spp. demonstrating the genetic basis for the efficiency and cross-protection of the A. davidanieli vaccine.

There is significant genome reduction in R. salmoninarum genome, which is 1.44 Mb smaller than the chromosome of TC1 and 1.55 Mb smaller than the chromosome of FB24. The two Arthrobacter strains have several large plasmids that are not present in the ATCC 33209 strain. In addition, these plasmids do not have high levels of similarity to sequences in the R. salmoninarum chromosome [51].

The presence of many IS elements, pseudogenes, and genomic islands in R. salmoninarum genome coupled with a lack of restriction-modification systems contribute to the extensive disruption of ORFs as a strategy to reduce many pathways in the bacteria. Moreover, the highly homogeneous nature of R. salmoninarum with respect to the overall genomic structure, biochemical properties, and surface antigens [114, 115] points to the evolution of this pathogen towards a strictly intracellular life style.

Several virulence factors including capsular synthesis genes, heme acquisition operons, genes encoding possible hemolysins, and the poorly characterized msa genes identified in the R. salmoninarum genome seems to be horizontally acquired. Arthrobacter spp. lacks most of these gene sequences, thus underlining the differential evolution and adaptation of these two very closely related species to contrasting ecological niches.

9. STREPTOCOCCUS AND LACTOCOCCUS

Gram-positive cocci belonging to the genera Streptococcus and Lactococcus are increasingly being recognized as important fish pathogens all over the world [116]. There are several different species of Gram-positive cocci, including Streptococcus parauberis, S. iniae, S. agalactiae (syn. Streptococcus difficilis), S. phocae [117, 118], Lactococcus garvieae (syn. Enterococcus seriolicida) [119], L. piscium [120–123], Vagococcus salmoninarum, and Carnobacterium piscicola [124], implicated in infectious diseases of warm water as well as cold water fishes.

Streptococcosis appears to have very few limitations in regard to geographic boundaries or host range, with outbreaks occurring in aquaculture facilities worldwide and in many different cultured species. S. iniae, S.parauberis, S. agalactiae, and L. garvieae are known as the major pathogens of streptococcosis and lactococcosis in Oncorhynchus mykiss, Seriola quinqueradiata, Siganus canaliculatus, and Tilapia spp. [125].Recently, S. iniae and L. garvieae are also recognized as emerging zoonotic pathogens, causing diseases in both fish and human beings [23, 126].

S. iniae is a β-haemolytic, Gram-positive coccus that causes generalized septicaemia and meningoencephalitis in a variety of warm water fishes [127], whereas S. parauberis is an α-hemolytic, Gram-positive coccus, mainly pathogenic in cultured turbot (Scophthalmus maximus) and olive flounder, Paralichthys olivaceus. L. garvieaecauses a hyperacute and haemorrhagic septicemia in fishes particularly during the summer time. General pathological symptoms of streptococcosis and lactococcosis in fishes are hemorrhage, congestion, lethargy, dark pigmentation, erratic swimming, and exophthalmos with clouding of the cornea [117, 128].

Complete genome sequences of different strains of S. parauberis and L. garvieae, important pathogenic species isolated from both fish and human, have been published [129–132].

S. parauberis is recognized as the dominant etiological agent of streptococcosis in fish [117], whereas both S. parauberis and S. uberis are involved the causation of bovine mastitis in dairy cow [133, 134].

S. parauberis is closer to S. uberis than with other Streptococcus spp. and is biochemically and serologically indistinguishable from S. uberis [135]. Both species were earlier considered as type I and II of S. uberis, but later shown to be phylogenetically distinct and renamed the type I as S. uberis and type II as S. parauberis [134].

The S. parauberis strain KCTC11537BP genome size falls in the middle of the 1.8 to 2.3 Mb range of streptococcal genomes sequenced to date and the average G+C content of 35.6% is significantly lower than those of S. pyogenes [132]. About 78% of genes are shared between the genomes of S. parauberis strain KCTC11537BP and S. uberis NC_012004, but they differ significantly at two regions of the genome, demonstrating the genomic basis for their separation into two species.

S. parauberis genome encodes an M-like protein of S. iniae (SiM), which is an important virulence factor in S. iniae [136]. It also encodes hasA and hasB genes that may be involved in capsule production for resistance against phagocytosis. The genome analysis indicates that S. parauberis could possibly possess the ability to regulate the metabolism of more carbohydrates than other Streptococcus species and to synthesize all the aminoacids and regulatory factors required to adapt and survive in a highly hostile host environment.

Complete genome sequences of L. garvieae strain UNIUD074, isolated from diseased rainbow trout in Italy, a virulent strain Lg2 (serotype KG2) and a nonvirulent strain ATCC 49156 (serotype KG+), both isolated from diseased yellowtail in Japan have recently been published [130, 131]. In addition, genome sequence of L. garvieae strain 21881, isolated from a man suffering from septicemia has been published [129].

The strains Lg2 and ATCC 49156 have 99% sequence identity and share 1944 orthologous genes, but are different in 24 Lg2-specific genes that were absent in the ATCC 49156 genome. One of the Lg2-specific genes is a 16.5 kb capsule gene cluster, which confirms the earlier transmission electron microscopic finding that Lg2 is encapsulated, and ATCC 49156 is nonencapsulated [137]. In fact, the capsule gene cluster has the features of a horizontally acquired genomic island conferring virulence to the Lg2 strain but might have been lost from the ATCC 49156 strain while subculturing in the laboratory [131]. Both genomes carried three types of IS elements, prophage sequences, and integrase genes and were found smaller than those of at least five sequenced L. lactis genomes. The Lg2 genome lacks several aminoacid biosynthesis genes, which is a characteristic feature of pathogenic bacteria with reduced genomes. The Lg2 strain contains hemolysins, NADH oxidase and superoxide dismutase (SOD), adhesins and sortase, which are known virulence factors [137–139]. It also encodes a gene for phosphoglucomutase, a virulence factor conferring the resistance to peptide antimicrobials in S. iniae [140].

Although L. garvieae and L. lactis genomes share 75% CDS, about 25% genes are Lg2-specific hypothetical proteins and proteins of unknown functions, which may be involved in the virulence of the Lg2 strain. These findings indicate that L. garvieae and L. lactis have significantly diverged from the common ancestor, and the L. garvieae is evolving into a pathogenic species equipped with virulence features suitable for living in the host environment.

10. MYCOBACTERIA

Chronic infections in fish caused by different species of mycobacteria have been well recognized [23, 141, 142]. Several slow growing as well as fast growing species of mycobacteria such as Mycobacterium marinum, M. fortuitum, M. chelonae, and M. avium have been isolated from wild and cultured fish suffering from mycobacteriosis in different parts of the world

[143–145]. Among them, M. marinum is the most important fish pathogen, frequently isolated from a variety of fish species with granulomas [146]. It is also a known zoonotic pathogen, transmitted to man though fish handling in aquariums and aquaculture tanks, producing superficial and self-limiting lesions called "fish tank or aquarium tank granuloma" involving the cooler parts of the body such as hands, forearms, elbows, and knees [147, 148]. Although strain variation has been reported [149], there is significant intraspecies sequence homogeneity among different M. mrinum strains [150]. However, it is hypothesized that only certain strains of M. marinum have zoonotic potential [151]. Phylogenetic studies have shown that M. marinum is most closely related to M. ulcerans followed by M. tuberculosis [150].Owing to this, M. marinum and M. tuberculosis share many virulence factors and significant pathological features and respond to similar antibiotics [152, 153]. Hence, M. marinum is also an important model organism to study the pathogenesis of tuberculosis [152, 153].

Interestingly, the genome of M. marinum is 50% bigger than that of M. tuberculosis and seems to have acquired a number of genes encoding NRPSs and the huge repertoire of PE, PPE, and ESX systems probably by HGT [154]. Both species might have evolved differently from a common environmental mycobacteria. M. tuberculosis might have adapted to its host intracellular life by extensive genome reduction and M. marinum, by and large retained or obtained genes required for its dual lifestyle and broad-host range.

11. GENOME SEQUENCING TO FIND NOVEL VACCINE AND DRUG TARGETS IN FISH PATHOGENS

Our understanding of the molecular basis of virulence of certain well-studied fish bacterial pathogens has increased dramatically during the past decade. This has resulted from the application of recombinant DNA technology and cell biology to investigate bacterial infections, and the development of genetic techniques for identifying virulence genes.

More recently, genome sequence information of several bacterial fish pathogens has become available from genome sequencing projects. There is strong reason to believe that this understanding will be exploited to develop new interventions against fish bacterial infections.

The relevance of sequencing projects for drug and vaccine discovery is obvious. During the "pregenomic" era, the vaccine candidate genes were individually identified by tedious gene knockout studies and virulence attenuation. But now, the complete genome sequencing provides information on every virulence gene and all potential vaccine candidates, and the sequence databases will become indispensable for research in fish vaccinology and drug development.

After sequencing, the open reading frames (ORFs) are searched against available databases for sequence similarity with genes of known functions in other organisms. There are several strategies for gene annotation employing the tools of predictive bioinformatics programs combined with analyses of the published literature.

Multiple target vaccine candidate genes can be chosen and deleted simultaneously by various strategies including global transposon mutagenesis and gene replacement techniques [155, 156] to study their effect on virulence and essentiality. A number of important virulence determinants identified in the sequenced genome can be targeted. For example, the sortase enzyme in Gram-positive fish pathogens would be a very attractive universal vaccine and therapeutic drug target, as it mediates covalent anchoring of many surface displayed antigenic and/or virulence related proteins in Gram-positive bacteria [139]. The inactivation or inhibition of the sortase enzyme can simultaneously prevent the surface display of a number of virulence factors, thus effectively attenuating the virulence of the pathogen [110, 157].

The availability of sequences of the complete surface antigenic repertoire of pathogens, including protein and noprotein antigens would facilitate strategies for rational design of vaccines and drugs. In addition, the recent availability of large collections of the "virulogenome" of fish bacterial pathogens will provide enormous virulence sequence information for DNA vaccination studies. The whole complement of IS elements, prophages, and pathogenicity islands that can harbor virulence, and antimicrobial resistance gene clusters can be easily identified in the genomes. The comparison of genomes of different strains of the same bacteria or closely related species can reveal how these strains or species behave differently while infecting fish hosts, thus opening exciting opportunities for functional genomic analysis of infection processes and pathogenesis. However, experimental validation of predicted functions of genes identified from sequencing projects has lagged far behind the speed of annotation, and the major challenge of researchers in the field today is to understand the functional framework of the sequenced genomes.

12. CONCLUSIONS

There has been a steady increase in the number of species of bacteria implicated in fish diseases. The common fish pathogenic bacterial species belong to the genera Vibrio, Aeromonas, Flavobacterium, Yersinia, Edwardsiella, Streptococcus, lactococcus, Renibacterium, and Mycobacterium [23]. However, there is growing indications that the pathogenic species spectrum as well as the geographic and host range is widening among fish pathogens [158–161], leading to the emergence of new pathogens. Unlike the situation in human and animal medicine, fish diseases pose unique and daunting challenges. Fish are always bathed in a continuous medium of water, and fish disease treatment is essentially a population medicine. In addition, the current treatment methods are largely ineffective, and the biology and genetics of most fish bacterial pathogens are poorly understood, limiting the application of modern science-based pathogen intervention strategies.

Rapid growth and expansion of genome sequencing of human and animal pathogens enabled better understanding of their biology, evolution, and host adaptation strategies, and helped in combating many major diseases. Unfortunately, such developments and progress in the genomics and functional genomics of fish pathogenic bacteria have been very slow. However, recent availability of cost-effective high-throughput sequencing technologies has set the pace of sequencing of more fish pathogenic bacteria. Genome sequencing of a number of important bacterial pathogens of fish has helped us to better understand their biology and genetics. The sequencing projects have unearthed exciting new information on the adaptive evolution of fish pathogens, for example, how the nonpathogenic and ubiquitous soil bacteria such as Arthrobacter sp. has evolved into a strictly obligate fish pathogen, R. salmoninarum, by shedding functional genes through genomic reduction to lead to a very cosy intracellular life style.

On the other hand, phenotypically similar strains of the same species differ in certain set of virulence gene clusters, acquired through HGT and become highly virulent. The capsule gene cluster in the L. garvieae Lg2 strain confers virulence compared to noncapsulated ATCC 49156, which lacks the gene cluster. Nonpathogenic strains acquire genomic islands from distantly related pathogenic species and emerge as new pathogens of fish.

Comparative pathogenomics of closely related bacteria has increased our knowledge of how they vary in their virulence and their ability to adapt to different ecological niches. This is clearly evident in the difference in virulence of various strains of V. anguillarum and V. vulnificus, and among the closely related species of the genus Flavobacterium. As more strain-specific sequence information on bacterial pathogens of fish becomes available, we will have a better understanding of the subtle genomic differences among strains with varying virulence characteristics.

The typical pathogen evolutionary strategy of acquiring, shuffling and shedding genes mediated by IS elements, pseudogenes, prophage sequences, and HGT is also observed in most bacterial pathogens of fish. It is certain that the new genomic information will bring paradigm changes in bacterial pathogenesis and should provide new perspectives to our current thinking on the evolutionary and adaptive strategies of aquatic bacteria and how they colonize and establish in wider ecological niches and new host species. Moreover, the identification of key virulence factors in pathogenic strains should help us design efficient drugs and vaccines to combat major bacterial pathogens of fish.

However, it should be stressed that the genomic information will provide only a snapshot of the microorganism. Highly virulent clones armed with one or more acquired virulence factors can suddenly develop from the existing harmless microorganisms in the face of environmental, antibiotic, and host-induced selective pressures.

More intriguingly, about 40% of the genes in sequenced bacterial genomes constitute new putative genes and hypothetical proteins with mysterious functions and are conserved among several different species of bacteria. Even in Escherichia coli, the most studied of all bacteria, only 54% genes have currently been functionally characterized based on experimental evidence [162]. A close scrutiny of the sequenced genomes of fish pathogens reveals that the above situation is essentially true for these pathogens as well. Although current advances in functional genomics, structural genomics and bioinformatics have contributed immensely to deciphering and extracting useful biological information from the vast genomic data, understanding and assigning functionality to the unique and new gene sequences discovered in the genomes will be the major task of genome biologists in the coming years.

ACKNOWLEDGMENTS

This work was financially supported by various research fundings from the Ministry of Agriculture and Fisheries Wealth, Oman.

REFERENCES

1. J. M. Rothberg and J. H. Leamon, "The development and impact of 454 sequencing," Nature Biotechnology, vol. 26, no. 10, pp. 1117–1124, 2008.
2. J. Zhang, R. Chiodini, A. Badr, and G. Zhang, "The impact of next-generation sequencing on genomics," Journal of Genetics and Genomics, vol. 38, no. 3, pp. 95–109, 2011.
3. E. Pennisi, "Genome data shake tree of life," Science, vol. 280, no. 5364, pp. 672–674, 1998.
4. W. F. Doolittle, "Phylogenetic classification and the universal tree," Science, vol. 284, no. 5423, pp. 2124–2128, 1999.
5. W. F. Doolittle, "Lateral genomics," Trends in Cell Biology, vol. 9, no. 12, pp. M5–M8, 1999.
6. E. Pennisi, "Is it time to uproot the tree of life?" Science, vol. 284, no. 5418, pp. 1305–1307, 1999.
7. W. F. Doolittle, "Uprooting the tree of life," Scientific American, vol. 282, no. 2, pp. 90–95, 2000.
8. E. V. Koonin, K. S. Makarova, and L. Aravind, "Horizontal gene transfer in prokaryotes: quantification and classification," Annual Review of Microbiology, vol. 55, pp. 709–742, 2001.
9. N. A. Moran, "Microbial minimalism: genome reduction in bacterial pathogens," Cell, vol. 108, no. 5, pp. 583–586, 2002.
10. Mira, L. Klasson, and S. G. E. Andersson, "Microbial genome evolution: sources of variability," Current Opinion in Microbiology, vol. 5, no. 5, pp. 506–512, 2002.
11. L. Holm, "Codon usage and gene expression," Nucleic Acids Research, vol. 14, no. 7, pp. 3075–3087, 1986.
12. L. S. Frost, R. Leplae, A. O. Summers, and A. Toussaint, "Mobile genetic elements: the agents of open source evolution," Nature Reviews Microbiology, vol. 3, no. 9, pp. 722–732, 2005.
13. U. Dobrindt, B. Hochhut, U. Hentschel, and J. Hacker, "Genomic islands in pathogenic and environmental microorganisms," Nature Reviews Microbiology, vol. 2, no. 5, pp. 414–424, 2004.
14. U. Dobrindt and J. Hacker, "How bacterial pathogens were constructed," in Bacterial Virulence: Basic Principles, Models and Global Approaches, P. Sansonetti, Ed., pp. 3–15, Wiley-VCH, GmbH & Co. KGaA, Weinheim, Germany, 2010.
15. M. J. Pallen and B. W. Wren, "Bacterial pathogenomics," Nature, vol. 449, no. 7164, pp. 835–842, 2007.
16. J. G. Lawrence and J. R. Roth, "Genomic flux: genome evolution by gene loss and acquisition," in Organization of the Prokaryotic Genome, R. L. Charlebois, Ed., pp. 263–289, American Society for Microbiology, Washington, DC, USA, 1999.
17. F. P. Meyer, "Aquaculture disease and health management.," Journal of Animal Science, vol. 69, no. 10, pp. 4201–4208, 1991.
18. C. D. Harvell, K. Kim, J. M. Burkholder et al., "Emerging marine diseases—climate links and anthropogenic factors," Science, vol. 285, no. 5433, pp. 1505–1510, 1999.
19. R. Subasinghe, "Fish health and quarantine; review of the state of the World aquaculture," in FAO Fisheries Circular, no. 886, pp. 45–49, Rome, Italy, Food and Agriculture Organization of the United Nations, 1997.
20. B. Hjeltnes and R. J. Roberts, "Vibriosis," in Bacterial Diseases of Fish, R. J. Roberts, N. R. Bromag, and V. Inglis, Eds., pp. 109–121, Blackwell Scientific, Oxford, UK, 1993.
21. D. V. Lightner and R. M. Redman, "Shrimp diseases and current diagnostic methods," Aquaculture, vol. 164, no. 1–4, pp. 201–220, 1998.
22. O. Bergh, F. Nilsen, and O. B. Samuelsen, "Diseases, prophylaxis and treatment of the Atlantic halibut Hippoglossus hippoglossus: a review," Diseases of Aquatic Organisms, vol. 48, no. 1, pp. 57–74, 2001.

23. B. Austin and D. A. Austin, Bacterial Fish Pathogens: Disease in Farmed and Wild Fish, Springer, New York, NY, USA, 3rd edition, 1999.
24. B. Austin and X. H. Zhang, "Vibrio harveyi: a significant pathogen of marine vertebrates and invertebrates," Letters in Applied Microbiology, vol. 43, no. 2, pp. 119–124, 2006.
25. C. Amaro, E. G. Biosca, C. Esteve et al., "Comparative study of phenotypic and virulence properties in Vibrio vulnificus biotypes 1 and 2 obtained from a European eel farm experiencing mortalities," Diseases of Aquatic Organisms, vol. 13, pp. 29–35, 1992.
26. E. G. Biosca, H. Llorens, E. Garay, and C. Amaro, "Presence of a capsule in Vibrio vulnificus biotype 2 and its relationship to virulence for eels," Infection and Immunity, vol. 61, no. 5, pp. 1611–1618, 1993.
27. E. Benediktsdóttir, L. Verdonck, C. Spröer, S. Helgason, and J. Swings, "Characterization of Vibrio viscosus and Vibrio wodanis isolated at different geographical locations: a proposal for reclassification of Vibrio viscosus as Moritella viscosa comb. nov.," International Journal of Systematic and Evolutionary Microbiology, vol. 50, no. 2, pp. 479–488, 2000.
28. L. A. Actis, M. E. Tolmasky, and J. H. Crosa, "Vibriosis," in Fish Diseases and Disorders, Viral, Bacterial, and Fungal Infections, P. T. K. Woo and D. W. Bruno, Eds., vol. 3, pp. 570–605, CABI International, Oxfordshire, UK, 2nd edition, 2011.
29. E. Hjerde, M. Lorentzen, M. T. G. Holden et al., "The genome sequence of the fish pathogen Aliivibrio salmonicida strain LFI1238 shows extensive evidence of gene decay," BMC Genomics, vol. 9, article 616, 2008.
30. P. A. Gulig, V. D. Crécy-Lagard, A. C. Wright, B. Walts, M. Telonis-Scott, and L. M. McIntyre, "SOLiD sequencing of four Vibrio vulnificus genomes enables comparative genomic analysis and identification of candidate clade-specific virulence genes," BMC Genomics, vol. 11, no. 1, article 512, 2010.
31. H. Naka, G. M. Dias, C. C. Thompson, C. Dubay, F. L. Thompson, and J. H. Crosa, "Complete genome sequence of the marine fish pathogen Vibrio anguillarum harboring the pJM1 virulence plasmid and genomic comparison with other virulent strains of V. anguillarum and V. ordalii," Infection and Immunity, vol. 79, no. 7, pp. 2889–2900, 2011.
32. J. L. Larsen, K. Pedersen, and I. Dalsgaard, "Vibrio anguillarum serovars associated with vibriosis in fish," Journal of Fish Diseases, vol. 17, no. 3, pp. 259–267, 1994.
33. U. B. S. Sorensen and J. L. Larsen, "Serotyping of Vibrio anguillarum," Applied and Environmental Microbiology, vol. 51, no. 3, pp. 593–597, 1986.
34. E. Toranzo and J. L. Barja, "A review of the taxonomy and seroepizootiology of Vibrio anguillarum, with special reference to aquaculture in the NorthWest Spain," Diseases of Aquatic Organisms, vol. 9, pp. 73–82, 1990.
35. J. H. Crosa, "A plasmid associated with virulence in the marine fish pathogen Vibrio anguillarumspecifies an iron-sequestering system," Nature, vol. 284, no. 5756, pp. 566–568, 1980.
36. E. Toranzo, J. L. Barja, S. A. Potter et al., "Molecular factors associated with virulence of marine vibrios isolated from striped bass in Chesapeake Bay," Infection and Immunity, vol. 39, no. 3, pp. 1220–1227, 1983.
37. M. H. Schiewe, T. J. Trust, and J. H. Crosa, "Vibrio ordalii sp. nov.: a causative agent of vibriosis in fish," Current Microbiology, vol. 6, no. 6, pp. 343–348, 1981.
38. D. P. Ransom, C. N. Lannan, J. S. Rohovec, and J. L. Fryer, "Comparison of histopathology caused by Vibrio anguillarum and Vibrio ordalii in three species of Pacific salmon," Journal of Fish Diseases, vol. 7, pp. 107–115, 1984.
39. K. Makino, K. Oshima, K. Kurokawa et al., "Genome sequence of Vibrio parahaemolyticus: a pathogenic mechanism distinct from that of V. cholerae," The Lancet, vol. 361, no. 9359, pp. 743–749, 2003.
40. N. Okada, S. Matsuda, J. Matsuyama et al., "Presence of genes for type III secretion system 2 in Vibrio mimicus strains," BMC Microbiology, vol. 10, article 302, 2010.

41. Alam, K. A. Miller, M. Chaand, J. S. Butler, and M. Dziejman, “Identification of Vibrio cholerae type III secretion system effector proteins,” Infection and Immunity, vol. 79, no. 4, pp. 1728–1740, 2011.
42. R. Osorio, S. Juiz-Rio, and M. L. Lemos, “A siderophore biosynthesis gene cluster from the fish pathogen Photobacterium damselae subsp. piscicida is structurally and functionally related to the Yersiniahigh-pathogenicity island,” Microbiology, vol. 152, no. 11, pp. 3327–3341, 2006.
43. E. S. Yip, B. T. Grublesky, E. A. Hussa, and K. L. Visick, “A novel, conserved cluster of genes promotes symbiotic colonization and σ-dependent biofilm formation by Vibrio fischeri,” Molecular Microbiology, vol. 57, no. 5, pp. 1485–1498, 2005.
44. H. S. Kim, S. J. Park, and K. H. Lee, “Role of NtrC-regulated exopolysaccharides in the biofilm formation and pathogenic interaction of Vibrio vulnificus,” Molecular Microbiology, vol. 74, no. 2, pp. 436–453, 2009.
45. E. Sanjuan, B. Fouz, J. D. Oliver, and C. Amaro, “Evaluation of genotypic and phenotypic methods to distinguish clinical from environmental Vibrio vulnificus strains,” Applied and Environmental Microbiology, vol. 75, no. 6, pp. 1604–1613, 2009.
46. N. Bisharat, D. I. Cohen, M. C. Maiden, D. W. Crook, T. Peto, and R. M. Harding, “The evolution of genetic structure in the marine pathogen, Vibrio vulnificus,” Infection, Genetics and Evolution, vol. 7, no. 6, pp. 685–693, 2007.
47. M. B. Schrøder, S. Espelid, and T. Ø. Jørgensen, “Two serotypes of Vibrio salmonicida isolated from diseased cod (Gadus morhua L.); virulence, immunological studies and vaccination experiments,” Fish and Shellfish Immunology, vol. 2, no. 3, pp. 211–221, 1992.
48. M. Tomich, P. J. Planet, and D. H. Figurski, “The tad locus: postcards from the widespread colonization island,” Nature Reviews Microbiology, vol. 5, no. 5, pp. 363–375, 2007.
49. P. Siguier, J. Filee, and M. Chandler, “Insertion sequences in prokaryotic genomes,” Current Opinion in Microbiology, vol. 9, no. 5, pp. 526–531, 2006.
50. J. Parkhill, G. Dougan, K. D. James et al., “Complete genome sequence of a multiple drug resistant Salmonella enterica serovar Typhi CT18,” Nature, vol. 413, no. 6858, pp. 848–852, 2001.
51. G. D. Wiens, D. D. Rockey, Z. Wu et al., “Genome sequence of the fish pathogen Renibacterium salmoninarum suggests reductive evolution away from an environmental Arthrobacter ancestor,” Journal of Bacteriology, vol. 190, no. 21, pp. 6970–6982, 2008.
52. Martin-Carnahan and S. W. Joseph, “Aeromonadaceae,” in Bergey's Manual of Systematic Bacteriology, G. M. Garrity, Ed., vol. 2, pp. 556–580, Springer, New York, NY, USA, 2nd edition, 2005.
53. M. J. Figueras, “Clinical relevance of Aeromonas sM503,” Reviews in Medical Microbiology, vol. 16, no. 4, pp. 145–153, 2005.
54. R. Seshadri, S. W. Joseph, A. K. Chopra et al., “Genome sequence of Aeromonas hydrophila ATCC 7966T: jack of all trades,” Journal of Bacteriology, vol. 188, no. 23, pp. 8272–8282, 2006.
55. M. E. Reith, R. K. Singh, B. Curtis et al., “The genome of Aeromonas salmonicida subsp. salmonicidaA449: Insights into the evolution of a fish pathogen,” BMC Genomics, vol. 9, article 427, 2008.
56. Y. Li, Y. Liu, Z. Zhou et al., “Complete genome sequence of Aeromonas veronii strain B565,” Journal of Bacteriology, vol. 193, no. 13, pp. 3389–3390, 2011.
57. S. A. Beatson, M. D. G. de Luna, N. L. Bachmann et al., “Genome sequence of the emerging pathogen Aeromonas caviae,” Journal of Bacteriology, vol. 193, no. 5, pp. 1286–1287, 2011.
58. E. M. Bernoth, “Furunculosis: the history of the diseases and of disease research,” in Furunculosis, Multidisciplinary Fish Disease Research, E. M. Bernoth, A. E. Ellis, P. J. Midtlyng, G. Olivier, and P. Smith, Eds., pp. 1–20, Academic Press, London, UK, 1997.

59. M. Hiney and G. Olivier, "Furunculosis (Aeromonas salmonicida)," in Fish Diseases and Disorders III: Viral, Bacterial and Fungal Infections, P. T. K. Woo and D. W. Bruno, Eds., pp. 341–425, CAB Publishing, Oxford, UK, 1999.
60. C. Paniagua, O. Rivero, J. Anguita, and G. Naharro, "Pathogenicity factors and virulence for rainbow trout (Salmo gairdneri) of motile Aeromonas spp. isolated from a river," Journal of Clinical Microbiology, vol. 28, no. 2, pp. 350–355, 1990.
61. S. E. Burr, D. Pugovkin, T. Wahli, H. Segner, and J. Frey, "Attenuated virulence of an Aeromonas salmonicida subsp. salmonicida type III secretion mutant in a rainbow trout model," Microbiology, vol. 151, no. 6, pp. 2111–2118, 2005.
62. J. Sha, L. Pillai, A. A. Fadl, C. L. Galindo, T. E. Erova, and A. K. Chopra, "The type III secretion system and cytotoxic enterotoxin alter the virulence of Aeromonas hydrophila," Infection and Immunity, vol. 73, no. 10, pp. 6446–6457, 2005.
63. J. M. Janda and R. P. Kokka, "The pathogenicity of Aeromonas strains relative to genospecies and phenospecies identification," FEMS Microbiology Letters, vol. 90, no. 1, pp. 29–33, 1991.
64. J. Madetoja, I. Dalsgaard, and T. Wiklund, "Occurrence of Flavobacterium psychrophilum in fish-farming environments," Diseases of Aquatic Organisms, vol. 52, no. 2, pp. 109–118, 2002. View at Google Scholar ·View at Scopus
65. Nematollahi, A. Decostere, F. Pasmans, and F. Haesebrouck, "Flavobacterium psychrophilum infections in salmonid fish," Journal of Fish Diseases, vol. 26, no. 10, pp. 563–574, 2003.
66. J. G. Leahy and R. R. Colwell, "Microbial degradation of hydrocarbons in the environment," Microbiological Reviews, vol. 54, no. 3, pp. 305–315, 1990.
67. J. Lehmann, D. Mock, F. J. Stuerenberg, and J. F. Bernardet, "First isolation of Cytophaga psychrophilafrom a systemic disease in eel and cyprinids," Diseases of Aquatic Organisms, vol. 10, pp. 217–220, 1991.
68. R. A. Holt, Cytophaga psychrophila, the causative agent of bacterial cold-water disease in salmonid fish, Ph.D. thesis, Oregon State University, Corvallis, Ore, USA, 1987.
69. G. J. Heo, K. Kasai, and H. Wakabayashi, "Occurrence of Flavobacterium branchiophila associated with bacterial gill disease at a trout hatchery," Fish Pathology, vol. 25, pp. 99–105, 1990.
70. V. E. Ostland, J. S. Lumsden, D. D. MacPhee, and H. W. Ferguson, "Characteristics of Flavobacterium branchiophilum, the cause of salmonid bacterial gill disease in Ontario," Journal of Aquatic Animal Health, vol. 6, no. 1, pp. 13–26, 1994.
71. J. F. Bernardet, "Flexibacter columnaris: first description in France and comparison with bacterial strains from other origins," Diseases of Aquatic Organisms, vol. 6, pp. 37–44, 1989.
72. D. J. Speare and H. W. Ferguson, "Clinical and pathological features of common gill diseases of cultured salmonids in Ontario," Canadian Veterinary Journal, vol. 30, pp. 882–887, 1989.
73. J. F. Bernardet, P. Segers, M. Vancanneyt, F. Berthe, K. Kersters, and P. Vandamme, "Cutting a gordian knot: emended classification and description of the genus Flavobacterium, emended description of the family Flavobacteriaceae, and proposal of Flavobacterium hydatis nom. nov. (basonym, Cytophaga aquatilis Strohl and Tait 1978)," International Journal of Systematic Bacteriology, vol. 46, no. 1, pp. 128–148, 1996.
74. E. Duchaud, M. Boussaha, V. Loux et al., "Complete genome sequence of the fish pathogen Flavobacterium psychrophilum," Nature Biotechnology, vol. 25, no. 7, pp. 763–769, 2007.
75. M. Touchon, P. Barbier, J. F. Bernardet et al., "Complete genome sequence of the fish pathogen Flavobacterium branchiophilum," Applied and Environmentl Microbiology, vol. 77, no. 21, pp. 7656–7662, 2011.
76. M. L. Lawrence, A. Karsi, H. C. Tekedar et al., "Comparative genomics of Flavobacterium columnareisolates from two genetic divisions and with different

pathogenic potential for channel catfish (Ictalurus punctatus)," in Proceedings of the Conference on FLAVOBACTERIUM, Paris, France, September 2009.

77. C. Rodkhum, I. Hirono, J. H. Crosa, and T. Aoki, "Four novel hemolysin genes of Vibrio anguillarum and their virulence to rainbow trout," Microbial Pathogenesis, vol. 39, no. 4, pp. 109–119, 2005.

78. M. Palmer, "The family of thiol-activated, cholesterol-binding cytolysins," Toxicon, vol. 39, no. 11, pp. 1681–1689, 2001.

79. M. Shoji, D. B. Ratnayake, Y. Shi et al., "Construction and characterization of a nonpigmented mutant of Porphyromonas gingivalis: Cell surface polysaccharide as an anchorage for gingipanis," Microbiology, vol. 148, no. 4, pp. 1183–1191, 2002.

80. K. Sato, E. Sakai, P. D. Veith et al., "Identification of a new membrane-associated protein that influences transport/maturation of gingipains and adhesins of Porphyromonas gingivalis," The Journal of Biological Chemistry, vol. 280, no. 10, pp. 8668–8677, 2005.

81. S. L. Abbott and J. M. Janda, "The genus Edwardsiella," Prokaryotes, vol. 6, pp. 72–89, 2006.

82. J. E. Clarridge, D. M. Musher, V. Fanstein, and R. J. Wallace Jr., "Extra-intestinal human infection caused by Edwardsiella tarda," Journal of Clinical Microbiology, vol. 11, no. 5, pp. 511–514, 1980.

83. P. Nagel, A. Serritella, and T. J. Layden, "Edwardsiella tarda gastroenteritis associated with a pet turtle," Gastroenterology, vol. 82, no. 6, pp. 1436–1437, 1982.

84. W. A. Rogers, "Edwardsiellosis in fishes," in Antigens of Fish Pathogens. Les Antigenes des Microorganisms Pathogenes des Poissons, D. P. Anderson, M. Dorson, and Ph. Dubourget, Eds., pp. 153–159, Fondation Marcel Merieux, Lyon, France, 1983.

85. F. P. Meyer and G. L. Bullock, "Edwardsiella tarda, a new pathogen of channel catfish (Ictalurus punctatus)," Applied Microbiology, vol. 25, no. 1, pp. 155–156, 1973.

86. S. H. M. Ling, X. H. Wang, L. Xie, T. M. Lim, and K. Y. Leung, "Use of green fluorescent protein (GFP) to study the invasion pathways of Edwardsiella tarda in in vivo and in vitro fish models," Microbiology, vol. 146, no. 1, pp. 7–19, 2000.

87. J. C. Newton, L. G. Wolfe, J. M. Grizzle, and J. A. Plumb, "Pathology of experimental enteric septicaemia in channel catfish, Ictalurus punctatus (Rafinesque), following immersion-exposure to Edwardsiella ictaluri," Journal of Fish Diseases, vol. 12, no. 4, pp. 335–347, 1989.

88. Q. Wang, M. Yang, J. Xiao et al., "Genome sequence of the versatile fish pathogen Edwardsiella tardaprovides insights into its adaptation to broad host ranges and intracellular niches," PLoS One, vol. 4, no. 10, Article ID e7646, 2009.

89. M. L. Williams, A. F. Gillaspy, D. W. Dyer et al., "Genome sequence of Edwardsiella ictaluri 93–146, a strain associated with a natural channel catfish outbreak of enteric septicemia of catfish," Journal of Bacteriology, vol. 194, no. 3, pp. 740–741, 2012.

90. S. R. Partridge and R. M. Hall, "The IS1111 family members IS4321 and IS5075 have sub-terminal inverted repeats and target the terminal inverted repeats of Tn21 family transposons," Journal of Bacteriology, vol. 185, no. 21, pp. 6371–6384, 2003.

91. T. Tanabe, T. Funahashi, H. Nakao, S. I. Miyoshi, S. Shinoda, and S. Yamamoto, "Identification and characterization of genes required for biosynthesis and transport of the siderophore vibrioferrin in Vibrio parahaemolyticus," Journal of Bacteriology, vol. 185, no. 23, pp. 6938–6949, 2003.

92. Q. Wang, Q. Liu, Y. Ma, L. Zhou, and Y. Zhang, "Isolation, sequencing and characterization of cluster genes involved in the biosynthesis and utilization of the siderophore of marine fish pathogen Vibrio alginolyticus," Archives of Microbiology, vol. 188, no. 4, pp. 433–439, 2007.

93. M. D. Furones, C. J. Rodgers, and C. B. Munn, "Yersinia ruckeri, the causal agent of enteric redmouth disease (ERM) in fish," Annual Review of Fish Diseases, vol. 3, no. C, pp. 105–125, 1993.

94. R. R. Rucker, "Redmouth disease of rainbow trout (Salmo gairdneri)," Bulletin de l'Office International des Epizooties, vol. 65, no. 5, pp. 825–830, 1966.
95. E. Tobback, A. Decostere, K. Hermans, F. Haesebrouck, and K. Chiers, "Yersinia ruckeri infections in salmonid fish," Journal of Fish Diseases, vol. 30, no. 5, pp. 257–268, 2007.
96. M. T. Horne and A. C. Barnes, "Enteric redmouth disease (Y. ruckeri)," in Fish Diseases and Disorders: Viral, Bacterial and Fungal Infections, P. T. K. Woo and D. W. Bruno, Eds., vol. 3, pp. 455–477, CABI Publishing, Oxfordshire, UK, 1999.
97. J. Ross, R. R. Rucker, and W. H. Ewing, "Description of a bacterium associated with redmouth disease of rainbow trout (Salmo gairdneri)," Canadian Journal of Microbiology, vol. 12, no. 4, pp. 763–770, 1966.
98. R. L. Davies, "Virulence and serum-resistance in different clonal groups and serotypes of Yersinia ruckeri," Veterinary Microbiology, vol. 29, no. 3-4, pp. 289–297, 1991.
99. R. M. V. Stevenson and D. W. Airdrie, "Serological variation among Yersinia ruckeri strains," Journal of Fish Diseases, vol. 7, pp. 247–254, 1984.
100. J. L. Romalde, B. Magarinos, J. L. Barja, and A. E. Toranzo, "Antigenic and molecular characterization of Yersinia ruckeri proposal for a new intraspecies classification," Systematic and Applied Microbiology, vol. 16, no. 3, pp. 411–419, 1993.
101. R. M. Stevenson, "Immunization with bacterial antigens: yersiniosis," Developments in Biological Standardization, vol. 90, pp. 117–124, 1997.
102. R. L. Davies, "Yersinia ruckeri produces four iron-regulated proteins but does not produce detectable siderophores," Journal of Fish Diseases, vol. 14, pp. 563–570, 1991.
103. M. Kotetishvili, A. Kreger, G. Wauters, J. G. Morris, A. Sulakvelidze, and O. C. Stine, "Multilocus sequence typing for studying genetic relationships among Yersinia species," Journal of Clinical Microbiology, vol. 43, no. 6, pp. 2674–2684, 2005.
104. P. E. Chen, C. Cook, A. C. Stewart et al., "Genomic characterization of the Yersinia genus," Genome Biology, vol. 11, no. 1, article R1, 2010.
105. J. E. Sanders and J. L. Fryer, "Renibacterium salmoninarum gen. nov., sp. nov., the causative agent of bacterial kidney disease in salmonid fishes," International Journal of Systematic Bacteriology, vol. 30, no. 2, pp. 496–502, 1980.
106. D. L. Mitchum and L. E. Sherman, "Transmission of bacterial kidney disease from wild to stocked hatchery trout," Canadian Journal of Fisheries and Aquatic Sciences, vol. 38, pp. 547–551, 1981.
107. G. L. Bullock, Bacterial Kidney Disease of Salmonid Fishes Caused by Renibacterium salmoninarum, no. 60, U.S. Fish and Wildlife Service, Fish Disease Leaflet, Washington, DC, USA, 1980.
108. K. E. Wolf and C. E. Dunbar, "Test of 34 therapeutic agents for control of kidney disease in trout," Transactions of American Fisheries Society, vol. 88, pp. 117–124, 1959.
109. S. K. Gutenberger, J. R. Duimstra, J. S. Rohovec, and J. L. Fryer, "Intracellular survival of Renibacterium salmoninarum in trout mononuclear phagocytes," Disease of Aquatic Organisms, vol. 28, no. 2, pp. 93–106, 1997.
110. P. S. Sudheesh, S. Crane, K. D. Cain, and M. S. Strom, "Sortase inhibitor phenyl vinyl sulfone inhibits Renibacterium salmoninarum adherence and invasion of host cells," Diseases of Aquatic Organisms, vol. 78, no. 2, pp. 115–127, 2007.
111. E. Stackebrandt, U. Wehmeyer, H. Nader, and F. Fiedler, "Phylogenetic relationship of the fish pathogenic Renibacterium salmoninarum to Arthrobacter, Micrococcus and related taxa," FEMS Microbiology Letters, vol. 50, no. 2-3, pp. 117–120, 1988.
112. K. Salonius, C. Siderakis, A. M. MacKinnon, and S. G. Griffiths, "Use of Arthrobacter davidanieli as a live vaccine against Renibacterium

salmoninarum and Piscirickettsia salmonis in salmonids," Developments in Biologicals, vol. 121, pp. 189–197, 2005.
113. E. F. Mongodin, N. Shapir, S. C. Daugherty et al., "Secrets of soil survival revealed by the genome sequence of Arthrobacter aurescens TC1," PLoS Genetics, vol. 2, no. 12, article e214, 2006.
114. R. G. Getchell, J. S. Rohovec, and J. L. Fryer, "Comparison of Renibacterium salmoninarum isolates by antigenic analysis," Fish Pathology, vol. 20, pp. 149–159, 1985.
115. F. Fiedler and R. Draxl, "Biochemical and immunochemical properties of the cell surface of Renibacterium salmoninarum," Journal of Bacteriology, vol. 168, no. 2, pp. 799–804, 1986.
116. T. Kitao, "Streptococcal infections," in Bacterial Disease of Fish, V. Inglis, R. J. Roberts, and N. R. Bromage, Eds., pp. 196–210, Blackwell Scientific, Oxford, UK, 1993.
117. Domenech, J. F. Fernandez-Garayzabal, C. Pasqual et al., "Streptococcosis in cultured turbot, Scophthalmus maximus (L.), associated with Streptococcus parauberis," Journal of Fish Diseases, vol. 19, pp. 33–38, 1996.
118. J. L. Romalde, C. Ravelo, I. Valdés et al., "Streptococcus phocae, an emerging pathogen for salmonid culture," Veterinary Microbiology, vol. 130, no. 1-2, pp. 198–207, 2008.
119. D. Vendrell, J. L. Balcázar, I. Ruiz-Zarzuela, I. de Blas, O. Gironés, and J. L. Múzquiz, "Lactococcus garvieae in fish: a review," Comparative Immunology, Microbiology and Infectious Diseases, vol. 29, no. 4, pp. 177–198, 2006.
120. M. D. Collins, J. A. E. Farrow, B. A. Phillips, and O. Kandler, "Streptococcus garvieae sp. nov. and Streptococcus plantarum sp. nov.," Journal of General Microbiology, vol. 129, no. 11, pp. 3427–3431, 1983.
121. M. Williams, J. L. Fryer, and M. D. Collins, "Lactococcus piscium sp. nov. a new Lactococcus species from salmonid fish," FEMS Microbiology Letters, vol. 68, no. 1-2, pp. 109–113, 1990.
122. Domenech, J. Prieta, J. F. Fernandez-Garayzabal et al., "Phenotypic and phylogenetic evidence for a close relationship between Lactococcus garvieae and Enterococcus seriolicida," Microbiologia SEM, vol. 9, no. 1, pp. 63–68, 1993.
123. Eldar, C. Ghittino, L. Asanta et al., "Enterococcus seriolicida is a junior synonym of Lactococcus garvieae, a causative agent of septicemia and meningoencephalitis in fish," Current Microbiology, vol. 32, no. 2, pp. 85–88, 1996.
124. S. Wallbanks, A. J. Martinez-Murcia, J. L. Fryer, B. A. Phillips, and M. D. Collins, "16S rRNA sequence determination for members of the genus Carnobacterium and related lactic acid bacteria and description of Vagococcus salmoninarum sp. nov.," International Journal of Systematic Bacteriology, vol. 40, no. 3, pp. 224–230, 1990.
125. E. S. Bromage and L. Owens, "Infection of barramundi Lates calcarifer with Streptococcus iniae: effects of different routes of exposure," Diseases of Aquatic Organisms, vol. 52, no. 3, pp. 199–205, 2002.
126. J. A. Elliott, M. D. Collins, N. E. Pigott, and R. R. Facklam, "Differentiation of Lactococcus lactis and Lactococcus garvieae from humans by comparison of whole-cell protein patterns," Journal of Clinical Microbiology, vol. 29, no. 12, pp. 2731–2734, 1991.
127. R. P. Perera, S. K. Johnson, M. D. Collins, and D. H. Lewis, "Streptococcus iniae associated with mortality of Tilapia nilotica X T.aurea hybrids," Journal of Aquatic Animal Health, vol. 6, no. 4, pp. 335–340, 1994.
128. R. Kusuda, K. Kawai, F. Salati, C. R. Banner, and J. L. Fryer, "Enterococcus seriolicida sp. nov., a fish pathogen," International Journal of Systematic Bacteriology, vol. 41, no. 3, pp. 406–409, 1991.
129. M. Aguado-Urda, G. H. López-Campos, M. M. Blanco et al., "Genome sequence of Lactococcus garvieae21881, isolated in a case of human septicemia," Journal of Bacteriology, vol. 193, no. 15, pp. 4033–4034, 2011.

130. P. Reimundo, M. Pignatelli, L. D. Alcaraz, G. D'Auria, A. Moya, and J. A. Guijarro, "Genome sequence of Lactococcus garvieae UNIUD074, isolated in Italy from a lactococcosis outbreak," Journal of Bacteriology, vol. 193, no. 14, pp. 3684–3685, 2011.
131. H. Morita, H. Toh, K. Oshima et al., "Complete genome sequence and comparative analysis of the fish pathogen Lactococcus garvieae," PLoS One, vol. 6, no. 8, Article ID e23184, 2011.
132. S. W. Nho, J. I. Hikima, I. S. Cha et al., "Complete genome sequence and immunoproteomic analyses of the bacterial fish pathogen Streptococcus parauberis," Journal of Bacteriology, vol. 193, no. 13, pp. 3356–3366, 2011.
133. J. L. Watts, "Etiological agents of bovine mastitis," Veterinary Microbiology, vol. 16, no. 1, pp. 41–66, 1988.
134. M. Williams and M. D. Collins, "Molecular taxonomic studies on Streptococcus uberis types I and II. Description of Streptococcus parauberis sp. nov.," Journal of Applied Bacteriology, vol. 68, no. 5, pp. 485–490, 1990.
135. R. W. Bentley, J. A. Leigh, and M. D. Collins, "Intrageneric structure of Streptococcus based on comparative analysis of small-subunit rRNA sequences," International Journal of Systematic Bacteriology, vol. 41, no. 4, pp. 487–494, 1991.
136. J. B. Locke, R. K. Aziz, M. R. Vicknair, V. Nizet, and J. T. Buchanan, "Streptococcus iniae M-like protein contributes to virulence in fish and is a target for live attenuated vaccine development," PLoS One, vol. 3, no. 7, Article ID e2824, 2008.
137. M. Kawanishi, T. Yoshida, M. Kijima et al., "Characterization of Lactococcus garvieae isolated from radish and broccoli sprouts that exhibited a KG+ phenotype, lack of virulence and absence of a capsule," Letters in Applied Microbiology, vol. 44, no. 5, pp. 481–487, 2007.
138. T. J. Mitchell, "The pathogenesis of streptococcal infections: from tooth decay to meningitis," Nature Reviews. Microbiology, vol. 1, no. 3, pp. 219–230, 2003.
139. L. A. Marraffini, A. C. Dedent, and O. Schneewind, "Sortases and the art of anchoring proteins to the envelopes of gram-positive bacteria," Microbiology and Molecular Biology Reviews, vol. 70, no. 1, pp. 192–221, 2006.
140. J. T. Buchanan, J. A. Stannard, X. Lauth et al., "Streptococcus iniaephosphoglucomutase is a virulence factor and a target for vaccine development," Infection and Immunity, vol. 73, no. 10, pp. 6935–6944, 2005.
141. G. N. Frerichs, "Acid-fast fish pathogens," in Bacterial Diseases of Fish, V. Inglis, R. J. Roberts, and N. R. Bromage, Eds., pp. 217–233, Blackwell Scientific, Oxford, UK, 1993.
142. E. J. Noga, Fish disease diagnosis and treatment, Mosby, St. Louis, Mo, USA, 1996.
143. L. G. Wayne, G. P. Kubica et al., "Genus Mycobacterium Lehmann and Neumann 1896, 363AL," in Bergey's Manual of Systematic Bacteriology, P. H. A. Sneath, N. S. Mair, M. E. Sharpe, and J. G. Holt, Eds., vol. 2, pp. 1436–1457, Williams & Wilkins, Baltimore, Md, USA, 1986.
144. O. Thoen and T. A. Schliesser, "Mycobacterial infections in cold-blooded animals," in The Mycobacteria, G. P. Kubica and L. G. Wayne, Eds., part B, p. 1471, 1st edition, 1984.
145. G. E. Sanders and L. E. Swaim, "Atypical piscine Mycobacteriosis in Japanese medaka (Oryzias latipes)," Comparative Medicine, vol. 51, no. 2, pp. 171–175, 2001.
146. M. K. Stoskopf, Fish Medicine, WB Saunders, Philadelphia, Pa, USA, 1993.
147. T. Smith and H. P. Willett, "Other Mycobacterium species," in Zinsser Microbiology, K. Joklik, H. P. Willet, and D. B. Amos, Eds., pp. 674–698, Appleton-Century-Crofts, New York, NY, USA, 17th edition, 1980.
148. Huminer, S. D. Pitlik, C. Block et al., "Aquarium-borne Mycobacterium marinum skin infection. Report of a case and review of the literature," Archives of Dermatology, vol. 122, no. 6, pp. 698–703, 1986.

149. M. Ucko, A. Colorni, H. Kvitt, A. Diamant, A. Zlotkin, and W. R. Knibb, "Strain variation in Mycobacterium marinum fish isolates," Applied and Environmental Microbiology, vol. 68, no. 11, pp. 5281–5287, 2002.

150. M. J. Yip, J. L. Porter, J. A. M. Fyfe et al., "Evolution of Mycobacterium ulcerans and other mycolactone-producing mycobacteria from a common Mycobacterium marinum progenitor," Journal of Bacteriology, vol. 189, no. 5, pp. 2021–2029, 2007.

151. M. Ucko and A. Colorni, "Mycobacterium marinum infections in fish and humans in Israel," Journal of Clinical Microbiology, vol. 43, no. 2, pp. 892–895, 2005.

152. H. E. Volkman, H. Clay, D. Beery, J. C. W. Chang, D. R. Sherman, and L. Ramakrishnan, "Tuberculous granuloma formation is enhanced by a Mycobacterium virulence determinant," PLoS Biology, vol. 2, no. 11, article e367, 2004.

153. L. E. Swaim, L. E. Connolly, H. E. Volkman, O. Humbert, D. E. Born, and L. Ramakrishnan, "Mycobacterium marinum infection of adult zebrafish causes caseating granulomatous tuberculosis and is moderated by adaptive immunity," Infection and Immunity, vol. 74, no. 11, pp. 6108–6117, 2006.

154. T. P. Stinear, T. Seemann, P. F. Harrison et al., "Insights from the complete genome sequence of Mycobacterium marinum on the evolution of Mycobacterium tuberculosis," Genome Research, vol. 18, no. 5, pp. 729–741, 2008.

155. J. McFadden, "Recombination in mycobacteria," Molecular Microbiology, vol. 21, no. 2, pp. 205–211, 1996.

156. N. R. Salama, B. Shepherd, and S. Falkow, "Global transposon mutagenesis and essential gene analysis of Helicobacter pylori," Journal of Bacteriology, vol. 186, no. 23, pp. 7926–7935, 2004.

157. P. Cossart and R. Jonquières, "Sortase, a universal target for therapeutic agents against Gram-positive bacteria?" Proceedings of the National Academy of Sciences of the United States of America, vol. 97, no. 10, pp. 5013–5015, 2000.

158. M. Maher, R. Palmer, F. Gannon, and T. Smith, "Relationship of a novel bacterial fish pathogen to Streptobacillus monilifomris and the fusobacteria group, based on 16S ribosomal RNA analysis," Systematic and Applied Microbiology, vol. 18, no. 1, pp. 79–84, 1995.

159. J. L. Fryer and M. J. Mauel, "The rickettsia: an emerging group of pathogens in fish," Emerging Infectious Diseases, vol. 3, no. 2, pp. 137–144, 1997.

160. T. H. Birkbeck, L. A. Laidler, A. N. Grant, and D. I. Cox, "Pasteurella skyensis sp. nov., isolated from Atlantic salmon (Salmo salar L.)," International Journal of Systematic and Evolutionary Microbiology, vol. 52, no. 3, pp. 699–704, 2002.

161. Michel, J. F. Bernardet, P. Daniel, S. Chilmonczyk, M. Urdaci, and P. de Kinkelin, "Clinical and aetiological aspects of a summer enteritic syndrome associated with the sporulating segmented filamentous bacterium "Candidatus Arthromitus" in farmed rainbow trout, Oncorhynchus mykiss(Walbaum)," Journal of Fish Diseases, vol. 25, no. 9, pp. 533–543, 2002.

162. M. Riley, T. Abe, M. B. Arnaud et al., "Escherichia coli K-12: a cooperatively developed annotation snapshot—2005," Nucleic Acids Research, vol. 34, no. 1, pp. 1–9, 2006.

CHAPTER 17

Loop-Mediated Isothermal Amplification as an Emerging Technology for Detection of *Yersinia Ruckeri*the Causative Agent of Enteric Red Mouth Disease in Fish

Mona Saleh[1], Hatem Soliman[1,2], and Mansour El-Matbouli[*1]

[1]*Clinic for Fish and Reptiles, Faculty of Veterinary Medicine, University of Munich, Germany, Kaulbachstr.37, 80539 Munich, Germany*

[2]*Veterinary Serum and Vaccine Research Institute, El-Sekka El-Beda St., P.O. Box 131, Abbasia, Cairo, Egypt*

ABSTRACT

Background

Enteric Redmouth (ERM) disease also known as Yersiniosis is a contagious disease affecting salmonids, mainly rainbow trout. The causative agent is the gram-negative bacterium *Yersinia ruckeri*. The disease can be diagnosed by isolation and identification of the causative agent, or detection of the *Pathogen* using fluorescent antibody tests, ELISA and PCR assays. These diagnostic methods are laborious, time consuming and need well trained personnel.

Results

A loop-mediated isothermal amplification (LAMP) assay was developed and evaluated for detection of *Y. ruckeri* the etiological agent of enteric red mouth (ERM) disease in salmonids. The assay was optimised to amplify the *yruI/yruR* gene, which encodes *Y. ruckeri* quorum sensing system, in the presence of a specific primer set and *Bst* DNA polymerase at an isothermal temperature of 63°C for one hour. Amplification products were detected by visual inspection, agarose gel electrophoresis and by real-time monitoring of turbidity resulted by formation of LAMP amplicons. Digestion with *Hph*I restriction enzyme demonstrated that the amplified product was unique. The specificity of the assay was verified by the absence of amplification products when tested against related bacteria. The assay had 10-fold higher sensitivity compared with conventional PCR and successfully detected *Y. ruckeri* not only in pure bacterial culture but also in tissue homogenates of infected fish.

Conclusion

The ERM-LAMP assay represents a practical alternative to the microbiological approach for rapid, sensitive and specific detection of *Y. ruckeri* in fish farms. The assay is carried out in one hour and needs only a heating block or water bath as laboratory furniture. The advantages of the ERM-LAMP assay make it a promising tool for molecular detection of enteric red mouth disease in fish farms.

BACKGROUND

Yersiniosis or enteric red mouth disease (ERM) is a serious systemic bacterial infection of fishes which causes significant economic losses in salmonid aquaculture worldwide [1]. Although infection with this agent has been reported in other fish species, salmonids especially rainbow trout *Oncorhrynchus mykiss*, are highly susceptible to ERM [2, 3]. The disease was first described in the rainbow trout in the United State in 1958, from Hagerman Valley, Idaho by Rucker [4], and later the causative organism named *Yersinia ruckeri* [5]. The disease is endemic in North America [3] and widespread elsewhere. It was also described in 1981 in France, Germany and United Kingdom and has now been reported in most of Europe, Australia [6, 7] and South Africa [8].

The causative agent, *Yersinia ruckeri*, is a gram-negative, non-spore-forming rod-shaped bacterium with rounded ends and like the other members of the Enterobacteriaceae family is glucose-fermentative, oxidase-negative and nitrate-reductive [9, 10]. ERM outbreaks usually begin with low mortality, and then escalate to result in high losses. Characteristic symptoms of ERM are haemorrhages of the mouth and gills, though these are rarely seen in acute infections but may be present in chronic infections, diffuse haemorrhages within the swim bladder, petechial haemorrhage of the pyloric caecae, bilateral exophthalmia, abdominal distension as a result of fluid accumulation, general septicaemia with inflammation of the gut, the spleen is often enlarged and can be almost black in colour [4]. Transmission occurs by direct contact with carrier fish, other aquatic invertebrates and birds [4, 11]. The ability of *Y. ruckeri* to survive and remain infective in the aquatic environment is considered to be a major factor in spread of the disease. Furthermore, *Y. ruckeri* is able to form biofilms and grow on surfaces and solid supports in fish tanks, like many bacteria in aquatic environments, which lead to recurrent infections in rainbow trout farms [12]. Although vaccination has for a decade been very successful in the control of infections caused by *Y. ruckeri* in trout farms [13], cases of yersiniosis have been reported in trout farms where vaccination didn't provide enough protection against the infection [14] and due to carrier state [13]. Different diagnostic methods have been developed for detection of *Y. ruckeri* including culturing, serological and molecular techniques. Isolation and identification using agar media and the organism's biochemical characteristics are considered the gold standard for *Y. ruckeri*diagnosis. Serological methods for detection of *Y. ruckeri* have also been developed and these include ELISA, agglutination, and the immunofluorescence antibody technique (IFAT) [15]. Molecular techniques are able to detect low levels of the bacterium and facilitate detection of asymptomatic carriers, which is very important for prevention of ERM transmission and spread [16]. Restriction fragmentation-length polymorphism [17] and PCR assays [18, 19, 20] are widely used for detection of low levels of *Y. ruckeri* in infected trout tissues and blood and also for detection of asymptomatic carriers. Although PCR has been shown to be a powerful and sensitive tool in detection of *Y. ruckeri*, its requirements for expensive equipments, a precision thermocycler and laboratory training limit its use in the field as a routine diagnostic tool.

Alternate isothermal nucleic acid amplification methods, which require only a simple heating device, have been developed to offer feasible platforms for rapid and sensitive detection of a target nucleic acid. These include nucleic acid-based amplification (NASBA), loop-mediated isothermal amplification (LAMP) and ramification amplification [21, 22, 23]. LAMP is a nucleic acid amplification method that synthesises large amounts of DNA in a short period of time with high specificity [22, 24]. The strand displacement activity of *Bst*DNA polymerase impels auto-cyclic DNA synthesis with loop-forming primers to yield long-stem loop products under isothermal conditions: 60–65°C for about 60 min [22, 25]. The LAMP reaction requires four or six primers that target six or eight separate DNA sequences on the target and give the assay very high specificity [22, 25]. LAMP amplification products can be detected by gel electrophoresis, by real time monitoring of turbidity with a turbidimeter [24, 26] or with the naked-eye. Visual detection can be accomplished using different methods such as detection of a white precipitate (magnesium pyrophosphate), use of an intercalating DNA dye such as SYBR Green I gel stain [27], use of florescent detection reagent, FDR, [28], or use of oligonucleotide probes labelled with different fluorescent dyes and low molecular weight cationic polymers such as polyethylenimine, PEI [29].

LAMP-based assays have been developed for numerous aquaculture animal pathogens, including white spot syndrome virus [30], yellow head virus [31], *Edwardsiella tarda* [32] and *Nocardia seriolae* [33], *Tetracapsuloides bryosalmonae, Myxobolus cerebralis, Thelohania contejeani* [34, 35, 36], Koi herpes virus (CyHV-3) and viral hemorrhagic septicaemia (VHS) [27, 37]. The objective of this study was to develop and evaluate LAMP, as a simple, rapid and sensitive diagnostic tool for ERM disease.

METHODS

Bacteria

The bacterial strains used in this study were listed in (table 1). *Y. ruckeri* strains were cultured on trypticase-soy-agar [3]. The purity of the cultures was tested with Gram stain and confirmed biochemically with the API 20E rapid identification system.

Each strain from other bacterial strains was propagated on its specific medium and then tested by Gram stain and biochemically.

DNA extraction

DNA was extracted from bacterial cultures using QIAamp® DNA mini kit (QIAGEN, Hilden, Germany). Bacterial cells were harvested in a microcentrifuge tube by centrifugation at 5000 × *g* for 10 min. Cell pellets were re-suspended in 180 µl lysis buffer (20 mg/ml lysozym; 20 mM Tris-HCl, pH 8.0; 2 mM EDETA; 1.2% Triton) and incubated at 37°C for 30 min. Proteinase K and Buffer AL were then added and mixed by vortexing. After 30 min incubation at 56°C, ethanol was added and thoroughly mixed to yield a homogenous solution. DNA was then extracted as per manufacturer's instructions. DNA was extracted from tissue samples (liver, kidney, spleen) by QIAamp® DNA mini kit (QIAGEN, Hilden, Germany) according to the manufacturer's instructions following the animal tissues protocol.

Table 1 Bacterial species assayed in ERM-LAMP experiments

Bacterial Strains	**Source**
Y. ruckeri	DSMZ[1] 18506 (ATCC 29473)
Y. ruckeri	CECT[2] 955
Y. ruckeri	CECT 956
Y. ruckeri	Dr. Joachim Nils[3]
Y. aldovae	DSMZ 18303 (ATCC 35236)
Y. enterocolitica	DSMZ 4780 (ATCC 9610)
Y. frederiksenii	DSMZ 18490 (ATCC 33641)
Y. intermedia	DSMZ 18517 (ATCC 29909)
Y. kristensenii	DSMZ 18543(ATCC 33638)
Aeromonas salmonicida	Clinic for Fish and Reptiles
Aeromonas sorbia	Clinic for Fish and Reptiles
Renibacterium salmoninarum	Clinic for Fish and Reptiles
Flavobacterium columnare	Clinic for Fish and Reptiles
Pseudomonas aeroginosa	Clinic for Fish and Reptiles

1) DSMZ: Deutsche Sammlung von Mikroorganismen und Zellkulturen GmbH (German Collection of Micro-organisms and Cell Cultures) Braunschweig, Germany.
2) CECT: Colección Española de Cultivos Tipo (Spanish Type Culture Collection) Valencia, Spain.
3) Fischgesundheitsdienst im Staatlichen Untersuchungsamt, Veterinäruntersuchungsamt Mittelhessen, Giessen, Germany.

Oligonucleotide primers

ERM-LAMP primers were designed according to the published sequence of *yruI/yruR* (GenBank accession number AF274748, [20]) using Primer Explorer version 4 (Net Laboratory, Tokyo, Japan). Five primers were constructed; two outer primers F3 and B3, two inner primers: forward inner primer (FIP) backward inner primer (BIP) and loop forward primer (LF). FIP comprised the F1c sequence complementary to F1, a TTTT linker, and F2 sequence. BIP consisted of the B1c sequence complementary to B1, a TTTT Linker and B2 sequence. After modification of the 3' end with Rox, the loop forward primer LF was used as an Oligo DNA Probe (ODP). PCR specific primers IF-2 and IR-2 were used to amplify 1000 bp of *yruI/yruR* genes of *Y. ruckeri*[20]. Details of the LAMP and PCR primers are given in (Table 2).

Table 2 Details of oligonucleotide primers used for ERM-LAMP assay and PCR assay.

Primer name	Length	Sequence (5'-3')
F3	20-mer	TCGATATAGTTACCTTCCGG
B3	18-mer	ATGGGCAGTGAACTGTAG
FIP	46-mer	TGTTCGTTTATTGAACTTCACCGATTTTCGTCGAACTGAGCGTTAA
BIP	50-mer	AAGCTGATTTCCATAAATTCCGAGTTTTTAATGACATGGAGTTTGATGAG
Loop Forward(LF)	25-mer	AGGTATCGTGTGTTAGGATTATCGT
ODP	25-mer	AGGTATCGTGTGTTAGGATTATCGT-Rox
IF-2	24-mer	GAGCGCTACGACAGTCCCAGATAT
IR-2	24-mer	CATACCTTTAACGCTCAGTTCGAC

Optimization of ERM- LAMP condition

ERM-LAMP reactions were carried out in a Loopamp real-time turbidimeter (LA-200, Teramecs Co., Ltd., Kyoto, Japan) at 60, 63 and 65°C, for 30, 45 and 60 min, followed by 80°C for 2 min to terminate the reaction. The reaction mixture contained 40 pmol each of inner primers FIP and BIP, 5 pmol each of outer primers F3 and B3, 20 pmol of LF (forward loop primer), 1.4 mM of dNTP mix, 1.6 M betaine (Sigma-Aldrich, GmbH, Schnelldorf, Germany), 4.5 mM $MgSO_4$, 8 U of *Bst* DNA polymerase (New England Biolabs GmbH, Frankfurt, Germany), 1× of the supplied Thermopol buffer, and a specified amount of template DNA in a final volume of 25 µl. Reaction mix without DNA template was included as a negative control.

PCR amplification

Amplification was performed in a 50 µl reaction volume with 2× ready mix PCR Master mix (Thermo Scientific, Hamburg, Germany) which contained (75 mM Tris-HCl (pH 8.8), 20 mM $(NH_4)_2 SO_4$, 1.5 mM $MgCl_2$, 0.01% Tween-20, 0.2 mM each nucleotide triphosphate, 1.25 U thermoprime plus DNA polymerase, and red dye for electrophoresis), 1.5 µl of DNA template and 20 pmol each of forward and reverse primers. The amplification was carried out in Mastercycler Gradient thermocycler, Eppendorf, with the following cycling profile: 94°C for 2 min, then 40 PCR cycles of 92°C for 1 min (DNA denaturation), 65°C for 1 min (primer annealing) and 72°C for 1.5 min (DNA extension), with a terminal extension step of 72°C for 5 min.

Detection of the amplification products

Three detection methods were used: real-time turbidity detection, agarose gel analysis and visual detection. Changes in absorbance at 650 nm were measured for real-time turbidity detection with a Loopamp real-time turbidimeter (LA-200). A cut off value was determined based on the mean of the negative detection control optical density. Specimens with an optical density of less than 0.1 were determined to be negative for *Y. ruckeri* bacterial DNA. LAMP and PCR amplification products were analysed by gel electrophoresis stained with GelRed™ Nucleic Acid Gel Stain, 10,000× in water (BIOTREND Chemikalien GmbH, Köln, Germany) and then visualised under UV light. A TrackIt™ 100 bp DNA ladder (Invitrogen GmbH, Karlsruhe, Germany) was used as molecular weight marker. Visual detection of the LAMP products was carried out either by using 1 µl of Fluorescent Detection Reagent, FDR, (Eiken Chemical Co., Ltd) added before incubation of the reaction mixture at 63°C, or by addition of 1 µl of 1:10 diluted SYBR Green I nucleic acid gel stain 10,000 × concentration in DMSO (Cambrex BioSceince, Rockland, Inc., ME, USA) to the LAMP product after termination of the reaction. Any colour changes of the reaction mixture were noted. For detection with Rox- labelled probe, 0.2 µmol of low molecular weight PEI (Wako chemical GmbH, Neuss, Germany) was added to the LAMP product after centrifugation for 10 s at 6000 rpm to form insoluble PEI-amplicon complex, containing the Rox- labelled probe, which was precipitated by additional centrifugation at 6000 rpm for 10 s. Reaction tubes were then visualised under a conventional UV illuminator or by fluorescence microscopy.

Restriction analysis digestion of the ERM- LAMP products

To confirm the structure of the LAMP amplicons, it was purified using a High pure PCR purification kit (Roche Molecular Biochemicals, Mannheim, Germany) and then subjected to digestion with restriction enzyme *Hph*I (New England BioLabs GmbH, Frankfurt, Germany). Fragment sizes were analyzed by 2% agarose gels electrophoresis stained with GelRed™ Nucleic Acid Gel Stain, 10,000× in water (BIOTREND Chemikalien GmbH, Köln, Germany) and then visualised under UV light.

ERM- LAMP assay specificity

DNAs from *Y. ruckeri strains* and from other bacterial strains (*Y. aldovae, Y. enterocolitica, Y. frederiksenii, Y. intermedia, Y. kristensenii, Aeromonas salmonicida, Aeromonas sorbia, Pseudomonas aeruginosa, Renibacterium salmoninarum and Flavobacterium columnare*) were tested by ERM-LAMP assay to assess the specificity of the constructed primers. DNA from non-infected fish tissues and a negative LAMP reaction control were used to detect any non-specific amplification.

Sensitivity of the ERM-LAMP assay

One microgram genomic *Y. ruckeri* DNA was 10-fold serially diluted to assess the lower detection limit of the LAMP assay compared with conventional PCR. The products were analysed visually and by 2% agarose gel electrophoresis.

Feasibility of the ERM- LAMP assay

The use of the ERM-LAMP assay to detect *Y. ruckeri* DNA in clinical specimens was evaluated by testing 15 rainbow trout samples infected with ERM submitted to our clinic and 4 control fish samples. These fish were suffering from diffuse haemorrhages in the swim bladder and enlarged black spleen. The samples were tested by both ERM-LAMP assay and PCR assay.

RESULTS

Optimal amplification of the *Y. ruckeri yruI/yruR* gene by ERM-LAMP assay was obtained at 63°C for 60 min, as shown by both agarose gel electrophoresis and real time turbidity measurements. Amplified products exhibited a ladder-like pattern on the gel (Fig. 1). Specificity of the amplification was confirmed by digestion of the LAMP products using *Hph*I restriction enzyme (Fig. 1), the sizes of the resultant digestion products were as predicted (87 bp and 108 bp). Results obtained with the visual detection methods correlated with agarose gel electrophoresis results. When FDR used, a strong green fluorescence was emitted by LAMP positive reactions (F ig. 2, Tube No.3) when exposed to UV light and no fluorescence was evident for a negative reaction (Fig. 2, Tube No. 4). Likewise, after addition of SYBR Green I dye, the ERM-LAMP products appeared green (Fig. 2, Tube No. 5), while in the negative control tube the original orange colour of SYBR Green I did not change (Fig. 2, Tube No. 6). With Rox-labelled probe, a pellet formed emitted a red fluorescence for a positive reaction (Fig. 2, Tube No. 2), but there was neither pellet nor fluorescence observed in the negative control tube (Fig. 2, Tube No. 1).

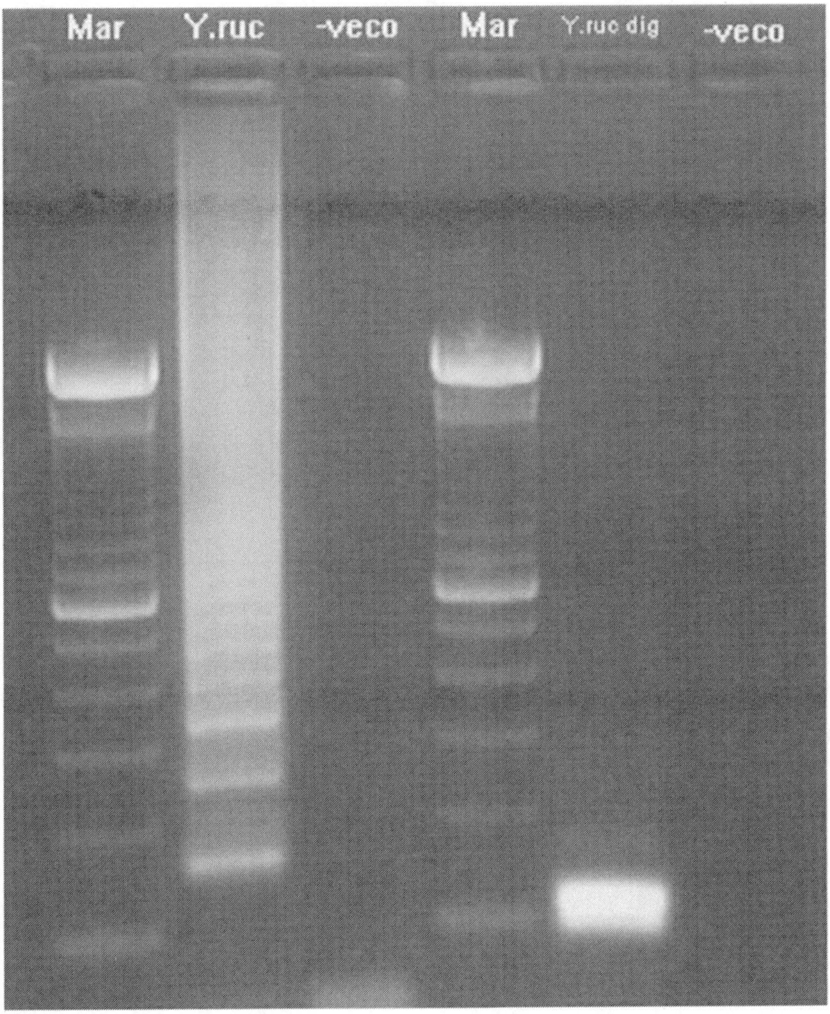

Figure 1 ERM-LAMP. *Yersinia ruckeri* loop-mediated isothermal amplification (ERM-LAMP) products and restriction analysis of ERM- LAMP product with *Hph*I enzyme. Lane Mar = 100-base-pair DNA ladder, lane Y. ruc = Amplified *Y. ruckeri*LAMP product shows a ladder-like pattern, lane Y. ruc dig = Digested *Y. ruckeri* LAMP product with *Hph*I with production of 87 bp and 108 bp bands, lane – veco = Negative (No template) control.

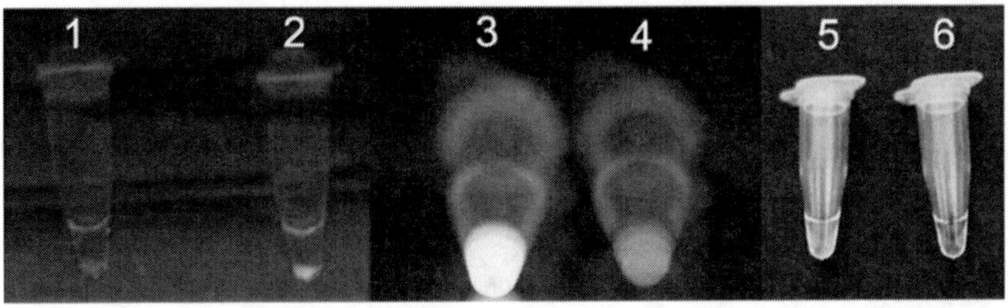

Figure 2 Visual detection of ERM-LAMP product. Using different naked eye detection methods: 1 = Negative control of ERM-LAMP reaction using Rox- labelled probe, there is neither pellet nor red fluorescence; 2 = Positive ERM-LAMP reaction using Rox- labelled probe, the pellet emitted red fluorescence; 3 = positive sample by using FDR, emitted strong green fluorescence when exposed to UV light; 4 = negative sample by using FDR, did not emitted strong green fluorescence under UV light; 5 = positive sample with green colour by using SYBR green I stain; 6 = negative sample with orange colour by using SYBR green I stain.

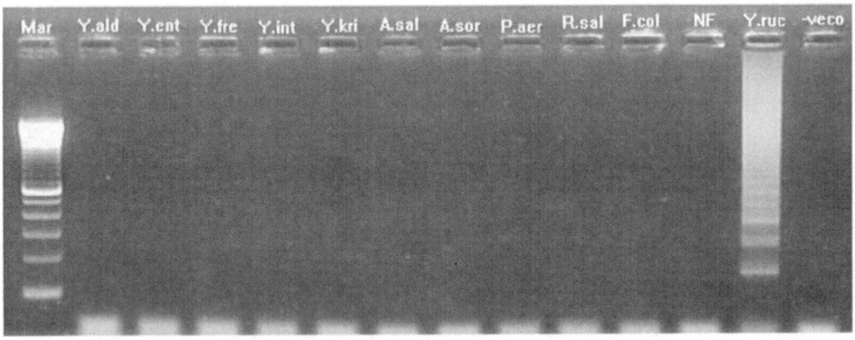

Figure 3 Specificity of ERM-LAMP primers for detection of *Y. ruckeri* DNA. Lane Mar = 100-base-pair DNA ladder, lane Y. ald = DNA from *Yersinia aldovae*, lane Y. ent = DNA from *Yersinia enterocolitica*, lane Y. fre = DNA from *Yersinia frederiksenii*, lane Y. int = DNA from *Yersinia intermedia*, lane Y. kri = DNA from *Yersinia kristensenii*, lane A. sal = DNA from *Aeromonas salmonicida*, lane A. sor = DNA from *Aeromonas sorbia*, lane P. aer = DNA from *Pseudomonas aeruginosa*, lane R. sal = DNA from *Renibacterium salmoninarum*, lane F. col = DNA from *Flavobacterium columnare*, lane NF = DNA from non-infected Fish tissues, lane Y. ruc = DNA from *Yersinia ruckeri*, lane – veco = Negative control.

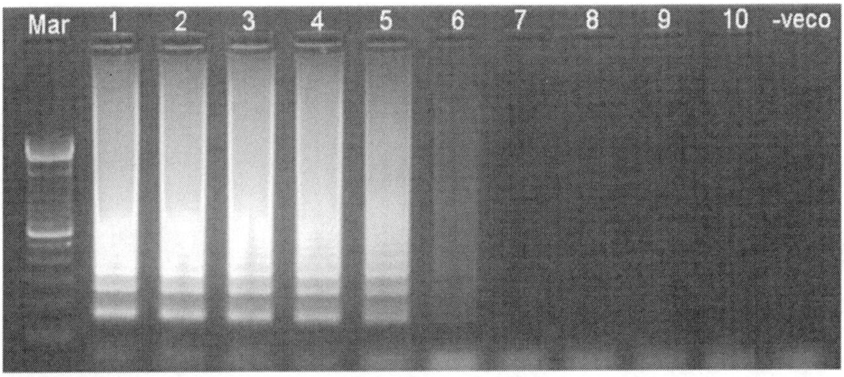

Figure 4 Sensitivity of ERM-LAMP assay. Lower detection limit of the *Yersinia ruckeri* DNA by LAMP assay. Lane Mar = 100-base-pair DNA ladder, lane 1–10 = 10-fold serial dilution of 1 μg *Yersinia ruckeri* DNA from 10^{-1}-10^{-10}; lane – veco = No template control.

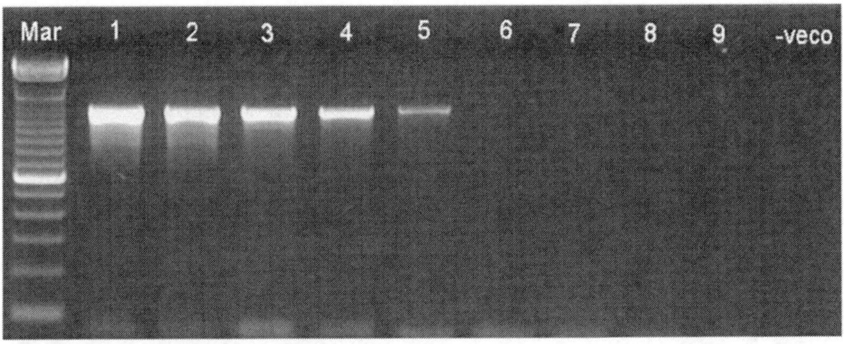

Figure 5 Sensitivity of ERM-PCR assay. Lower detection limit of (1000 bp fragment) *Yersinia ruckeri* DNA by PCR. Lane Mar = 100-base-pair DNA ladder, lane 1–9 = 10-fold serial dilution of 1 μg *Yersinia ruckeri* DNA from 10^{-1}-10^{-9}; lane – veco = No template control.

The specificity of ERM-LAMP primers was confirmed by amplification of *yruI/yruR* gene from all *Y. ruckeri*tested strains while there are no amplification products detected from the other bacterial species, non-infected fish tissues or negative (no template) LAMP reaction control (Fig. 3). Both agarose gel electrophoresis and visual detection methods showed that, the lower detection limit of the ERM- LAMP method is 10^{-6}dilution, which equal to 1 pg of the *Y. ruckeri* genomic DNA (Fig. 4), while PCR showed no amplification after a dilution of 10^{-5} which equal to 10 pg *Y. ruckeri* genomic DNA (Fig. 5). The LAMP assay detected *Y. ruckeri* DNA from

15 infected fish samples, which were also positive by PCR (Fig. 6 &7). Samples from all 4 control fish were negative in both assays.

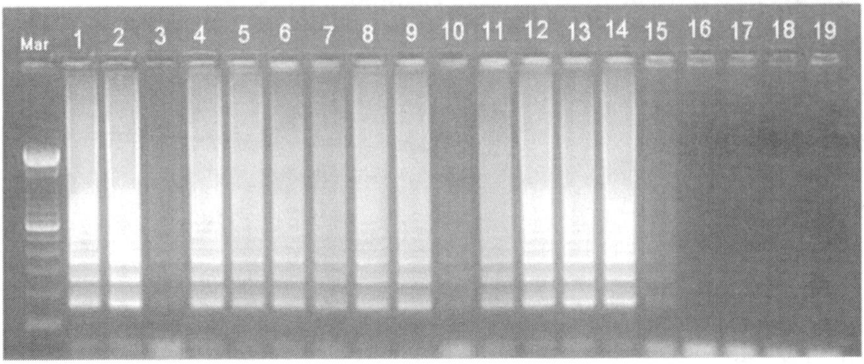

Figure 6 Feasibility of ERM-LAMP assay. Detection of *Yersinia ruckeri* DNA from 15 infected kidney samples by ERM-LAMP while there is no amplifications appeared with the non-infected kidney samples. Lane Mar = 100-base-pair DNA ladder, lanes 1–15 = DNA from infected kidney samples, lanes 16–19 = DNA from non-infected kidney samples.

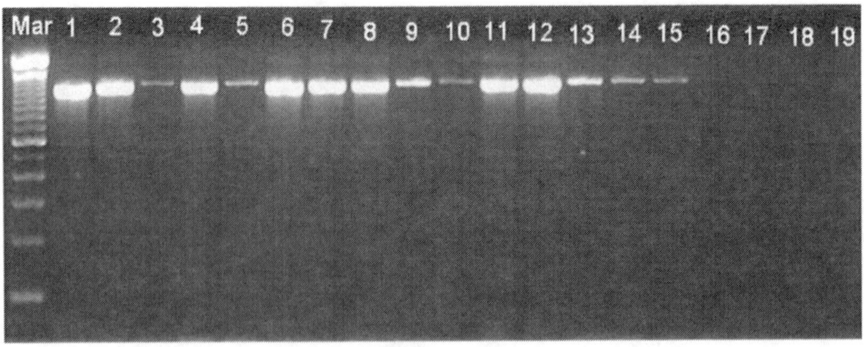

Figure 7 Feasibility of ERM-PCR assay. Detection of *Yersinia ruckeri* DNA from 15 infected kidney samples by ERM-PCR while there is no amplifications appeared with the non – infected kidney samples. Lane Mar = 100-base-pair DNA ladder, lanes 1–15 = DNA from infected kidney samples, lanes 16–19 = DNA from non-infected kidney samples.

DISCUSSION

Efficient, rapid and timely diagnosis is critical for successful management of diseases in aquaculture. For field diagnosis, the optimal detection system should be economical, quick, and easy to operate, moreover should meet the requirements of specificity and sensitivity [38]. ERM disease is a serious infection that causes sever economic losses in salmonid aquaculture. It usually occurs as an acute condition with high morbidity and mortality rates, which necessitates rapid and accurate methods for detection of its causative agent, *Y. ruckeri*[18]. A traditional microbiological approach for isolation and identification usually takes 2 to 3 days, and given that different numerical profiles for *Y. ruckeri* can be obtained with commercial multi-substrate identification systems, particularly the API 20E system, they must be interpreted with caution [3]. Although PCR assays are more accurate, specific, and faster than the microbiological approach [18, 19, 20], they require precision equipments which are beyond the capacity of most diagnostic sites to purchase, maintain and operate, and the complexity of the assay procedures obviates the possibility of point-of-care use.

In this study, a rapid and sensitive diagnostic system based on LAMP technology was developed to detect *Y. ruckeri*. The ERM-LAMP assay requires only a simple water bath or heating block to incubate the reaction mixture at 63°C for 1 hr before the reaction products are visualised. The assay utilizes a single DNA polymerase that is active at relatively high isothermal amplification temperatures, which diminishes the probability of non-specific priming [39]. The *yruI/yruR* quorum sensing system encoding gene of *Y. ruckeri*was chosen as a suitable target, as it controls virulence gene expression through cell to cell communication and has great potential for rapid and specific identification of this fish pathogen [20]. Although there is a serotypic diversity among *Y. ruckeri* strains [40, 41], *yruI/yruR* gene was amplified from all *Y. ruckeri* tested strains by PCR and produced one RFLP pattern which demonstrate a high degree of genotypic homogeneity among *Y. ruckeri* strains regarding this gene [20].

A LAMP assay requires at least 4 highly specific primers to distinguish six distinct regions on the target DNA [42]. In developing the ERM-LAMP assay, several primer sets were appraised, with the most effective set presented here. The assay was optimized to amplify *Y. ruckeri* at 63°C using a set of 4 or 5 primers. In initial trials of the assay, a characteristic ladder-like pattern of LAMP amplification is demonstrated upon gel electrophoresis [43] and confirmed the identity of the product by *Hph*I digestion. The ERM-LAMP assay was able to amplify the target *yruI/yruR* gene from all *Y. ruckeri* tested strains while it did not show any cross-reactivity with a panel of DNAs from other Yersinia species or from other related bacterial species, which confirm its specificity. Due to the isothermal nature of the LAMP assay, there is no time lost in temperature cycling, which leads to extremely high efficiency compared with regular PCR [22, 44]. Another advantage of LAMP is that real-time monitoring of the reaction is possible [24] and this decreases the time needed to get results and reduces the risk of carry-over contamination in the post-PCR process [45]. Alternatively, LAMP reaction products can be visualized using SYBR Green I nucleic gel stain which has high binding affinity to double stranded DNA and hence turns from orange to green as the LAMP amplicons are produced [46, 47]. LAMP product can also be monitored by placing a reaction tube directly on a UV transilluminator; when the FDR added into the reaction mixture. The calcein in FDR is initially combined with manganese ions and is quenched, but as amplification generates by-product pyrophosphate ions, these bind to and remove manganese from the calcein, resulting in fluorescence which is intensified further as calcein combines with magnesium ions [28, 45]. On the other hand, if low molecular weight PEI is used, this forms an insoluble complex with high molecular weight DNAs, like LAMP products, which then captures the hybridized Rox-labelled probe into a pellet which fluoresces red under UV light [29]. All of our data confirmed that visual detection of assay results was compatible with the real-time turbidity measurement and agarose gel electrophoresis. Hence simple visual detection facilitates use of the assay in basic laboratories and in fish farms.

Compared with biochemical, microbial culture methods and PCR assay (24–48 hrs, 3 hrs respectively); the ERM-LAMP is convenient, rapid, and sensitive. The ERM-LAMP assay is 10-fold more sensitive than regular PCR as it detected a very low concentration of *Y. ruckeri* genomic DNA (1 pg), while the PCR can detect only till 10 pg *Y. ruckeri* genomic DNA. The assay successfully detected *Y. ruckeri* DNA in infected fish samples and hence appears suitable for use with clinical specimens.

CONCLUSION

Loop mediated isothermal amplification assay as a new diagnostic tool for diagnosis of ERM disease in salmonids was developed and evaluated. The ERM-LAMP assay is rapid, as its result appeared after one hour, and sensitive than the conventional diagnostic method of ERM disease. The ERM-LAMP assay requires only a regular laboratory water bath and is hence suitable as a routine diagnostic tool in private clinics and field applications where equipment such as thermal cycling machines and electrophoresis apparatus are not available.

ACKNOWLEDGEMENTS

The Authors are grateful for Dr. Joachim Nils, Fischgesundheitsdienst im Staatlichen Untersuchungsamt, Veterinäruntersuchungsamt Mittelhessen, Giessen, Germany, for providing *Yersinia ruckeri* strain used in this endeavour. We would like also to thank Dr. Sieghard Frischmann, Mast Diagnostica Laboratoriumspräparate GmbH, Reinfeld, Germany, for providing us the real-time turbidimeter LA-200.

REFERENCES

1. Raida MK, Buchmann K: Bath vaccination of rainbow trout (*Oncorhynchus mykiss* Walbaum) against *Yersinia ruckeri*: Effects of temperature on protection and gene expression. Vaccine. 2008, 26: 1050-1062. 10.1016/j.vaccine.2007.12.029.
2. Austin B, Austin DA: Bacterial Fish Pathogens: Disease of farmed and wild fish. 3rd edition. Praxis Publishing Ltd, Chichester, UK, 1993.
3. Furones MD, Rodgers CJ, Munn CJ: *Yersinia ruckeri*, the causative agent of enteric redmouth disease (ERM) in fish. Ann Rev Fish Dis. 1993, 3: 105-125. 10.1016/0959-8030(93)90031-6.
4. Rucker R: Redmouth disease of Rainbow trout (Salmo gairdneri). Bull Off Int Epizoot. 1966, 65 (5): 825-830.
5. Ewing WH, Ross AJ, Brenner DJ, Fanning GR: *Yersinia ruckeri* sp. Nov., the redmouth (RN) bacterium. Int J Syst Bacteriol. 1978, 28: 37-44.
6. Bullock GL, Stuckkey HM, Shotts EB: Enteric redmouth bacterium: comparison of isolates from different geographic areas. J Fish Dis. 1978, 1: 351-356. 10.1111/j.1365-2761.1978.tb00039.x.
7. Llewellyn LC: A bacterium with similarities to the redmouth bacterium and *Serratia liquefaciens* (Grimes and Hennerty) causing mortalities in hatchery-reared salmonids in Australia. J Fish Dis. 1980, 3: 29-39. 10.1111/j.1365-2761.1980.tb00181.x.
8. Bragg RR: Health status of salmonids in river systems in Natal (South Africa). III. Isolation and identification of bacteria. Onderstepoort J Vet Res. 1991, 58: 67-70.
9. Ross AJ, Rucker RR, Ewing WH: Discription of a bacterium associated with redmouth disease of rainbow trout (Salmo gairdneri). Can J Microbiol. 1966, 12: 763-770.
10. Post G: Text book of fish health. TFH Publication, Inc, Neptune City, NJ 1987.
11. Willumsen B: Birds and wild fish as potential vectors of *Yersinia ruckeri*. J Fish Dis. 1989, 12: 275-277. 10.1111/j.1365-2761.1989.tb00313.x.
12. Coquet L, Cosette P, Junter GA, Beucher E, Saiter JM, Jouenne T: Adhesion of *Yersinia Ruckeri* to fish farm materials: influence of cell and material surface properties. Colloids and surfaces B: Biointerfaces. 2002, 26: 373-378. 10.1016/S0927-7765(02)00023-1.
13. Stevenson RMW: Immunization with bacterial antigens: Yersiniosis. Dev Biol Stand. 1997, 90: 117-124.
14. Austin DA, Robertson PAW, Austin B: Recovery of a new biogroup of *Yersinia ruckeri* from diseased rainbow trout (*Oncorhynchus mykiss*, Wahlbaum). Syst Appl Microbiol. 2003, 26: 127-131. 10.1078/072320203322337416.
15. Smith AM, Goldring OL, Dear G: The production and methods of use of polyclonal antisera to the pathogenic organisms *Aeromonas salmonicida, Yersinia ruckeri*, and *Renibacterium salmoninarum*. J Fish Biol. 1987, 31A: 225-226. 10.1111/j.1095-8649.1987.tb05317.x.

16. Tobback E, Decostere A, Hermans K, Haesebrouck F, Chiers K: *Yersinia ruckeri* infections in salmonid fish. J Fish Dis. 2007, 30: 257-268. 10.1111/j.1365-2761.2007.00816.x.
17. Garcia JA, Dominguez L, Larson JL, Pederson K: Ribotyping and plasmid profiling of *Yersinia ruckeri*. J Appl Microbiol. 1998, 85: 949-955.
18. Gibello A, Blanco MM, Moreno MA, Cutuli MT, Domenech A, Dominguez L, Fernandez-Garayzabal JF: Development of a PCR assay for detection of *Yersinia ruckeri* in Tissues of inoculated and naturally infected trout. Appl Environ Microbiol. 1999, 65: 346-350.
19. Altinok I, Grizzle JM, Liu Z: Detection of *Yersinia ruckeri* in rainbow trout blood by use of polymerase chain reaction. Dis Aquat Org. 2001, 44: 29-34. 10.3354/dao044029.
20. Temprano A, Yugueros J, Hernanz C, Sanchez M, Berzal B, Luengo JM, Naharro G: Rapid identification of *Yersinia ruckeri*by PCR amplification of yruI- yruR quorum sensing. J Fish Dis. 2001, 24: 253-261. 10.1046/j.1365-2761.2001.00261.x.
21. Compton J: Nucleic acid sequence-based amplification. Nature. 1991, 350: 91-92. 10.1038/350091a0.
22. Notomi T, Okayama H, Yonekawa T, Watanabe K, Amino N, Hase T: Loop- mediated isothermal amplification of DNA. Nucleic Acids Res. 2000, 28: 63-10.1093/nar/28.12.e63.
23. Zhang DY, Brandwein M, Hsuih T, Li HB: Ramification amplification: A novel isothermal DNA amplification method. Mol Diagn. 2001, 6: 141-150.
24. Mori Y, Kitao M, Tomita N, Notomi T: Real-Time turbidimetry of LAMP reaction for quantifying template DNA. J Biochem Biophys Methods. 2004, 59: 145-157. 10.1016/j.jbbm.2003.12.005.
25. Nagamine K, Hase T, Notomi T: Accelerated reaction by Loop-mediated isothermal amplification using loop primers. Mol Cell Probes. 2002, 16: 223-229. 10.1006/mcpr.2002.0415.
26. Mori Y, Nagamine K, Tomita N, Notomi T: Detection of loop-mediated isothermal amplification reaction by turbidity derived from magnesium pyrophosphate formation. Biochem Biophys Res Commun. 2001, 289: 150-154. 10.1006/bbrc.2001.5921.
27. Soliman H, El-Matbouli M: An inexpensive and rapid diagnostic method of the koi herpesvirus (KHV) infection by loop-mediated isothermal amplification. Virol J. 2005, 2: 83-10.1186/1743-422X-2-83.
28. Yoda T, Suzuki Y, Yamazaki K, Sakon N, Aoyama I, Tsukamoto T: Evaluation and application of reverse transcription loop-mediated isothermal amplification for detection of noroviruses. J Med Virol. 2007, 79: 326-334. 10.1002/jmv.20802.
29. Mori Y, Hirano T, Notomi T: Sequence specific visual detection of LAMP reactions by addition of cationic polymers. BMC Biotechnol. 2006, 6: 3-10.1186/1472-6750-6-3.
30. Kono T, Savan R, Sakai M, Itami T: Detection of white spot syndrome virus in shrimp by loop-mediated isothermal amplification. J Virol Methods. 2004, 115: 59-65. 10.1016/j.jviromet.2003.09.015.
31. Mekata T, Kono T, Svan R, Sakai M, Kasornchandra J, Yoshida T, Itami T: Detection of yellow head virus in shrimp by loop-mediated isothermal amplification. J Virol Methods. 2006, 135: 151-156. 10.1016/j.jviromet.2006.02.012.
32. Savan R, Igarashi A, Matsuoka S, Sakai M: Sensitive and rapid detection of edwardsiellosis in fish by a loop-mediated isothermal amplification method. Appl Environ Microbiol. 2004, 70: 621-624. 10.1128/AEM.70.1.621-624.2004.
33. Itano T, Kawakami H, Kono T, Sakai M: Detection of fish nocardiosis by loop-mediated isothermal amplification. J Appl Microbiol. 2005, 100: 1381-1387. 10.1111/j.1365-2672.2006.02872.x.
34. El-Matbouli M, Soliman H: Rapid diagnosis of *Tetracapsuloides bryosalmonae*, the causative agent of proliferative kidney disease (PKD) in salmonid fish by a novel DNA amplification method loop mediated isothermal amplification (LAMP). Parasitol Res. 2005, 96: 277-284. 10.1007/s00436-005-1361-3.

35. El-Matbouli M, Soliman H: Development of a rapid assay for diagnosis of *Myxobolus cerebralis* in fish and oligochaetes using loop-mediated isothermal amplification. J Fish Dis. 2005, 28: 549-557. 10.1111/j.1365-2761.2005.00660.x.

36. El-Matbouli M, Soliman H: Development and evaluation of two molecular diagnostic methods for detection of *Thelohania contejeani* (Microsporidia), the causative agent of porcelain disease in crayfish. Dis Aquat Org. 2006, 69: 205-211. 10.3354/dao069205.

37. Soliman H, El-Matbouli M: Reverse transcription loop mediated isothermal amplification (RT-LAMP) for rapid detection of viral hemorrhagic septicaemia virus (VHS). Vet Microbiol. 2005, 114: 205-213. 10.1016/j.vetmic.2005.11.063.

38. Teng P, Chen C, Sung P, Lee F, Ou B, Lee P: Specific detection of reverse transcription-loop-mediated isothermal amplification amplicons for Taura syndrome virus by colorimetric dot-blot hybridization. J Virol Methods. 2007, 146: 317-326. 10.1016/j.jviromet.2007.07.027.

39. Boehme CC, Nabeta P, Henostroza G, Raqib R, Rahim Z, Gerhardt M, Sanga E, Hoelscher M, Notomi T, Hase T, Mark D, Perkins MD: Operational feasibility of using Loop-Mediated Isothermal Amplification for diagnosis of pulmonary tuberculosis in microscopy centres of developing countries. J Clin Microbiol. 2007, 45: 1936-1940. 10.1128/JCM.02352-06.

40. Davies RL: Virulence and serum-resistance in different clonal groups and serotypes of *Yersinia ruckeri*. Vet Microbiol. 1991, 29: 289-297. 10.1016/0378-1135(91)90136-4.

41. Davies RL: Outer membrane protein profiles of *Yersinia ruckeri*. Vet Microbiol. 1991, 26: 125-140. 10.1016/0378-1135(91)90049-L.

42. Enosawa M, Kageyama S, Sawai K, Watanabe K, Notomi T, Onoe S, Mori Y, Yokomizo Y: Use of loop-mediated isothermal amplification of the IS900 sequence for rapid detection of cultured *Mycobacterium avium* subsp. *Paratuberculosis*. J Clin Microbiol. 2003, 9: 4359-4365. 10.1128/JCM.41.9.4359-4365.2003.

43. Thai HTC, Le MQ, Vuong CD, Parida M, Minekawa H, Tsugunori N, Hasebe F, Morita K: Development and evaluation of a novel loop-mediated isothermal amplification method for rapid detection of sever acute respiratory syndrome Coronavirus. J Gen Virol. 2004, 36: 93-109.

44. Nagamine K, Watanabe K, Ohtsuka K, Hase T, Notomi T: Loop-mediated isothermal amplification reaction using a nondenaturated template. Clin Chem. 2001, 47: 1742-1743.

45. Imai M, Ninomiya A, Minekawa H, Notomi T, Ischzaki T, Van Tu P, Tien NT, Tashiro M, Odagiri T: Rapid diagnosis of H5N1 avian influenza virus infection by newly developed influenza H5 hemagglutinin gene-specific loop-mediated isothermal amplification method. J Virol Methods. 2007, 141: 173-180. 10.1016/j.jviromet.2006.12.004.

46. Karleson F, Steen H, Nesland J: SYBR green I DNA staining increases the detection sensitivity of viruses by polymerase chain reaction. J Virol Methods. 1995, 55: 153-156. 10.1016/0166-0934(95)00053-W.

47. Iwamoto T, Sonobe T, Hayashi K: Loop-mediated isothermal amplification of *Mycobacterium tuberculosis* complex, *M. avium*, and *M. intracellulare* in sputum samples. J Clin Microbiol. 2003, 41: 2616-2622. 10.1128/JCM.41.6.2616-2622.2003.

CHAPTER 18

Systematic Identification and Characterization of Cardiac Long Intergenic Noncoding Rnas in Zebrafish

Lei Wang[1,2], Xiao Ma[3,4,5], Xiaolei Xu[3,4] & Yuji Zhang[1,2]

[1]*Department of Epidemiology & Public Health, University of Maryland School of Medicine, Baltimore, MD, 21201, USA.*

[2]*Division of Biostatistics and Bioinformatics, University of Maryland Marlene and Stewart Greenebaum Comprehensive Cancer Center, Baltimore, MD, 21201, USA.*

[3]*Department of Biochemistry and Molecular Biology, Mayo Clinic, Rochester, MN, 55905, USA.*

[4]*Division of Cardiovascular Diseases, Mayo Clinic, Rochester, MN, 55905, USA.*

[5]*Clinical and Translational Science Track, Mayo Graduate School, Mayo Clinic College of Medicine, Rochester, MN, 55905, USA.*

ABSTRACT

Long intergenic noncoding RNAs (lincRNAs) are increasingly recognized as potential key regulators of heart development and related diseases, but their identities and functions remain elusive. In this study, we sought to identify and characterize the cardiac lincRNA transcriptome in the experimentally accessible zebrafish model by integrating bioinformatics analysis with experimental validation. By conducting genome-wide RNA sequencing profiling of zebrafish embryonic hearts, adult hearts, and adult muscle, we generated a high-confidence set of 813 cardiac lincRNA transcripts, 423 of which are novel. Among these lincRNAs, 564 are expressed in the embryonic heart, and 730 are expressed in the adult heart, including 2 novel lincRNAs, TCONS_00000891 and TCONS_00028686, which exhibit cardiac-enriched expression patterns in adult heart. Using a method similar to a fetal gene program, we identified 51 lincRNAs with differential expression patterns between embryonic and adult hearts, among which TCONS_00009015 responded to doxorubicin-induced cardiac stress. In summary, our genome-wide systematic identification and characterization of cardiac lincRNAs lays the foundation for future studies in this vertebrate model to elucidate crucial roles for cardiac lincRNAs during heart development and cardiac diseases.

INTRODUCTION

Cardiovascular development is a dynamic, spatiotemporally coordinated, and transcriptionally regulated process involving thousands of genes[1]. Even subtle perturbations of this process can lead to a spectrum of heart diseases, such as congenital heart defects[2], coronary artery disease[3, 4], cardiac hypertrophy[5, 6], and arrhythmias[7]. Traditional laboratory studies have focused on protein-coding genes and their regulatory roles in heart development[8, 9], but these genes account for less than 3% of the human genome. Recent applications of high-throughput sequencing technologies to analyze the transcriptome in human and other model organisms have revealed a wide spectrum of transcripts that do not encode proteins: the so-called noncoding RNAs (ncRNAs)[10, 11]. Several consortium projects, such as the Encyclopedia of DNA Elements (ENCODE)[12] and Functional Annotation of the Mammalian Genome (FANTOM)[13] have reported that ncRNAs substantially outnumber protein-coding genes and perform various specific roles in controlling gene expression and function[14]. Thus, these ncRNAs have emerged as new key players in the cardiovascular regulatory network[15]. Based on the size of their mature transcripts, ncRNAs can be broadly classified as small (<200 nucleotide [nt]) RNAs, large (>200 nt) ncRNAs (lncRNAs), and regulatory RNAs (e.g., rRNA and tRNA). As the largest group of mammalian ncRNA transcripts[16], lncRNAs have been shown to exert nonredundant roles in diverse biological processes, including chromosome X inactivation, imprinting, splicing, and transcriptional regulation[17,18,19]. To complete the cardiac transcriptome landscape, investigating the biological roles of lncRNAs during cardiac-related biological processes in a systematic genome-wide manner is essential.

Recent cardiovascular mechanism studies have detected and characterized the unique expression of several lncRNAs under both normal physiological conditions and disease states[20,21,22,23,24,25], and the results suggest that lncRNAs substantially affect gene regulation in cardiovascular-related biological processes. For instance, *chast* was shown to promote cardiac remodeling and influence cardiomyocyte hypertrophy[26], whereas *CARMEN* is a super enhancer–associated lncRNA controlling cardiac specification, differentiation, and homeostasis[27]. Long intergenic noncoding RNAs (lincRNAs), a subclass of lncRNAs that do not overlap with annotated coding regions[28], are especially interesting because their specific genomic locations facilitate experimental manipulation and computational analysis. LincRNAs have been shown to govern cardiac development and stress responses[22, 23], to orchestrate cardiac transcriptional programs[29], and to serve as potential biomarkers and novel therapeutic targets for cardiovascular diseases[25]. For instance, both *Bvht* and *Fendrr* interact with the epigenetic silencing complex polycomb repressive complex 2 (PRC2) to regulate key genes during early cardiac development[22, 23]. However, because comprehensive lincRNA annotation is lacking, only a few lincRNAs have been studied in the context of heart development and cardiac diseases[30]. Existing annotations of lincRNAs in human and other model organisms are usually derived from large-scale sequencing studies[31]. Such annotations, however, may miss lincRNAs that are only expressed during narrow spatiotemporal developmental windows[32,33,34]. To better understand the potential roles of cardiac lincRNAs (i.e., lincRNAs expressed during heart development), performing genome-wide identification, characterization, and analysis of lincRNAs in the context of heart tissue is necessary.

In this study, we sought to systematically identify and characterize cardiac lincRNAs using the zebrafish model. Specifically, we focused on lincRNAs that are expressed in embryonic and adult hearts. First, we reconstructed *de novo* most of the annotated zebrafish genes in the RefSeq and Ensembl databases. Through an integrative bioinformatics filtering approach, we defined a high-confidence set of 813 cardiac lincRNA transcripts, 423 of which were novel. Both *cis*- and *trans*-level analyses suggested that several cardiac lincRNAs were potentially functionally associated with heart development and cardiac diseases. Independent experimental validation also revealed unique activation patterns of the novel cardiac lincRNA TCONS_00009015 in the doxorubicin (DOX)-induced cardiomyopathy model. Our investigation of cardiac lincRNAs in a vertebrate developmental model revealed that these cardiac lincRNAs have crucial biological roles during heart development and may serve as potential key regulators in biological processes closely related to heart diseases. Our cardiac lincRNA catalog will facilitate future experimental and computational studies to uncover lincRNA functions in heart-associated biological processes and disease mechanisms.

RESULTS

Assembly of a high-confidence transcriptome in zebrafish

Based on our genome-wide cDNA sequencing experiments in 3 tissue types (embryonic heart, adult heart, and adult muscle), which were described in our previous work[35], we obtained ≈318 million reads. More than 75% of these reads passed the initial quality control step and were aligned to the zebrafish reference genome (Zv9). By applying our stepwise transcriptome assembly pipeline (Fig. 1A), we reconstructed 103,442 transcripts from 43,338 loci across the 3 tissue types; of these transcripts, 87,669 had at least 2 exons, and 15,773 had 1 exon.

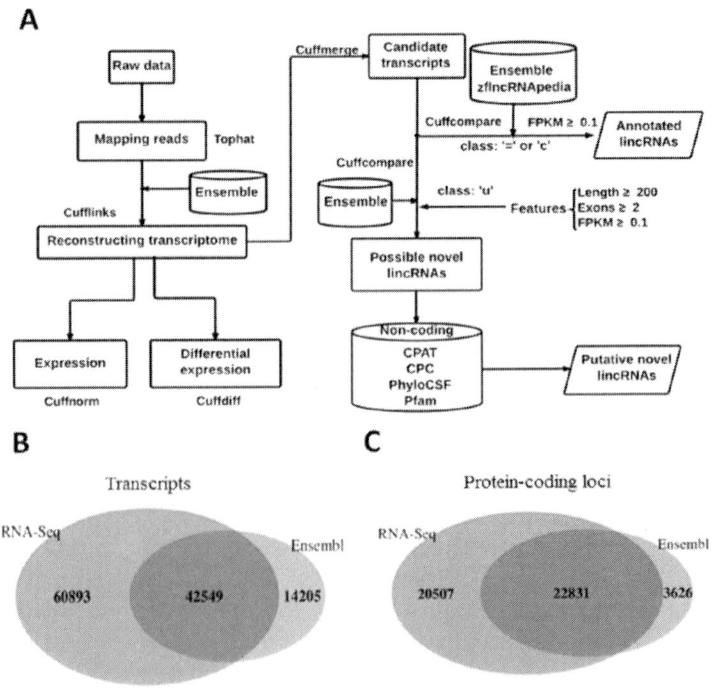

Figure 1 Overview of the RNA-seq–based cardiac transcriptome assembly. (**A**) The computational bioinformatics workflow to assemble the cardiac transcriptome and identify lincRNAs. 'u', intergenic transcript in Cufflinks; '=', complete match of intron chain in Cufflinks (see Methods). (**B**) Overlap between transcripts from the RNA-seq–based cardiac transcriptome assembly (blue) and Ensembl annotated transcripts (red). (**C**) Overlap between RNA-seq–based transcripts in protein-coding loci (blue) and Ensembl annotated protein-coding transcripts (red).

To assess the quality and coverage of our reconstructed zebrafish transcriptome, we compared it to the RefSeq (Zv9) and Ensembl annotations (Danio_rerio.Zv9.79). Of 16,146 zebrafish RefSeq transcripts, 12,582 mRNAs (78%) were identified in our zebrafish transcriptome. Similarly, evidence for transcripts from 42,549 of 56,754 Ensembl genes (75%) with comparable transcript structures was identified in our cardiac transcriptome (Fig. 1B). Notably, 22,831 protein-coding genes (86%) in the Ensembl annotation were recovered in our transcriptome (Fig. 1C). In addition, 26,300 transcripts in our zebrafish transcriptome are variants of known Ensembl genes (i.e., novel isoforms or partial transcripts). This high degree of overlap provides independent confirmation for a large fraction of our zebrafish transcriptome.

Identification of a stringent set of zebrafish lincRNAs

Using our transcriptome derived from zebrafish heart and muscle, we applied 6 filters in a stringent bioinformatics filtering pipeline to identify putative lincRNAs (Fig. 1A). First, we removed candidate transcripts that overlapped with known protein-coding genes in the zebrafish Ensembl annotation (Danio_rerio.Zv9.79.gtf). Second, we removed candidate transcripts with only 1 exon, a length less than 200 nt, or a fragments per kilobase of transcript per million mapped reads (FPKM) value less than 0.1. For the remaining putative noncoding transcripts, 343 were known lincRNA genes in the Ensembl annotation, and 100 were annotated by the zflncRNApedia annotation database[36]. Third, for the remaining 1,739 novel candidate transcripts, we removed transcripts with coding potential using the Cording Potential Assessing Tool (CPAT)[37] based on the size of the open-reading frames, open-reading frame coverage, Fickett TESTCODE statistics, and hexamer bias. This filter retained 863 candidate putative noncoding transcripts. Fourth, we removed candidate transcripts that had high similarity to known proteins or protein domains in the Pfam database using the Hidden Markov Model-based sequence alignment tool (HMMER)[38]. Candidate transcripts with a significant Pfam hit ($P < 1 \times 10^{-5}$) were removed, which resulted in 707 candidate noncoding transcripts. Fifth, we evaluated the protein-coding potential of these 707 transcripts with the Coding Potential Calculator (CPC) algorithm based on 6 biologically meaningful sequence features[39]. This filter retained 565 putative noncoding transcripts. Sixth, we used the Phylogenetic Coding Substitution Frequency (PhyloCSF) tool to score the coding potential of transcripts using multispecies nucleotide sequence alignment[40]. We chose a PhyloCSF threshold of less than 20 because it retained most of the RefSeq ncRNAs[41]. The resulting set contained 495 novel lincRNAs. In total, we identified 938 lincRNAs with detectable expression (i.e., FPKM value >0.1), 443 of which were known lincRNAs (i.e., 343 lincRNAs from the

Ensembl annotation Zv9 and 100 from the zflncRNApedia database) and 495 of which were novel lincRNAs (Fig. 2A).

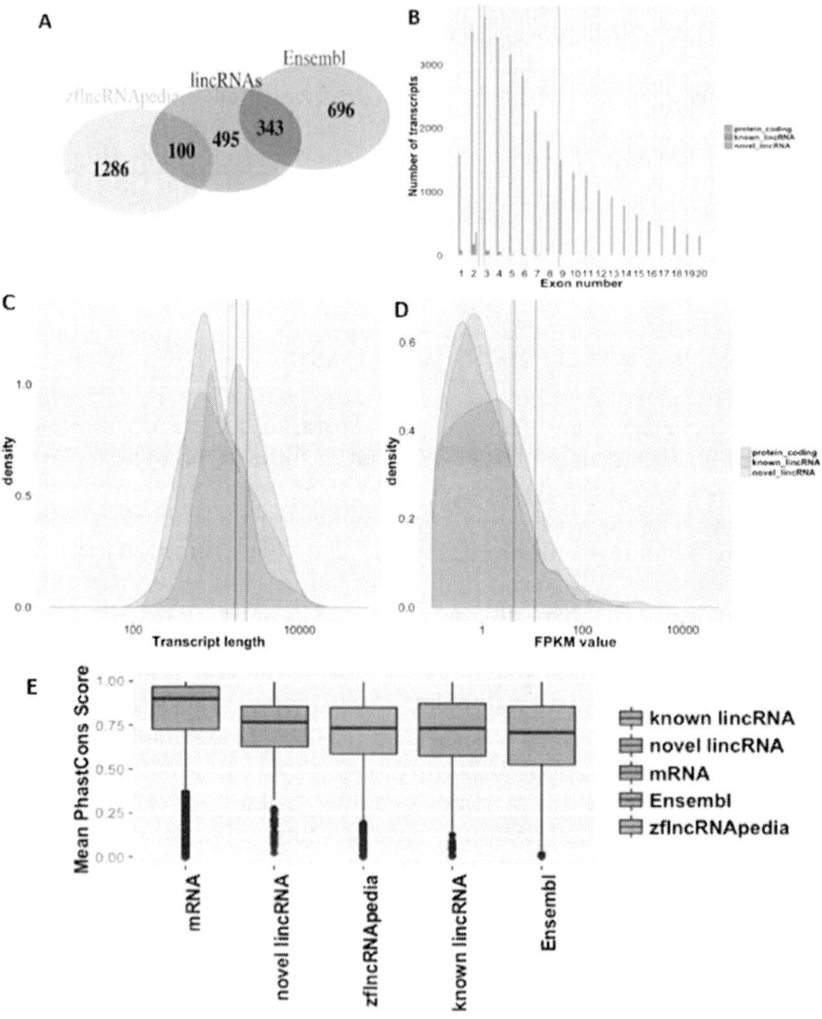

Figure 2 Zebrafish lincRNAs are shorter, less conserved, and expressed at lower levels than protein-coding genes. (**A**) Overlap among cardiac lincRNAs (dark blue) and other existing annotations: Ensembl (dark green), and zflncRNApedia (orange). Distribution of exon numbers (**B**), transcript length (**C**), expression level (**D**), and phastCons scores (**E**) for protein-coding genes, previously annotated lincRNAs, and novel lincRNAs in our RNA-seq dataset.

To assess the gene structure of lincRNAs reconstructed by our bioinformatics pipeline, we experimentally tested 5 randomly selected lincRNAs, including 2 novel predicted lincRNAs identified by our pipeline (TCONS_00022936 and TCONS_00000891) and 3 lincRNAs that overlap with known transcripts (TCONS_00048108, TCONS_00009015, and TCONS_00053013). We cloned all 5 lincRNAs from the zebrafish cardiac cDNA library and confirmed all exon junction regions via Sanger sequencing (Fig. 3A and Supplemental Figure 1). These data proved the fidelity of the gene structure of the lincRNA transcripts identified by our bioinformatics pipeline.

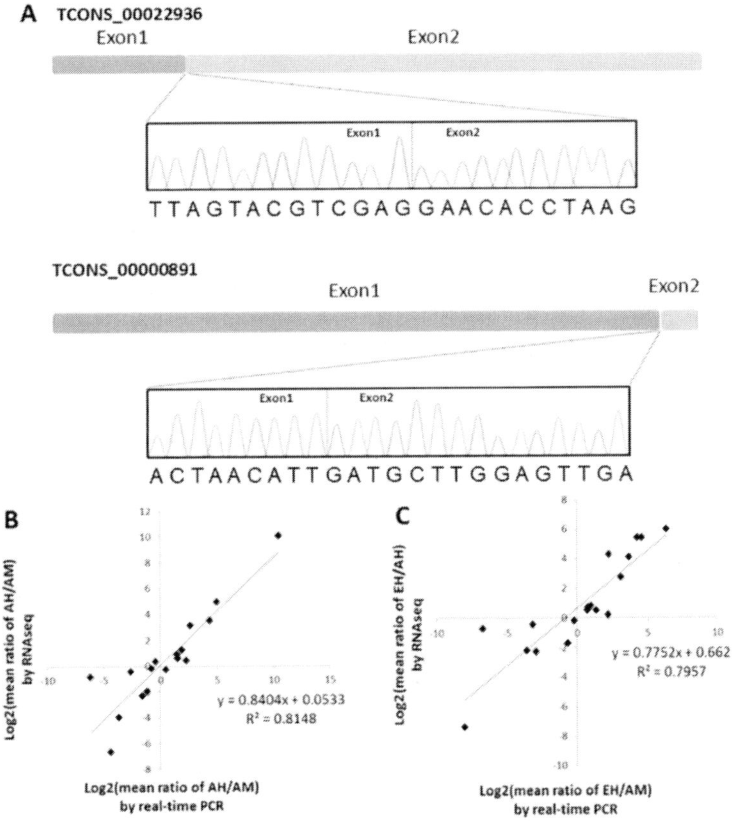

Figure 3 Validation of RNA-seq–predicted lincRNAs. (**A**) Sanger sequence validation of exon junction regions of 2 novel lincRNAs, TCONS_00022936 and TCONS_00000891. Sequence of TCONS_00000891 is shown in reverse complement form. (**B**) Quantitative real-time PCR validation of expressions of 17 lincRNAs between adult fish heart (AH) and adult fish muscle (AM). (**C**) Quantitative real-time PCR validation of expressions of 18 lincRNAs between embryonic fish heart (EH) and adult fish heart.

To assess the expression level of lincRNAs detected by our RNA-seq dataset, we used quantitative reverse-transcription polymerase chain reaction (qRT-PCR) with a selected set of 17 lincRNAs to compare their relative expression levels between adult heart and adult muscle. We observed a correlation coefficient of 0.81 between readouts from RNA-seq and qRT-PCR experiments (Fig. 3B). Similarly, comparing the relative expression levels of 18 lincRNAs between embryonic heart and adult heart resulted in a correlation coefficient of 0.79. These results proved the fidelity of the expression levels of the lincRNAs from our RNA-seq data set (Fig. 3C).

Molecular and expression properties of zebrafish lincRNAs

Many genome-wide studies have demonstrated that lincRNAs are shorter, less conserved, and expressed at significantly lower levels than protein-coding genes[32, 42, 43]. Analyzing the structures, expression levels, and conservation of 938 lincRNAs in our RNA-seq dataset showed that these lincRNAs had fewer exons per transcript (2.8 exons for known lincRNAs; 2.4 exons for novel lincRNAs) than protein-coding transcripts (8.7 exons; Fig. 2B). The novel lincRNAs also exhibited a length density distribution similar to that of known lincRNAs (a mean length of 1,343 nt for known lincRNAs, and a mean length of 1,692 nt for novel lincRNAs), both of which are shorter than protein-coding transcripts (mean length, 2,324 nt) (Fig. 2C). Among the lincRNA transcripts with mean FPKM values higher than 0.1 in at least 1 of the 3 tissue types, lincRNAs were expressed at lower levels than protein-coding transcripts (on average, 8.7 FPKM for protein-coding transcripts, 3.8 FPKM for known lincRNAs, and 4.2 FPKM for novel lincRNAs; Fig. 2D). Based on the phastCons scores, the conservation level of novel lincRNAs was similar to those of known lincRNAs in the Ensembl and zflncRNApedia databases[36], which are substantially lower than those of protein-coding transcripts (Fig. 2E).

Because lincRNAs in the human genome are ubiquitously distributed across all chromosomes[44] but have tissue-specific expression[32], we investigated the genomic distribution of 938 lincRNAs and their expression patterns in the 3 muscular tissue types in zebrafish. These lincRNAs were distributed across all chromosomes in the zebrafish reference genome (Fig. 4A). Overall, 601 lincRNAs were expressed in both heart and muscle tissues, and 212 lincRNAs displayed a putative enriched expression pattern in heart tissues but not muscle. Thus, we termed the 813 heart-expressed lincRNAs the *cardiac lincRNA catalog*. Among the 212 cardiac-specific lincRNAs, 99 were expressed only in the adult heart, and 43 were expressed only in the embryonic heart (Fig. 4B). Notably, 62 of the 99 adult heart-specific lincRNAs and 11 of the 43 embryonic heart-specific lincRNAs were novel (Fig. 4C and D). The remaining 125 lincRNAs showed more enriched expression in the muscle tissue instead of in the cardiac tissue (Fig. 4B). Thus, approximately 36% of the lincRNAs (212 in the heart and 125 in the muscle) showed a tissue-enriched expression pattern.

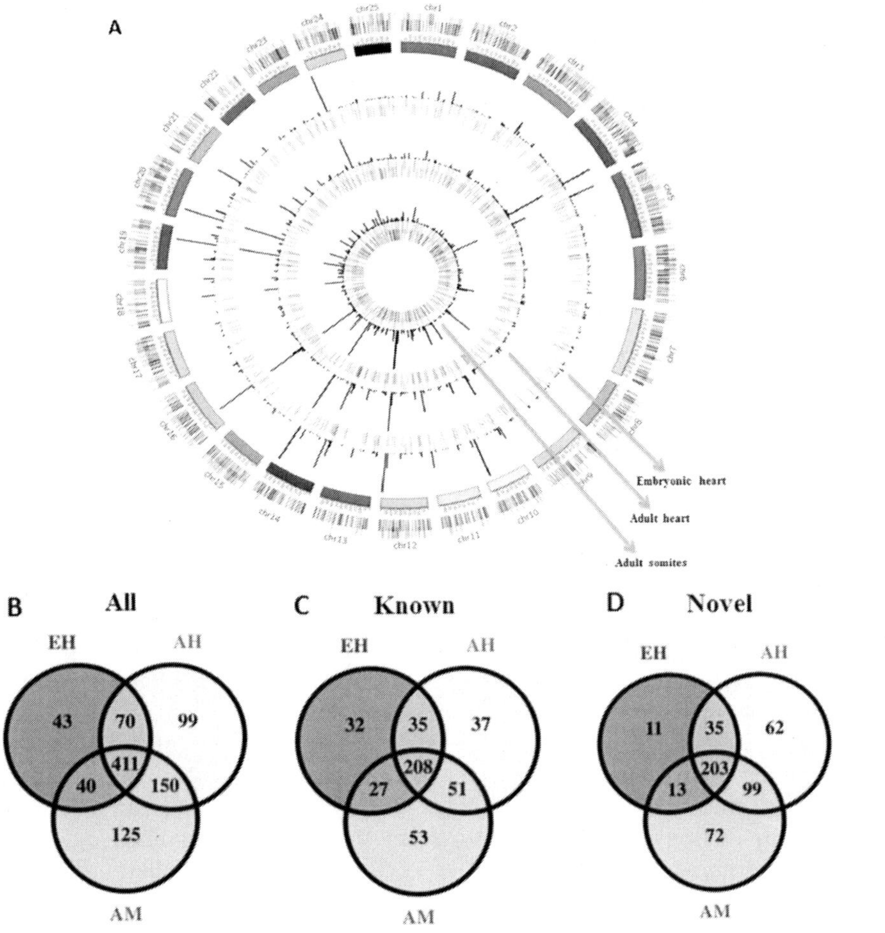

Figure 4 Overview of lincRNA distribution across the reference genome and different tissue types. (**A**) A ubiquitous distribution of cardiac lincRNAs in the zebrafish genome and spatiotemporal expression patterns (putative novel lincRNAs, red; annotated lincRNAs, green; the FPKM value (FPKM <50) for histogram). Also shown is the number of (**B**) all lincRNA transcripts; (**C**) known lincRNA transcripts; and (**D**) novel lincRNA transcripts expressed in 3 tissue types and their overlaps. AH, adult heart; AM, adult muscle; EH, embryonic heart.

Analysis of embryonic heart-enriched lincRNAs

Among the 813 cardiac lincRNAs, 564 were expressed in the embryonic heart (i.e., mean FPKM value >0.1); of these, 302 (53.5%) were known (i.e., annotated in Ensembl or the zflncRNApedia database), and 262 (46.5%) were putative novel lincRNAs (Supplemental Table 1). In the gene functional enrichment analysis by the Ingenuity Pathway Analysis (IPA) tool (http://www.ingenuity.com/), the neighboring protein-coding genes of embryonic heart-enriched lincRNAs were significantly enriched in development-related biological processes, such as embryonic development, organismal development, and tissue morphology (Supplemental Table 2). Since lincRNAs have a greater likelihood to be functionally associated with their nearest neighboring protein-coding genes[45, 46], it is likely that these lincRNAs are also involved in same biological processes. Notably, 21 embryonic heart-enriched lincRNAs were bidirectional lincRNAs (Supplemental Table 3). Based on a gene set analysis (GSA) between embryonic heart-enriched lincRNAs and functional gene sets associated with heart development and disease (Supplemental Table 4), the embryonic heart-enriched lincRNAs were significantly associated with mitogen-activated protein kinase (MAPK) signaling, heart function, heart contraction, and sarcomere organization (Fig. 5A).

For the 11 lincRNAs with the highest expression levels in embryonic heart (i.e., mean FPKM value >25 in all 3 biological replicates), their expression levels were between 10% and 60% of that of *gapdhs*, a housekeeping gene with an average FPKM of 127 in embryonic heart[47](Fig. 5B). In whole-mount *in situ* hybridizations targeting these lincRNAs, cardiac expression was confirmed for 5 lincRNAs in both the atrium and ventricle of 2dayspostfertilization (dpf) embryos. Instead of a cardiac or muscle restricted expression pattern, these lincRNAs were also expressed in other tissues, including brain, muscle, pectoral fin, and tail (Fig. 5C). Dynamic expression was noted for some lincRNAs during embryogenesis. For example, longitudinal studies of TCONS00009015 revealed no cardiac expression at 1 dpf, weak cardiac expression at 2 dpf, and strong cardiac expression at 4 dpf (Supplemental Figure 2).

Analysis of adult heart-enriched lincRNAs

In adult zebrafish heart, 730 lincRNAs had detectable expression levels (i.e., mean FPKM value >0.1); of these, 331 (45.3%) were known, and 399 (54.7%) were putative novel lincRNAs (Supplemental Table 1). In the gene functional enrichment analysis (i.e., IPA) of adult heart-enriched

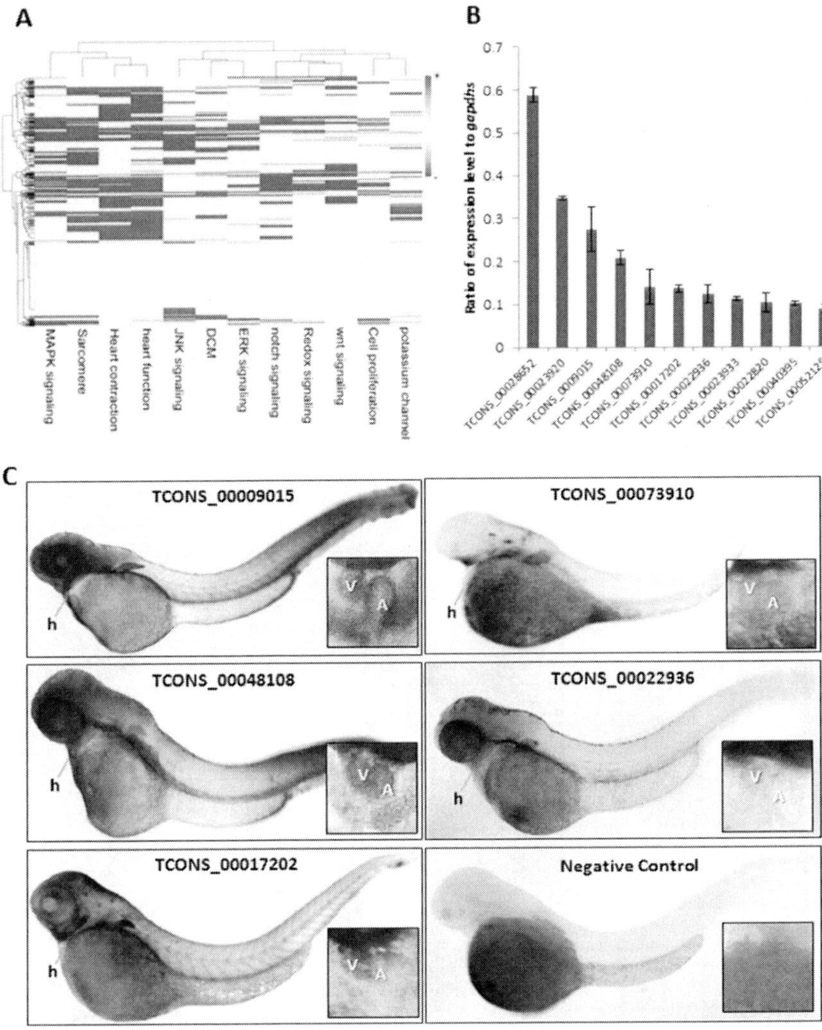

Figure 5 Analysis of embryonic heart–enriched lincRNAs. (**A**) Expression-based association matrix of 564 embryonic heart–enriched lincRNA transcripts (rows) and functional gene sets (columns). Red, positive correlation; blue, negative correlation; yellow, no correlation. (**B**) Expression of embryonic heart-enriched lincRNAs, as revealed by Quantitative real-time PCR. *gapdhs* (FPKM = 127) was used as reference gene. Means ± S.D. N = 3. (**C**) Validation of 5 selected lincRNAs with high embryonic cardiac expression by *in situ* hybridization at 2-days post fertilization (2 dpf), and negative control is a lincRNA candidate with low FPKM in embryonic heart. A, atrium; h, heart; V, ventricle.

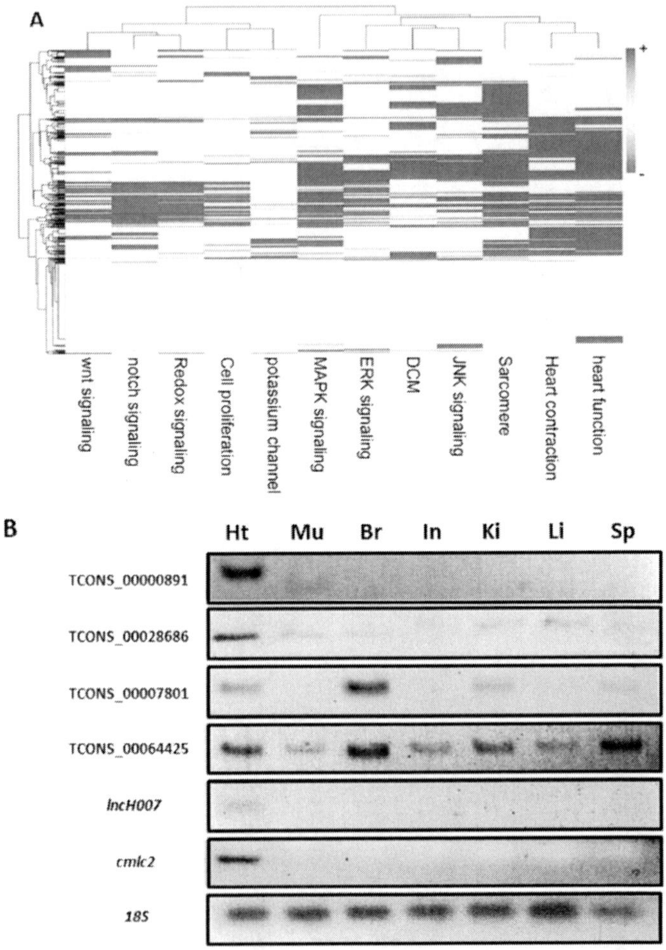

Figure 6 Analysis of adult heart–enriched lincRNAs. (**A**) Expression-based association matrix of 730 adult heart–enriched lincRNA transcripts (rows) and functional gene sets (columns). Red, positive correlation; blue, negative correlation; yellow, no correlation. (**B**) Tissue-specific expression pattern of adult heart-enriched lincRNAs across 7 tissues. Semi-quantitative RT-PCR of adult cardiac–enriched lincRNAs in 7 fresh tissue samples. PCR products were electrophoresed in 3% agarose gel. 18 S served as internal control and *lncH007* and *cmlc2*served as cardiac-specific positive controls. Br, brain; Ht, heart; In, intestine; Ki, kidney; Li, liver; Mu, muscle; Sp, spleen. Gel images are cropped around 100 bp-300 bp region. Product size information is listed in Supplemental Table 8.

lincRNAs, the neighboring protein-coding genes of these lincRNAs were significantly enriched in basic biological processes, such as RNA postmodification and cardiovascular system development and function (Supplemental Table 2). Notably, 29 of the adult heart-enriched lincRNAs were bidirectional lincRNAs (Supplemental Table 3). Based on a GSA between adult heart-enriched lincRNAs and functional gene sets associated with heart development and disease, the adult heart-enriched lincRNAs were associated not only with MAPK signaling, heart function, heart contraction, and sarcomere organization, but also with extracellular signal-regulated kinase (ERK) signaling, c-Jun N-terminal protein kinase (JNK) signaling, and dilated cardiomyopathy (DCM) (Fig. 6A).

We selected 21 lincRNAs with the highest expression ratios between adult heart and adult muscle for downstream experimental validation. The FPKM ratios of these lincRNAs between the heart and muscle groups ranged from 1,071 to 1.3 (Supplemental Table 5). We determined the expression levels of these lincRNAs in 7 adult zebrafish tissues—heart, muscle, brain, intestine, kidney, liver, and spleen—using semi-quantitative PCR. Tissue-specific expression patterns were found for several candidates. TCONS_00000891 was only expressed in heart, similar to the cardiac-restricted expression patterns of *lncH007*, a known cardiac lincRNA, and *regulatory myosin light chain* (*cmlc2*), a known cardiac gene[48] (Fig. 6B). TCONS_00028686 was expressed predominantly in heart rather than the other tissue types; TCONS_00040108 was expressed in heart, brain, and kidney; TCONS_00022820 was expressed in heart and spleen; TCONS_00007801 was expressed in heart and brain; and TCONS_00064425 was expressed in heart, brain, kidney, and spleen (Fig. 6B and Supplemental Figure 3).

Analysis of fetal lincRNAs

In the cardiac transcriptome, a subgroup of developmentally active genes, dubbed the *fetal gene program*, is particularly interesting. These genes are characterized by differential expression between embryonic and adult hearts. It is hypothesized that many of these quiescent genes have important roles during cardiac diseases, especially those that are reactivated during pathogenesis upon stress. Similar to the fetal gene program, the fetal lincRNA program has been explored recently in mouse models49. Using our definition of the fetal lincRNA program in a zebrafish heart (i.e., the lincRNA expression level is at least 4-fold different between embryonic heart and adult heart50), we identified 51 fetal lincRNAs (Supplemental Table 6). To investigate biological processes associated with these putative fetal lincRNAs, we did a GSA between fetal lincRNAs and all 8047 biological processes in the ZebrafishMine database (http://www.zebrafishmine.org/). However, only few biological processes showed correlation relationship with fetal lincRNAs (Supplemental Figure 4). To further investigate whether these putative fetal lincRNAs are associated with cardiac-remodeling biological processes, we manually collected a list of biological processes associated with heart development and diseases (Supplemental Table 4) and performed a GSA between the expression level of fetal lincRNAs and that of genes in these biological processes. The expression level of most fetal lincRNAs was positively or negatively correlated with our manually curated heart-related biological processes, including MAPK signaling, heart function, heart contraction, potassium channel, sarcomere functional gene sets, and the notch signaling gene set (Fig. 7A).

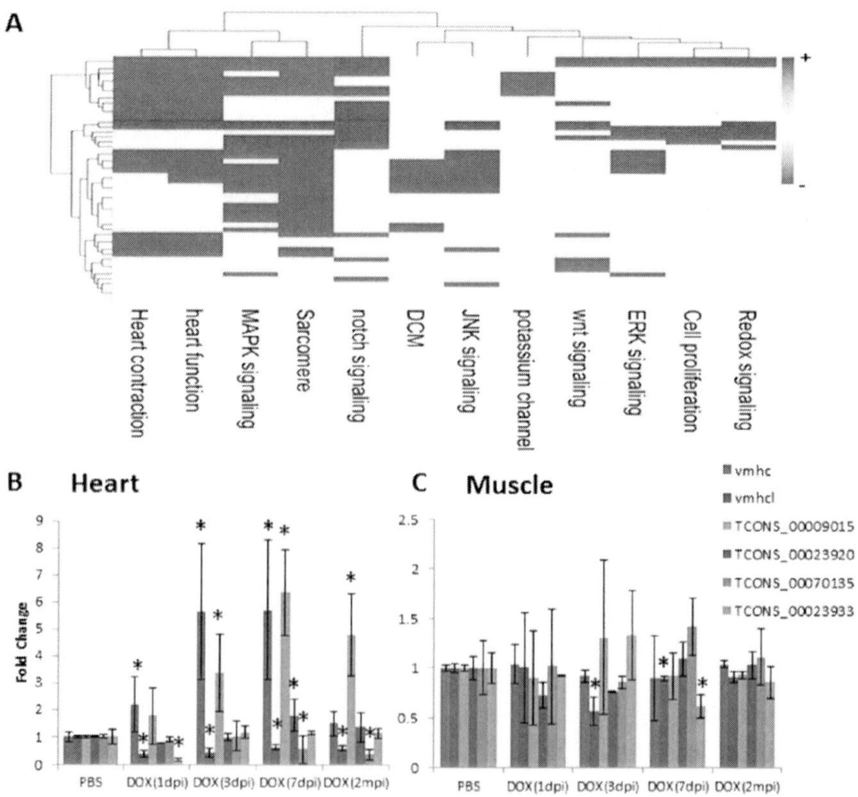

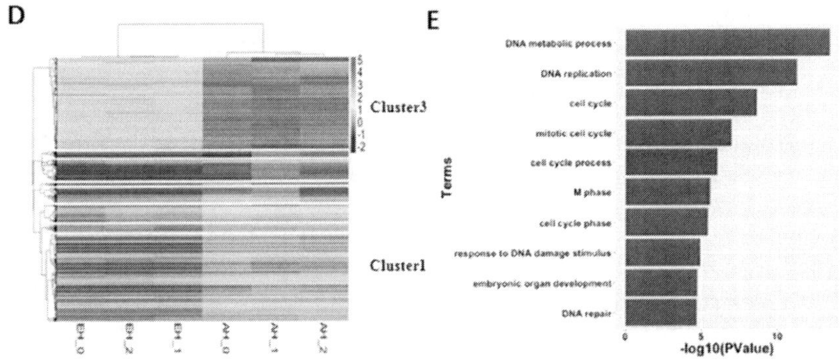

Figure 7 Analysis of fetal lincRNAs. (**A**) Expression-based association matrix of 51 fetal lincRNA transcripts (rows) and functional gene sets (columns). Red, positive correlation; blue, negative correlation; yellow, no correlation. Expression of fetal lincRNAs upon DOX stress in adult heart and muscle. Quantification of expression level change of lincRNAs upon DOX stress in the heart (**B**) and in the muscle (**C**). Means ± S.D. N = 5. $^*P < 0.05$ compared to control (PBS) group. Ratio of *vmhc* to *vmhcl* served as cardiac remodel marker. (**D**) K-means clustering for all differentially expressed transcripts including protein-coding transcripts and lincRNAs. (**E**) Gene functional enrichment analysis of protein-coding genes in Cluster 3 suggested that fetal lincRNAs (all grouped into Cluster 3) were significantly enriched in several biological processes including DNA metabolic processes, DNA replication, and cell cycle (FDR <0.01).

To identify fetal lincRNAs that can be reactivated under cardiac stress, we investigated the expression change of 15 fetal lincRNAs with higher expression levels in embryonic heart in an established DOX-induced cardiomyopathy (DIC) model in zebrafish[51]. Four fetal lincRNAs exhibit changes of expression upon DOX stress, while other 11 fetal lincRNAs remain unchanged. TCONS_00009015 was activated by at least 4-fold upon DOX stress at 1 day post injection (dpi), which remained highly expressed until 2 months post injection. TCONS_00023920 was only induced temporarily at 1 week after the DOX injection. By contrast, expression of TCONS_00070135 started to decrease at 7 dpi of DIC, and expression of TCONS_00023933 was reduced at 3 dpi and recovered at later time (Fig. 7B). We did not observe similar inductions in the muscle, which suggests that the transcriptional activation is specific to cardiac stress induced by DOX (Fig. 7C).

Unsupervised K-means clustering for all differentially expressed transcripts (i.e., protein-coding transcripts and lincRNAs) showed that all 16 fetal lincRNAs that were highly expressed in embryonic heart were grouped into

Cluster 3, whereas 25 of the 35 (71%) lincRNAs that were highly expressed in adult heart were in Cluster 1 (Fig. 7D). Gene functional enrichment analysis of the protein-coding genes in Cluster 3 by the Database for Annotation, Visualization and Integrated Discovery (DAVID) v6.8[52] suggested that the transcripts in Cluster 3 were significantly enriched in several biological processes, including DNA metabolic processes, DNA replication, and cell cycle (false discovery rate [FDR] <0.01; Fig. 7E). In contrast, as shown by IPA network enrichment analysis, the transcripts in Cluster 1 are significantly involved in various biological functions, including cardiac hypertrophy, cardiovascular disease, and developmental disorders, which suggest that the lincRNAs highly expressed in adult heart are involved in similar biological processes (Supplemental Table 2).

DISCUSSIONS

To systematically identify and characterize the lincRNAs associated with heart development and cardiac diseases, we generated a reference catalog of 813 cardiac lincRNAs in the zebrafish model by integrating RNA-seq data from 3 tissues with publicly available transcript annotations. In all, 423 (52%) of the transcripts in the cardiac catalog are novel lincRNAs identified using our RNA-seq dataset. Characterizing these lincRNAs via structural, expression, and evolutionary analyses indicated that cardiac lincRNAs share similar features with other lincRNA annotations in the public domain. Our cardiac lincRNA catalog will shed new light on future experimental and computational studies investigating lincRNA function in heart-associated biological processes and disease mechanisms.

Similar genome-wide lncRNA studies in the heart have been conducted recently in human, mouse, and zebrafish. Using RNA-seq, Chunjiang *et al.* identified the contrasting coordination of *cis-* and *trans-* molecular regulation through lncRNAs expressed in human fetal and adult hearts[50]. Scot *et al.* used comprehensive quantitative RNA-seq data from mouse hearts, livers, and skin cells to identify both cardiac-expressed and cardiac-specific lncRNAs[53]. Kriti *et al.* combined RNA-seq with computational analysis to identify tissue-restricted lncRNA signatures in the heart, liver, muscle, brain, and blood of adult zebrafish[48]. Compared with these studies, our study on zebrafish cardiac lincRNAs is unique in several ways. First, we focused on lincRNAs because their specific genomic locations facilitate downstream experimental manipulation and computational analysis. Our independent experimental validation, which included Sanger sequencing, qRT-PCR, *in situ* hybridization, and DOX stress experiments, suggested that our cardiac lincRNA catalog is a high-quality lincRNA set that will facilitate experimental studies and further functional classification of these lincRNAs, some of which might serve as key regulators in biological processes involved in cardiac diseases. Second, we used more comprehensive reference annotation resources, including RefSeq, Ensembl, and zflncRNApedia, to identify the known lincRNAs expressed in our dataset, which helped improve the accuracy of our lincRNA identification. Third, we developed an integrative yet conservative bioinformatics filtering workflow that enabled us to identify new cardiac lincRNAs by generating a high-confidence cardiac lincRNA catalog. Cardiac lncRNA studies in mouse[53] and human[50] have focused only on known lncRNAs, which constitute only a small subset of lncRNAs[14]. Kriti *et al.* identified both known and novel lncRNAs from 5 different tissues of adult zebrafish. In contrast, our study used a more comprehensive and stringent bioinformatics filtering strategy to identify putative cardiac lincRNAs, several of which were further independently confirmed by biological experiments. Finally, we identified several novel lincRNAs expressed specifically in heart tissues. Among the 813 cardiac lincRNAs, 564 were expressed in embryonic heart, and 730 were expressed in adult heart. We also identified 16 fetal lincRNAs with much higher expression levels in embryonic heart than adult heart and 35 fetal lincRNAs with much higher expression levels in adult heart than embryonic heart. Specifically, TCONS_00000891 was expressed only in heart but not in other organs, whereas TCONS_00009015 was a unique fetal lincRNA that might play important biological roles after cardiac stresses, similar to known cardiac-remodeling hallmarks, such as the ratio between *ventricular myosin heavy chain* (*vmhc*) and *ventricular myosin heavy chain-like* (*vmhcl*), functional zebrafish orthologues for human genes *myosin heavy chain 6*(*MYH6*) and *MYH7* (Fig. 7)[35]. In contrast with *mhrt*, a cardiac protective ncRNA that is also an antisense transcript of *MYH7*[54], the location of TCONS_000009015 does note overlap with the loci of either *vmhc* or *vmhcl* on the zebrafish genome, indicating that it might be involved in cardiac remodeling process via a different mechanism. Further investigation of these cardiac lincRNAs is warranted and might lead to the discovery of novel disease biomarkers or targets for therapeutic intervention.

Mechanistically, lincRNAs can regulate gene expression in both *cis* and *trans* manners32. For instance, lincRNAs have been shown to perform *cis*-acting regulation of target genes located at or near the same genomic locus in human heart[50]. Mammalian lincRNAs are enriched near protein-coding genes encoding for transcription factors and genes involved in nervous system development[55,56,57]. The neighboring genes of zebrafish lincRNAs have also been found to be enriched in transcriptional processes[43]. In the present study, the neighboring genes of zebrafish lincRNAs from all 3 tissues were enriched in transcription regulatory processes: "transcription regulator activity" (GO:0030528) and "transcription factor activity" (GO:0003700). We found that the neighboring genes of cardiac-specific lincRNAs are enriched in muscle component categories, such as troponin complex and striated muscle thin filament. Among these cardiac-specific lincRNAs, 9 are bidirectional lincRNAs located within 1,000 bases of a neighboring protein-coding gene on the opposite strand. LincRNAs of this specific subclass have been shown to serve as bidirectional promoters for neighboring protein-coding genes[58, 59]. Within this subclass, TCONS_00016569 is a novel bidirectional lincRNA, and its counterpart protein-coding gene, *cacna1g*, is involved in calcium channel activity based on GO annotation. TCONS_00014804 is a known lincRNA (ENSDART00000152911) with an unknown function, and its counterpart protein-coding gene, *Zcchc4*, is associated with methyltransferase activity based on GO annotation. Thus, our cardiac lincRNA catalog should facilitate the generation of new hypotheses regarding the potential roles of these lincRNAs in cardiac processes and disease mechanisms.

In addition to *cis* effects, lincRNAs have also been shown to primarily regulate gene expression in *trans* by loss-of-function studies of hundreds of lincRNAs expressed in mouse embryonic stem cells[19]. Based on the "guilt-by-association" rule[56, 60], we observed that some cardiac lincRNAs were coexpressed with several heart-associated functional gene sets (Supplemental Table 4), which indicates that cardiac-specific lincRNAs are associated with heart development and regulation and may be potential regulators in heart diseases. Hierarchical clustering analysis of differentially expressed transcripts (including both lincRNAs and protein-coding transcripts) showed that most fetal lincRNAs were grouped into Clusters 1 and 3 with protein-coding transcripts that are significantly enriched in several biological processes, including DNA metabolic processes, DNA replication, and cell cycle (Fig. 7D and E). Thus, these cardiac lincRNAs may exert their functions by participating in biological processes similar to those of their coexpressed protein-coding partners. For instance, TCONS_00028686 in Cluster 1 had high expression in adult heart, and its genomic location overlaps with that of *lncH_009*, a known lincRNA with expression only in the adult heart in zebrafish[48]. Although the detailed function of TCONS_00028686 is not known, it shares a similar expression pattern with homeobox-containing genes, which implies that it could be involved in similar biological processes.

Despite their evolutionary distance, orthologues for zebrafish lincRNAs can be found in humans and mice[30]. Using multispecies genome alignments, we identified 72 human orthologous regions for 65 zebrafish lincRNAs and 67 mouse orthologous regions for 57 zebrafish lincRNAs (Supplemental Table 7). Among these orthologous regions, 14 human orthologous regions have annotations in human GENCODE (Release 19), and 11 mouse orthologous regions have annotations in mouse GENCODE (Release M1). Notably, the lincRNA TCONS_00071932 mapped to the human lincRNA

LINC00672. Low-level conservation across different organisms may be attributable to the importance of transcription from a specific genomic location, reduced selective pressure on the primary sequence of ncRNAs, or the rapid evolution of new functions. Our analysis was limited by the available transcript annotations in other species, which will be improved as more transcriptomes are sequenced in other organisms in the future.

In summary, our study provides a high-confidence catalog of cardiac lincRNAs in the zebrafish model, laying a solid foundation for future studies. First, changes in these cardiac lincRNAs during the pathogenesis of different cardiac disease models can be examined and reveal candidate lincRNAs that play vital roles during pathogenesis. Second, the specific functions of promising lincRNAs must be elucidated via loss-of-function and gain-of-function studies. We anticipate that these future studies will significantly improve our understanding of the biological functions of cardiac lincRNAs.

MATERIALS AND METHODS

Zebrafish husbandry

Wild-type WIK zebrafish (Danio rerio) were maintained under a 14-hour light/10-hour dark cycle at 28.5 °C. This study was approved by the Mayo Clinic Institutional Animal Care and Use Committee (A17515). All study procedures were performed in accordance with the Guide for the Care and Use of Laboratory Animals[61].

RNA-seq dataset

We used the RNA sequencing (RNA-seq) dataset from our previous study[35] for zebrafish cardiac transcriptome reconstruction. Nine biological samples from 3 groups (3 samples of adult heart tissue, 3 samples of adult muscle tissue, and 3 samples of embryonic heart tissue) were sequenced using the HiSeq 2000 platform (Illumina, San Diego, CA, USA) with a 50-bp pair-end sequencing protocol in the Mayo Clinic DNA Sequencing Core Facility.

Transcriptome assembly

Raw RNA-seq reads for each sample were aligned by TopHat (Version 2.0.12) to the zebrafish genome assembly (Zv9) using the Ensembl annotation Zv9 (Danio_rerio.Zv9.79.gtf)[62]. Each transcriptome was assembled by Cufflinks (Version 2.2.1)[63]. Transcripts were considered expressed if they had an FPKM value greater than 0.1 in at least 2 of 3 samples in the same group according to the Cuffnorm script from Cufflinks. Transcripts were considered to be differentially expressed across different groups if they exhibited a greater than 2-fold change and an FDR of less than 0.05 according to the Cuffdiff script from Cufflinks.

Bioinformatics filtering workflow to identify high-confidence lincRNAs

In the reconstructed transcriptome, transcripts designated in the 'u' category by Cufflink (i.e., unknown intergenic transcripts) were obtained using the Cuffcompare script with the Ensembl annotation file (Danio_rerio.Zv9.79.gtf). The annotated lincRNAs in Ensembl were also obtained in this step. The annotated lincRNAs in the zflncRNApedia database were obtained using the Cuffcompare script with the zflncRNApedia annotation[36]. The remaining novel candidate transcripts were then filtered in 5 steps: 1) FPKM value greater than 0.1; 2) sequence length longer than 200 nt; 3) exon number greater than 1; 4) evaluation of coding potential using CPAT, CPC, and PhyloCSF; and 5) filtration of candidates with known protein domains using Pfam. An overview of this process is shown in Fig. 1A.

Conservation analysis

To examine the sequence conservation of lincRNAs, we used the phastCons scores derived from the University of California, Santa Cruz (UCSC) 8-way vertebrate genome alignment seeded with the zebrafish reference genome (Zv9)[64]. The phastCons scores for the whole genome were downloaded from the UCSC database (http://hgdownload.cse.ucsc.edu/goldenPath/danRer7/phastCons8way). The phastCons score of each exon was defined as the average phastCons score of each nucleotide of that exon. Exons with no phastCons scores were ignored.

Identification of orthologous loci across zebrafish, mouse, and human

The lincRNAs in zebrafish were mapped to the human reference genome hg19 and the mouse reference genome mm9 through the UCSC database[65] using the bedToPsl, pslMap, pslToBed, and pslCDnaFilter scripts. This process was guided by the chain datasets (danRer7.hg19.all.chain and danRer7.mm9.all.chain) from the UCSC database (http://hgdownload.cse.ucsc.edu/downloads.html).

Clustering analysis

Unsupervised hierarchical clustering was performed with Pearson correlation and complete linkage options using the pheatmap R package (https://github.com/raivokolde/pheatmap). To infer potential functions of differentially expressed lincRNAs across tissue types, K-means clustering analysis was applied to all differentially expressed transcripts, including lincRNAs and protein-coding transcripts. The optimal k value was determined based on the average silhouette width using the fpc R package[66].

GSA

GSA was performed to infer the potential functions of cardiac lincRNAs. The Pearson correlation coefficient between each lincRNA and each protein-coding gene in each functional gene set was calculated by the GSA R package[67], as described previously[56, 68]. The functional gene sets included heart-associated gene sets annotated in the Zebrafishmine database (www.zebrafishmine.org; Supplemental Table 4) and our previous DCM gene set[35]. The GSA scores of all functional gene sets were transformed to a ternary scale (−1, 0, 1): Gene sets with FDR >0.5 were assigned a value of 0, gene sets with FDR <0.5 and GSA score >0 were assigned a value of 1, and gene sets with FDR <0.5 and GSA score <0 were assigned a value of −1. The score matrix was clustered and visualized with Pearson correlation distance metric and complete linkage options using the pheatmap R package.

Functional enrichment analysis

Functional enrichment analysis was performed using the Database for Annotation, Visualization and Integrated Discovery (DAVID) tool[52].

Pathway and network enrichment analysis

The gene lists of interest were annotated by IPA (QIAGEN) (http://www.ingenuity.com/). We queried IPA with the gene list of interest to map and generate putative biological processes/functions, networks, and pathways based on the manually curated knowledge database of molecular interactions extracted from the public literature. The enriched pathways and gene networks were generated using both direct and indirect relationships/connectivity. These pathways and networks were ranked by their enrichment score, which measures the probability that the genes were included in a network by chance.

qRT-PCR

RNA was extracted and purified from freshly dissected tissues using TRIzol Reagent (Thermo Fisher Scientific) following the manufacturer's protocol. In total, 500 ng of RNA from each tissue was used to prepare cDNA using the SuperScript III First-Strand Synthesis System (Thermo Fisher Scientific). A qRT-PCR was conducted using a LightCycler 480 QPCR apparatus in 96-well QPCR plates (Roche Life Science). The expression levels of the lincRNAs of interest were calculated by the $-\Delta\Delta$Ct (cycle threshold) method using glyceraldehyde-3-phosphate dehydrogenase (gapdh) as an internal reference. For each tissue type, at least 3 biological replicates were measured and analyzed. The primers for the lincRNAs of interest are listed in Supplemental Table 8.

In situ hybridization

In situ hybridization riboprobes were generated from 500- to 1,000-bp PCR products using T7-tagged primer pairs. Fish embryos were collected at 1, 2, and 4 dpf. Whole-mount *in situ* hybridization was conducted as described previously[69]. Images were recorded using an Axioplan 2 microscope (Zeiss) or a Leica III dissecting microscope equipped with a Nikon digital camera.

DOX stress experiment

Fish were exposed to DOX stress as described previously[70]. Adult WIK zebrafish were injected with DOX (Sigma) at a dose of 20 mg/g. More than 5 fish were included in each experimental group.

REFERENCES

1. Kathiriya, I. S., Nora, E. P. & Bruneau, B. G. Investigating the transcriptional control of cardiovascular development. Circulation research 116, 700–714, doi:10.1161/CIRCRESAHA.116.302832 (2015).
2. Bruneau, B. G. The developmental genetics of congenital heart disease. Nature 451, 943–948, doi:10.1038/nature06801 (2008).
3. McPherson, R. et al. A common allele on chromosome 9 associated with coronary heart disease. Science 316, 1488–1491, doi:10.1126/science.1142447 (2007).

4. Helgadottir, A. et al. A common variant on chromosome 9p21 affects the risk of myocardial infarction. Science 316, 1491–1493, doi:10.1126/science.1142842 (2007).
5. Heineke, J. & Molkentin, J. D. Regulation of cardiac hypertrophy by intracellular signalling pathways. Nature reviews. Molecular cell biology 7, 589–600, doi:10.1038/nrm1983 (2006).
6. Sayed, D., He, M., Yang, Z., Lin, L. & Abdellatif, M. Transcriptional regulation patterns revealed by high resolution chromatin immunoprecipitation during cardiac hypertrophy. The Journal of biological chemistry 288, 2546–2558, doi:10.1074/jbc.M112.429449 (2013).
7. Duygu, B., Poels, E. M. & da Costa Martins, P. A. Genetics and epigenetics of arrhythmia and heart failure. Frontiers in genetics 4, 219, doi:10.3389/fgene.2013.00219 (2013).
8. Brand, T. Heart development: molecular insights into cardiac specification and early morphogenesis. Developmental biology 258, 1–19, doi:10.1016/S0012-1606(03)00112-X (2003).
9. Olson, E. N. Gene regulatory networks in the evolution and development of the heart. Science 313, 1922–1927, doi:10.1126/science.1132292 (2006).
10. Mattick, J. S. & Makunin, I. V. Non-coding RNA. Human molecular genetics 15, R17–R29, doi:10.1093/hmg/ddl046 (2006).
11. Mercer, T. R., Dinger, M. E. & Mattick, J. S. Long non-coding RNAs: insights into functions. Nature Reviews Genetics 10, 155–159, doi:10.1038/nrg2521 (2009).
12. Raney, B. J. et al. ENCODE whole-genome data in the UCSC genome browser (2011 update). Nucleic acids research 39, D871–875, doi:10.1093/nar/gkq1017 (2011).
13. Consortium, F. et al. A promoter-level mammalian expression atlas. Nature 507, 462–470, doi:10.1038/nature13182 (2014).
14. Djebali, S. et al. Landscape of transcription in human cells. Nature 489, 101–108, doi:10.1038/nature11233 (2012).
15. Rayner, K. J. & Liu, P. P. Long Noncoding RNAs in the Heart: The Regulatory Roadmap of Cardiovascular Development and Disease. Circulation. Cardiovascular genetics 9, 101–103, doi:10.1161/CIRCGENETICS.116.001413 (2016).
16. Chowdhury, D., Choi, Y. E. & Brault, M. E. Charity begins at home: non-coding RNA functions in DNA repair. Nature reviews. Molecular cell biology 14, 181–189, doi:10.1038/nrm3523 (2013).
17. St Laurent, G., Wahlestedt, C. & Kapranov, P. The Landscape of long noncoding RNA classification. Trends in genetics: TIG 31, 239–251, doi:10.1016/j.tig.2015.03.007 (2015).
18. Sun, M. & Kraus, W. L. From discovery to function: the expanding roles of long noncoding RNAs in physiology and disease. Endocrine reviews 36, 25–64, doi:10.1210/er.2014-1034 (2015).
19. Rinn, J. L. & Chang, H. Y. Genome regulation by long noncoding RNAs. Annual review of biochemistry 81, 145–166, doi:10.1146/annurev-biochem-051410-092902 (2012).
20. Rizki, G. & Boyer, L. A. Lncing epigenetic control of transcription to cardiovascular development and disease. Circulation research 117, 192–206, doi:10.1161/CIRCRESAHA.117.304156 (2015).
21. Devaux, Y. et al. Long noncoding RNAs in cardiac development and ageing. Nat Rev Cardiol 12, 415–425, doi:10.1038/nrcardio.2015.55 (2015).
22. Klattenhoff, C. A. et al. Braveheart, a long noncoding RNA required for cardiovascular lineage commitment. Cell 152, 570–583, doi:10.1016/j.cell.2013.01.003 (2013).
23. Grote, P. et al. The tissue-specific lncRNA Fendrr is an essential regulator of heart and body wall development in the mouse. Developmental cell 24, 206–214, doi:10.1016/j.devcel.2012.12.012 (2013).

24. Ounzain, S. et al. Genome-wide profiling of the cardiac transcriptome after myocardial infarction identifies novel heart-specific long non-coding RNAs. Eur Heart J 36, 353–368a, doi:10.1093/eurheartj/ehu180 (2015).
25. Husso, T., Yla-Herttuala, S. & Turunen, M. P. A New Gene Therapy Approach for Cardiovascular Disease by Non-coding RNAs Acting in the Nucleus. Mol Ther Nucleic Acids 3, e197, doi:10.1038/mtna.2014.48 (2014).
26. Viereck, J. et al. Long noncoding RNA Chast promotes cardiac remodeling. Sci Transl Med 8, 326ra322–326ra22, doi:10.1126/scitranslmed.aaf1475 (2016).
27. Ounzain, S. et al. CARMEN, a human super enhancer-associated long noncoding RNA controlling cardiac specification, differentiation and homeostasis. J Mol Cell Cardiol 89, 98–112, doi:10.1016/j.yjmcc.2015.09.016 (2015).
28. Ma, L., Bajic, V. B. & Zhang, Z. On the classification of long non-coding RNAs. RNA Biology 10, 6–5, doi:10.4161/rna.24604 (2013).
29. Ounzain, S., Burdet, F., Ibberson, M. & Pedrazzini, T. Discovery and functional characterization of cardiovascular long noncoding RNAs. J Mol Cell Cardiol 89, 17–26, doi:10.1016/j.yjmcc.2015.09.013 (2015).
30. Uchida, S. & Dimmeler, S. Long noncoding RNAs in cardiovascular diseases. Circ Res 116, 737–750, doi:10.1161/CIRCRESAHA.116.302521 (2015).
31. Xu, J. et al. A comprehensive overview of lncRNA annotation resources. Brief Bioinform. doi:10.1093/bib/bbw015 (2016).
32. Ulitsky, I. & Bartel, D. P. lincRNAs: Genomics, Evolution, and Mechanisms. Cell 154, 26–46, doi:10.1016/j.cell.2013.06.020 (2013).
33. Quinn, J. J. & Chang, H. Y. Unique features of long non-coding RNA biogenesis and function. Nat Rev Genet 17, 47–62, doi:10.1038/nrg.2015.10 (2016).
34. Schwalb, B. et al. TT-seq maps the human transient transcriptome. Science 352, 1225–1228, doi:10.1126/science.aad9841 (2016).
35. Shih, Y. H. et al. Cardiac transcriptome and dilated cardiomyopathy genes in zebrafish. Circulation. Cardiovascular genetics 8, 261–269, doi:10.1161 /CIRCGENETICS.114.000702 (2015).
36. Dhiman, H., Kapoor, S., Sivadas, A., Sivasubbu, S. & Scaria, V. zflncRNApedia: A Comprehensive Online Resource for Zebrafish Long Non-Coding RNAs. PloS one 10, e0129997, doi:10.1371/journal.pone.0129997 (2015).
37. Wang, L. et al. CPAT: Coding-Potential Assessment Tool using an alignment-free logistic regression model. Nucleic Acids Research (2013).
38. Bateman, A. et al. The Pfam protein families database. Nucleic acids research 32, D138–141, doi:10.1093/nar/gkh121 (2004).
39. Kong, L. et al. CPC: assess the protein-coding potential of transcripts using sequence features and support vector machine. Nucleic Acids Res 35, W345–349, doi:10.1093/nar/gkm391 (2007).
40. Lin, M. F., Jungreis, I. & Kellis, M. PhyloCSF: a comparative genomics method to distinguish protein coding and non-coding regions. Bioinformatics 27, i275–282, doi:10.1093/bioinformatics/btr209 (2011).
41. Pauli, A. et al. Systematic identification of long noncoding RNAs expressed during zebrafish embryogenesis. Genome research 22, 577–591, doi:10.1101 /gr.133009.111 (2012).
42. Cabili, M. N. et al. Integrative annotation of human large intergenic noncoding RNAs reveals global properties and specific subclasses. Genes Dev 25, 1915–1927, doi:10.1101/gad.17446611 (2011).
43. Ulitsky, I., Shkumatava, A., Jan, C. H., Sive, H. & Bartel, D. P. Conserved function of lincRNAs in vertebrate embryonic development despite rapid sequence evolution. Cell 147, 1537–1550, doi:10.1016/j.cell.2011.11.055 (2011).
44. Hangauer, M. J., Vaughn, I. W. & McManus, M. T. Pervasive Transcription of the Human Genome Produces Thousands of Previously Unidentified Long Intergenic Noncoding RNAs. PLoS genetics 9, e1003569, doi:10.1371/journal.pgen.1003569 (2013).

45. Wang, K. C. et al. A long noncoding RNA maintains active chromatin to coordinate homeotic gene expression. Nature 472, 120–124, doi:10.1038/nature09819 (2011).
46. Orom, U. A. et al. Long noncoding RNAs with enhancer-like function in human cells. Cell 143, 46–58, doi:10.1016/j.cell.2010.09.001 (2010).
47. Casadei, R. et al. Identification of housekeeping genes suitable for gene expression analysis in the zebrafish. Gene Expr Patterns 11, 271–276, doi:10.1016/j.gep.2011.01.003 (2011).
48. Kaushik, K. et al. Dynamic expression of long non-coding RNAs (lncRNAs) in adult zebrafish. PLoS One 8, e83616, doi:10.1371/journal.pone.0083616 (2013).
49. Matkovich, S. J., Edwards, J. R., Grossenheider, T. C., de Guzman Strong, C. & Dorn, G. W. 2nd Epigenetic coordination of embryonic heart transcription by dynamically regulated long noncoding RNAs. Proc Natl Acad Sci USA 111, 12264–12269, doi:10.1073/pnas.1410622111 (2014).
50. He, C. et al. Systematic Characterization of Long Noncoding RNAs Reveals the Contrasting Coordination of Cis- and Trans-Molecular Regulation in Human Fetal and Adult Hearts. Circulation. Cardiovascular genetics 9, 110–118, doi:10.1161/CIRCGENETICS.115.001264 (2016).
51. Ding, Y. et al. Haploinsufficiency of target of rapamycin attenuates cardiomyopathies in adult zebrafish. Circ Res 109, 658–669, doi:10.1161/CIRCRESAHA.111.248260 (2011).
52. Huang, D. W., Sherman, B. T. & Lempicki, R. A. Systematic and integrative analysis of large gene lists using DAVID bioinformatics resources. Nature protocols 4, 44–57, doi:10.1038/nprot.2008.211 (2009).
53. Matkovich, S. J., Edwards, J. R., Grossenheider, T. C., de Guzman Strong, C. & Dorn, G. W. 2nd Epigenetic coordination of embryonic heart transcription by dynamically regulated long noncoding RNAs. Proceedings of the National Academy of Sciences of the United States of America 111, 12264–12269, doi:10.1073/pnas.1410622111 (2014).
54. Han, P. et al. A long noncoding RNA protects the heart from pathological hypertrophy. Nature 514, 102–106, doi:10.1038/nature13596 (2014).
55. Mercer, T. R., Dinger, M. E., Sunkin, S. M., Mehler, M. F. & Mattick, J. S. Specific expression of long noncoding RNAs in the mouse brain. Proceedings of the National Academy of Sciences 105, 716–721, doi:10.1073/pnas.0706729105 (2008).
56. Guttman, M. et al. Chromatin signature reveals over a thousand highly conserved large non-coding RNAs in mammals. Nature 458, 223–227, doi:10.1038/nature07672 (2009).
57. Ponjavic, J., Oliver, P. L., Lunter, G. & Ponting, C. P. Genomic and transcriptional co-localization of protein-coding and long non-coding RNA pairs in the developing brain. PLoS genetics 5, e1000617, doi:10.1371/journal.pgen.1000617 (2009).
58. Engstrom, P. G. et al. Complex Loci in human and mouse genomes. PLoS genetics 2, e47, doi:10.1371/journal.pgen.0020047 (2006).
59. Kornienko, A. E., Guenzl, P. M., Barlow, D. P. & Pauler, F. M. Gene regulation by the act of long non-coding RNA transcription. BMC biology 11, 59, doi:10.1186/1741-7007-11-59 (2013).
60. Su, X. et al. Comprehensive analysis of long non-coding RNAs in human breast cancer clinical subtypes. Oncotarget 5, 9864–9876, doi:10.18632/oncotarget.2454 (2014).
61. Council, N. R. Guide for the Care and Use of Laboratory Animals: Eighth Edition. (The National Academies Press, 2011).
62. Kim, D. et al. TopHat2: accurate alignment of transcriptomes in the presence of insertions, deletions and gene fusions. Genome Biol 14, R36, doi:10.1186/gb-2013-14-4-r36 (2013).
63. Trapnell, C. et al. Transcript assembly and quantification by RNA-Seq reveals unannotated transcripts and isoform switching during cell differentiation. Nature biotechnology 28, 511–515, doi:10.1038/nbt.1621 (2010).

64. Siepel, A. et al. Evolutionarily conserved elements in vertebrate, insect, worm, and yeast genomes. Genome research 15, 1034–1050, doi:10.1101/gr.3715005 (2005).
65. Meyer, L. R. et al. The UCSC Genome Browser database: extensions and updates 2013. Nucleic acids research 41, D64–69, doi:10.1093/nar/gks1048 (2013).
66. Hennig, C. fpc: Flexible procedures for clustering. R package version 2, 0–3 (2010).
67. Efron, B. & Tibshirani, R. On Testing the Significance of Sets of Genes. Annals of Applied Statistics 1, 107–129, doi:10.1214/07-Aoas101 (2007).
68. Subramanian, A. et al. Gene set enrichment analysis: a knowledge-based approach for interpreting genome-wide expression profiles. Proc Natl Acad Sci USA 102, 15545–15550, doi:10.1073/pnas.0506580102 (2005).
69. Xu, X. et al. Cardiomyopathy in zebrafish due to mutation in an alternatively spliced exon of titin. Nature genetics 30, 205–209, doi:10.1038/ng816 (2002).
70. Ding, Y. et al. Haploinsufficiency of target of rapamycin attenuates cardiomyopathies in adult zebrafish. Circulation research 109, 658–669, doi:10.1161/CIRCRESAHA.111.248260 (2011).

Index